Atomic-Scale Modelling of Electrochemical Systems

Atomic-Scale Modelling of Electrochemical Systems

Edited by

Marko M. Melander
University of Jyväskylä
Department of Chemistry
Jyväskylä
Finland

Tomi T. Laurila
Aalto University
Electrical Engineering and Automation
Espoo
Finland

Kari Laasonen
Aalto University
Chemistry and Materials Science
Espoo
Finland

WILEY

This edition first published 2022

Registered Offices
John Wiley & Sons, Inc., 111 River Street, Hoboken, NJ 07030, USA
John Wiley & Sons Ltd, The Atrium, Southern Gate, Chichester, West Sussex, PO19 8SQ, UK

Editorial Office
The Atrium, Southern Gate, Chichester, West Sussex, PO19 8SQ, UK

For details of our global editorial offices, customer services, and more information about Wiley products visit us at www.wiley.com.

Wiley also publishes its books in a variety of electronic formats and by print-on-demand. Some content that appears in standard print versions of this book may not be available in other formats.

Library of Congress Cataloging-in-Publication Data applied for
Hardback ISBN: 978-1-119-60561-4

Cover Design: Wiley
Cover Images: Background © krishh/iStock/Getty Images. Cover Illustration courtesy of Marko Melander.

Set in 9.5/12.5pt STIXTwoText by Straive, Chennai, India
Printed and bound by CPI Group (UK) Ltd, Croydon, CR0 4YY

C9781119605614_250821

Contents

List of Contributors

Andreussi, Oliviero
Department of Physics
University of North Texas
Denton, TX, USA

Arias, Tomás A.
Department of Physics, Laboratory of Atomic and Solid State Physics, and Cornell Center for Materials Research (CCMR), Cornell University, Cornell
University, Ithaca, NY, USA

Bhowmik, Arghya
Department of Energy Conversion and Storage
Technical University of Denmark
Kgs. Lyngby, Denmark

Bouzid, Assil
Institut de Recherche sur les Céramiques
Centre Européen de la Céramique
Limoges

Bölle, Felix Tim
Department of Energy Conversion and Storage
Technical University of Denmark
Kgs. Lyngby, Denmark

Castelli, Ivano E.
Department of Energy Conversion and Storage
Technical University of Denmark
Kgs. Lyngby, Denmark

Chang, Jin Hyun
Department of Energy Conversion and Storage
Technical University of Denmark
Kgs. Lyngby, Denmark

Fletcher, Stephen
The Fletcher Consultancy
Loughborough, Leicestershire
United Kingdom

Groß, Axel
Institute of Theoretical Chemistry
Ulm University

and

Helmholtz Institute Ulm (HIU)
Electrochemical Energy Storage
Ulm, Germany

Hörmann, Nicolas Georg
Chair of Theoretical Chemistry
Technische Universität München
Garching, Germany

Kastlunger, Georg
School of Engineering, Brown University
Providence, Rhode Island
USA

Laasonen, Kari
Department of Chemistry and Materials Science, Aalto University,
Espoo, Finland

Lastra, García
Department of Energy Conversion and Storage
Technical University of Denmark
Kgs. Lyngby, Denmark

Laurila, Tomi T.
Department of Electrical Engineering and Automation, Aalto University
Espoo, Finland

Lindgren, Per
School of Engineering
Brown University
Providence, Rhode Island
USA

Maria, Juan
Department of Energy Conversion and Storage
Technical University of Denmark
Kgs. Lyngby, Denmark

Mathiesen, Nicolai Rask
Department of Energy Conversion and Storage
Technical University of Denmark
Kgs. Lyngby, Denmark

Melander, Marko M.
Nanoscience Center, Department of Chemistry,
University of Jyväskylä, Jyväskylä, Finland

Nattino, Francesco
Theory and Simulations of Materials (THEOS) and
National Centre for Computational Design and Discovery of Novel Materials (MARVEL)
École Polytechnique Fédérale de Lausanne
Lausanne, Switzerland

Nazmutdinov, Renat R.
Kazan National Research Technological University
Kazan, Republic of Tatarstan
Russian Federation

Pasquarello, Alfredo
Chaire de Simulation à l'Echelle Atomique (CSEA)
Ecole Polytechnique Fédérale de Lausanne (EPFL)
Lausanne, Switzerland

Peterson, Andrew A.
School of Engineering
Brown University
Providence, Rhode Island
USA

Sundararaman, Ravishankar
Materials Science and Engineering,
Rensselaer Polytechnic Institute, Troy, NY, USA

Tygesen, Alexander Sougaard
Department of Energy Conversion and Storage
Technical University of Denmark
Kgs. Lyngby, Denmark

Ulstrup, Jens
Kazan National Research Technological University
Kazan, Republic of Tatarstan
Russian Federation

and

Department of Chemistry
Technical University of Denmark
Kgs. Lyngby, Denmark

Vegge, Tejs
Department of Energy Conversion and Storage
Technical University of Denmark
Kgs. Lyngby, Denmark

Part I

1

Introduction to Atomic Scale Electrochemistry

Marko M. Melander, Tomi T. Laurila, and Kari Laasonen

1.1 Background

Electrochemistry and electrocatalysis are at the forefront of many technological fields related to solving the grand challenges encountered in advanced energy solutions, personalized medicine, and environmental issues. Electrochemical technologies of interest include, among others, batteries, CO_2 mitigation, various sensor technologies, water purification, molecular electronics, fuel-cells, hydrogen powered energies, and solar-powered renewable technologies.

To improve upon existing electrochemical technologies in a rational way, understanding and controlling the atomic scale properties of the electrochemical interface is vital. In particular, the connection between atomic scale surface chemistry and the electrocatalytical performance needs to be established. Rational design of better electrocatalysts working in complex electrochemical environments needs insight from experiments, computational methods, as well as theoretical approaches. While experimental electrochemical and spectroelectrochemical methods are well-established and can often be routinely applied, theoretical and computational methods have not yet reached the same level of maturity. The lack of generally accepted and applicable computational and theoretical tools is due to the high complexity of the electrochemical interface which provides a number of challenges for atomic scale theory and modelling. Specific challenges include; (i) inclusion of the electrode potential, (ii) the need for several time and length scales to assess both thermodynamic and kinetic properties of the solid-liquid interface, and (iii) a quantum mechanical treatment to describe chemical bond making and breaking.

The field of atomistic modelling in electrochemistry has made impressive progress during the last 15 years. In this book, we will the review state-of-the-art computational and theoretical methods for modelling, understanding, and predicting the properties of electrochemical interfaces. Specifically, we discuss different ways of (i) including the electrode potential in the computational setup and fixed potential calculations within the framework of grand canonical density functional theory, (ii) quantum mechanical models for the solid-liquid interface and the formation of an electrochemical double-layer using molecular dynamics and/or continuum descriptions, (iii) thermodynamic description of the interface and reactions taking place at the interface as a function of the electrode potential, (iv) novel ways of

Atomic-Scale Modelling of Electrochemical Systems, First Edition.
Edited by Marko M. Melander, Tomi T. Laurila, and Kari Laasonen.

describing rates for heterogeneous electron transfer (both with the outer and inner sphere redox couples), proton-coupled electron transfer, and other electrocatalytic reactions as a function of the electrode potential, and (v) multiscale modelling where atomic level information is used for predicting experimental observables to enable direct comparison with experiments, to rationalize experimental results and to predict the electrochemical performance. We will also highlight several applications in electrocatalysis and electrochemistry using state-of-the-art methods.

This book will provide a comprehensive view on the current theoretical and computational methods and their application for understanding, predicting, and optimizing the properties of electrochemical interfaces starting from the atomic scale. While several books have been devoted to either experimental electrochemistry or computational chemistry, there are no books on atomistic computational electrochemistry! Hence, we hope that this volume contributes to fill this gap in the literature.

1.2 The thermodynamics of electrified interface

In Fig. 1.1, a simplified view of a typical electrochemical interface in aqueous electrolyte is shown. On the far left is the electrode, which provides the source/sink of electrons at a constant electrochemical potential as well as a substrate for any chemical reactions accompanying redox reactions. The surface of the electrode typically contains specifically adsorbed ions in addition to solvent molecules and reaction intermediates and products. In addition, the surface chemistry of the electrode is heavily dependent on the value of the applied potential.

The structure of the solvent adjacent to the electrode is significantly affected by the charge on the electrode. On the solvent side we have one compact layer where the change in the

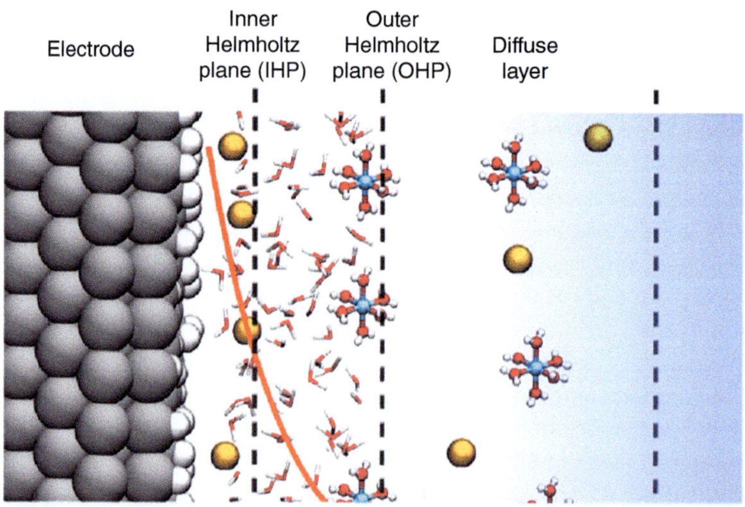

Figure 1.1 Schematic presentation of the electrochemical double layer showing the IHP, OHP, and diffuse layer as well as how the potential changes as a function of distance from the electrode surface. See text for further details. Source: Courtesy of Nico Holmberg.

potential is approximately linear and a more spread out region where the change is more or less exponential (diffuse layer). These layers together constitute the so-called electrical double layer (EDL). Note that the EDL is formed at the interface of any type of different phases owing to the electrostatic interactions and is by no means restricted to solid/liquid interfaces. However, the structure of this interphasial region can be quite different at different type of interfaces. For example, if one is working with ionic liquids the double layer region differs significantly from the one shown in Fig. 1.1 [1]. The compact layer is further composed of two sublayers as shown in Figure 1.1. The one closest to the electrode is called the inner Helmholtz layer (IHL) and it contains, in addition to the solvent molecules, specifically adsorbed ions (typically anions or large cations). These ions have lost at least partially their solvation sheet and have direct contact to (also partly desolvated) electrode. The so-called inner Helmholtz plane (IHP) is typically defined to go through the centers of these specifically adsorbed ions. In the other part of the compact layer further away from the electrode surface, in the outer Helmholtz layer (OHL), in addition to solvent, ions that have retained their solvation sheets are located. Thus, within OHL there are no specific chemical interactions between the electrode and the redox species and the interaction is thus purely electrostatic. The so-called outer Helmholtz plane (OHP) is typically defined as the plane going through the centers of the these non-specifically adsorbed ions. Thus, the OHP can be thought of as the distance of closest approach of surface inactive ions. After the compact layer, before the bulk of the solution, comes the diffuse layer, where there is a dynamic equilibrium between the ordering tendency caused by the electric field from the electrode and the disordering thermal motion. At this point, it is helpful to mention something about the dimensions of these layers. The thickness of the compact layer (consisting of inner and outer Helmholtz layers) is typically in the order of 0.5 nm or less. The thickness of the diffuse layer depends heavily on the total ionic concentration and is less than 10 nm for concentrations greater than 0.01 M. Note also that the inner layer and diffuse layer together have a net electrical charge equal in magnitude to that of the electrode surface but of opposite polarity. As a result, the complete structure is electrically neutral.

In addition to the ion distribution and potential profile, important properties of the double layer include its capacitance. The double layer capacitance is an important factor in electroanalytical measurements as well as in sensor technologies and electrocatalytical applications. The double layer capacitance defines the electric field experienced by the species at the surface and within the EDL. The surface capacitance also directly influences how the charge state of the electrode surface depends on the applied potential. In addition, when one drives reactions on the electrode surface by changing the potential, there will always be a contribution from the charging current of the double layer (so-called non-faradic current) to the total current that has to be somehow subtracted from it to obtain the faradic current corresponding to the actual redox reaction under investigation. The charging current arises from the rearrangement of the species constituting the double layer as the potential changes. In fact, there are several electroanalytical techniques, such as differential pulse voltammetry (DPV) and square wave voltammetry (SWV), that have been developed in order to minimize the contributions arising from this background current. This double layer capacitance can be measured with various different ways the simplest being cyclic voltammetry (CV) – if there is a suitable double layer region in the voltammogram. It is to be noted that when measured with CV the double layer capacitance usually contains some fraction of pseudocapacitance from the parasitic faradic reactions occurring during the potential cycling. Another feasible way is to use Electrochemical Impedance Spectroscopy (EIS) where the pseudocapacitance is typically neglible and one obtains smaller (perhaps more correct) values for the capacitance than in the case of CV.

Well-established classical models for the double layer capacitance exist and are reviewed for example in [2] and we will therefore not discuss those here. We just want to emphasize two things of importance: (i) the effect of the electrode on the magnitude of the double layer capacitance and (ii) variation of the value of the dielectric constant in the interfacial region. In the context of the first point the electronic structure of the electrode will significantly influence the capacitance of the compact part of the double layer and thus must be appropriately taken into account. For example, metals typically have higher double layer capacitances than graphite owing to the differences in their electronic structure [3]. Regarding point (ii), it is perhaps good to take water as an example. Based on extensive experimental studies it appears that the dielectric constant for highly oriented water immediately next to the electrode (in IHP) is about 5 and it then steadily grows toward the bulk value as one moves away from the electrode/solvent interface [4]. As the nature of the electrode will have an effect on the ordering of water at the interface points (i) and (ii) are not independent of each other but are heavily interrelated.

There is a third layer close to the electrode that has not been shown in Fig. 1.1 called diffusion layer. This is a region, which is electrically neutral, but where concentration (or more precisely chemical potential) profiles exist created by the redox reactions at the interface. This layer has a significant role in many electrochemical reactions and its thickness develops as a function of time being zero right at the start of the measurement and growing steadily up to millimeters during the duration of the experiment (although natural convection starts to interfere with its growth around thicknesses of 0.5 mm or so). The thickness of this layer can be minimized by accelerating mass transport from the bulk towards the interface (preventing the relaxation of the diffusion layer into the bulk of the solution). Acceleration can be achieved, for example, by using convection as in the case of rotating disk electrode (RDE) or by increasing cycling speed during cyclic voltammetry (CV) measurements. By using electroanalytical techniques under typical conditions (for example not using extremely fast cycling speed of hundreds of thousands of V/s or exceeding rotating speeds of 10000 rpm in RDE) the thickness of the diffuse layer cannot be pushed lower than to about 5μm or so. Thus, its dimensions during practical experiments or measurements greatly exceed the dimensions of the EDL.

The electrified interface together with the diffusion layer sets the stage for the electrochemical reactions and as seen from Fig. 1.1 is relatively complex. Thus, it is a good idea to divide the complete electrochemical interface into smaller parts and see what atomistic simulations can contribute to our understanding in each case.

1.2.1 Electrode

If the electrode material used in the experiments is a single crystal metal, such as Pt(111) or Pt(100), its structure is relatively easily simulated by slab calculations with periodic boundary conditions. However, as electrodes used in practise are often polycrystalline (especially when applications are considered) these single crystal studies can only act as model systems. For example, work function, which is needed when the value of electrode potential is calculated, is different for various lattice planes of the metal, and some kind of average must be used for a polycrystalline surface. In addition, the effect of surface oxidation, defects, roughness and impurities at the electrode surfaces will have a significant effect on both faradic and non-faradic currents in practice. In the case of surface oxidation, the surface composition can also change significantly. The starting material can be an almost pure metal, but after several consecutive cycles in basic environment the surface can be completely oxidized. The real value of the high purity and well controlled single crystal electrochemical studies is that they provide a convenient experimental benchmark for the simulations.

In case of non-metallic materials, the situation is even more challenging. For example, with carbonaceous materials the surface contains different functional groups of which relative fractions depend heavily on the applied potential and the history of the sample. In addition, the standard (or in practice formal) potential of a metal with respect to its reduction reaction is obtained as a steady state with respect to dissolution and deposition (in aqueous solutions), whereas the formal potential of carbonaceous materials is typically established with respect to some surface reactions occurring on a carbon surface under experimental conditions (owing to the restricted solubility of the carbon to aqueous solutions). Moreover, the structure of the solvent next to the electrode can be expected to be different with metals and carbonaceous materials due to differences in location of the image plane [3] and chemical interactions. The chemical nature of the surface heavily influences the so-called inner sphere redox reactions where breaking and making of chemical bonds take place. This process is also accompanied by the simultaneous loss of solvation sheet of the redox active species as well as that of the electrode. Thus, in addition to the direct chemical interactions between the redox probe and the electrode, all factors influencing the solvent structure around the redox species and next to the electrode, will have a strong effect on inner sphere reactions (ISR). This is in contrast to the so-called outer sphere redox couples where no bond breaking or formation during the reactions take place and the redox species retain their solvation sheet to a large degree. If one considers the double layer picture above one can roughly state that outer sphere reaction (OSR) species never enter the IHL and the electron transfer takes place at a fixed distance from the OHP without direct contact with the electrode. More discussion on the differences between OSR and ISR probes is provided below.

1.2.2 Electrical double layer

To complement the classical macroscopic models of the EDL discussed above, many computational models have been introduced, these models are often based on modified Poisson–Boltzmann approaches and are thus continuum models similar to the classical ones [5, 6]. They have proven themselves practical tools to account for the presence of the electrical double layer at the electrochemical interfaces. However, as these are simplifications of the actual complex electrochemical interface, there are issues and open questions, such as which features must be included into the models to achieve an adequate description of the EDL and which features can be left out to simplify the approach. In addition, as one of the goals of applying molecular simulation to the EDL region is to provide a molecular interpretation of the experimental capacitance data, continuum models are not appropriate for this. Thus, recently much activity has been devoted to combining the continuum models with explicit DFT based (grand canonical) simulations to provide an accurate description of the interface at the most important locations complemented with an adequate continuum description in the less critical regions. In this book, several examples of various recent approaches striving to accomplish this task are presented.

1.2.3 Solvation sheets

As already discussed, solvation of the redox active species is highly important for electrochemical reactions. It is well known that the structure of the solvation sheet is typically time dependent and thus requires explicit calculations to simulate thermodynamic averages. In a recent paper a computationally feasible methodology to estimate solvation free energies from dynamically evolving solvation sheets was presented [7]. The main outcome from the study was that only by explicitly treating water molecules dynamically (using *ab initio* molecular dynamics (AIMD)) can the whole complexity of the solvation process be addressed. As many of the electrochemical reactions of practical importance include ion transfer, the changes occurring in the solvation sheets of the redox active species as they approach the electrode must be taken into account. This has been mainly done by utilizing potential of mean force approach (PMF) [8]. However, the approach does not provide details of the changes of the solvation sheet structure during ion movement and thus new approaches need to be developed. One effect omitted (to our knowledge) from all the solvation effect calculations so far is the explicit effect of ionic strength on the process. As already discussed, the ionic strength has a strong effect on the structure of the EDL and likewise on the solvation sheet. According to the Debye–Huckel theory all ions are surrounded by an atmosphere of counter ions in the solution and this should be taken into account when solvation effects are simulated. During dynamic changes in the solvation sheet there is likely to be also changes in the local concentrations of these counter ions. This charge in/outflow can be modelled using a grand canonical approach and will have an effect when for the electron transfer considerations (see below) as changes in the local ionic atmosphere should accompany the fundamental act of electron transfer. There are various molecular level solution models that take explicitly into account the presence of ions and their interactions in the electrolyte [9, 10], but to our knowledge a combination of these type of models with explicit solvation calculations are missing. One link between computational studies and experimental information on solutions that has not been extensively utilized with topics central to the present book is provided by the Kirkwood–Buff theory [11, 12]. Application of Kirkwood–Buff theory and especially the approach developed by Ben-Naim [12] within the context of this book is briefly touched upon in Chapter 3. The theory provides relationships between particle distribution functions accessible by computational techniques in the grand canonical ensemble and derivatives of the chemical potentials of all species involved thus enabling direct comparison between experiment and simulations.

1.2.4 Electrode potential

The inner or Galvani (ϕ) potential of a phase is divided into outer (Volta, ψ) potential and surface (χ) potential (Fig. 1.2). These formal definitions are based on the following thought

Figure 1.2 Schematic presentation showing outer, surface and inner electrode potentials. See text for further details.

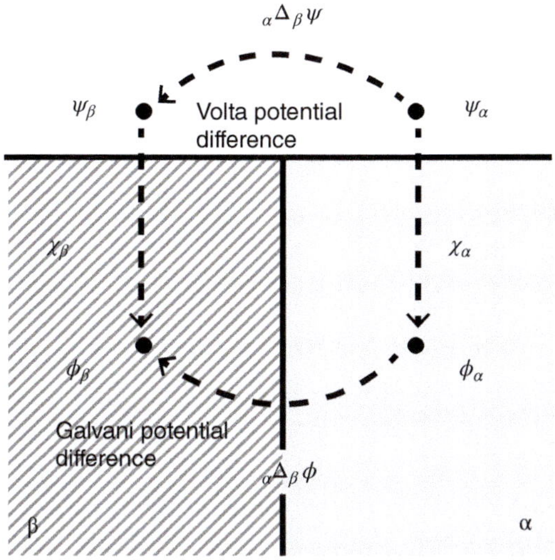

experiments. In the case of outer potential, the electrode and the electrical double layer (solvent) are dismantled and both are placed in contact with the vacuum. A test charge of unit positive charge is brought from infinity (in vacuum) close to the electrode surface to a location that is just outside the reach of image forces arising from the electrode surface. The work done in bringing a test unit charge from infinity up to this point as well as the potential at this point, is determined solely by the charge on the electrode and is not influenced by any image interactions (chemical terms) between the test charge and the electrode. This potential just outside the charged electrode is defined as the outer potential of an electrode. A similar thought experiment is then carried out for the solution phase to obtain its outer potential. After these two steps have been completed, one may reassemble the original interface to obtain the outer (Volta) potential difference. It is to be noted that the outer potential is defined on charged but dipole free surface. Further, it should be emphasized that the Volta potential difference for an electrode can be measured experimentally [13], which is not the case for the inner and surface potentials as will be discussed.

To define the surface potential another thought experiment is carried out. The electrode-solvent interface is again dismantled and placed in contact with the vacuum and additionally all excess charge is eliminated from both phases (electrode and solvent). However, the surface dipoles on both surfaces remain intact. Again a test charge (unit positive charge) is brought from infinity towards the dipole layer and then made to cross it. The work done during this process defines a potential, which is not related to the excess charge on the solution/electrode phase as the excess charges on the phases are maintained at zero. Since the work is related to penetrating a surface layer on the electrolyte/electrode, the corresponding potential is called a surface potential (or a χ potential). If, now, the electrode and solution phases are brought together, there will be dipole layers in the two phases and the work done to take the test charge through these layers defines the surface potential difference. As the surface potentials above were defined separately in contact with vacuum there will be some changes in both of them when we reassemble the system (electrons overspill less from the metal in the presence of solvent than at vacuum interface and solvent molecules are also differently oriented at electrode/solvent than at vacuum/solvent interface), which have to be taken into account in the calculations. It is also important to realize that surface potentials cannot be measured (as the potential and its reference state are situated in different phases and the interactions are not purely electrostatic). This means that the inner potential of an electrode cannot be measured either and only differences between two inner potentials can be measured.

The charge free phase used while defining the surface potential gives us access to an important property of an electrode/electrolyte system called potential of zero charge. This is the potential where there is no excess charge on the surface of the electrode in contact with an electrolyte. This quantity can be measured and is typically (on clean metal electrodes) associated with the minimum in the double layer capacitance. However, as most of the electrolytes adsorb on the surface of the electrodes at least to some degree the measured quantity is typically considered as the potential of zero total charge as it contains in addition to free charge also any charge associated with the adsorbed species [14]. As stated above, out of the constituents contributing to inner potential, only outer potential can be measured in practice. This means that also the inner potential cannot be measured and only differences in inner potentials can be measured. Therefore, an arbitrary reference has been established that is used as the reference point for all other redox reactions. This is called the standard hydrogen electrode (SHE) and is defined under standard conditions (activities of the species taking part into the reaction are all equal to one, pressure is 1 atm and pH is zero) and put as zero at all temperatures. If we relax the requirements of unit activity a bit and use 1 M acid solution instead of unit activity we obtain a quantity called the normal hydrogen electrode (NHE) whose value is practically the same as that of SHE. Another variant of of hydrogen electrodes used in practice is the reversible hydrogen electrode (RHE) where the reference electrode is immersed into the test solution used for the the electroanalytical experiment (and hydrogen is bubbled through the cell). Other widely used reference electrodes in aqueous solutions include saturated calomel electrodes (SCE) as well as silver/silver chloride (Ag/AgCl), whose potentials are then defined against the SHE reference.

The types of references described above are not ideal ones to be used in computational studies, where the vacuum level (either absolute or just outside solution) or deep in the electrolyte would perhaps be more feasible. Thus, there has been number of studies how to relate these scales to each other in computational studies. This is intimately related to the incorporation of the electrode potential to the calculations in the atomistic simulations and therefore there are many examples about this work also in this book. The theoretical basis for the connection between SHE and the vacuum level reference was established already in the 1960s by Bockris et al. [15]. The concept of an "absolute electrode potential" raised significant discussion among electrochemists during the 1970s and 1980s. The disputes were settled in 1990 [16] when it was shown that no universally general "absolute electrode potential" exists – instead a "single electrode potential" scale should be used. Besides semantics, the use of a single – rather than absolute – electrode potential scale opens up the possibility of choosing the potential scale conveniently. What is convenient depends on whether one utilizes the single potential experimentally or computationally. For experiments the most useful scale is set by the electron at rest in vacuum while in a computational work the most convenient reference might an electron in the solvent interacting only electrostatically with other species as is the case for Poisson–Boltzmann models [5]. Indeed, it can be shown [16] that independent of the reference state, all single electrode potentials are related to each other by

$$E(abs) = E(red) + K \tag{1.1}$$

where K is a constant depending on the reference scale and the reduced absolute potential is given by

$$E(red) = \delta\phi_M^S - \mu_e, \tag{1.2}$$

with a Galvani potential difference $\delta\phi_M^S$ between the electrode and solvent and μ_e the chemical potential of electrons, i.e. the Fermi-level (see Flecther). Therefore, depending on the definition of K various single electrode potentials can be used and inter-converted flexibly. Nevertheless, it should not be forgotten that the experimental determination of single inner electrode potential is not possible.

When we have redox active species in the system (or a metal in contact with aqueous solvent) there will be faradic reactions taking place and the reaction rates are directly proportional to the measured current (density). When forward and backward rates are equal and no net current is flowing in the system, a steady state has been established and the corresponding potential is called the standard electrochemical potential for that particular reaction. If the species participating in the reaction are not in their standard states (activities are less than one when pure element reference state is used), the quantity is called the formal potential. In some cases the measurement of formal potentials for especially sluggish reactions can be difficult. Therefore, in a recent paper, a computationally feasible methodology to calculate formal potentials of various species without any experimental input was proposed [7]. This opens up new possibilities to predict beforehand these values for molecules without appropriate electrochemical data as well as providing us with the means to assess formal potentials of molecules of which redox reactions are kinetically so sluggish that they do not take place inside the solvent electrochemical window despite the (possible) thermodynamic feasibility. The solvents' thermodynamic electrochemical window defines the interval of potentials between the anodic and cathodic background processes arising from the decomposition of the solvent. In aqueous solvents this means the formal potential difference between hydrogen evolution reaction (HER) at the cathodic end and oxygen evolution reaction (OER) at the anodic end. The kinetic electrochemical window used in practice is defined as the potential range where the currents arising from these reactions do not yet significantly interfere with any other currents arising from the faradic reactions of interest. The kinetic electrochemical window can be significantly wider than the thermodynamic one. As discussed, the formal potential is associated with a steady state electron transfer (no net current) and the current corresponding to the situation is called the exchange current density, which is directly related to the heterogeneous electrochemical rate constant and thus to the intrinsic kinetic feasibility of the reaction under given conditions (at the formal potential). As this current density can be measured (by Tafel analyses or electrochemical impedance spectroscopy (EIS), for example) and, based on that, the heterogeneous electrochemical rate constant at formal potential calculated, the latter provides a natural variable that can be compared with the computational results.

1.3 Chemical interactions between the electrode and redox species

Many of the practically important electrochemical reactions proceed *via* an inner sphere route and therefore involve specific interactions between the electrode and the redox species in question. One of the most important processes is the adsorption of either reactants, intermediate species or products on the electrode surface. This may lead to electrocatalysis or blocking of the electrode surface depending on the nature and extent of adsorption. This phenomenon can also be utilized to improve the performance of the electrochemical devices. For example in sensing applications, in a biological environment, one of the greatest challenges is to increase the selectivity towards the analyte and expel interfering elements. Specific adsorption of various species on the electrode surface can be used to improve selectivity and to shift the oxidation or reduction peaks anodically or cathodically depending on the details of interaction of the electrode surface with the reactant and product species. If one can induce preferential adsorption of product species of a given oxidation reaction on the electrode surface this will result into reduction in the activation energy for the reaction and corresponding cathodic shift in the oxidation peak or a appearance of a pre-peak depending on the solute concentration. Favoring the adsorption of a reactant species naturally provides an opposite effect. Another manifestation of adsorption and subsequent surface reactions on the electrode is the passivation of the electrode surface which is typical for many biological reactions [17]. It is obvious that computational studies must play a key role in understanding these types of phenomena and in defining the optimal surface chemistries for a given analyte as shown in Refs. [18, 19].

In the area of chemical catalysis there is a long history of utilizing computational schemes to understand catalytic behavior at the microscopic level [20]. However, in the electrochemical case, owing to the complexity of the electrified interface, the situation is considerably more difficult. Some semi-quantitative guidelines, such as volcano plots exist, but the validity of these rather *ad hoc* correlations remain a controversial issue and therefore they have only a limited predictive value [21]. Much of the work in this area has been concentrating in defining suitable descriptors to access interfacial interactions and electrocatalytical reactions with variable success [22]. However, to our knowledge there are no studies where adsorption studies on electrode surface would have been combined with explicit solvation/-electrolyte models and interfacial charge transfer.

Most of these correlations and guidelines have been directly adopted from the field of heterogeneous catalysis. However, it needs to be stressed that there is no *a priori* guarantee that concepts derived for solid-gas interfaces can be directly utilized at electrified solid-liquid interfaces.[23] For instance, the celebrated d-band model [24] in heterogeneous catalysis is derived from a Newns–Anderson–Grimley (NAG)[25, 26] Hamiltonian and the use of Greens function methods to define chemisorption functions. The role of the d-band center in the interaction between the molecule and surface is then inferred from these chemisorption functions. At electrochemical interfaces the NAG model needs to be extended to include the electrode potential and solvent interactions [27–29]. Both the electrode potential (Fermi-level) and the solvent enter the chemisorption functions and explicitly determine the interaction strength between the surface and molecule. Therefore, directly adopting the d-band model and its predictions from heterogenous catalysis to electrocatalytic environments should be done with care. The same care should also be exercised when transferring descriptors, volcano plots, Brønstedt–Evans–Polanyi free energy relationships between binding energies and reaction kinetics etc. from heterogeneous catalysis to the electrified interface.

1.4 Reaction kinetics at electrochemical interfaces

Arguably the most interesting and wide-spread electrochemical and electrocatalytic reactions are electron transfer (ET), proton transfer (PT), and proton-coupled electron transfer reactions (PCET). These reactions constitute the elementary steps in e.g. oxygen reduction and evolution reactions, hydrogen evolution, electrosynthesis of CO_2-based fuels and chemicals, organic electrochemistry, ammonia synthesis, and so forth. Understanding the chemistry of these reactions ET, PT, and PCET is at the very core of many electrochemical and electrocatalytic applications.

The traditional approach to quantifying electron transfer reactions is to use the Butler–Volmer (B-V) analysis, which is the linearized version of the more advanced Marcus–Hush theory of electron transfer [30]. The B-V equation provides a connection between the measured current density and the overpotential used. It is strictly valid only for pure electron transfer reactions (outer sphere), but it works surprisingly well also for ion transfer reactions (inner sphere) and electrocatalytic reactions. It has three important special solutions: (i) with overpotential of zero it reduces to the Nernst equation, (ii) with small overpotential (around 10 mV) it provides a linear relationship that enables a convenient way of assessing charge transfer resistance (and thus heterogeneous electron transfer coefficient), and (iii) at large anodic or cathodic overpotentials (more than 50 mV) it becomes the famous Tafel equation. The latter can be used to obtain mechanistic information about the redox reactions under study by analyzing the so-called Tafel slopes and transfer coefficients [31]. For OSR reactions the transfer coefficients are always close to ½ and they sum up to unity (they are in fact often associated with the name symmetry factor). They are also in general independent of temperature. On the contrary, in ion transfer/inner sphere (ITR/ISR) reactions the single electron transfer step is not necessary the rate determining step (RDS). There may be other steps than simple electron transfer that constitute the RDS or there are electron transfer steps before the RDS. In these cases the simple symmetry factor must be replaced by transfer coefficient, which includes symmetry factor (kinetics), but contains also stoichiometric and rate determining step considerations [32]. Thus, there are no grounds to assume that transfer coefficient in these cases would be close to 0.5 or anodic and cathodic parts would add up to one. It should also be noted that in the case of inner sphere reactions where adsorption plays a significant role, the close proximity of the electrode and the redox probe enable much higher interaction strengths than in the OSR case and thus electrocatalysis becomes possible.

1.4.1 Outer and inner sphere reactions

Traditionally, ET, PT, and PCET reactions at electrochemical interfaces have been divided to inner- and outer-sphere reactions. Outer-sphere reactions are rather independent of the electrode material and the role of electrode is mainly to provide a sink/source for the charge transfer. In outer-sphere reactions there is no bond breaking or making during the redox reaction (which of course does not rule out changes in the bond lengths or in the solvation sheet). Inner-sphere reactions on the other hand involve bond breaking and/or making, at least partial loss of their solvation sheet as well as chemical (specific) interaction between the reactants and the electrode – in fact the electrode should be considered as a reactive "species" .

Outer-sphere reactions on unoxidized metals are usually taken to be adiabatic (examples include such redox couples as $Ru(NH3)_6$ and $IrCl_6$) as they proceed extremely rapidly [33]. On oxidized metals, semiconductors as well as on insulators on other hand OSR reactions are considered as non-adiabatic transitions between initial and final states. Hence, the kinetics are based on variations of Fermis Golden Rule; the prefactor is determined by the non-adiabatic transition probability while the reaction energy barrier is determined by medium reorganization facilitating the non-adiabatic tunneling event. There are only a limited number of simple OSR couples in practice. In addition to the two species mentioned above, a ferrocene(methanol), some aromatic compounds and anthracene-based redox couples behave as OSR probes on most electrode surfaces (in appropriate solvents). On the other hand, ferrocyanide couple previously taken to be an OSR probe has been shown to experience strong chemical interactions with especially oxygen-based functional groups on the electrode surfaces [34, 35]. As the electronic interaction between the electrode and the redox couple decays exponentially as a function of distance it is possible to change an adiabatic OSR reaction into a non-adiabatic one by increasing the distance of the redox species from the electrode surface by using for example layers of self-assembled monolayer (SAMs) as shown experimentally in [36].

As pointed out in the chapter by Fletcher, one of the known shortcoming of the traditional Marcus–Hush theory is that it does not explicitly take into account effects of ionic strength (ions do not have the ionic atmosphere required by the Debye–Huckel theory). These effects can be included into expression for the reaction rate by changing the Coulomb interactions involved in the binary correlation function into the screened Coulomb interactions (in case of outer-sphere reacions between ions) as well as including the additional contribution from the reorganization of the ionic atmosphere (to reach the transition configuration) to the activation free energy, which will affect the probability of the electron transfer. With a suitable choice of reaction coordinate (by using diabatic vertical energy gap – although the validity to use this as a reaction coordinate has been challenged (see again chapters by Fletcher and Melander) and with a proper solvent model it may be feasible to take ionic strength into account directly also in the computational studies.

Most electrocatalytic reactions of practical interest take place *via* the inner-sphere route. In this case one requires at least two reaction coordinates out of which one is likely to be the distance of the incoming species from the electrode surface. This is in contrast to the outer OSR reaction that takes place from a fixed distance (from OHP typically), and thus one reaction coordinate is typically sufficient. Under these circumstances the problem becomes significantly more complicated as the chemical nature of the electrode surface and changes in the solvation sheets of the redox species and the electrode become decisive. There are many approaches that have tried to tackle this problem, but there are still many open questions related to these issues as exemplified also by this book. For example, approaches based on grand canonical DFT might in the near future become feasible tools to study practically interesting electron transfer reactions and be able to replace approaches based on model Hamiltonians (see below).

In many cases, especially in biological redox and electrocatalytic reactions, electron transfer is coupled with proton transfer. One example is oxidation of an important neurotransmitter dopamine, where in addition to two electrons two protons are also transferred. There are a few investigations where it has been claimed that the dopamine (DA) oxidation reaction is enhanced on electrode surface, which are capable of forming hydrogen bonds. However, no quantitative data has been provided to back up these claims. From the theory of PCET reactions it is known that if there is a tendency for strong hydrogen bonding from the acceptor (either molecule or electrode), this tends to make the proton transfer more adiabatic and thus necessarily reduces the activation energy for tunneling. This might explain the experimental observations regarding dopamine.

Simultaneous PCET reactions have received significant interest also from the electrocatalytic community. The computational electrocatalysis field in particular has focused almost exclusively on the study of simultaneous PCET. Usually decoupled ET/PT have been neglected because the electron transfer is considered very fast and coupled to the proton motion in inner-sphere electrocatalytic reactions. Also simultaneous transfer of electrons and protons seems thermodynamically favorable because formation of unstable charged species is avoided. It should, however, be kept in mind that the decoupled PCET might be kinetically favored over the simultaneous pathway. Furthermore, whether the simultaneous or decoupled PCET is the dominant pathway depends on the solvent, pH, the electrode material, reaction etc. For instance, strongly binding catalysts such as platinum seem to favor simultaneous PCET while a decoupled pathway is observed on weakly binding catalysts such as carbon or gold. Also the lack of protons in non-protonated solvents or under alkaline conditions favors the decoupled pathway [37].

One example of a biologically important non-concerted proton electron transfer can be found from the redox reactions associated with the family of catechols. The structures of catechol, methylcatechol, and dopamine are closely related and are shown in Fig. 1.3.

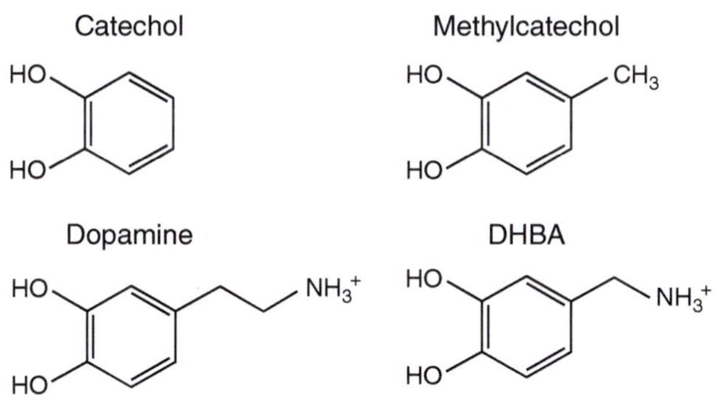

Figure 1.3 The structures of catechol, methylcatechol, dopamine and DHBA (3,4-dihydroxybenzylamine) at physiological pH [17]. Source: From Tommi Palomki, Sara Chumillas, Sami Sainio, Vera Protopopova, Minna Kauppila, Jari Koskinen, Vctor Climent, Juan M. Feliu, and Tomi Laurila. Electrochemical reactions of catechol, methylcatechol and dopamine at tetrahedral amorphous carbon (ta-c) thin film electrodes. Diamond and Related Materials, 59:30–39, 2015.

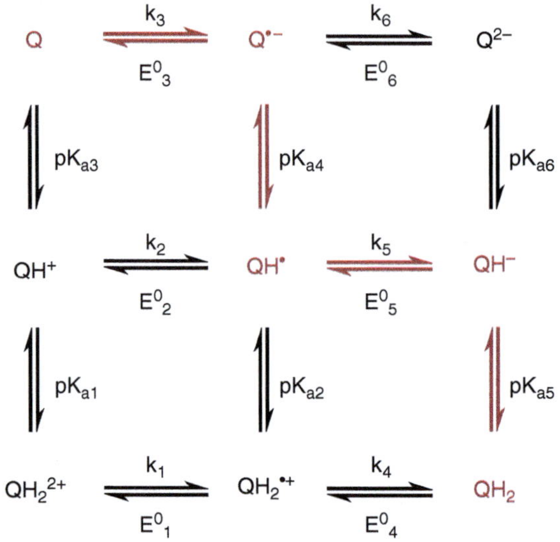

Figure 1.4 The scheme of squares represents the possible electron and proton transfer sequences of catechols. The formal potentials (E^0) and heterogeneous reaction rate constants (k) affect electron transfer whereas pKa values relate to the oxidation state of the molecule. The suggested transfer sequence at neutral pH on GC is illustrated in red [17]. Source: From Tommi Palomki, Sara Chumillas, Sami Sainio, Vera Protopopova, Minna Kauppila, Jari Koskinen, Victor Climent, Juan M. Feliu, and Tomi Laurila. Electrochemical reactions of catechol, methylcatechol and dopamine at tetrahedral amorphous carbon (ta-c) thin film electrodes. Diamond and Related Materials, 59:30–39, 2015.

These three catechols oxidize to the corresponding orthoquinones in a two-electron, two-proton transfer process as discussed above. The order of proton and electron transfer follows the scheme of squares, presented in Fig. 1.4, and is heavily dependent on pH. This model assumes that protonations are at equilibrium and that the electron transfer step is rate-determining [38–40].

The proposed electron and proton transfer sequence at neutral pH $H^+ \rightarrow e^- \rightarrow H^+ \rightarrow e^-$ (sequence starting from QH_2 at the lower right corner) and it is illustrated in red in the scheme of squares in Fig. 1.4. This transfer sequence has been suggested for hydroquinone on platinum [38] and for methylcatechol, 3,4-dihydroxybenzylamine (DHBA), and 3,4-dihydroxyphenylacetic acid (DOPAC) on carbon paste electrodes [39, 40]. Since DA is structurally very similar to DHBA (Fig. 1.4), it was suggested in [17] that it most likely follows the same reaction sequence during its oxidation.

1.4.2 Computational aspects

The separation of electrochemical reaction into inner- and outer-sphere reactions is also reflected in the quantum mechanical treatment used for computing the reaction energetics. Outer-sphere reactions are typically treated using model Hamiltonians parametrized in terms of the environment reorganization and interaction strength (coupling element) between the initial and final states. In an electrochemical setting the electrode potential is usually added as an external parameter controlling the reaction energy in a predefined way. While the model Hamiltonians are often analytically solvable, parametrizing these models in a meaningful way is difficult. For instance, the solvent contributions are often treated using simple (Born) continuum models, which overestimates the solvation energies and Poisson–Boltzmann double layer models. As the OSR reactions take place from a fixed position (redox species is assumed to be localized at OHP) a single reaction/solvent coordinate is typically used. The electrode electronic structure is usually based on semi-elliptic d-band density of states models or more complicated analytic models [41] and the role of electrode potential is to shift the Fermi-level of an otherwise static electronic structure. Transition probabilities are computed as orbital-to-orbital transitions between localized orbitals of the electrode and a molecular redox-level; computing these matrix elements from e.g. density function theory depends heavily on the orbital localization [42]. Due to a large number of free and tunable parameters, it is possible that various parameter sets can fit the same experimental data [43]. To date, studies on outer-sphere kinetics with general Hamiltonians including the solvent, electrode potential, reactive species, and general rate theory have not yet been accomplished (to our knowledge).

Unlike outer-sphere models, studies on inner-sphere reactions are routinely carried out using general first principles models, usually based on large-scale density functional theory calculations. The advantage of these models is that they can accurate model the electronic structure to capture the intricate bonding between the electrode and reactants. Nowadays, also the electrode potential and solvent can be included in the model. Solving the electronic structure problem, however, requires significant computing facilitates and (both human and machine) time. Analytically solvable models of inner-sphere reactions have proven more difficult and the most successful theory is Schmickler's extension of the Newns–Anderson–Grimley chemisorption theory to electrochemical systems [44]. In fact, using the Schmickler–Newns–Anderson–Grimley model for ET and its extension to PCET reactions can provide a general model Hamiltonian for both adiabatic (inner-sphere) and non-adiabatic (outer-sphere) reactions [28, 29]. The inner-sphere reaction kinetics are usually addressed using adiabatic transition state theory (TST) [28, 29, 42]. Unlike in the outer-sphere case, the reaction coordinate may include the bonds being broken/formed. In the case of PT and PCET, for instance, movement of the transferring proton provides the reaction coordinate. If the solvent polarization/reorganization is regarded important, also the solvent coordinate should be included in the adiabatic TST model [27–29] .

1.4.3 Challenges

Given the analytical and numerical tools to compute various energetic quantities of an electrocatalytic reaction, it should still be kept in mind that computing the reaction barrier is not enough (be it the adiabatic transition state barrier or the non-adiabatic barrier); computing the reaction pre-factor determining the transition probability is equally important. For instance, is the reaction non-adiabatic? Is nuclear tunneling for PT or PCET important? Another uncertainty is the preconceived choice of the mechanism: Does the reaction take place in a concerted PCET step or do ET/PT follow each other sequentially?

Another crucial question regards the choice of the reaction coordinate(s) and the way of computing the reaction rate: should the reaction be described in terms of a non-adiabatic transition and environment reorganization or is it more appropriate to a transition state theory description of bond forming and breaking? The latter has been widely used in the field computational heterogeneous catalysis and practically all first principles work on electrocatalytic reaction barriers. On the other hand, the non-adiabatic formalism has been the preferred way in the traditional theoretical electrochemical and electrocatalytic fields utilizing model Hamiltonians. Finally, in multistep reactions care should be taken to verify which step is the rate determining step (RDS) and can the preceding steps be taken to be under steady state conditions etc.

1.5 Charge transport

In addition to interfacial electron transfer, redox reactions may be controlled in some cases also by electron transport inside the electrode materials. This is the case e.g. for semiconducting electrodes encountered in some (photo)electrocatalytic applications, battery materials, or, as shown in a recent investigation, when using ultra thin tetrahedral amorphous carbon (ta-C) films as electrodes [45]. It can be expected that similar phenomena can be observed also in other dielectric materials with highly heterogeneous spatial distribution of more conductive and more insulating parts. Depending on the material, the nature of the charge conduction may vary. As discussed in Chapter 11 by Vegge et al., for example, in metal oxide materials hopping-like electron transfer between different redox centers has been demonstrated as the conduction mechanism. It has also been recently demonstrated that non-adiabatic effects in electron transfer may have considerable role in the poor conduction of electrons in lithium sulphur batteries. [46] Again in the field of lithium battery materials, slow lithium diffusion may become the rate-determining step. While the above examples are based on an equilibrium transition state theory -like picture, a broad class of non-equilibrium transport phenomena may also be encountered in electrochemical systems. To investigate these types of processes at the DFT level (or beyond) requires the use of techniques based on non-equilibrium Green's function or similar.

1.6 Mass transport to the electrode

Mass transfer at electrodes takes place due to electron transfer reactions, where the species involved in the reactions are either depleted or produced at the electrode surface. As a consequence, (electro)chemical potential gradients will develop in the system leading to mass transfer. The modes of mass transport are:

1. Diffusion, which is the movement of a chemical species caused by a chemical gradient i.e. a difference in chemical potential (quite often approximated by concentration differences).
2. Migration, which is the movement of a charged species under the effect of an electric field.
3. Convection, resulting from a fluid flow that occurs naturally as a consequence of density gradients or forced by the cell design (for example a rotating disk electrode).

The mass transfer to an electrode due all these effects is coined in the Nernst–Planck equation. An exact solution of the Nernst–Planck equation that takes into account all of the three modes of mass transfer is difficult. Therefore, electrochemical systems are usually designed so that the mass transfer of an electroactive species at the electrode can be restricted to diffusion alone.

Diffusive transport from the bulk to the electrode surface takes place within the diffusion layer whose thickness is time dependent and varies from practically zero at the start of the experiment to several mm as discussed in section 2. Typically the mass transport by diffusion is approached by continuum methods by: (i) first using Fick's II law, with appropriate initial and boundary conditions to obtain time dependent concentration profile and

(ii) then using this profile (at a fixed time) in Fick's I law combined with the relationship between current and diffusion flux at the surface of the electrode. This classical continuum approach works well when one is working with electrodes larger than nanometer size, but when the diffuse double layer and diffusion layer start to overlap owing to the reduced size of the electrode the classical treatment might not be valid anymore. Naturally, also all kind of atomic-level information about the transfer processes are missing from this treatment.

There have been some attempts to use molecular dynamics to obtain a more microscopic view of the diffusion processes in electrolytes [47, 48]. However, we are unaware of any investigations where detailed atomistic studies would have been carried out where diffusion to the electrode would have been coupled to the redox reactions taking place at the electrode surface. This type of investigation undoubtedly requires multilevel approaches.

1.7 Summary

As described above, the electrochemical process can be divided into several steps that include in the case of inner sphere redox species (at least):

1. mass transfer to the electrode
2. possible adsorption on the surface
3. diffusion on the surface to find a reactive site
4. electron transfer and other chemical reactions
5. transport of the electron through the electrode to the outside circuitry
6. diffusion away from the reactive site
7. desorption from the surface
8. diffusion away from the surface region.

In addition to these steps related to the faradic reaction there is always the non-faradic current present, which is associated with the changes in the fine structure of the electrical double layer as a function of potential. In this book, there are many different examples where parts of the electrochemical interface have been assessed from the atomic perspective using different levels of theory and computation. These studies have provided a wealth of very important atomic level information that can be used to rationalize the experimental results as well as to provide a more fundamental understanding of the phenomena taking place at the electrified interface. There are also examples in this book where multilevel approaches have been utilized in order to model the whole interfacial area. This is an area which undoubtedly will grow in importance in the near future and requires new innovative approaches to coarse grain the electrified interface during simulations to keep the computational burden bearable.

There are many challenges in experimental electrochemistry that will benefit greatly from the increased understanding arising from these computational studies. One particularly important topic is electrocatalysis, whose understanding holds the key to better advanced energy solutions, superior sensor devices, as well as to CO_2 mitigation technologies, to name just a few. Taking into account the advancement of the field during the last 15 years, we can expect a plethora of major breakthroughs in the next coming 15 years in this field. This book tries to show the current state of the art of computational and theoretical atomic-scale electrochemistry. As the reader hopefully notices, there are various different flavours in how the same problems are approached, but that also unifying concepts and methods are emerging that can hopefully in the future bridge over various levels of theory and time scales that are needed in computational electrochemistry.

References

1 Martin Z. Bazant, Brian D. Storey, and Alexei A. Kornyshev. Double layer in ionic liquids: Overscreening versus crowding. *Phys. Rev. Lett.*, 106:046102, Jan 2011.

2 J. Bard Allen, M. Stratmann, E Gileadi, and M Urbakh. *Thermodynamics and Electrified Interfaces*, volume I. Wiley, 2002.

3 N.B. Luque and W. Schmickler. The electric double layer on graphite. *Electrochimica Acta*, 71:82–85, 2012.

4 J. O'M. Bockris, E. Gileadi, and K. Mller. Dielectric relaxation in the electric double layer. *The Journal of Chemical Physics*, 44(4):1445–1456, 1966.

5 Marko M. Melander, Mikael J. Kuisma, Thorbjørn Erik Køppen Christensen, and Karoliina Honkala. Grand-canonical approach to density functional theory of electrocatalytic systems: Thermodynamics of solid-liquid interfaces at constant ion and electrode potentials. *The Journal of Chemical Physics*, 150(4):041706, 2019.

6 Nicolas G. Hörmann, Oliviero Andreussi, and Nicola Marzari. Grand canonical simulations of electrochemical interfaces in implicit solvation models. *The Journal of Chemical Physics*, 150(4):041730, 2019.

7 Miguel A. Caro, Olga Lopez-Acevedo, and Tomi Laurila. Redox potentials from ab initio molecular dynamics and explicit entropy calculations: Application to transition metals in aqueous solution. *Journal of Chemical Theory and Computation*, 13(8):3432–3441, 2017.

8 Leandro M. C. Pinto, Paola Quaino, Mauricio D. Arce, Elizabeth Santos, and Wolfgang Schmickler. Electrochemical adsorption of oh on pt(111) in alkaline solutions: Combining dft and molecular dynamics. *ChemPhysChem*, 15(10):2003–2009, 2014.

9 Nav Nidhi Rajput, Xiaohui Qu, Niya Sa, Anthony K. Burrell, and Kristin A. Persson. The coupling between stability and ion pair formation in magnesium electrolytes from first-principles quantum mechanics and classical molecular dynamics. *Journal of the American Chemical Society*, 137(9):3411–3420, 2015.

10 Kim D. Collins, George W. Neilson, and John E. Enderby. Ions in water: Characterizing the forces that control chemical processes and biological structure. *Biophysical Chemistry*, 128(2):95–104, 2007.

11 John G. Kirkwood and Frank P. Buff. The statistical mechanical theory of solutions. I. *The Journal of Chemical Physics*, 19(6):774–777, 1951.

12 A. Ben-Naim. Inversion of the kirkwood–buff theory of solutions: Application to the water–ethanol system. *The Journal of Chemical Physics*, 67(11):4884–4890, 1977.

13 M. Gamboa-Aldeco J. O. M. Bockris, A. K. Reddy. *Modern Electrochemistry: Fundamentals of Electrodics*, volume 2a. Plenum Press, 1998.

14 V. Climent, N. García-Araez, E. Herrero, and J. Feliu. Potential of zero total charge of platinum single crystals: A local approach to stepped surfaces vicinal to Pt(111). *Russian Journal of Electrochemistry*, 42(11):1145–1160, Nov 2006.

15 J. O'M. Bockris and S. D. Argade. Work function of metals and the potential at which they have zero charge in contact with solutions. *The Journal of Chemical Physics*, 49(11):5133–5134, 1968.

16 Sergio Trasatti. The "absolute" electrode potential—the end of the story. *Electrochimica Acta*, 35(1):269–271, 1990.

17 Tommi Palomäki, Sara Chumillas, Sami Sainio, Vera Protopopova, Minna Kauppila, Jari Koskinen, Víctor Climent, Juan M. Feliu, and Tomi Laurila. Electrochemical reactions of catechol, methylcatechol and dopamine at tetrahedral amorphous carbon (ta-C) thin film electrodes. *Diamond and Related Materials*, 59:30–39, 2015.

18 Anja Aarva, Tomi Laurila, and Miguel A. Caro. Doping as a means to probe the potential dependence of dopamine adsorption on carbon-based surfaces: A first-principles study. *The Journal of Chemical Physics*, 146(23):234704, 2017.

19 Miguel A. Caro, Anja Aarva, Volker L. Deringer, Gábor Cásnyi, and Tomi Laurila. Reactivity of amorphous carbon surfaces: Rationalizing the role of structural motifs in functionalization using machine learning. *Chemistry of Materials*, 30(21):7446–7455, 2018.

20 Jeffrey Greeley. Theoretical heterogeneous catalysis: Scaling relationships and computational catalyst design. *Annual Review of Chemical and Biomolecular Engineering*, 7(1):605–635, 2016.

21 Cyrille Costentin and Jean-Michel Savéant. Heterogeneous molecular catalysis of electrochemical reactions: Volcano plots and catalytic Tafel plots. *ACS Applied Materials & Interfaces*, 9(23):19894–19899, 2017.

22 Federico Calle-Vallejo, David Loffreda, Marc T M Koper, and Philippe Sautet. Introducing structural sensitivity into adsorption–energy scaling relations by means of coordination numbers. *Nature Chemistry*, 7:403, Apr 2015.

23 Paola Quaino, Fernanda Juarez, Elizabeth Santos, and Wolfgang Schmickler. Volcano plots in hydrogen electrocatalysis–uses and abuses. *Beilstein Journal of Nanotechnology*, 5:846–854, 2014.

24 B. Hammer and J.K. Nørskov. Electronic factors determining the reactivity of metal surfaces. *Surface Science*, 343(3):211–220, 1995.

25 S. G. Davison and K.W. Sulston. *Green-Function Theory of Chemisorption*. Springer Netherlands, 2006.

26 J K Nørskov. Chemisorption on metal surfaces. *Reports on Progress in Physics*, 53(10):1253–1295, oct 1990.

27 Wolfgang Schmickler. A theory of adiabatic electron-transfer reactions. *Journal of Electroanalytical Chemistry and Interfacial Electrochemistry*, 204(1):31–43, 1986.

28 Wolfgang Schmickler. Adiabatic and non-adiabatic electrochemical electron transfer in terms of green's function theory. *Russian Journal of Electrochemistry*, 53(10):1182–1188, Oct 2017.

29 Yan-Choi Lam, Alexander V. Soudackov, Zachary K. Goldsmith, and Sharon Hammes-Schiffer. Theory of proton discharge on metal electrodes: Electronically adiabatic model. *The Journal of Physical Chemistry C*, 123(19):12335–12345, 2019.

30 Martin Z. Bazant. Theory of chemical kinetics and charge transfer based on nonequilibrium thermodynamics. *Accounts of Chemical Research*, 46(5):1144–1160, 2013.

31 Guidelli Rolando, Compton Richard G, Feliu Juan M, Gileadi Eliezer, Lipkowski Jacek, Schmickler Wolfgang, and Trasatti Sergio. Defining the transfer coefficient in electrochemistry: An assessment (IUPAC Technical Report), 2014.

32 J. O'M Bockris and Z. Nagy. Symmetry factor and transfer coefficient. a source of confusion in electrode kinetics. *Journal of Chemical Education*, 50(12):839, 1973.

33 T. Iwasita, W. Schmickler, and J. W. Schultze. The influence of the metal on the kinetics of outer sphere redox reactions. *Berichte der Bunsengesellschaft fr physikalische Chemie*, 89(2):138–142, 1985.

34 Richard L. McCreery. Advanced carbon electrode materials for molecular electrochemistry. *Chemical Reviews*, 108(7):2646–2687, 2008.

35 Tomi Laurila, Sami Sainio, and Miguel A. Caro. Hybrid carbon based nanomaterials for electrochemical detection of biomolecules. *Progress in Materials Science*, 88:499–594, 2017.

36 Dimitri E. Khoshtariya, Tina D. Dolidze, Leonid D. Zusman, and David H. Waldeck. Observation of the turnover between the solvent friction (overdamped) and tunneling (nonadiabatic) charge-transfer mechanisms for a $Au/Fe(CN)_6^{3-/4-}$ electrode process and evidence for a freezing out of the marcus barrier. *The Journal of Physical Chemistry A*, 105(10):1818–1829, 2001.

37 Marc T. M. Koper. Theory of multiple proton–electron transfer reactions and its implications for electrocatalysis. *Chem. Sci.*, 4:2710–2723, 2013.

38 E. Laviron. Electrochemical reactions with protonations at equilibrium: Part X. the kinetics of the p-benzoquinone/hydroquinone couple on a platinum electrode. *Journal of Electroanalytical Chemistry and Interfacial Electrochemistry*, 164(2):213–227, 1984.

39 Mark R. Deakin, Paul M. Kovach, K. J. Stutts, and R. Mark. Wightman. Heterogeneous mechanisms of the oxidation of catechols and ascorbic acid at carbon electrodes. *Analytical Chemistry*, 58(7):1474–1480, 1986.

40 Mark R. Deakin and R.Mark Wightman. The kinetics of some substituted catechol/o-quinone couples at a carbon paste electrode. *Journal of Electroanalytical Chemistry and Interfacial Electrochemistry*, 206(1):167–177, 1986.

41 Renat R. Nazmutdinov, Michael D. Bronshtein, and Elizabeth Santos. Electron transfer across the graphene electrode/solution interface: Interplay between different kinetic regimes. *The Journal of Physical Chemistry C*, 123(19):12346–12354, 2019.

42 Marko M. Melander. Grand canonical rate theory for electrochemical and electrocatalytic systems i: General formulation and proton-coupled electron transfer reactions. *Journal of The Electrochemical Society*, 167(11):116518, Jul 2020.

43 Zachary K. Goldsmith, Yan Choi Lam, Alexander V. Soudackov, and Sharon Hammes-Schiffer. Proton discharge on a gold electrode from triethylammonium in acetonitrile: Theoretical modeling of potential-dependent kinetic isotope effects. *Journal of the American Chemical Society*, 141(2):1084–1090, 2019.

44 E. Santos, A. Lundin, K. Pötting, P. Quaino, and W. Schmickler. Model for the electrocatalysis of hydrogen evolution. *Phys. Rev. B*, 79:235436, Jun 2009.

45 Tommi Palomäki, Niklas Wester, Miguel A. Caro, Sami Sainio, Vera Protopopova, Jari Koskinen, and Tomi Laurila. Electron transport determines the electrochemical properties of tetrahedral amorphous carbon (ta-C) thin films. *Electrochimica Acta*, 225:1–10, 2017.

46 Haesun Park, Nitin Kumar, Marko Melander, Tejs Vegge, Juan Maria Garcia Lastra, and Donald J. Siegel. Adiabatic and nonadiabatic charge transport in Li–S batteries. *Chemistry of Materials*, 30(3):915–928, 2018.

47 Dean R. Wheeler and John Newman. Molecular dynamics simulations of multicomponent diffusion. 1. Equilibrium method. *The Journal of Physical Chemistry B*, 108(47):18353–18361, 2004.

48 Dean R. Wheeler and John Newman. Molecular dynamics simulations of multicomponent diffusion. 2. Nonequilibrium method. *The Journal of Physical Chemistry B*, 108(47):18362–18367, 2004.

Part II

2

Retrospective and Prospective Views of Electrochemical Electron Transfer Processes: Theory and Computations

Renat R. Nazmutdinov[1] *and Jens Ulstrup*[1,2]

[1] *Kazan National Research Technological University, K. Marx Str., 68, 420015 Kazan, Republic of Tatarstan, Russian Federation*
[2] *Department of Chemistry, Building 207, Technical University of Denmark, 2800 Kongens Lyngby, Denmark*

2.1 Introduction – interfacial molecular electrochemistry in recent retrospective

2.1.1 An electrochemical renaissance

Prompted by concepts and methodologies of solid state and surface physics, a new era of physical and theoretical electrochemistry, almost like an electrochemical renaissance, began in the 1970s, transforming electrochemistry into interdisciplinary, highly sophisticated condensed matter physical science [1–4]. Notions were "clean", i.e. single-crystal, atomically planar electrochemical surfaces, and a range of surface spectroscopies including UV/Vis, infrared, Raman, surface enhanced Raman, and X-ray photoelectron spectroscopy.

From the same time statistical physics [5–7] and electronic structure theories [8–10] were introduced, with the palatable jellium or free electron gas model in initial focus [10]. Prior to the electrochemical renaissance the electrochemical surface was viewed as a hard wall with all physical events on the solution side. A jellium "tail" interfacing the electrolyte, exposed to surface charging and molecular "pseudopotentials", could incorporate exciting features of metal dependent capacitance [8, 9], solvent structure [6–9, 11], adsorption [12], and effects such as surface plasmonics [13] and second harmonic generation [14]. Well-defined electrochemical microenvironments [15–17] paved the way for electrochemical scanning tunneling (STM) and atomic force microscopy (AFM) (*in situ* STM and AFM) [4, 17–20], with both surface and tip under electrochemical control. *In situ* STM/AFM brought electrochemical surface mapping to atomic and adsorbates to sub-molecular resolution [3, 20–23] by a subtle phenomenon, *quantum mechanical tunneling* [24, 25]. New approaches to electrochemical *processes* were opened, creating a new area suitably denoted as single-molecule electrochemistry.

2.1.2 A bioelectrochemical renaissance

Similar boundary-traversing efforts were introduced in interfacial electrochemistry of redox metalloproteins [4, 26, 27], DNA-based molecules [4, 28–30], and even whole protein

Atomic-Scale Modelling of Electrochemical Systems, First Edition.
Edited by Marko M. Melander, Tomi T. Laurila, and Kari Laasonen.

and enzyme complexes [31]. It is in fact remarkable that molecules as large and fragile as metalloproteins can now be mapped to the single-molecule level directly in aqueous biological media. Single-molecule electrochemistry has prompted new concepts of interfacial electron transfer (ET) through complex molecules, relating both to the finite-size system nature (stochastic as opposed to statistical) [32] and to new single-molecule ET phenomena [23, 32–35]. Phenomenological theory offers powerful frames, but major computational efforts are now needed. For example, even the prettiest *in situ* STM images of the best organized self-assembled molecular monolayers (SAMs) reduce to "blobs", unless backed up by large-scale electronic structure computations. Such computations convert the images to electronic conductivity through the molecular fragments, with amazing detail and sometimes surprises, never deciphered just by looking at the images (see section 2.5).

We start with a retrospective view on quantum electrochemical *processes*, including less-known elements of electrochemical ET theory. We then proceed to the new nanoscale and single-molecule electrochemistry. Along the way, and in our third part, we overview computational methods and their application to some electrochemical systems.

2.2 Analytical theory of molecular electrochemical ET processes

2.2.1 A Reference to molecular ET processes in homogeneous solution

As a reference, we provide first a brief overview of simple ET processes in homogeneous solution. Core concepts carry over to electrochemical processes. Following pioneering work of Marcus [36, 37], Hush [38, 39], Levich and Dogonadze [40, 41], and their associates from the late 1950s, condensed matter ET theory developed to comprehensive sophisticated chemical rate theory throughout following decades. The core notion of ET as a quantum mechanical transition between donor and acceptor states, coupled to both solvent and local nuclear modes was recognized in this early phase. The quantum mechanical approach of Levich and Dogonadze, and of Hush drew inspiration from other condensed matter electronic processes, i.e. optical intervalence transitions [42], nonradiative ionization of impurity atoms in solids, electronic energy transfer, and excess electron localization in solids or liquids (polaron formation) [24, 25]. Proton and proton coupled ET, also introduced early, is another major area with close physical and formal relations to chemical ET.

The following attractive equation, first reported by Levich and Dogonadze [40] (1959), already summarizes these early achievements and is still a broad conceptual basis for chemical, electrochemical, and biological ET theory [24, 25]:

$$W_{AD} = \kappa_{AD}^{el} \frac{\omega_{eff}}{2\pi} S_{wv} \exp\left\{ -\frac{\Delta G_{wv}^{\neq}}{k_B T} \right\} \tag{2.1}$$

k_B is Boltzmann's constant, T the temperature, and ω_{eff} the effective vibrational frequency of all nuclear modes reorganized. The other symbols are defined below. We note:

- The ET process rests on Born-Oppenheimer separation of the electronic and nuclear system. Relaxation of this view is discussed elsewhere [43, 44]. The electronic subsystem is inherent in the transmission coefficient ("tunneling factor"), κ_{AD}^{el}, the nuclear system in

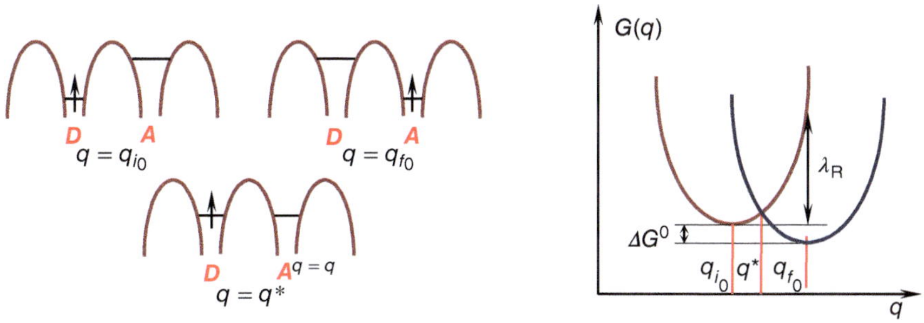

Figure 2.1 Left: The electronic energies at the electron donor (D) and acceptor (A). The electronic energies in the reactants' and products' state at different nuclear coordinate q values are indicated. Right: Potential (free) energy surfaces. ΔG^0 is the reaction free energy (driving force) and λ_R the total reorganization free energy.

the rest of Eq. (2.1). This view is the basis for the notion of potential (free) energy surfaces, Figure 2.1.

- In the diabatic limit of weak donor-acceptor electronic coupling, κ_{AD}^{el} takes the form

$$\kappa_{AD}^{el} = [T_{AD}]^2 \sqrt{\frac{4\pi^3}{\lambda_R k_B T \hbar^2 \omega_{eff}^2}} << 1; \quad \lambda_R = \lambda_{sol} + \lambda_{loc} \tag{2.2}$$

λ_{sol} and λ_{loc} are the reorganization free energies of the collective (polarization and other) nuclear modes and of local, still low-frequency metal-ligand stretching and other modes. The most important factor in Eq. (2.2) is the electronic coupling T_{AD}, with all information about donor-acceptor interactions, wave functions, tunneling and spin, intermediate "matter" etc. $\hbar = h/2\pi$ where h is Planck's constant.
$\kappa_{AD}^{el} \rightarrow 1$ in the opposite, adiabatic limit of strong interaction. ET is still an electron tunneling process, now reflected in a lower activation free energy, $\Delta G_{AD}^{\neq} \rightarrow \Delta G_{AD}^{\neq} - ½ T_{AD}$. The two limits are commonly bridged by the Landau-Zener correlation, taking the following form for motion along a pair of one-dimensional potential energy surfaces, $U_D(q)$ and $U_A(q)$ spanned by the nuclear coordinate, q

$$\kappa_{AD}^{el} = 1 - \exp(-2\pi\lambda_{el}); \qquad \gamma_{el} = \frac{|T_{AD}|^2}{h\nu|U_A'(q^*) - U_D'(q^*)|} \tag{2.3}$$

ν is the nuclear velocity, and $q = q^*$ at the potential surface crossing. $|U_A'(q) - U_D'(q)|$ is replaced by grad$|U_A'(\{q_k^*\}) - U_D'(\{q_k^*\})|$ in multidimensional $\{q_k\}$-space. Landau-Zener nuclear tunneling is discussed elsewhere [45]. An attractive form alternative to Eq. (2.3) is

$$\kappa_{el} = \frac{2|T_{AD}|^2}{T_{AD,critical}^2}; \qquad T_{AD,critical} = \sqrt{\frac{\lambda_R k_B T (\hbar\omega_{eff})^2}{4\pi^3}} \tag{2.4}$$

where $T_{AD,critical}$ is a "critical value" of T_{AD} separating the diabatic and adiabatic limits. Practical boundaries are $T_{AD,critical} \approx 0.05$ eV, i.e. roughly $\approx k_B T$ ($= 0.025$ eV, $T = 298$ °K).

- The activation free energy is

$$\Delta G^{\neq}_{wv} = \frac{1}{4\lambda_R}(\lambda_R + \Delta G^0 + \varepsilon_w - \varepsilon_v + w_A - w_D)^2 \tag{2.5}$$

where w_D and w_A are work terms equivalent to electrochemical double layer effects. ΔG^0 is the only parameter directly accessible (the reduction potentials). The other parameters can only be extracted indirectly, or by electronic structure or molecular dynamics computations.
- The high-frequency modes are represented by the Franck Condon overlap factor, S_{wv} coupling the vibrational states ε_v and ε_w, and modifying the driving force as $\Delta G^0 \rightarrow \Delta G^0 + \varepsilon_w - \varepsilon_v$. High-frequency modes are reorganized by *nuclear* tunneling, particularly important in proton and hydrogen atom transfer processes. Vibrational motion e.g. in organic skeletal, C-O etc. modes are also high enough that nuclear tunneling prevails.

Eqs. (2.1) and (2.5) do not specify the particular vibrational wave functions, and incorporate not only displaced harmonic modes but also mode deformation, anharmonicity etc. High- and low-frequency notions depend on the temperature. Proton stretching ($\omega \approx 3000\,\mathrm{cm}^{-1}$) is well in the tunneling regime at room temperature, metal-ligand stretching modes ($\omega \approx 3$–$400\,\mathrm{cm}^{-1}$) borderline cases. As temperature is lowered, increasingly lower frequency modes are quantized. *All* chemical processes are solely by nuclear tunneling as $T \rightarrow 0$.

2.2.2 Brief discussion of contemporary computational approaches

We provide here a prelude to our discussion of computational approaches to ET kinetics in section 2.6. Modern computational approaches rest on state-of-the-art quantum chemistry, mostly density functional theory (DFT) and molecular dynamics (MD) simulations [46–51]. The latter employ classical force field (pair potentials), or *ab initio* MD (AIMD), as well as Carr-Parrinello MD, CPMD) methods. All forces in both AIMD and CPMD are calculated from the position of the nuclei and the total electronic density. Quantum chemical calculations offer detailed information on geometry and electronic structure of the reactants and products. Wave functions are used to estimate the electronic transmission coefficient (weak coupling) or activation barrier lowering (strong coupling). Electrostatic reactants' potential, inner-sphere reorganization, potential energy surfaces for dissociative ET, and standard electrode potentials can be calculated as well. Solvation is addressed, either explicitly by including individual solvent molecules, or implicitly using continuum theories (for example, Polarized Continuum Model, PCM or conductor like Screening Model, COSMO). Mixed models are effective, when the first coordination shells include solvent molecules, while long-range solvation is treated implicitly. A hybrid scheme is frequently used for complex (biochemical) systems, when part of the solvent molecules is addressed by molecular mechanics (QM/MM, *e.g.* the review [51].

MD offers other insight. The following scheme is widely adopted [46]. The reaction coordinate (X) is defined as the fluctuational non-equilibrium energy difference between the equilibrium solvation free energies in the reactants' (R) and products' (P) states. Histograms (fitted to smooth the probability functions, $P_{R(P)}(X)$) describing such fluctuations are built in the next step. The free energy surfaces are then calculated as

$G_{R(P)}(X) = -k_B T \ln P_{R(P)}(X)$. The $G_R(X)/G_P(X)$crossing defines the activation barrier. In the strong-coupling limit $G_R(X)$ and $G_P(X)$ are combined using the resonance integral (cf. below). An important reorganization asymmetry effect (when the solvent reorganization energies of forward and reverse ET are different) was investigated [47], cf. section 2.2.4.1. The $P_{R(P)}(X)$s are usually well described by Gaussians. $G_{R(P)}(X)$ are then parabolic, but in some environments (for example protein/water interfaces [48] non-Gaussian fluctuations prevail, deviating from the quadratic free energy relation. Still another issue is high-frequency quantum solvent mode cutting off, relocating this free energy barrier contribution to nuclear tunneling. This correction can be estimated as [24, 25]

$$\xi = \frac{2}{\pi C} \int_0^{\omega^*} \frac{\mathrm{Im}\varepsilon(\omega)}{\omega \|\varepsilon(\omega)\|^2} \, d\omega, \tag{2.6}$$

C is the Pekar factor ($= \varepsilon_o^{-1} - \varepsilon_{st}^{-1}$ where ε_o is the optical and ε_{st} the static dielectric constant). $\varepsilon(\omega)$ is the complex dielectric spectrum and ω^* a vibrational frequency ($\approx k_B T/\hbar$) roughly separating nuclear tunneling and thermal activation. For water, $\xi \approx 0.8$–0.9 from dielectric spectra. As in MD simulations, nuclear tunneling is not considered directly, but can be addressed by path integral methods [49]. MD is also used to calculate the reactants' and products' potentials of mean force, i.e. the free energy surface [50]. The PMF is crucial, as it defines the distance of closest approach. PMFs usually take the form of a double-well potential but can also reduce to a single-well potential.

As a final note, resting on linear response theory [24, 25, 36, 52], the attractive quadratic free energy formalism is very general and has met with great success, but the generality is at the same time a limitation. The nature of the electronic-vibrational interactions cannot, for example, be extracted. Pressure-density response in polar media is equally represented [53], as long as the response is linear, but mode distortion (frequency changes), anharmonicity, mode mixing, and other non-linear effects common in chemical processes are not covered.

With the fundamental ET phenomenon broadly understood, present-day efforts seem to move along at least three major lines. One is to find new sophisticated target systems. Chemical and biological ET systems, as well as multifarious catalytic and electrocatalytic systems [4, 23, 54, 55], are examples. A second line is that new ET *phenomena* are under discovery in single-molecule electrochemistry [4, 23] (cf. section 2.5). Multi-ET, interplay between tunneling and hopping [56], and stochastic molecular systems are examples [32]. The third line is to integrate first principle electronic structure and MD computational efforts with ET parameters into the complex interfaces and single-molecule electrochemistry [57].

2.2.3 Molecular electrochemical ET processes and general chemical rate theory

The introduction of quantum mechanical approaches to the fundamental electrochemical process dates back to Gurney (1931) [58], whose description of the electrochemical dihydrogen evolution reaction already included core concepts of later theoretical work. Although qualitative – and for a long time overlooked – Gurney introduced electron tunneling, the continuous electronic spectrum of the electrode, and electronic-vibrational

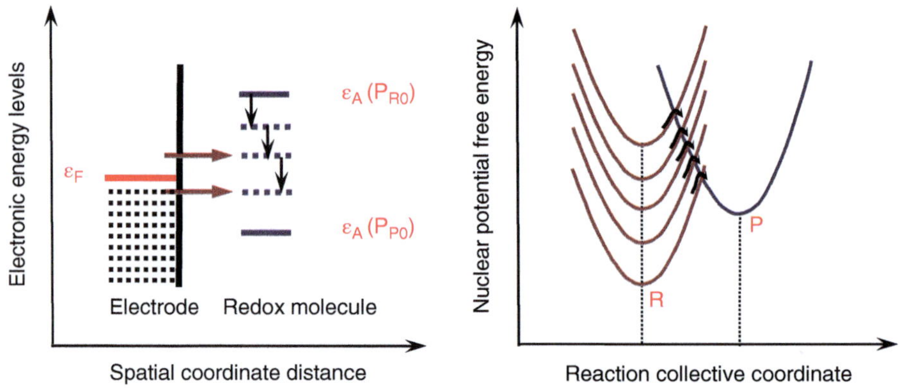

Figure 2.2 Left: Electronic energies of electrode and redox molecule at different solvent polarization P. P_{R0} and P_{P0} are the equilibrium values of P in the reactants' and products' states, respectively. Right: Potential free energy surfaces for electrochemical (cathodic) electron transfer process.

coupling. Gurney's work was followed nearly 30 years later by Gerischer [59], and slightly later, in more stringent form by Dogonadze, Chizmadzhev, Levich, and their associates. In spirit and reality these early studies, summarized by Dogonadze and Kuznetsov [60, 61], and by others [24, 25], constitute the basis also of present day electrochemical rate theory.

Electrochemical ET bears obvious relations to ET in solution. A strongly solvated molecular reactant is involved in both cases, with nuclear reorganization and driving force as core notions. Both classes are quantum mechanical transitions, with diabatic and adiabatic notions in common. The core *difference* is the continuous electronic manifold in the metal (or semiconductor) electrode. For, say, a cathodic process the electron is transferred from a continuous distribution of delocalized "quasiparticle" states to a localized state on the molecule, Figure 2.2. In general the electronic transition is accompanied by redistribution of all the other electrons. A continuum of potential (free) energy surfaces (PFESs) in both reactants' (electron in the metal electrode) and products' state (electron on the molecule) therefore in general represents the process (see section 2.6.2). In a single-electron view the simpler representation in Figure 2.2 applies, which we shall first discuss.

Figure 2.2 illustrates the electrochemical process. A single reactive attempt at the crossing of a pair of reactants' and products' surfaces occurs in homogeneous solution. Frictional forces in the transition region can modify this view [62]. Electrochemical ET involves multiple attempts as the nuclear system begins to relax after the first transition around the Fermi level. Most of these are abortive in the diabatic limit, but a multitude of forward and reverse transitions operate in the adiabatic limit, giving a random walk pattern across the crossing region. These have been analyzed [57, 63–65] (cf. section 2.6) and lead to more tricky diabaticity/adiabaticity distinction than for ET in homogeneous solution.

The differences are summarized in the relation between current density and overpotential, equivalent to the free energy relation for ET in homogeneous solution, Eqs. (2.1)–(2.3) [24, 25, 60, 61]:

$$j(\eta) = \int d\varepsilon f(\varepsilon)\rho(\varepsilon)j(\varepsilon;\eta); \qquad j(\varepsilon;\eta) = e\Gamma_{ox}^{(1-\alpha)}\Gamma_{red}^{\alpha}W_{AD}(\varepsilon;\eta) \tag{2.7}$$

$W_{AD}(\varepsilon;\eta)$ is the rate constant (s^{-1}) and $j(\varepsilon;\eta)$ the (infinitesimal) current density from a given electrode level ε, $f(\varepsilon)$ the Fermi function, $\rho(\varepsilon)$ the electrode electronic level density, Γ_{ox} and Γ_{red} the population of the oxidized and reduced molecule close to the surface, and e the electronic charge. α is the observable electrochemical transfer coefficient

$$\alpha = -k_B T d \ln j(\eta)/d(e\eta) \tag{2.8}$$

The following equation, equivalent to Eqs. (2.1)–(2.3), applies broadly:

$$W_{AD}(\varepsilon;\eta) = \kappa_{el}\frac{\omega_{eff}}{2\pi}\exp\left\{-\frac{[\lambda_R - e\eta + w_P^{el} - w_A^{el} - (\varepsilon - \varepsilon_F)]^2}{4\lambda_R k_B T}\right\} \tag{2.9}$$

$\kappa_{el}(\varepsilon;\eta)$ is the transmission coefficient, where the electron exchange energy, $T_{\varepsilon A}(\varepsilon;\eta)$, couples the molecular acceptor level (A) with the metallic, "quasiparticle" level ε. Levels around the Fermi level $\varepsilon \approx \varepsilon_F$ dominate, as long as $|e\eta| << \lambda_R = \lambda_{sol} + \lambda_{loc}$. Eqs. (2.4)–(2.9) give

$$j = j(\eta) = e\Gamma_{ox}^{(1-\alpha)}\Gamma_{red}^{\alpha}W_{DA};\quad W_{DA} \approx \kappa_{eff}\frac{\omega_{eff}}{2\pi}\exp\left[-\frac{(\lambda_R - e\eta + w_P^{el} - w_A^{el})^2}{4\lambda_R k_B T}\right] \tag{2.10}$$

$$\kappa_{eff} = 4\pi\kappa_{el}(\varepsilon_F;\eta)\rho(\varepsilon_F)k_B T,\ \text{if}\ \kappa_{eff} << 1,\ \text{otherwise}\ \kappa_{eff} \to 1$$

Eqs. (2.4)–(2.10) summarize similarities and differences between ET processes in solution and electrochemical ET processes, and underlines computational challenges:

- The electronic continuum is specific to metal, semiconductor, semi-metal (such as carbon), or superconducting materials. Different conduction bands contribute differently.
- The current/overpotential relation, equivalent to free energy correlations in solution, holds greater computational challenges. The "work terms", w_R and w_P are now the electrochemical double layer effects, first recognized by Frumkin [66, 67]. Present and forthcoming theoretical efforts were initiated by Kornyshev, Badiali, Schmickler, Henderson, Nazmutdinov and others [6–13, 57], and will be discussed below.
- Eqs. (2.4)–(2.10) constitute a frame of the Tafel relation. A quadratic overpotential dependence turns linear with $\alpha = 1/2$ for small overpotentials, $|e\eta| << \lambda_R = \lambda_{sol} + \lambda_{loc}$, observed for simple ET processes [68–70], but there are also deviations. Eqs. (2.4)–(2.10) rest on nuclear motion as a set of displaced harmonic oscillators. Vibrational frequency changes give limiting α-values different from $1/2$. In other cases such as electrochemical dihydrogen evolution, the linear Tafel range is "too wide" (see section 2.5.3). Anharmonic modes [71], charge redistribution [72], and excited vibrational states (cf. section 2.6.2) account for such behaviour ultimately, however, to be resolved by computational efforts. As $|e\eta|$ approaches the reorganization free energy, levels below the Fermi level take over, and "activationless" ET prevails. There is therefore no "inverted" free energy region in electrochemical ET.
- ET in homogeneous solution rests on electronic *wave functions*, while theoretical approaches to interfacial structural properties, capacitances etc. mostly rest on electronic *densities* [6–9]. Correlations between *densities* and *wave functions* are inherent in Kohn-Sham schemes, but transparent correlations, where for example interfacial capacitance is combined with ET properties, are missing. Suggestions as to how jellium

density can be incorporated in electrochemical ET were offered [74], but have not been followed up.

- Analytical theory of *adiabatic* electrochemical ET is available [74, 75, 78]. Eq. (2.10) with $\kappa_{\text{eff}} \to 1$ and *electronic* activation (free) energy lowering, but is more sophisticated than for single-electron levels, including for example spin interaction. Attempts to bridge diabatic and adiabatic electrochemical ET processes analytically have been offered with noted limitations [74], and would be a field for novel theoretical efforts.
- The transmission coefficient, κ_{eff}, Eqs. (2.7)–(2.10) can be recast as energy broadening, Δ, rooted in Anderson-Newns theory of atomic adsorption on metal surfaces [76, 77]

$$\Delta(\varepsilon) = \pi \sum_k |V_{AD}^k(\varepsilon)|^2 \delta(\varepsilon - \varepsilon_k) \to \pi |V_{AD}(\varepsilon)|^2 \rho(\varepsilon) \tag{2.11}$$

"k" refers to the momentum of the electron in the electrode. In the diabatic limit,

$$4\pi \frac{\Delta k_B T}{\hbar} \sqrt{\pi/(\lambda_R k_B T)} << 1; \;\; j(\eta) \approx 4\pi e \Gamma_{ox}^{1-\alpha} \Gamma_{red}^{\alpha} \frac{\Delta}{\hbar} k_B T \sqrt{\frac{\pi}{\lambda_R k_B T}} \exp \left[-\frac{(\lambda_R - e\eta + w_P^{el} - w_R^{el})^2}{4\lambda_R k_B T} \right] \tag{2.12}$$

In the opposite, adiabatic limit, $k_B T < \Delta < \lambda_R$ electron delocalization prevails, leading to the notion of electronic occupation number of the molecular reactant, n_A $(0 < n_A < 1)$ [12, 57].

It is important to note that the activation barrier is lowered, when the donor and acceptor orbital overlap is significant. This is illustrated most simply by homogeneous ET. The free energy surface $E(q)$ along the reaction coordinate q is recast as

$$E(q) = \frac{1}{2}[U_R(q) + U_P(q) - \sqrt{(U_R(q) - U_P(q))^2 + 4T_{RP}^2}] \tag{2.13}$$

where T_{RP} is the resonance integral (measure of the orbital overlap). $U_R(q) = \lambda_R q^2$ and $U_P(q) = \lambda_R(q-1)^2 + \Delta G^0$, where λ_R is the reorganization energy along the dimensionless q-mode and ΔG^0 the reaction free energy. If the resonance integral is small, then $E(q) = U_R(q)$ or $U_P(q)$ for the reactants' and products' states, respectively.

However, this simple model is not suitable for electrochemical ET, because of the manifold of donor (acceptor) levels. As first noted by Schmickler [78], Anderson-Newns theory is instead a convenient frame. In the spin-less, correlation energy-free, and "wide band" limit (s- and p-metals), the activation free energy can be brought to the form $(\eta = 0)$ [23, 57, 78]

$$\Delta G_{el}^{\neq} = \frac{1}{4}\lambda_R(1 - 2n_A) + \frac{\Delta}{2\pi} \ln \frac{\Delta^2}{\lambda_R^2(1 - 2n_A)^2 + \Delta^2}; n_A = \frac{1}{\pi} \text{arccot} \frac{\varepsilon_A - \varepsilon_F}{\Delta} \tag{2.14}$$

ε_A being the configurationally fluctuating molecular level (counted from the Fermi level). $\Delta \to 0$ and $n_A \to 0$ in the diabatic limit, reducing $\Delta G_{el}^{\neq}$ to Eqs. (2.9), (2.10), and (2.12).

Experimental studies point to the nature of most electrochemical processes, whether simple ET or complex processes such as dihydrogen evolution, as adiabatic or weakly diabatic. Strongly diabatic processes prevail at surfaces modified by self-assembled molecular monolayers (SAMs). We shall discuss cases of both these limits below.

2.2.4 Some electrochemical ET systems at metal electrodes

Our discussion above offers a frame for simple electrochemical ET at metal electrodes. Electrochemical ET theory extends, however, much wider to include ET at semiconductor and semimetal electrodes, complex processes such as proton and atom group transfer, dissociative ET, and biological processes [24, 25, 79–84], some of which we shall discuss further.

2.2.4.1 Some outer sphere electrochemical ET processes

Transition metal complexes have long been primary probes for theoretical ET patterns in solution and electrochemical ET. In recent times computational focus has been on the core rate parameters including double layer effects ("work terms") [85, 86], and other effects such as chirality [87], and solvent controlled electronic delocalization between metal complex and electrode surface [88]. We illustrate here some results.

Outer sphere ET between transition metal complexes have been core probes since earliest ET science [30]. In a major review (1983) [89], E.D. German analyzed a wealth of data like

$$[Fe(OH_2)_6]^{3+} + [Fe(OH_2)_6]^{2+} \rightleftharpoons [Fe(OH_2)_6]^{2+} + [Fe(OH_2)_6]^{3+} \tag{2.15}$$

and with other metals and ligands. The analysis was based on calculation of work terms and reorganization free energies, using detailed solvation models. The strategy was to insert calculated activation free energy parameters in the rate constant forms, compare with experimental data, and then see "if anything was left" for a transmission coefficient. A notable outcome was that nearly all simple ET processes belong to the adiabatic limit of strong intermolecular interaction. Co-complexes were an exception, substantiated by much later studies, and likely associated with accompanying electronic spin changes.

The number of reported cases of computational analysis of electrochemical ET of robust transition metal complexes is much more limited but offer a subtle view. As illustration, we address first robust Os, Ru, Fe, and Co polypyridine complexes, for which detailed voltammetry and *in situ* STM are available [23, 90]. A DFT study of the mixed bipy (2,2′-bipyridine) and p0p (4,4′-bipyridine) complexes of Os and Co ($[M(bipy)_2(p0p)Cl]^{2+/+}$, M = Os, Co) in free and adsorbed state on a 6 × 6 atoms model Au(111) and Pt(111)-surface was reported [88]. The (cationic) complexes were combined with stoichiometric counter anions (chloride), and the GPAW (real-space Grid-based Projector-augmented wave) DFT code combined with Verlet MD of cells of 198 water molecules (see Figure 2.3). The complexes were adsorbed on a "mined" adatom, as this gives stronger adsorption in spite of the mining energy spent (see section 2.5.2.1). Stoichiometric Os/Co complex and counter anions in electrostatically neutral form were allowed to relax. As expected but now computationally substantiated, inclusion of the solvent takes the neutral $[M(bipy)_2(p0p)Cl]\cdot Cl$ or $[M(bipy)_2(p0p)Cl] \bullet Cl_2$ unit from neutral complex and chlorine *atom* charge state to a chemically meaningful $[M(bipy)_2(p0p)Cl]^{n+}\bullet\ (Cl^-)_n$, n = 1, 2 solvated complex. The same applies to the complexes adsorbed on Au(111) or Pt(111). The neutral forms undergo unphysical charge delocalization into the metal electrodes, but solvation leads to charge separation and $[M(bipy)_2(p0p)Cl]^{2+/+}$ adsorption of chemical cationic species.

A second illustration relates to electrochemical ET of hexa-ammine Co(III), Cr(III), and Ru(III) complexes at mercury electrodes. Theoretical analysis includes solvent and intramolecular reorganization, the transmission coefficient, and electrochemical double

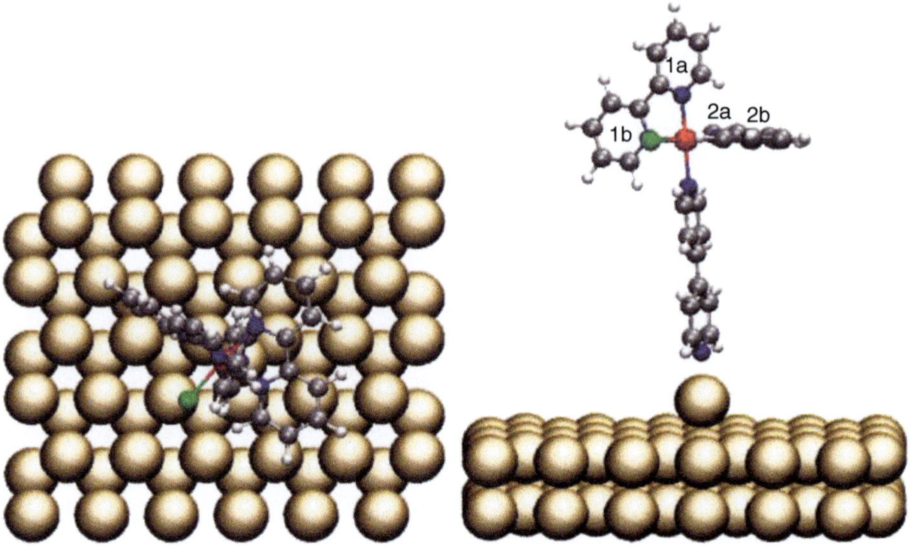

Figure 2.3 Left: Top view of [M(bipy)$_2$(p0p)Cl], M = Os, Co on 6 × 6 × 2 Au or Pt surface. Grey: carbon; blue: nitrogen; white: hydrogen; green: chloride; red: metal center. Right: Side view showing the perpendicular orientation on a mined metal atom. The pyridine rings are labelled. Adapted from [89].

layer effects [85, 86]. $[Co(NH_3)_6]^{2+/3+}$ and $[Co(NH_3)_5F]^{+/2+}$ display spin interconversion. Ligand substitution lability of Co(II) and Cr(II) is another complicating factor.

The molecular orbital (MO) structure is crucial. For example, in reduction of low-spin $[Fe(CN)_6]^{3-}$ $T_{2g}^5E_g$ an electron is transferred to the bonding orbital T_{2g} which strengthens the metal-ligand bonds. Similarly, the electron is transferred to T_{2g} in low-spin $[Ru(NH_3)_6]^{3+}$, $T_{2g}^5E_g$, with negligible intramolecular reorganization. Other examples illustrate the opposite. $[Co(CN)_6]^{3-}$ is low-spin $T_{2g}^6E_g$. As the triply degenerate MO T_{2g} is fully occupied, two anti-bonding E_g MOs participate in the reduction, leading to weakened $[Co(CN)_6]^{4-}$ metal-ligand bonds, with the further Jahn-Teller distortion. $[Cr(NH_3)_6]^{3+}$ is high-spin $T_{2g}^3E_g$. Despite only half-occupied T_{2g}, reduced $[Cr(NH_3)_6]^{2+}$ is $T_{2g}^3E_g^1$ with Cr-ligand bond weakened and further Jahn-Teller distortion. Significant inner sphere reorganization therefore accompanies $[Co(CN)_6]^{3-}$ and $[Cr(NH_3)_6]^{3+}$ reduction.

As a robust second row transition metal couple with no intramolecular reorganization, $[Ru(NH_3)_6]^{2+/3+}$ is the least complex of the couples. The electrochemical rate constants, 0.3–2 cm s^{-1} are orders of magnitude larger than for the other couples, and the Tafel coefficient, $\alpha = dlnj/d(e\eta)$ close to ½ [86]. Solvent reorganization free energies based on a conducting sphere close to an adlayer of low dielectric constant [91] are 0.4–0.6 eV for all the couples. The most challenging $[Ru(NH_3)_6]^{2+/3+}$ feature is the transmission coefficient, Eq. (2.16)

$$\kappa_{eff}(z) \approx \kappa_{eff,0} \exp(-\beta z) \tag{2.16}$$

Ridge and *face* adsorption, Figure 2.4 are by far the most efficient, $\kappa_{eff,0}^{ridge}, \kappa_{eff,0}^{face} >> \kappa_{eff,0}^{apex}$ (400–600 vs. 0.3–0.4). The electronic factors are an order or two larger for $[Ru(NH_3)_6]^{2+/3+}$ than for the other couples, due to better overlap of the 4d metal orbitals. $\kappa_{eff} << 1$. All the

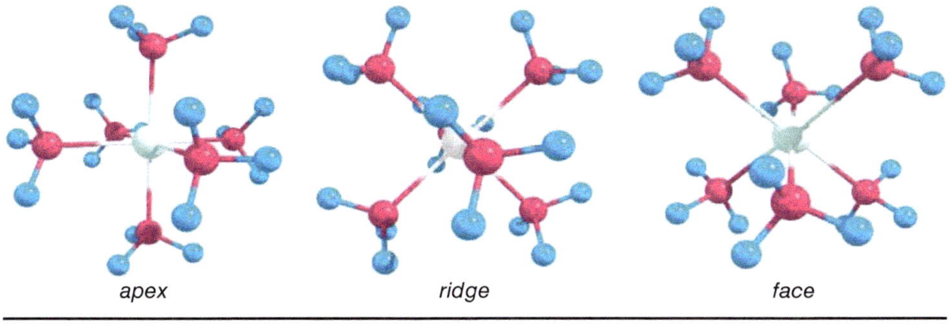

Figure 2.4 $[Ru(NH_3)_6]^{2+/3+}$, $[Cr(NH_3)_6]^{2+/3+}$ or $[Co(NH_3)_6]^{2+/3+}$ in different orientations on the electrode surface. The horizontal line indicates the projection of the surface perpendicular to the plane.

couples are therefore diabatic, although weakly for $[Ru(NH_3)_6]^{2+/3+}$ with κ_{eff} of 0.05–0.06, an order of magnitude higher than for the other couples. The parameters substantiate that intramolecular reorganization and transmission coefficient are controlling factors.

Other issues remain for $[Co(NH_3)_6]^{2+/3+}$ and $[Cr(NH_3)_6]^{2+/3+}$. Both reduced forms, with high-spin d^7 and d^4 configurations are, first highly substitution labile, e.g.

$$[Co(NH_3)_6]^{2+} + 6\,H_2O \rightarrow [Co(OH_2)_6]^{2+} + 6NH_3 \tag{2.17}$$

An equilibrium potential therefore requires "extrathermodynamic" assumptions. Both processes also involve antibonding e_g orbitals, subject to vibrational frequency changes, with the lower frequencies, Ω_{loc} in the reduced form, $\Omega_{loc}^{red} < (<)\Omega_{loc}^{ox}$. This offers a clue to the transfer coefficients of 0.7–0.75 strikingly exceeding the "iconic" value of 0.5.

Potential surfaces spanned both by a set of low-frequency environmental modes, $\{q_k\}$ with frequencies $\{\omega_k\}$, and a local mode Q, with frequencies Ω_{loc}^{red} and Ω_{loc}^{red} in the reduced and oxidized state are represented as

$$U_R = \frac{1}{2}\sum_k \hbar\omega_k q_k^2 + \frac{1}{2}\hbar\Omega_{loc}^{ox}Q^2$$
$$U_P = \frac{1}{2}\sum_k \hbar\omega_k (q_k - q_{k0})^2 + \frac{1}{2}\hbar\Omega_{loc}^{red}(Q - Q_0)^2 + \Delta G^0 \tag{2.18}$$

with "0" representing the coordinate displacements. Following available procedures [24, 25], the quadratic form emerges when the local mode is subject to displacement only,

$$\Delta G^{\neq}\} = \frac{(\lambda_{sol} + \lambda_{loc} - e\eta)^2}{4(\lambda_{sol} + \lambda_{loc})}; \quad \alpha = 1/2, \text{ at } |e\eta| << \lambda_{solv} + \lambda_{loc}, \; \Omega_{loc}^{red} = \Omega_{loc}^{ox} \tag{2.19}$$

The following form emerges, when $\Omega_{loc}^{red} < (<)\Omega_{loc}^{ox}$

$$\Delta G^{\neq} = \frac{(\lambda_{sol} + \lambda_{loc}^{red} - e\eta)^2}{4\lambda_{sol}}; \quad \alpha = \frac{1}{2}\frac{\lambda_{sol} + \lambda_{loc}^{red}}{\lambda_{sol}} > 0.5, \text{ when } \Omega_{loc}^{red} <(<)\Omega_{loc}^{ox} \tag{2.20}$$

λ_{loc}^{red} is the local mode reorganization energy in the low-frequency, reduced state. When $\Omega_{loc}^{ox} <(<)\Omega_{loc}^{red}$, then

$$\Delta G^{\neq}\} = \lambda_{loc}^{ox} + \frac{(\lambda_{sol} - \lambda_{loc}^{ox} - e\eta)^2}{4\lambda_{sol}}; \quad \alpha = \frac{1}{2}\frac{\lambda_{sol} - \lambda_{loc}^{red}}{\lambda_{sol}} < 0.5 \tag{2.21}$$

The transfer coefficient thus assumes other values than ½ at low overpotentials, when kinematic details beyond the displaced modes are included.

2.2.4.2 Dissociative ET: the electrochemical peroxodisulfate reduction

The view of electrochemical ET in section 2.2.2 needs obvious extension beyond the linear response notions, when mode distortion and anharmonicity as in dissociative ET is important. Although inherent in the ET formalism, other factors such as charge dependence of the double layer or the reaction zone, are commonly assigned a secondary role but prevail for strongly charged reactant molecules at high excess electrode charges [92, 93]. The electrochemical reduction of the peroxodisulfate ion

$$S_2O_8^{2-} + 2e^- \rightarrow 2SO_4^{2-} \tag{2.22}$$

explored over decades [93], strikingly illuminates these challenges (Figure 2.5).

Reduction requires high negative overpotentials, with the (Hg) surface strongly negatively charged, reflected in a voltammetric "pit" at the verge of activationless ET where a current plateau is expected. Tsirlina, Petrij and associates explored the process [9, 92],

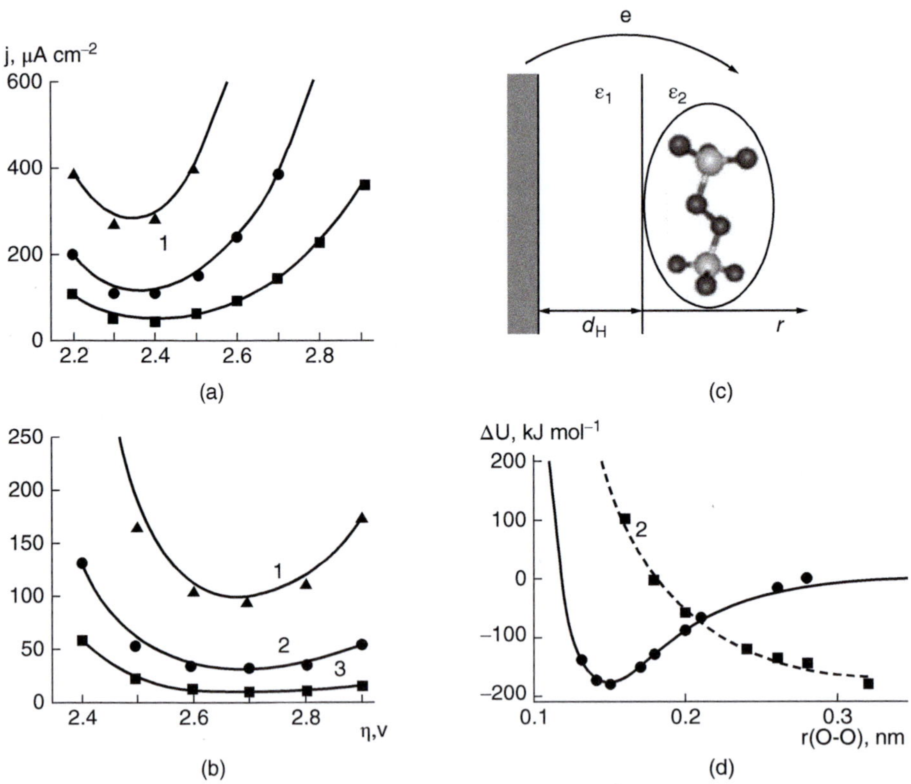

Figure 2.5 Left, top: Model current/potential relations of peroxodisulfate at a mercury electrode. Left, bottom: Experimental current/potential curves. The characteristic "dip" is noted. Right, top: Peroxodisulfate at an electrochemical surface with an inner (ε_1) and a bulk dielectric constant (ε_2). Right, bottom: Energy surfaces of oxidized (1) and reduced (2) peroxodisulfate along the O-O coordinate. Adapted from [93].

leading to understanding of how such a pattern can emerge. The strongly negatively charged electrode repels the $S_2O_8^{2-}$ ion. As the negative overpotential rises, the current first drops due to increasingly unfavorable repulsion and distance dependent transmission coefficient that outwin the driving force. As the overpotential increases further, a subtle combination of increasing driving force, expansion of the Hg-electrode electronic density overlap, and reaction zone, Δz, extension, Eq. (2.11) take over. This outcome follows naturally from both the properties of the $S_2O_8^{2-}$/Hg system and the ET formalism when charge dependent ET parameters not so commonly in focus are included.

2.2.5 d-band, cation, and spin catalysis

The high catalytic activity of some *d*-metals compared with less active *s*- and *p*-metals is intriguing. An approach resting on the Anderson-Newns (AN) model [76, 77] in the "narrow band" approximation was developed by Schmickler and Santos [94] to elucidate this challenge. The most successful applications of this theory are targeted at ET reactions with an adsorbed product. In spin-less formalism (i.e. for a one-electron step) the reaction free energy surface $E(q)$ along the dimensionless solvent coordinate q is then written as

$$E(q) = \lambda_s q^2 + \int_{-U_0}^{0} \varepsilon \rho_a(\varepsilon, q)\, d\varepsilon, \tag{2.23}$$

$\rho_a(\varepsilon, q)$ is the electronic density of states (DOS) of the reactant level broadened by coupling with the solid (in what follows a metal electrode), and U_0 the bottom of the conduction band ($\approx$ 12 eV), the Fermi level (upper integration limit) is taken as zero.

The second term in Eq. (2.23) describes the electronic part of the free energy surface from interaction between reactant and electrode, assuming that the Fermi-Dirac distribution function is replaced by a step function The reactant DOS then takes the form [77]

$$\rho_a(\varepsilon, q) = \frac{1}{\pi} \frac{\Delta(\varepsilon)}{\Delta^2(\varepsilon) + (\varepsilon - \varepsilon_a(q) - \Lambda(\varepsilon))^2}, \tag{2.24}$$

where $\Delta(\varepsilon)$ describes the effective reactant level broadening in generalized form, $\Delta(\varepsilon) = \pi V^2 \rho(\varepsilon)$, where $\rho(\varepsilon)$ is the metal DOS and V is an electronic coupling constant.

The DOS for transition metals *d*-bands is frequently described by a semi-ellipse as first suggested by Newns [77]. ε_a is the reactant energy level which, at zero overpotential, depends on the solvent coordinate q as $\varepsilon_a = -2\lambda_{ssol} q$. $\Lambda(\varepsilon)$ is the Hilbert transform of $\Delta(\varepsilon)$, the physical meaning of which is a shift of the reactant energy level. If $\Delta(\varepsilon)$ is approximated by a semi-ellipse, then $\Lambda(\varepsilon)$ is a linear function in the energy interval corresponding to the horizontal axis. The activation barrier ΔE_a is defined as the difference between values of $E(q)$ related to the transition (q =1/2) and initial (q = 0) states,

$$\Delta E_a \approx \frac{\lambda_s}{4} - \frac{1}{\pi} \int_{-U_0}^{0} \left(\frac{\varepsilon\, \Delta(\varepsilon)}{\Delta^2(\varepsilon) + (\varepsilon - \Lambda(\varepsilon))^2} - \frac{\varepsilon\, \Delta(\varepsilon)}{\Delta^2(\varepsilon) + (\varepsilon - \lambda_s - \Lambda(\varepsilon))^2} \right) d\varepsilon. \tag{2.25}$$

ΔE_a is sensitive, first to $\Delta(\varepsilon)$ (Eq. (2.25)), which in turn is proportional to the squared coupling constant V and the metal DOS. For metals such as Pt, Pd and the Cu group the *d*-electron DOS is larger than for a wide *sp*-band. Additional analysis of Eq. (2.25) shows that the closer the narrow metal *d*-band to the Fermi level (to $\varepsilon = 0$), the larger the integral and, therefore the more pronounced the barrier decrease (catalytic effect). This explains why Pt and Pd are good catalysts, as the *d*-band (the center of the semi-ellipse describing $\Delta(\epsilon)$, ϵ_0) for these metals is closer to the Fermi level than in Au, Ag, Cu and other metals. If ϵ_0 is close, then a difference in coupling constants becomes crucial, and the *sp*-band of *d*-metals begins to play a more important role in the ET process.

Electrochemical reduction of anions in the presence of counter-ions (cations) can also result in catalysis, induced by ion pair formation. The electrostatic repulsion from a negatively charged electrode surface for such ionic associates is then smaller and they approach closer to the electrode. As the ion pair is larger than the anion, the solvent reorganization energy is also smaller. If cations from the electrolyte solution are adsorbed specifically, they can induce a local potential which facilitates the approach of anionic reactants [95].

The dioxygen reduction offers particular interest regarding spin catalysis. Molecular dioxygen in the excited singlet state is significantly more reactive than the triplet ground state, facilitating rate controlling O-O bond break in the elementary ET act. Singlet O_2 is formed on adsorption of molecular dioxygen on some electrode surfaces, nanoparticles, doped carbon nanotubes etc., which then serve as spin catalysts [96]. The ET rate with quantum dots (small transition metal clusters) as electron donor or acceptor, immobilized on solid surfaces, can also be controlled by the spin conservation rule.

2.2.6 New solvent environments in simple electrochemical ET processes – ionic liquids

Heterogeneous ET processes in a new environment – Room Temperature Ionic Liquids (RTIL) in particular – are new theoretical and computational targets. Outer sphere ET reactions in homogeneous RTII solution have been extensively investigated using classical MD simulations [97–99]. A combined approach resting on quantum chemical modeling and MD was employed in [100] to describe ET across the Au(111)/RTIL interface. No unusual features were found compared with molecular solvents. In fact, surprisingly, linear response theory (Gauss fluctuations) is still valid so that ET can be described in terms of the solvent reorganization free energy. A large contribution of solvent high-frequency quantum modes (the reducing factor nearly 0.7, Eq. (2.6)) was, however, found for $[\text{bmim}^+][\text{BF}_4{}^-]$ using the experimental dielectric spectrum [100].

Major progress in the detailed understanding of the electrochemical metal/ionic liquid interface has been achieved over the last decade by Kornyshev and associates, experimental progress by Mao and associates, and by Nichols and associates. Progress covers both interfacial structural properties, say the double-hump "camel-like" capacitance and kinetic parameters such as the reorganization free energy. We refer to recent authoritative articles and reviews by Kornyshev, Mao, and Nichols and their associates on this novel and non-traditional area of the new electrochemistry [101–105].

2.2.7 Proton transfer, proton conductivity, and proton coupled electron transfer (PCET)

Next to ET [106], proton transfer (PT) and proton coupled electron transfer (PCET) are the most ubiquitous "elementary" chemical processes. PT is the core in general acid-base catalysis, proton conductors, and electrochemical, for example fuel cell processes. PT/PCET are central in hydrolytic and redox enzyme processes, for example conversion of dioxygen to water in cytochrome *c* oxidase or blue Cu-oxidases [107]. PCET extends all the way from separate ET and PT to fully synchronous hydrogen atom transfer (HAT) [109–111]. In the former case ET and PT are still coupled, by the field changes that accompany each step, for example in charge-induced driving force in proton pumping. PT and HAT have been in experimental and theoretical focus for a long time and is reviewed extensively [108–110].

By involving transfer of a "light" particle between two heavy molecular fragments, and with conspicuous quantum mechanical (tunnelling) features already at room temperature, *PT* and *HAT* display close analogies with *electron transfer*. The analogies include strong coupling to the environment and vibrationally assisted tunneling. A particular perspective is the kinetic deuterium or tritium isotope effects (KIEs) [24, 25, 110–117]. Important *differences* are the bond breaking and formation in PT/HAT, whereas the ET donor and acceptor retain their integrity. Also, while electron tunneling over long distances, i.e. tens of Ångströms, is feasible, PT/HAT is only feasible over a fraction of an Ångström or so [109, 114].

This difference holds at least three implications. One is that "gating" is much more important than for ET. Thermal activation along donor-acceptor translational modes is an important contribution to the activation (free) energy [109, 114] and the dominating contribution to the temperature coefficient of the KIEs, caused by the shorter tunneling distance of the heavier isotopes. A second implication is that the donor-acceptor interaction at the short distances is much stronger than in ET. PT/HAT therefore belongs to the limit of so strongly *electronically* adiabatic processes that "explicit" tunneling features are concealed. The HAT donor-acceptor interactions are weaker with larger HAT distances and conspicuous quantum mechanical features [118–120], particularly large KIEs. A third implication is that "long-range" PT as in trans-membrane proton pumping is entirely different from the notion of long-range ET. The latter is rooted in a clearly defined single-ET step over distances which exceed the structural extension of the donor and acceptor groups. "Long-range" PT is a sequence of individual short-range proton hops. Donor and acceptor groups along a proton transport channel are amino or carboxylate groups or water molecules. "Gating", i.e. conformational fluctuations of the lining proton transport groups, is therefore crucial. This is analogous to proton conductivity in Nafion and other synthetic membranes [118].

Large, i.e. > 10 KIEs, indicative of conspicuous quantum mechanical tunneling are long known for poorly hydrogen bonded C-donor and –acceptor acids and bases [119]. Even larger values, up to orders of magnitude have been reported for HAT in frozen glasses [116] and in enzyme processes at room temperature [11, 115]. In either case the large KIEs are associated with "freezing" of the gating mode. In most cases of O- and N-acids and -bases with strong hydrogen bonding the KIEs are small, i.e. weakly in excess of the stretching vibrational frequency ratio of $\sqrt{2}$ [4, 119, 120]. Since the *equilibrium* PT distance is still significant, say ≈ 0.8 Å or so, this reflects the strongly gated nature of the process. The PT/HAT scenario is summarized in Figures 2.6 and 2.7 [24, 25, 109, 118–122]. More recent approaches to single- and multi-PT processes in chemical and biological systems are given elsewhere [121, 123, 124]. We note:

- PT/HAT is coupled to *two* nuclear mode sets with quite different timescales, proton and solvent modes. The proton/H-atom is trapped on the donor at the equilibrium solvent

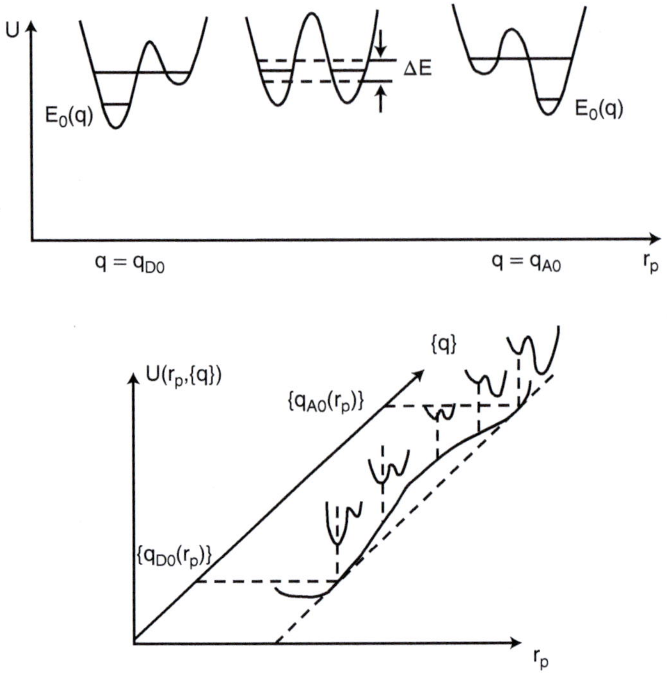

Figure 2.6 Two views of double-well potential free energy projections on the proton coordinate, r_p, in a PT/HAT process. The environmental coordinate(s) represented by q. The proton is trapped near the donor and near the acceptor, when q takes its equilibrium values in the reactants' state, q_{DO}, and product's state, q_{AO}, respectively. Adapted from [120].

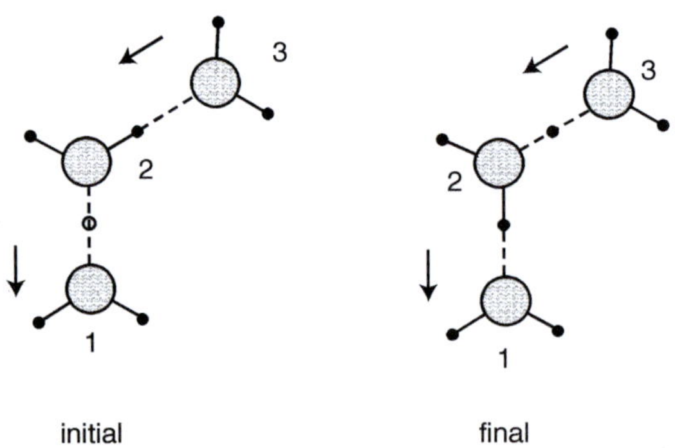

Figure 2.7 Double-proton transfer in proton conduction via the Zundel complex $H_5O_2^+$. The proton is initially located symmetrically between the water molecules 1 and 2. In the final state the proton has been translocated to a symmetrical position between water molecules 2 and 3. Mechanisms analogous to these operate in proton conduction through biological and synthetic membranes. Adapted from [107, 118]

configuration in the reactants' state. *Fluctuations* induce a state of resonance between the *proton/H-atom vibrational* levels in the reactants' and products' states. Gating is part of this process. Reorganization of the proton and electronic system proceeds at this dynamic resonance, followed by proton/H-atom trapping in the products' state (Figure 2.6).

- The electronic system follows "adiabatically" the PT/HAT motion, but PT/HAT may be obstructed by a tunneling barrier. This limit is denoted as "partially adiabatic" [24, 25]. Or the proton/H-atom may follow "adiabatically" the solvent nuclear dynamics if the

PT/HAT barrier is small. Barrier attenuation is part of the gating, and the barrier shallower, the closer the gating modes take the donor and acceptor fragments.

The partially adiabatic limit applies to *HAT*, in redox enzymes or frozen glasses. The following formalism is immediately palatable, cf. Eqs. (2.1)–(2.3) [24, 25, 113, 114]

$$W_{RP} = Z_P^{-1} \sum_{v,w} \exp\left(-\frac{\varepsilon_{Rv}^P}{k_B T}\right) W_{RP}^{vw} \tag{2.26a}$$

$$W_{RP}^{vw} = \kappa_{vw}^P \frac{\omega_{eff}}{2\pi} \exp\left(-\frac{G_{vw}^{\neq}}{k_B T}\right) \; ; \qquad Z_P = \sum_{v,w} \exp\left(-\frac{\varepsilon_{Rv}^P}{k_B T}\right) \tag{2.26b}$$

$$G_{vw}^P = \frac{[\lambda_R + \Delta G^0 + (\varepsilon_{Pw}^P - \varepsilon_{Pw=0}^P) - (\varepsilon_{Rv}^P - \varepsilon_{Rv=0}^P)]^2}{4\lambda_R} ; \kappa_{vw}^P = \left(\frac{1}{2}\Delta E_{vw}^P\right)^2 \sqrt{\frac{\pi^3}{\lambda_R k_B T \hbar^2 \omega_{eff}^2}} \tag{2.26c}$$

$$\Delta\varepsilon_{vw}^P = \frac{1}{\pi}\hbar\Omega_P \exp\left\{-\frac{1}{\hbar}\int_{left}^{right} \left\{2m_P[U_{vw}(r_P) - \varepsilon_{vw}^P]^{\frac{1}{2}}\right\}\right\} \tag{2.26d}$$

m_P is the proton mass. The integration limits represent the proton tunneling barrier. We note:

1. The same quadratic free energy relation, as for ET.
2. Averaging over all reactants', ε_{Rv}^P, and summation over the products' proton/H-atom vibrational states, ε_{Pw}^P are included. This is important for the KIEs but often reduces to the ground vibrational states $v = w = 0$ (see section 2.2.6.2).
3. The transmission coefficient, κ_{vw}^P, is determined by the energy splitting of the proton vibrational levels at resonance, $\Delta\varepsilon_{vw}^P$. $\Delta\varepsilon_{vw}^P$ is determined by the proton vibrational frequency, Ω_P, and the tunneling barrier, $U_{vw}(r_P)$ along r_P, at the fluctuational environmental resonance configuration, Eq. (2.26d).

The PT/HAT *character* changes entirely when the donor-acceptor interactions are strong. The rate constant reduces to

$$W_{RP}^{vw} = \frac{\omega_{eff}}{2\pi} \exp\left(-\frac{G_{vw}^{\neq}}{k_B T}\right) \; ; \qquad G_{vw}^{\neq} \to G_{vw}^{\neq} - \frac{1}{2}\Delta\varepsilon_{vw}^P \tag{2.27}$$

Proton tunneling is thus reflected by a lower activation barrier rather than a pre-exponential tunneling factor (see sections 2.2.2 and 2.2.3). The KIEs are caused by a larger activation free energy, rooted in smaller splitting, $\Delta\varepsilon_{vw}^P$, for the heavier isotope.

Proton conductors, proton pumping or fuel cell membranes such as Nafion, belong to the class of strongly adiabatic PT [110, 118, 121, 122], illustrated by the excess proton conductivity in aqueous solution (Figure 2.7) [4, 118, 122]. The dominating room temperature species is the Zundel ion, $H_5O_2^+$, with the proton located symmetrically between two strongly solvating but labile water molecules. Only tiny structural shifts are needed for proton translocation [122, 123] with nearly all activation free energy in the gating mode(s) and virtually no PT barrier left. Such views carry over to proton transport in biological proton transport channels, the proton transporting sulfonate groups in Nafion, and multi-PT in hydrolytic enzymes [118, 121, 125].

2.2.7.1 Some further notes on the nature of PT/PCET processes

PT/PCET processes are different from ET processes by the two-step character, i.e. coupled ET and PT of the elementary process and by the unique tool in the form of the KIEs. The KIEs reflect directly the quantum mechanical nature of the PT/HAT dynamics, but *all* nuclear modes including solvent librational modes are subject to a KIE. Significant values, $\gg \sqrt{2}$, are therefore required for unambiguous interpretation. Secondly, PT/PCET processes are "gated" in the sense that PT/HAT over the crystallographic distances, up to an Å or so, would both be virtually blocked and display KIEs far exceeding observed values. Gating takes the PT/HAT distances to smaller values, 0.3–0.5 Å or so that accord with observed KIEs [109, 114]. Gating is obstructed in HAT processes in rigid protein environment and between radicals (methanol, acetonitrile) in frozen glasses, reflected in huge KIEs.

Most recent PCET analysis is based on the diabatic, electron/proton and proton/solvent diabatic formalism, Eqs. (2.26) and (2.27) [24, 25, 112, 115, 126]. This is analytically convenient, but the partially (Eq. (2.26)) or totally adiabatic mechanisms (Eq. (2.27)) are much more likely. The last few decades have witnessed major progress in PT/PCET processes extended to stochastic chemical rate theory and path integral approaches [121, 125], and to coupled multi-PT as in hydrolytic enzyme processes. Among new system types now better understood are:

- Elementary steps in transition metal complex catalysis such as gated elementary processes in hydride and H-atom insertion reactions [126].
- Heterogeneous catalysis as in methane refining, Fischer-Tropsch processes, and other processes of technological interest [127].
- Photochemical PCET between transition metal complexes and organic molecules [55, 111].
- Elementary HAT steps in enzyme systems. Lipoxygenase, alcohol dehydrogenase, and amine deoxygenases are core cases [115, 128].
- Novel perspectives for single-molecule PCET in electrochemical STM, with prospects for directly recording molecular conductivity and PT/PCET distances [129].

2.2.7.2 The electrochemical hydrogen evolution reaction, and the Tafel plot on mercury

The hydrogen evolution reactions (HERs) have attracted experimental and theoretical attention over decades. The mechanism of these two-step processes is complex and involves both electron and proton transfer across electrode/aqueous solution interfaces. A most striking effect at a mercury electrode is a linear $\ln j/\eta$ (Tafel) plot (with a transfer coefficient close to ½) over a surprisingly wide overpotential range [130–132]. We consider here the HER from acid solution, where the first stage (discharge of a hydronium ion with formation of an adsorbed intermediate) can be represented as (Volmer step),

$$H_3O^+ + e \rightarrow H_{ads} + H_2O. \tag{2.28}$$

In alkali solution the proton donor is a water molecule ($H_2O + e \rightarrow H_{ads} + OH^-$).

Since the Volmer step involves an adsorbed hydrogen atom, ET belongs to the adiabatic limit. In the Levich-Dogonadze-Kuznetsov (LDK) theory [113] the dimensionless current j can be calculated as a sum of normalized partial contributions j_{vw}

$$j = \frac{1}{Z}\sum_{v,w} j_{vw}; \qquad Z = \sum_{v} \exp\{\Delta\varepsilon_{1v}^{(R)}/k_B T\}. \tag{2.29}$$

$\Delta\varepsilon_{1v}^{(R)}$ is the excitation energy from the ground vibrational level (1) to the level v. Further

$$j_{vw} = \kappa_p^{(vw)} \exp\{-\Delta\varepsilon_{1v}^{(R)}/k_B T\} \exp\left\{-\frac{(\lambda_R + \Delta H_{diss.ads}^{H_2} + \varepsilon_w^{(P)} - \varepsilon_v^{(R)} - e\eta)^2}{4\lambda k_B T}\right\} \quad (2.30)$$

cf. Eq. (2.26), which corresponds to PT from the reactants' state v to the products' state w. $\kappa_p^{(vw)}$ are the partial transmission coefficients, $\varepsilon_v^{(R)}$ И $\varepsilon_w^{(P)}$ the vibrational energy levels in the reactants' (R) and products' (P) states, λ_R the solvent reorganization free energy and $\Delta H_{diss.ads}^{H_2}$ the dissociative adsorption energy of a dihydrogen molecule on mercury.

ET and PT constitute the same elementary act, i.e. a PCET. $\Delta H_{diss.ads}^{H_2}$ is important, as this quantity makes the Volmer step rate determining for mercury and Hg-like metals (Pb, Cd, Ag) with weak Me-H_{ads} interaction. The Heyrovskii mechanism, $H_{ads} + H_3O^+ = H_2 + H_2O$, can also be considered. $\Delta H_{diss.ads}^{H_2}$ in Eq. (2.30) should here be taken with "minus". As for Hg-like metals, $\Delta H_{diss.ads}^{H_2}$ is large and positive, the activation barrier is significantly reduced, which results in increasing the Heyrovskii step rate. In contrast, $\Delta H_{diss.ads}^{H_2}$ is small or even negative for other metals (like Ni, Pt, Pd) for which the Heyrovskii stage may be rate determining. The larger v and w, the closer is $\kappa_p^{(vw)}$ to unity, but the smaller the contribution to the partial current. If the reactant is a solvated hydronium ion ($H_7O_3^+$), λ_R holds both inner and outer sphere contributions.

Quantum chemical modeling combined with the LDK theory [133] shows that partial currents depend non-linearly on the mercury electrode overpotential, whereas the resulting current vs η was found to demonstrate perfect linear dependence in a wide overpotential range with a Tafel slope close to ½. The interplay between partial current contributions thus leads to a linear Tafel plot. The LDK theory also explains qualitatively the KIE [134], with the ratio $j_{D(T)}/j_H$ increasing with increasing overpotential. This is corroborated by DFT calculations [133]. Notably, molecular modeling for a mercury electrode [133] does not accord with the concept of a barrierless discharge region [132]. Other recent approaches to the KIE of the HER are reported elsewhere [135].

The proton transmission coefficient is sensitive to the distance of closest approach of the hydrated proton to the electrode surface, x_0. For hydrophilic electrodes with strong H_{ads}–metal interaction, i.e. $\Delta H_{diss.ads}^{H_2}$ notably smaller than for mercury (e.g. Pt, Pd), PT is expected to be adiabatic ($\kappa_p^{(nm)} \approx 1$) and H-tunneling not to be important. H-atom motion can here be treated as classical [120, 136]. Moreover, the strong Me–H_{ads} orbital overlap reduces notably the activation barrier [137]. The Volmer step cannot then be rate determining and the origin of the linear Tafel plot [131] is related to other mechanisms (Heyrovski step or surface recombination). The HER and H-oxidation were investigated recently by Nørskov and associates [138] using DFT. The metal surface was modeled by periodical slabs, but modern quantum mechanical charge transfer theory was not employed.

2.3 Ballistic and stochastic (Kramers-Zusman) chemical rate theory

Condensed matter chemical charge transfer involves strong coupling to the environment. In most approaches the solvent is regarded as equilibrated, and the electronic transitions as friction-free, or "ballistic". These views are broadly adequate but solvent equilibration

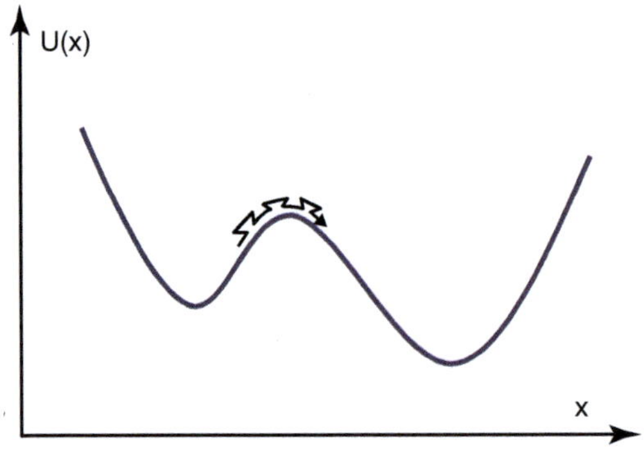

Figure 2.8 Adiabatic potential surface with frictional motion along a diffusion-like nuclear mode.

imposes constraints in ultra-fast ET processes. Non-equilibrated solvation dynamics implies that part of the coupling between solvent and reacting molecules is omitted, when potential surfaces are constructed, and re-introduced as friction along the surfaces, Figure 2.8. These views originate from Kramers [139], dating further back to Einstein's renowned study on Brownian motion [140], and has today developed to what is known as stochastic chemical rate theory, lead e.g. by Hynes [141], Zusman [142], Kuznetsov [143], and by Sumi and Marcus [144].

Solvent dynamics can control adiabatic ET for shallow barriers. The heart of the stochastic approach is that the barrier passage resembles Brownian diffusion with an apparent diffusion coefficient, D and by the Stokes-Einstein relation, a friction coefficient ς and a shear viscosity, η. In the "overdamped limit" the rate constant is proportional to the inverse solvent relaxation time playing the role of a frequency factor. In this limit the higher the viscosity, the longer the relaxation time, τ_L

$$W_{AD} \approx \frac{1}{\tau_L} \exp\left(-\frac{\Delta G^{\neq}}{k_B T}\right); \qquad \tau_L \approx \frac{a^2}{2D} \approx \frac{a^2}{2k_B T}\varsigma \approx \frac{3\pi a^3}{k_B T}\eta \tag{2.31}$$

where a is a molecular-scale length. These correlations stem from macroscopic kinematics and classical theory of dielectrics, now with molecular theoretical frames.

The reduction of the electrochemical target molecule $S_2O_8^{2-}$ on mercury, section 2.2.4.2 from variable-viscosity water-ethylene glycol (EG) solutions was recently investigated [145]. The current in the "polarization pit" region discloses a challenging non-monotonous viscosity dependence. The current first decreases with increasing viscosity, reaches a plateau, but then increases again. As $S_2O_8^{2-}$ reduction is accompanied by -O-O- dissociation, intramolecular reorganization is crucial and a three-dimensional free energy surface needed. Coupling to both a local, frictionless mode and a solvent relaxational mode was treated in detail by Zusman [142], Kuznetsov [143], Sumi and Marcus [144], and others. A "pragmatic" version of Eq. (2.35) for the two-mode system, suited for experimental data analysis [144] is

$$W_{AD} \approx \frac{1}{(\tau_L)^{\alpha}}\left(\frac{\omega_{eff}}{2\pi}\right)^{1-\alpha} \exp\left(-\gamma\frac{\Delta G^{\neq}}{k_B T}\right); \ \alpha \approx 1 - (\lambda_{loc}/\lambda_{sol}); \ \gamma \approx \frac{\lambda_{sol}}{\lambda_{sol} + \lambda_{loc}} \tag{2.32}$$

$\lambda_{loc}/\lambda_{sol} << 1$, and $2 - (\lambda_{loc}/\lambda_{sol}) < 1$.

The rate constant (W_{AD}) dependence on solution viscosity (or relaxation time τ_L) is usually the experimental target, which forms a basis for widely accepted estimations of the "degree of adiabaticity" θ ($= -dln\ k/dln\ \tau_L$) [145]. $\theta = 1$ in the adiabatic limit, $\theta = 0$ in diabatic ET. Solvent dynamics, i.e. the effects of dynamic electrolyte characteristics on the rate constant are therefore important in adiabatic ET, but disappear for diabatic ET. The kinetics is then well described by "ballistic" transition state theory.

A linear ln$k$ vs ln$\tau_L$dependence for simple ET processes can be described by the Kramers-Zusman theory [142–144], with an effective frequency factor $\approx \tau_L{}^{-1}$, but non-linear ln$k$ vs ln$\tau_L$correlations are frequently observed, when ET is accompanied by intramolecular reorganization or bond fission. The correlations can then be addressed by Sumi-Marcus theory [144], with "diffusive" and slow motion along the solvent coordinate, while the fast intramolecular motion crosses a Transition State barrier, Figure 2.9. The equation

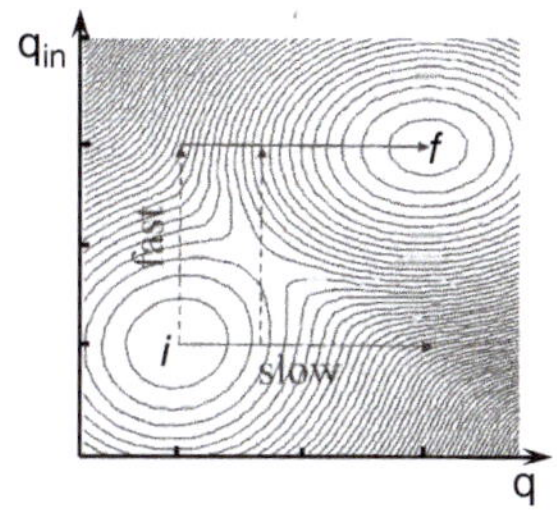

Figure 2.9 Section of a three-dimensional free energy surface along a dimensionless solvent (q) and an intramolecular (q_{in}) coordinate illustrating the Sumi-Marcus model [144].

$$\frac{\partial P(q,\tau)}{\partial \tau} = \hat{L}P(q,\tau), \tag{2.33}$$

must be solved. $\hat{L}$ is the Smoluchowski operator supplemented by a sink term $k_{in}(q)$:

$$\hat{L}P(q,\tau) = D\left\{\frac{\partial^2}{\partial q^2} + \frac{1}{kT}\frac{dU(q)}{dq}\frac{\partial}{\partial q}\right\}P(q,\tau) - k_R(q)P(q,\tau), \tag{2.34}$$

$P(q,\tau)$ is the probability density of finding a reactant in the reactants' well and D the diffusion coefficient along q, $D = k_BT/2\lambda_{sol}\tau_L$. $U(q)$ is a section of the reaction free energy surface. The sink term, $k_R(q)$, is written as

$$k_R = \nu_R \exp\{-\Delta E_a^*(q)/kT\}, \tag{2.35}$$

where ν_R is an effective frequency factor. In the Sumi-Marcus model only forward reaction along q_R is considered and the reverse process neglected. The energy barrier along the intra-molecular mode, ΔE_a^*, depends on the solvent coordinate q as

$$\Delta E_a^*(q) = U(q, q_{saddle}^*(q)) - U(q, q_R = 0), \tag{2.36}$$

$q_{saddle}^*(q)$ denotes the saddle line on the three-dimensional free energy surface $E(q, q_R)$ and is defined by a transcendental equation. $P(q,\tau)$ is described by the initial-boundary problem,

$$P(q,\tau)|_{\tau=0} = P^*(q) = \frac{1}{N^*}\exp(-\lambda_{sol}q^2/kT); \tag{2.37}$$

$$\frac{\partial P(q,\tau)}{\partial q} + \frac{1}{kT}\frac{\partial U(q)}{\partial q}P(q,\tau)|_{q=a} = 0$$

where $N^*$is the normalization coefficient $N^* = \sqrt{\lambda_{sol}/\pi k_BT}$ and $a = q_L$ or q_R (i.e. the q values at the left and right boundaries, respectively). Two different average survival times apply in the reactants' state:

$$\tau_a = \int_0^\infty d\tau \int_{q_L}^{q_R} P(q,\tau)dqd\tau \text{ and } \tau_b = \frac{1}{\tau_a}\int_0^\infty \tau d\tau \int_{q_L}^{q_R} P(q,\tau)dqd\tau \tag{2.38}$$

The ET rate constant W_{AD} can be defined as $1/\tau_a$ or $1/\tau_b$. τ_a can be considered as a short-time survival time, while τ_b is more relevant to describe the reaction kinetics at longer time scales. Both might contribute to the observable reaction rate.

We cannot here review the broad area of stochastic chemical and optical charge transfer theory, e.g. [24, 25, 143]. We illustrate instead what can be extracted from data analysis of a specific process, the electrochemical peroxodisulfate reduction (section 2.3), for which new data for viscous solvents have disclosed unexpected – non-monotonous – viscosity dependence of the current [145, 147], most intriguingly an ascending branch at high viscosity. This was also observed for $[Cr(EDTA)]^{-/2-}$ [145], i.e. for ET with no bond fission. This behaviour cannot be framed by equilibrium theories, but is accommodated by stochastic chemical rate theory, which also enables extracting "hidden" solvation properties.

Non-monotonous rate constant vs solvent viscosity dependence accords qualitatively with the Sumi-Marcus model, with the assumption that the Pekar factor decreases with increasing viscosity. This assumption is more or less realistic for syrups [146] but not for EG-water mixtures [147]. The non-traditional solvent dynamics was, however, rationalized at the molecular level with a constant Pekar factor [147] by introducing the time (τ) correlation function $M(\tau) = <\varepsilon_A(\tau)\varepsilon_A(0)>$ which describes the fluctuating reactant energy level ε_A. This function can be calculated by MD simulations, but also be obtained from the complex solvent dielectric spectrum $\varepsilon(\omega)$ [148]:

$$M(\tau) = 2k_B T\lambda_{sol}\frac{Q(\tau)}{Q(0)}, \quad \text{with } Q(\tau) = \frac{i}{2\pi}\int_{-\infty}^{\infty}\exp(-i\omega\tau)\left[\frac{1}{\varepsilon(\omega)} - \frac{1}{\varepsilon_{st}}\right]\frac{d\omega}{\omega} \tag{2.39}$$

with $\varepsilon_{st} = \varepsilon(\omega \to 0)$ being the static dielectric constant. The frequency dependent function $\varepsilon(\omega)^{-1} - \varepsilon_0{}^{-1}$ can be regarded as a generalized Pekar factor, (section 2.3).

The solvent reorganization energy for $S_2O_8{}^{2-}$ reduction, Eq. (2.26) refers to the bond break of the first, rate determining ET step

$$S_2O_8{}^{2-} + e = S_2O_8{}^{3-} = SO_4{}^{2-} + SO_4{}^{-}. \tag{2.40}$$

The dielectric spectrum for water-EG mixtures displays 3-D behavior [147]. Following Zusman [142] we can expand the correlation function

$$M(\tau) = 2k_B T\lambda\sum_{i=1}^{3}\delta_i \exp(-\tau/\tau_i^*), \tag{2.41}$$

Δ_i are the contributions to the solvent reorganization free energy from the i-th solvent mode $\left(\sum_{i=1}^{3}\delta_i = 1\right)$ and δ_i^* their relaxation times.

The solvent modes do not behave "chaotically" but interplay like "an orchestra", each mode assigned one instrument. Such adiabatic free energy surfaces can be recast as

$$E_R(\vec{q}, r) = U_R(\vec{q}) + U_R^*(r); \qquad E_P(\vec{q}, r) = U_P(\vec{q}) + U_P^*(r) - e\eta \tag{2.42}$$

$U_R^*(r)$ and $U_P^*(r)$ represent intramolecular reorganization, r the peroxodisulfate O-O bond length, $\vec{q}$ a vector which describes selected dimensionless solvent coordinates, R and P refer to the reactants' and products' states, respectively, and η is the overpotential.

According with Eq. (2.41) the "solvent" parts of Eqs. (2.42), $U_R(\vec{q})$ and $U_P(\vec{q})$, take the form

$$U_D(\vec{q}) = \sum_{i=1}^{3} \delta_i \lambda_{sol} q_i^2 \qquad (2.43)$$
$$U_A(\vec{q}) = \sum_{i=1}^{3} \delta_i \lambda_{sol} (q_i - 1)^2$$

The classical Langevin (Brownian) molecular dynamics was employed to calculate the reaction rate (W_{AD}) [143, 147]. Four differential equations describe the motion in five-dimensional reaction space along three solvent and one intramolecular degree of freedom:

$$\frac{\partial q_i}{\partial \tau} = -\frac{1}{2\delta_i \lambda \tau_i^*}\frac{\partial U_D}{\partial q_i} + \frac{dq_i^{rand}}{d\tau}, \quad i = 1, 2, 3$$
$$\frac{\partial^2 r}{\partial \tau^2} = -\frac{1}{m}\frac{\partial U_D^*}{\partial r} - \gamma_r \frac{dr}{d\tau} + \frac{dv^{rand}}{d\tau}, \qquad (2.44)$$

where m is the reduced mass of $S_2O_8{}^{2-}$, dq_i^{rand} random solvent coordinate increments, dv^{rand} a random velocity increment along r, and γ_r the intramolecular friction coefficient.

The first three equations describe the solvent dynamics in the overdamped limit [143], while the intramolecular motion is fast. The random increments were generated as reported [149]. A set of reactant life times ($\tau_{life,\,i}$) was computed by generating several thousand trajectories, the lifetime for the i-th trajectory defined as the time required to reach the saddle line starting from the bottom of the reactants' well. The resulting lifetime is the average:

$$< \tau_{life} >= \frac{1}{N}\sum_{i=1}^{N} \tau_{life,i} \qquad (2.45)$$

where N is the number of trajectories. The rate constant can then be estimated as $1/ < \tau_{life}>$.

The rate constants vs. x(EG) at several overpotentials are shown in Figure 2.10. Starting from $\eta = 2.9$ V the rate constant vs. x(EG) in fact *increases* with increasing x(EG), with the ascending branches more pronounced at higher overvoltages, according with the data [147].

The computed results can be understood in terms of saddle point avoidance (SPA), which is one of the most remarkable manifestations of solvent dynamics. Usually the larger this

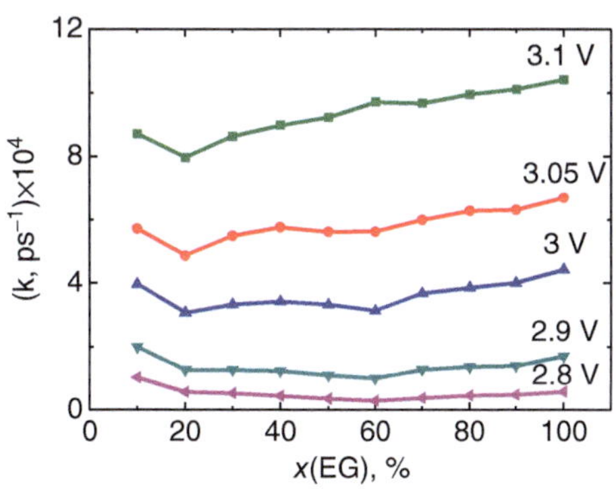

Figure 2.10 Rate constant of the electrochemical $S_2O_8{}^{2-}$ reduction vs. x(EG) in water-EG mixtures at different electrode overpotentials (2.8 V – 3.1 V). Adapted from [147].

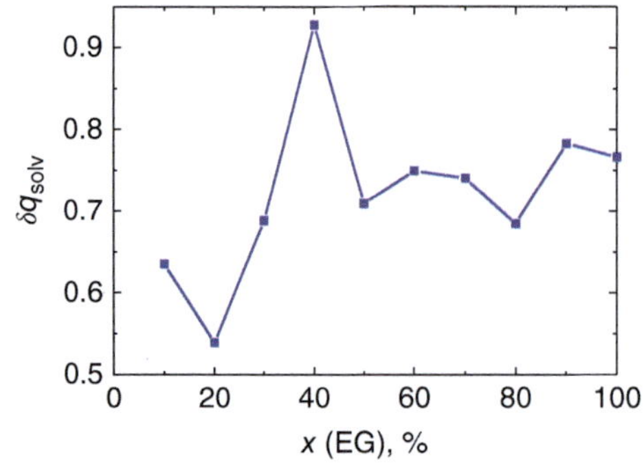

Figure 2.11 Saddle point avoidance effect as a function of x(EG) at $\eta = 3.1$ V. Adapted from [147].

effect, the slower the reaction rate. As the intramolecular coordinate is here fast, the SPA can be assigned to a sum of contributions from all three solvent modes:

$$\delta q_{solv} = \sum_{i=1}^{3} (q_i^* - q_i^{saddle}), \tag{2.46}$$

q_i^* are the solvent mode values corresponding to the reaction window where the barrier is overcome, q_i^{saddle} the solvent coordinates at the saddle point of the 5D free energy surface.

The Δq_{solv} dependence of x(EG) at $\eta = 3.1$ V is shown in Figure 2.11. SPA was largest in the vicinity of the $W_{AD}(x(\text{EG}))$ minima. On further viscosity increase, SPA is less pronounced, leading to increasing rate constant due to decreasing Arrhenius factor. It is, however, still an open and challenging question, how the multimodal "orchestra" concept can be extended to non-Debye solvents (like syrups, glycerol etc.), where an exact expansion of the solvent correlation function, Eq. (2.41) is not possible.

The formalism above rests on using the experimental solvent complex dielectric spectrum. We then lose, however, information about the solvent dynamics in the reaction layer (so-called problem of solvent "local viscosity"). The "local" solvent relaxation times can differ significantly from the bulk values. The corresponding correlation function $M_{loc}(\tau)$ can be calculated from MD simulations [99] enabling more reliable predictions.

2.4 Early and recent views on chemical and electrochemical long-range ET

The concept "long-range" ET initiated from inner sphere ET between transition metal complexes as pioneered by Henry Taube [150]. "Superexchange" is a concept introduced by Kramers already in the 1930s [151] to account for magnetic interactions between transition metal ions in crystalline oxides, and McConnell [152] introduced superexchange in chemistry. This attractive formalism based on high-order quantum mechanical perturbation theory is still the core of much present analysis, now enveloped in "Green Function" and other terminology [153, 154]. Figure 2.12 illuminates the role of intermediate "matter" in long-range ET. The donor and acceptor group can be coupled electronically via intermediate LUMOs or HOMOs. These states are then not physically populated, but long-range coupling outwins direct ET due to favorable nearest neighbor interactions.

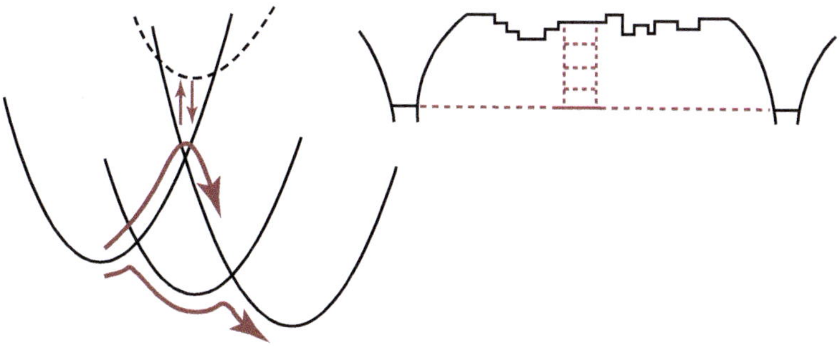

Figure 2.12 Left: Free energy surfaces, schematic for three-level bridge-assisted ET. Critical energy quantities are indicated. The upper vertical arrows show the superexchange pattern with purely electronic coupling to a high-energy intermediate state indicated by the dashed line at the crossing of the reactants' and products' potential surfaces. The upper bold arrow shows the barrier crossing in this limit. The lower bold arrow shows the pattern for ET through a low-lying intermediate state. Right: A scheme for the donor and acceptor levels along with a set of bridge group LUMOs in long-range superexchange ET (cf. Figure 2.1).

A single bridge group, B, converts the direct electron exchange factor T_{DA}, Eqs. (2.2)–(2.4) into

$$T_{DA}^{(2)} \rightarrow \sum_{i} \frac{V_{DB} V_{BA}}{\Delta_{DB}} \tag{2.47}$$

V_{DB} and V_{BA} are the electron exchange factors between the donor and the intermediate bridge group, and between the bridge group and the acceptor, respectively, Δ_{DB} the energy gap between the donor (redox) level and intermediate bridge group energies. With N intermediate states, higher-order expansion takes Eq. (2.47) into forms such as

$$T_{AD}^{N} = \sum_{i,i\prime} \frac{V_{DB1} V_{B1B2} \cdots \ldots V_{BNA}}{\Delta_{DB1} \Delta_{DB2} \cdot \ldots \Delta_{DBN}} \tag{2.48}$$

If all V_{BiBi}'s and all Δ_{Bi}'s are the same, V_B and Δ_{Bi}, respectively and a single ET path dominates, then

$$V_{DA}^{(M)} \rightarrow \frac{V_D V_A}{V_B} \left(\frac{V_B}{\Delta_B} \right)^M \tag{2.49}$$

where V_D and V_A are the electronic couplings between the donor and the nearest bridge group, and between the acceptor and the bridge group nearest to the acceptor, respectively. M is the number of bridge groups. Eq. (2.49) can then be converted to

$$V_{DA} \rightarrow \frac{V_D V_A}{V_B} \exp \left[-\frac{1}{a} \left(\ln \frac{\Delta_B}{V_B} \right) z \right] \tag{2.50}$$

a is the structural extension of each intermediate group, $z = Ma$ the donor-acceptor distance. Eq. (2.50) reflects exponential distance decay, i.e. the larger the donor-bridge energy gap, and the weaker the electronic couplings, the faster the distance fall-off. Superexchange is also a formal basis for "interference" in dual conductivity pathways, I and II, generically

$$|T_{AD}^{dual}|^2 = |T_{AD}^{I} + T_{AD}^{II}|^2 \tag{2.51}$$

Duality can either reinforce or counteract each other depending on the signs of the coupling factors and energy gaps. Electrochemical single-molecule junctions with negative interference in the form of "antiresonances" in the conductivity/overpotential correlations have been reported recently, by Tao [155], Nichols [156], and their associates.

The other limit is ET via individual ET steps, each vibrationally relaxed on electron or hole population also denoted as "hopping" or "incoherent" long-range ET, each step characterized by the formalism in section 2.2. Sequential ET requires low-lying electronic states in the ET chain. Such mechanisms follow formal consecutive multi-step kinetics. This alone invokes much weaker distance dependence than for superexchange. "Incoherent" hopping through N (> 1) intermediate states is characterized by rate constants, conductivities, charge separation times, or yield from the injection site to the Nth site [157–159]. Rate constants and other observables become prohibitively voluminous, unless applied to repetitive units, as for example for oligophenylenevinylene derivatives [159] and DNA fragments [157, 158]. One example of such a conductivity/distance correlation is [160]

$$g \to \left(\frac{|e|^2}{k_BT}\right)\frac{k_{av}^{bridge}\exp\left(-\dfrac{E_A^{bridge}}{k_BT}\right)}{\dfrac{k_{av}^{bridge}}{k_{inj}+k_{exit}}+(N-1)} \tag{2.52}$$

$k_{\rm inj}$ and $k_{\rm exit}$ are rate constants for electron (hole) injection from the negatively biased electrode to the nearest bridge group, and from the bridge group in the opposite end of the chain to the positively biased electrode (or vice versa for hole transfer). k_{av}^{bridge} is the "average" rate constant for ET or hole transfer between nearest neighbor bridge groups, and E_A^{bridge} the average energy gap between the bridge group states and the Fermi energy of the injecting electrode. Equations such as (2.52) illustrate the weak, approximately inverse polynomial distance dependence.

As illustrated by Figure 2.12 and recognized early [161–163], ultra-fast ET processes can proceed through "dynamically populated intermediate states" (DPIS). These are transiently populated, but do not relax before proceeding to the next step in the sequence. Such processes are still quantum mechanically second order as reflected by two transmission coefficients, Eq. (2.53), rather than a single one as in vibrationally relaxed processes

$$W_{AD}^{(2)}(DPIS) = \frac{2\pi^{3/2}(V_{DB})^2(V_{BA})^2}{\hbar^2\omega_{eff}\lambda_R^{DB}(k_BT|G_{DB}^{\neq}-G_{BA}^{\neq}|)^{1/2}}\exp\left[-\frac{\Delta G^{\neq}}{k_BT}\right] \tag{2.53}$$

$\Delta G^{\neq}$ is the activation free energy and coincides with the higher of the energies at the two crossings, $G_{\rm DB}{}^{\neq}$ and $G_{\rm BA}{}^{\neq}$. DPISs have aroused interest in contexts of ultra-fast chemical and biological processes (say the primary photosynthetic ET processes) [164].

As yet another issue, new resonances arise when the energies of two or all three crossings coincide and the energy gaps formally vanish, Eqs. (2.47)–(2.49) and (2.53). These effects are analogous to other three-level processes such as resonance Raman scattering. Formal divergence in the latter is often relaxed by adding lifetime broadening, but excited state coupling to a *continuum* of nuclear modes itself lifts the divergence [165]. Resonance peaks could thus appear in overpotential regions, where transition between superexchange and hopping occurs, but no such effects in electrochemistry seem to have been reported.

2.5 Molecular-scale electrochemical science

In "nanoscience" target objects display novel size-dependent properties. Quantum mechanical phenomena such as electron tunneling or electronic interference are conspicuous, and physical laws such as the renowned Ohm's law take new forms. Nanoscience offers technological perspectives [166, 167]. Metal and semiconductor nanostructures and nanoparticle arrays are also part of the new electrochemistry. 2D and 3D graphene and graphene composites are other examples of non-traditional materials in electrocatalysis, new battery types, and (bio)fuel cells [166–169]. New areas of redox metalloenzyme and DNA based electrochemistry also qualify as nanoscale (bio)electrochemistry [169–173].

Single-molecule electrochemistry is not so firmly established as some other single-molecule science. Also, single-molecule optics monitors truly single-molecule signals, but single-molecule electrochemical ET strictly speaking monitors the effects of successive ET events as in STM [171]. Single ET events may be observed in collision electrochemistry where signals from individual colliding NPs or even enzyme molecules exchange electrons [170]. Single-molecule electrochemistry can be combined with optical processes, if the different oxidation states of the target molecules have different optical properties. Fluorescence, electrochemiluminescence, Raman and tip or surface enhanced Raman scattering have then led to single-event electrochemistry. We refer to recent reviews of these exciting developing areas [171–174] and focus presently on *in situ* STM (and AFM).

2.5.1 Electrochemical in situ STM and AFM

Electrochemical STM and AFM in the *in situ* ("*in operando*") mode (Figure 2.13) [4, 23, 24] offers *two* tunneling current potential correlations, the current/overpotential and the current/bias voltage correlation. *Ex situ* STM in air or vacuum only offers the latter [18, 20]. *In situ* STM has become a powerful single-molecule tool to map small organic and inorganic molecules [3, 21, 22, 24, 105, 175–179], larger molecules such as transition metal complexes [23, 175, 176, 178], and even biomolecules such as redox (metallo)proteins [18–182] and DNA-based molecules [176, 177, 182, 183]. Conversion of *in situ* STM currents to molecular structure requires theoretical and computational support. Electrochemical *in situ* STM has disclosed new ET phenomena, for example "collective"

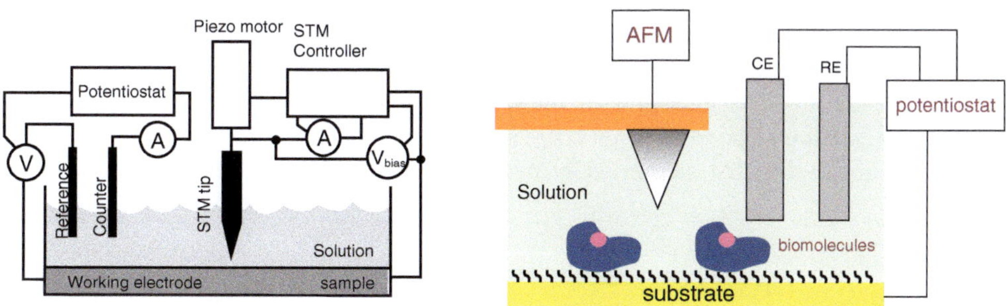

Figure 2.13 Left: *In situ* STM, schematic. The electrode/solution and the tip/electrode potentials are controlled independently relative to a common reference electrode. The tip is coated with an insulating material except at the very end. Right: Analogous electrochemical *in situ* AFM.

ET, possibly a rationale for often observed very high *in situ* STM current densities (per molecule) [23, 181]. Single-ET charging of inorganic particles [184, 185] are other new phenomena. Ionic liquid media are gaining importance [104, 105], and "nanoimpacts" [170] still other electrochemical single-entity phenomena. New electrode materials, with 2D and 3D graphene, graphene/NP hybrid, and perovskite materials [169, 186, 188], core-shell nanoparticles [187, 189], single carbon nanotubes [171, 217–219] (cf. below), and nanoporous electrodes [191, 192] as examples have further become prominent.

2.5.2 Nanoscale mapping of novel electrochemical surfaces

Modified electrochemical electrode surfaces, either in the metallic surface structure as in nanoporous gold [190, 191] or by self-assembled molecular monolayers (SAMs) have developed broadly, ultimately aiming at tailored electrochemical surfaces [192–195]. Characterization by surface spectroscopies (XPS, EDS, IR and Raman), transmission (TEM), and scanning electron microscopy (SEM) as well as STM/AFM have been paramount, enabling unprecedented structural resolution. We select here a particular system class, which has come to enjoy almost iconic status, viz. functionalized thiol SAMs on Au-surfaces.

2.5.2.1 Self-assembled molecular monolayers (SAMs) of functionalized thiol [192–194]

Solid gold and gold nanoparticles (AuNPs) have been coveted for thousands of years for investment, in jewelry, and for cult and decorative purposes. AuNPs and Au-surfaces must be protected by SAMs. Other functional groups in the SAM molecules can tailor the Au-surfaces for catalytic, molecular electronics, and other desirable properties. The electronically "soft" -S· ligand binds well to the "soft" Au-surface, in a Au(0)-S(0) strong van der Waals unit. The R-residues can be hydrophobic, hydrophilic, electrostatically charged or neutral, and structurally large or small. Au-thiyl based SAMs are determined by the Au-S bond and by lateral interactions, in addition to the solvent and electrolytes [192–195]. The latter are equally important, when R- is charged. Single-crystal electrochemistry, XPS and other surface spectroscopy, and *in situ* STM have been paramount in high-resolution mapping ranging from the single molecule to the whole SAM formation process [194], with theoretical and computational support essential.

Figures 2.14 and 2.15 illustrate the variety of packing modes of functionalized simple aliphatic thiols resolved to single-molecule or even sub-molecular resolution by *in situ* STM [23, 195–197]. Packing is determined not only by the molecules, but also by the atomic structure of the Au-surfaces, strikingly illuminated by the widely different packing of the amino acid cysteine on Au(111)-, Au(110)-, and Au(100)-electrode surfaces. Spectroscopy, diffraction, and *in situ* STM, supported by theoretical work has led to the adsorption process

$$\mathrm{Au(0) + HS(-I) - R \rightleftharpoons Au(0) - \cdot S(0) - R + \tfrac{1}{2}H_2} \tag{2.54}$$

with S bound as a thiyl radical, ·S(0) [192, 193]. The reverse process of Eq. (2.54) is electrochemical reductive desorption, associated with sharp voltammetric peaks crucial in determining the surface coverage of the ·S(0)-R SAMs.

Many studies have shown that unless sterically hindered, adsorption is generally via thiol "mining" of a gold atom from the surface, cf. section 2.2.4.1, followed by sideway

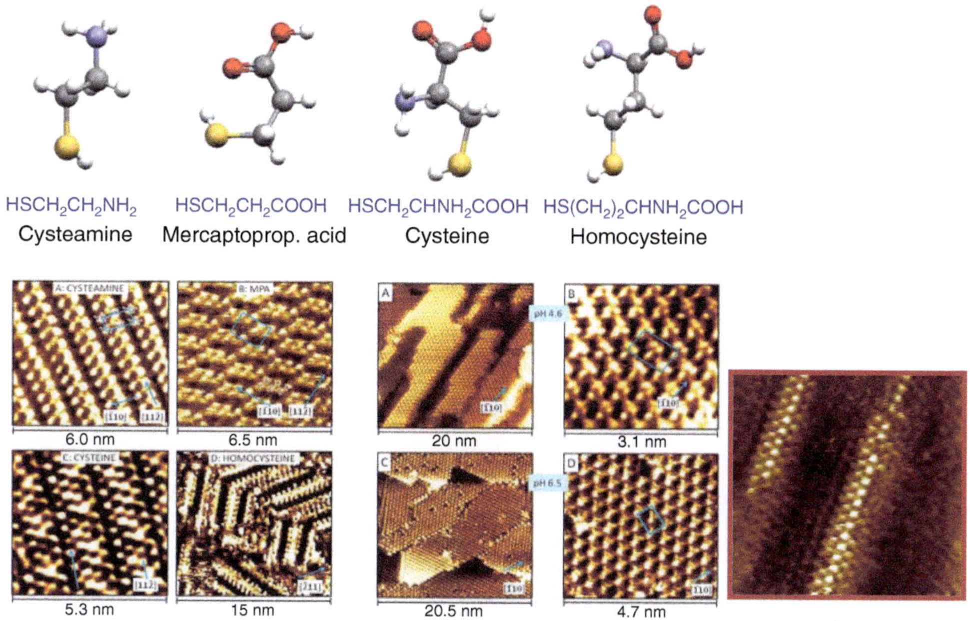

Figure 2.14 Top row: *In situ* STM target functionalized thiol molecules. Left image quartet, clockwise: *In situ* STM images of Cysteamine, mercaptopropionic acid, homocysteine, and L-cysteine all on Au(111). Middle image quartet: L-cysteine on Au(110), different magnification. Right Racemic L,D-cysteine on Au(100). Adapted from [23, 193–196].

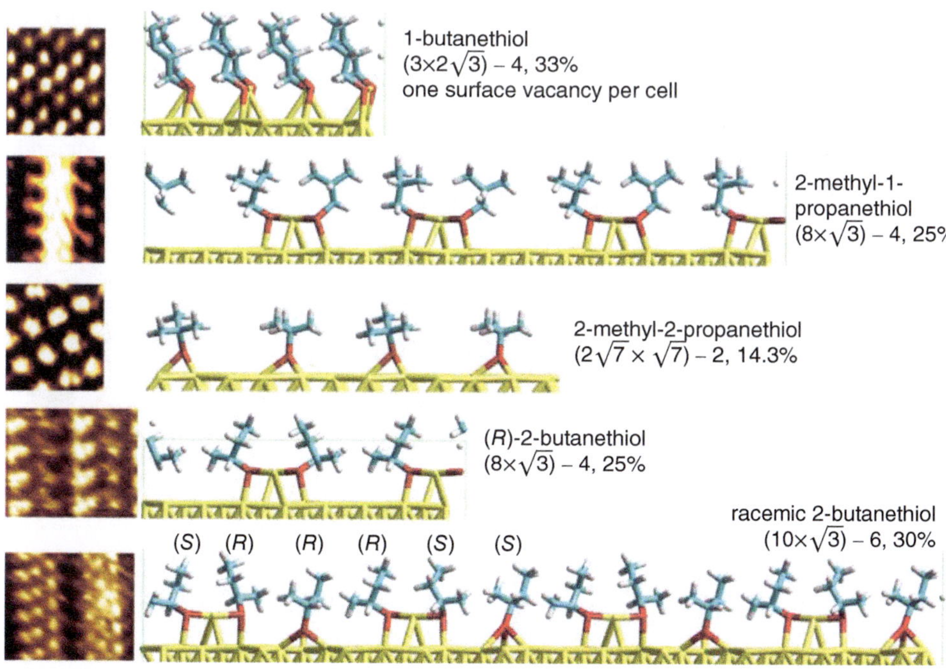

Figure 2.15 From top to bottom, binding of the four isomeric butanethiols on a Au(111)-surface. Column to the left: High-resolution *in situ* STM images on which the SAM interpretations are based. Adapted from [194].

binding of two adsorbate molecules. Among the butanethiols the bulky *tert*-butanethiol, 2-methyl-2-propanethiol is thus the only one binding on a flat surface [195, 198]. The high-density phase of racemic 2-butanethiol is another unique case for two binding modes on the same surface, both directly and via Au-atom mining. Even the simple thiols thus exhibit amazingly multifarious binding modes, increasingly well understood by electrochemistry, high-resolution spectroscopy and microscopy, and large-scale computations.

2.5.3 Electrochemical single-molecule ET and conductivity of complex molecules

Functionalized alkanethiol SAMs display a prodigious variety of binding patterns [194, 195], but are still "simple" molecules in the sense that they are small with no Faradaic currents except for reductive and oxidative desorption. Sharp capacitive peaks associated with a two-dimensional phase transition are, however, observed [199]. Redox molecules are instead what we shall denote as "complex". Even at the single-molecule level, they can be brought to undergo chemical processes. These are "simple" ET processes which are also probes for tunneling mechanisms, say sequential vs. superexchange, but also cover organic and inorganic redox probes [90, 178, 179], enzyme-substrate interactions [200, 215], and inter-protein and protein-DNA interactions [182, 202]. To this adds sophisticated quantum interference [155, 156], statistical approaches [32], ionic liquids [104, 105, 178], and the effects of the ionic atmosphere [23]. We first discuss in general terms *in situ* STM of complex molecules and some specific molecular targets, and then proceed briefly to redox metalloproteins.

Figure 2.16 shows an energy diagram of the *in situ* STM process of a molecule with a local redox level in a polar (aqueous, ionic liquid) medium, specifically reduction with negative bias (V_{bias}). The redox level is vacant at equilibrium, and located well above both Fermi levels, where superexchange prevails. As the overpotentials of the working electrode and

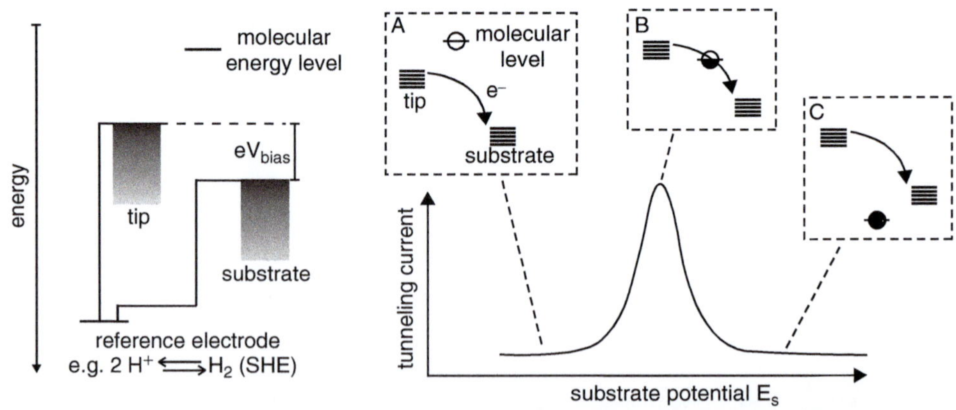

Figure 2.16 Left: Energy diagram showing energy levels of an oxidized redox molecule and of a pair of enclosing, negatively biased electrodes. Right: Tunnelling current/overpotential correlation on parallel variation of the substrate and tip potential. Adapted from [90].

the tip are shifted negatively in parallel, i.e. both Fermi levels are raised, the redox level "gets within thermal reach" of occupied levels of the negatively biased working electrode. ET then switches from superexchange to "hopping" with the resonance effects discussed in section 2.4 potentially at work. The redox level becomes physically populated and begins to relax. If coupling to the second electrode is weak, relaxation continues all the way across both Fermi levels, with renewed activation for the second step. If instead, by the condition $\hbar/|V_{BA}|^2 < \tau_{relax}$ [23, 204] where τ_{relax} is the nuclear relaxation time, the coupling is strong, multiple ET proceeds, until relaxation is achieved and renewed thermal activation sets in.

This view is qualitative but general and accounts for the "spectroscopic" tunneling current/overpotential feature. A phenomenological frame for the simplest two-step, diabatic ET via the molecular target is summarized in Eq. (2.55)

$$j_{AD} \approx 2en_{el}\frac{W_{BD}W_{AB}}{W_{BD}+W_{AB}}; \qquad n_{el} \approx \frac{|eV_{bias}|}{\delta\varepsilon^*} \tag{2.55a}$$

$$W_{BD} = \frac{\omega_{eff}}{2\pi}\frac{\kappa_{BD}}{\delta\varepsilon^*}\int_{-\infty}^{0}\exp\left[-\frac{(\lambda_R-\varepsilon-\gamma\varphi-\xi\eta)^2}{4\lambda k_BT}\right]d\varepsilon \tag{2.55b}$$

$$\vec{k}_2 = \frac{\omega_{eff}}{2\pi}\frac{\kappa_{AB}}{\delta\varepsilon^*}\int_{0}^{\infty}\exp\left[-\frac{(\lambda_R+\varepsilon-(1-\gamma)\varphi+\xi\eta)^2}{4\lambda k_BT}\right]d\varepsilon \tag{2.55c}$$

where V_{bias} is the bias voltage, ξ and γ parameters that represent the overpotential and bias voltage at the site of the molecule, and $\delta\varepsilon^*$ the effective energy integration range, $(\delta\varepsilon^*)^{-1}$ equivalent to energy density. n_{el} is the number of electrons coherently transferred. $\delta\varepsilon^*$ is wide (level broadening δ small) and $n_{\text{el}} \rightarrow 1$ in this limit. Small $\delta\varepsilon^*$ (large δ) and large $n_{\text{el}} \gg 1$ applies in the opposite, adiabatic limit.

Levels around the Fermi levels dominate at small η and V_{bias}, reducing Eq. (2.55) to

$$j_{AD} = e\frac{\omega_{eff}}{2\pi}\kappa\exp\left(\frac{\lambda_R-e|V_{bias}|}{k_BT}\right)\left[\cosh\left(\frac{\left(\frac{1}{2}-\gamma\right)e|V_{bias}|-e\xi\eta}{2k_BT}\right)\right]^{-1} \tag{2.56}$$

with $\kappa \approx \kappa_{\text{BD}} \approx \kappa_{\text{AB}}$, and a maximum in the overpotential correlation at

$$\eta_{\max} = \frac{1}{\xi}\left(\frac{1}{2}-\gamma\right)V_{bias} \tag{2.57}$$

i.e. at equilibrium, $\eta = 0$ for a symmetric junction, $\gamma = ½$. The maximum current is

$$j_{AD}^{\max} = en_{el}\frac{\omega_{eff}}{2\pi}\kappa\exp\left(\frac{\lambda_R-e|V_{bias}|}{k_BT}\right) \qquad n_{\text{el}} \rightarrow 1 \tag{2.58}$$

The following should be noted:

- The formalism is successful up to a point, cf. below, but needs adjustment. The j_{AD}/η and $j_{\text{AD}}/V_{\text{bias}}$ correlations are, however, well suited for data analysis, as long as $|e\eta, eV_{bias}| < \lambda$.
- The correlations in the *adiabatic* limit are less straightforward, studied in great detail by Medvedev [23, 205] based on Newns-Anderson formalism and presently under scrutiny. "Pragmatic" data analysis can be based on Eqs. (2.55)–(2.58) with κ simply set to unity.
- Attention to the electrolyte and to ionic liquid media should be given. The potential distribution parameters in the nanogap electrolyte medium are, for example correlated [23, 202] and do not accord with the generic pair values of $\xi = 1$and $\gamma = ½$.

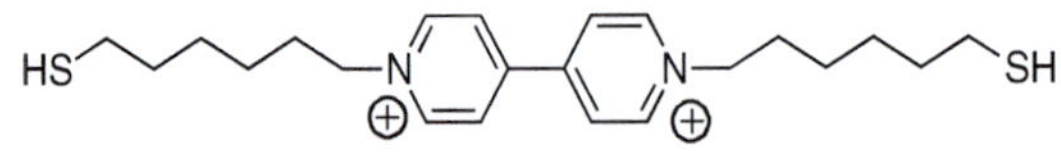

6V6 (as iodide or *p*-toluenesulfonate salt)

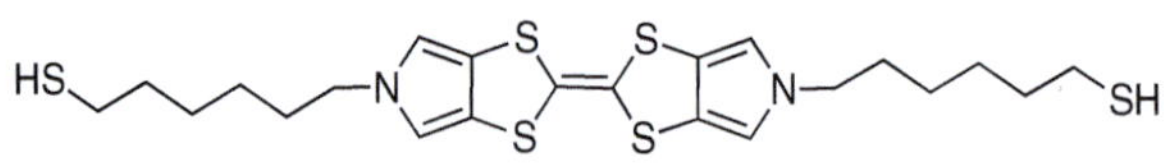

6PTTF6

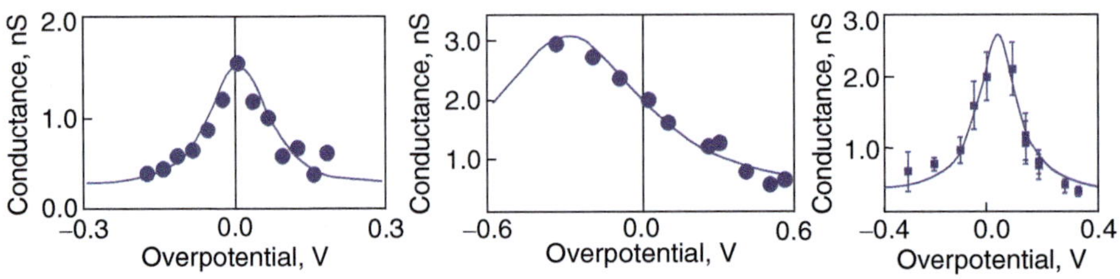

Figure 2.17 Top: The target molecules viologen (6V6) and 6,6′-pyrrolo-tetrathiafulvalene (6PTTF6). Bottom: conductivity/overpotential correlations, from left to right: 6V6 in ionic liquid (1-butyl-3-methylimidazolium trifluoromethanesulfonate), 6V6, and 6PTTF6 both in aqueous electrolyte. Adapted from [104, 178].

- High bias voltages are often needed. The solvent reorganization free energy in the tunneling gap is, however, small, 0.3 eV or less, and bias voltages needed may exceed the activationless limit in the individual ET steps, raising other questions.

2.5.4 Selected cases of in situ STM and STS of organic and inorganic redox molecules

Tao's study of Fe-protoporphyrin IX was the first case of *in situ* STS j_{tunn}/η showing a maximum around the equilibrium potential [175]. Single-molecule reports on other complex organic and inorganic molecules, with up to two orders of magnitude on-off current ratios have followed [90, 178, 179]. Mechanical properties of the molecules [206], the solvent [104, 105, 178], quantum interference [155, 156], and stochastic effects [32] have also been addressed. The comprehensive studies of Nichols and associates on viologen, Figure 2.17, illuminate both the potential and challenges of these powerful single-molecule approaches.

2.5.4.1 The viologens

The viologens (Figure 2.17) have come to stand forward as iconic single-molecule redox targets. New technology is built up such as break-junction, $j_{tunn}(s)$ and $j_{tunn}(t)$ (s distance, t time) with current-distance and current-time fluctuations recorded. The viologens offer strong support for two-step incoherent ET but has disclosed challenges that need attention in forthcoming efforts. 6V6 viologen between a Au tip and a Au(111)-electrode surface via two 6-carbon thiyl linkers can be compared with tetrathiafulvalene (6PTTF6). The current/overpotential correlations in aqueous electrolyte of the rigid 6PTTF6 displays a sharp current maximum. In contrast, the "floppy" 6V6 displays a sigmoidal dependence. Two-step formalism with "gated ET" via a "soft" torsional mode could reproduce latter behaviour. Another view is that the sigmoidal correlation simply reflects higher conductivity of the equilibrated oxidized than the equilibrated reduced form (cf. below).

6V6 displays, however, a pronounced maximum in 1-butyl-3-methylimidazolium triflate (BMIT). This is a subtle single-molecule challenge. Bulk dielectric differences cannot account for this pattern, but comparison with 6PTTF6 may hold a clue. Ionic liquid constituents might cling to the electrostatically charged 6V6 target, imposing rigidity compared with aqueous electrolyte. The 6V6/6PTTF6 pair thus displays multifarious achievements and challenges of *in situ* STM-based single-molecule electrochemistry.

2.5.4.2 Transition metal complexes as single-molecule in operando STM targets

Robust transition metal complexes of Os, Ru, and Co, particularly Os-complexes of 2,2′-bypyridine and linking 4,4′-bipy have been single-molecule *in situ* STM/STS targets [90, 207], Figure 2.18. The complexes show a strong maximum in the j_{tunn}/η correlations at the equilibrium potential, with on-off ratios up to two orders of magnitude, following the equilibrium potentials of two different Os-complexes, and the bias voltage in accordance with Eqs. (2.55)–(2.58).

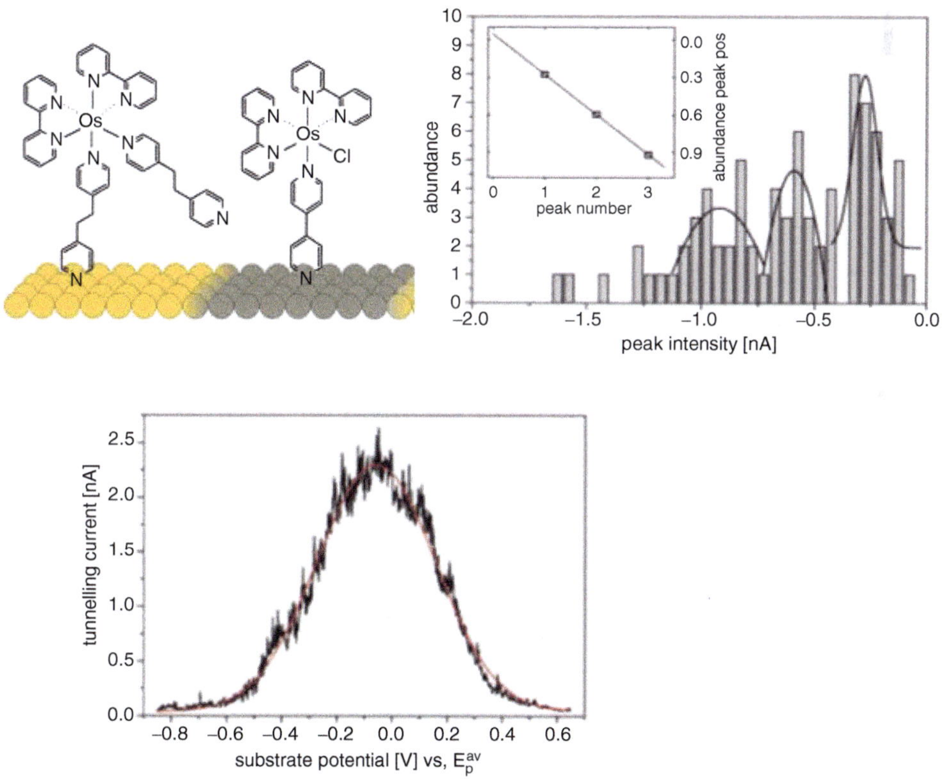

Figure 2.18 Top left: Two Os-complexes on Au(111) and Pt(111), schematic. Top right: Abundance of tunneling peak current densities. Equidistant peak abundances indicative of discrete single-, dual, etc. molecular contributions. Bottom left: Tunnelling current/overpotential correlations in an ionic liquid. The flat peak region possibly indicative of coherent electron transfer at the large bias voltages used (0.7 V). Adapted from [90, 207].

The Os-complexes illuminate other aspects. Repetitive *in situ* STS shows equidistant peaks, assigned to single-, dual- and triple- etc. molecular conductivity. *In situ* STS in an ionic liquid also shows a clear maximum with an on-off ratio of about 50 [207]. The maximum is broad, associated with the large bias voltage used (0.7 V). Os-polypy complexes offer direct comparison between electrochemical ET and the two-step *in situ* STM current, Eqs. (2.59a)–(2.59c), concluding that several hundred electrons are transferred in a single *in situ* STM event. *In situ* STM and STS of other polypy-based transition metal complex units open for electrochemical single-molecule spintronics [208].

As noted, transition from single-molecule to macroscopic, "thermodynamic" behaviour is a core in condensed matter nanoscience and nanoscale electrochemistry. Fan and Bard's single-molecule study at ultra-microelectrodes marks a first electrochemical approach [209]. In a quite recent study, Li, Tao, and associates introduced a novel approach to stochastic vs. bulk molecular conductivity [32]. The target system was ferrocene, Fc in acetonitrile solution linked to Au(111) and a Au-tip by variable-length alkanethiols, Figure 2.19. The data report single-molecule STS the average of which displays a sigmoidal pattern as for viologen in aqueous solution. It was argued that redox equilibrium prevails, without soft mode gating, but the sigmoidal average shows subtle single-molecule

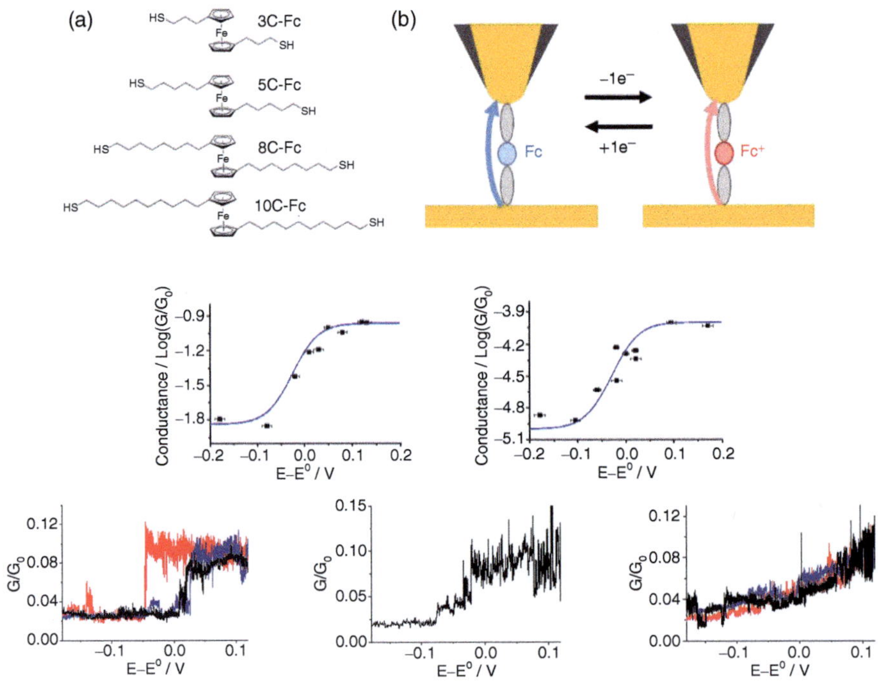

Figure 2.19 Top row, left: Ferrocene target compounds; right: Scheme of charge transport through a ferrocene molecule. Middle row: Averaged conductance vs. overpotential, 3-Carbon linker, left and 10-Carbon linker, right. The blue curves are Nernstian fittings. Bottom row: Single-molecule conductance vs. potential for 3-Carbon-Fc. Different fluctuation patterns around the equilibrium potential: (a) Discrete two level, (b) intermediate and (c) continuous variations of conductance with potential. Red, blue and black colors represent different potential sweeps. Only one curve shown in the middle plot. Adapted from [32].

tunneling current fluctuations. Figure 2.19, middle row shows two averaged STS correlations, 3C-Fc and 8C-Fc, the bottom row three kinds of single-molecule traces. The fluctuational transitions between low and high conductivity begin close to the equilibrium potential and range between abrupt and smooth. The higher the ET rate, the smoother the transition, so that the pattern to the left is dominated by Fc with the longest linker and lowest rate, the faster reacting Fc with the short linker by the pattern to the right. The study illuminates strikingly the transition from stochastic single-molecule behaviour to macroscopic patterns, which is exactly the heart of nanoscience.

2.5.5 Other single-entity nanoscale electrochemistry

Electrochemistry of molecular-scale entities covers highly diverse targets, developed over the last decade. In addition to single molecules, molecular-scale entities include "0D"-3D inorganic particle, wire, rod, "flower", layer etc. structures understood to different degrees. We refer to a recent Faraday Discussion [210] for overviews of single-entity electrochemistry and discuss two selected electrochemical nanoscale materials.

2.5.5.1 Electrochemistry in low-dimensional carbon confinement

The new "wonder" material graphene has come to stand forward in electrochemical materials science. Graphene is attractive for unique electronic properties (e.g. the Dirac point, and nearly linear ascending, or descending DOS near the Fermi level). Introducing graphene in current-voltage relations of ET processes is a challenge. An experimental challenge is to identify the graphene sheet charge, for example whether graphene is affected by a support electrode. Graphene holds a wealth of catalytic and other promises, attributed to surface functionalities and defects [213–217], the ET role of which is in need of theoretical exploration. Redox processes at the graphene edge are the most studied (see references in [222]). Graphene edge activity can be understood from scanning electron spectroscopy, which discloses local electronic states below the Fermi level at *zigzag*-type [213] but not at *chair*-type edges. Graphene sheet vacancies also result in local catalytic [222] electronic states. The FcMeOH/FcMeOH [+] rate constant was thus found to increase by two orders of magnitude on increasing the vacancy concentration to 5×10^{12} cm^{-2} (≈ 2 nm between the defects). Catalytic activity can be enhanced by nitrogen doping or by oxygen surface functionalities [215, 216].

"Nanotubes" are extensions from "0-D" NPs to 1-D nanoscale structures. Nano- and picoampere voltammetry of single-wall carbon nanotubes (SWCNT) is reported [211, 217–219], and experimental electrochemistry findings now warrant molecular level theoretical analysis, general principles of which are formulated, along with double layer modeling [220, 221]. Electrolyte structure and transport *inside* nanotubes are reviewed but harder to access [223]. Pure water structure and hydrated ion and proton transport inside SWCNTs was explored by X-ray diffraction and NMR technique, and by MD simulations. CNT mobility of water and hydrated protons is higher than in bulk solution, due to the ordered hydrogen bonds and weak interaction with the CNT walls. Cationic coordination is less ordered than in the bulk, but with only minor anion coordination shell changes. Understanding chemical *reactions* is highly challenging as reviewed recently [224], with primary attention to catalytic hydrogenation, oxidation, and other processes, and CNTs shown to modify the electronic properties of the metallic nanoparticle catalysts. Electron transfer inside a carbon nanotube filled with water was investigated in the framework of Marcus theory with the help of MD simulations [212].

2D and 3D graphenes such as reduced graphene oxide (rGO) and graphene "paper" have, finally, been introduced in both pure enzyme electrochemistry and, notably in working devices such as enzyme biofuel cells [169]. The new biocompatible electrode materials ascertain that enzymes are immobilized in robust, working states with high analytical sensitivity and power density, and visible perspectives for implantable devices. The materials are well characterized by SEM, TEM, and XPS. AFM and STM offer other high-resolution characterization. Altogether the new multifariously structured graphene-based materials offer prodigious new and exciting theoretical and computational challenges.

2.5.5.2 Electrochemistry of nano- and molecular-scale metallic nanoparticles

The vast area of metallic and non-metallic molecular- and nanoscale particles (NPs) has developed into sophisticated nanoscience with impressive progress over the last couple of decades, and strong impact in electrochemistry, all the way from macroscopic assemblies to pre-designed single NPs. We cannot here review this broad evolving area, but refer to meticulous recent reviews [225–227]. We discuss instead a few recent studies of AuNP-based electrocatalytic systems, where new computational efforts are warranted.

Metallic NPs represent exactly the nanoscience, where target systems are taken all the way from stochastic to macroscopic size. Theoretical approaches to NP-catalysis mostly address small molecules say dioxygen, water, dinitrogen, ammonia, methanol etc. and their intermediates in chemical conversion, and involve strong NP-molecule interactions [227]. 1–2 nm coated AuNPs display condensed matter single-ET (SET) charging [184], for which electrochemical *in situ* STM offers a further platform [185]. Electrochemical SET fades out when the NP size exceeds ≈ 2.5 nm still the size range of transition metal complexes or small redox metalloproteins. Contrary to molecular targets, the electronic spectrum of metallic NPs, however, rapidly approaches a continuum, still with catalytic enhancement of simple electrochemical and bioelectrochemical ET processes [228–233].

Thiol-protected AuNPs catalyze simple interfacial ET processes of transition metal complexes [228], quinones [229], the redox proteins cyt *c* [230] and azurin [231], as well as redox enzymes [232, 233]. In spite of ET distance extension by up to several nm, more than an order of magnitude rate constant *increase* is observed. Similarly extended molecular matter would close the ET channel entirely. Figure 2.20 illustrates here such a case. There is, however, presently no obvious single physical origin of electrochemical AuNP-triggered interfacial ET rate enhancement, and electronic structure computations of NPs that include both the electrochemical interface and the dynamic solvent, as well as complex redox molecules are not available. There is, therefore, surely a need for computational efforts directed at the catalytic effects (electronic transmission coefficients) of a single AuNP. Questions to be addressed would be AuNP size effects, catalytic efficiency of different AuNP surface structural facets, and the role of the protecting SAM. Such efforts would be a prodigious guide to AuNP electrocatalysis of both simple electrochemical ET processes and complex processes in fuel cell and other technologically interesting processes.

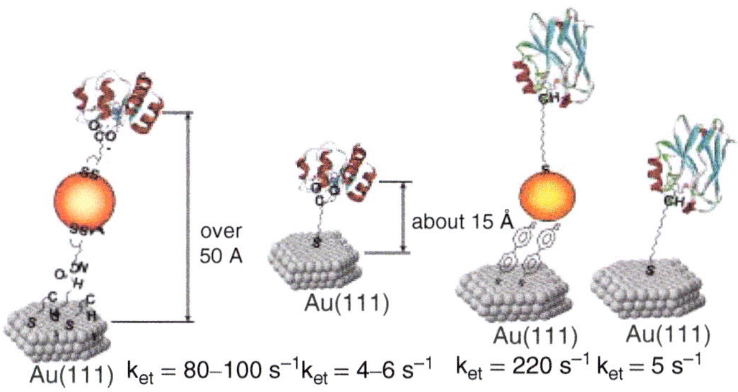

Figure 2.20 Left: The heme protein cyt *c* immobilized via a 3 nm coated AuNP and two molecular linkers on a Au(111)-electrode surface and the same molecule linked directly via a single molecular linker. Right: Similar linking of the blue copper ET protein azurin. The rate constants are also given. Adapted from [70, 230, 231].

2.5.6 Elements of nanoscale and single-molecule bioelectrochemistry

Single-molecule *in situ* STM and STS electrochemistry was early brought to include large biomolecules, with horse heart cytochrome *c* [234] and horseradish peroxidase [235] as focus proteins. What was later denoted a "paradigmatic" *in situ* STM metalloprotein, the blue copper protein azurin (*Pseudomonas aeruginosa*) was introduced slightly later [236, 237], followed by studies of protein immobilization, single-crystal electrochemistry, and *in situ* STM/STS [70, 238], subsequently picked up by other groups [180, 182, 239]. We discuss here a few single-molecule metalloprotein electrochemistry issues and note also some "puzzles".

2.5.6.1 A single-molecule electrochemical metalloprotein target – *P. aeruginosa* azurin

P. aeruginosa azurin (Mw ≈ 14,000 amu) is probably the most comprehensively studied single-molecule redox metalloprotein target. The strategy behind this choice is specific properties favorable for azurin as a high-resolution *in situ* STM target, Figure 2.21:

- The redox potential of the Type 1 Cu-center of the protein is comfortably accessible in both electrochemistry and *in situ* STM. The β-sheet protein structure is robust enough to withstand biologically "challenging" environments in STM, and even XPS.
- Azurin (Figure 2.21) holds two surface groups suitable for Au-surface linking. One is a disulfide group suitable for direct linking to "soft" metal surfaces such as Au(111). The other group is a hydrophobic amino acid patch around the Type 1 Cu-center opposite to the disulfide group, suitable for gentle linking to hydrophobically terminated thiol SAMs [70]. The strategic choice of azurin has impacted *in situ* STM based single-molecule bioelectrochemistry and other nanoscale azurin conductivity studies. Azurin in hydrophobically based surface orientation via variable length thiol-SAMs, have, for example, been mapped to single-molecule resolution (Figure 2.21). Azurin has overall emerged as a prototype electrochemical single-molecule metalloprotein target, but still poses questions as also for other metalloproteins, for example:

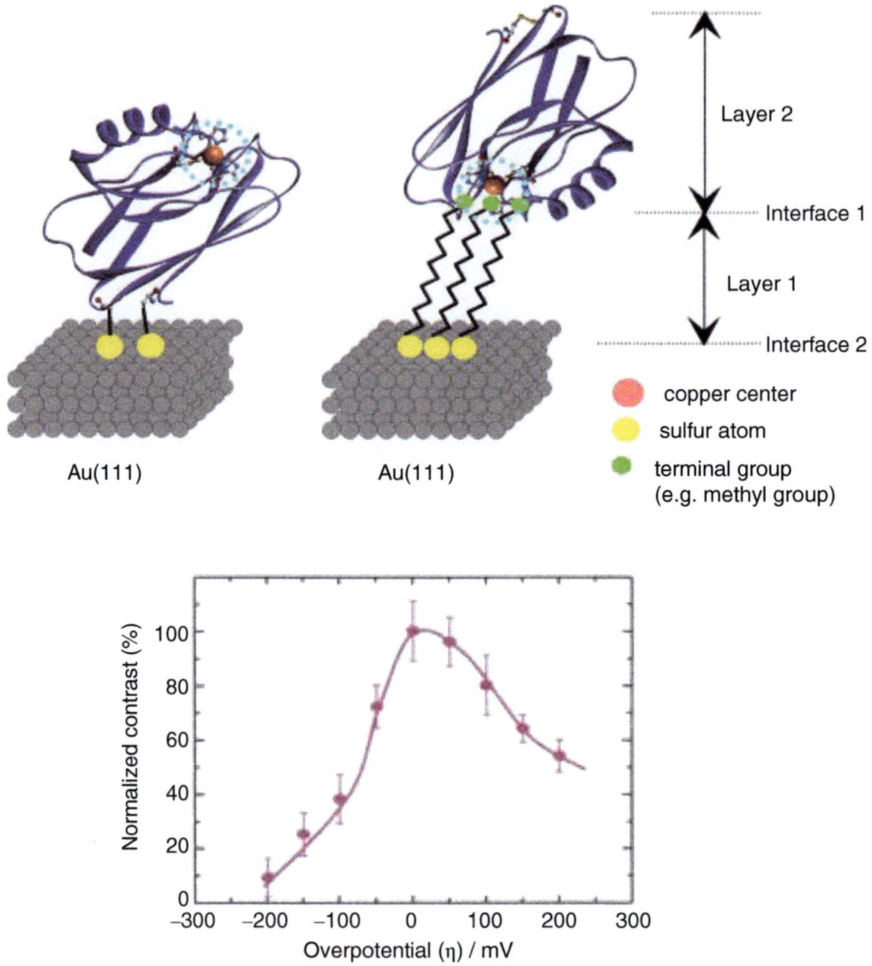

Figure 2.21 Top: *P. aeruginosa azurin* immobilized on Au(111), either directly via the disulfide group, or via the hydrophobic patch around the Cu-center, on variable-length alkanethiol SAMs. Bottom: Recorded tunneling current/overpotential STS of azurin on octanethiol SAM. Adapted from [71].

- The STS peak potential for azurin at alkanethiol SAMs is close to the equilibrium potential, but shifted negatively on azurin disulfide bonding to a gold tip [180, 239]. From the voltammetric quality, the former binding mode seems by far the most gentle.
- The STS correlations are asymmetric, inherently offering single-molecule "rectification" understandable from the location of the Cu-center, More rigorous bandshape analysis requires attention to the protein orientation, electrostatic potential distribution, ionic strength, and electronic coupling to the enclosing electrodes.
- The protein structure discloses other issues. Many studies of ET between the disulfide radical generated by pulse radiolysis and the Cu-center have pointed to "slow", millisecond intramolecular ET [240], but the nA single-molecule current densities

are only compatible with much higher rates, a recurrent *in situ* STM issue of complex molecules.

- This point addresses the diabatic or adiabatic nature of the interfacial ET transitions. High tunneling currents imply strong electronic coupling, possibly accommodated by adiabatic electrochemical ET theory, section 2.2.2. The strong *in situ* STM contrast of azurin molecules on the background of the high-resolution SAM images could reflect azurin embedding in the SAM [70]. Adsorption isotherms and distance independent ET rate constants at SAMs up to six-to-eight carbon atoms could be other reflections of strong electronic coupling.

Many *in situ* STM studies of the azurin core redox metalloprotein probe, introduced two decades ago have thus opened new single-molecule bioelectrochemistry, at the same time disclosing challenges for forthcoming single-molecule efforts.

2.5.6.2 Electrochemical SPMs of metalloenzymes, and some other "puzzles"

The introduction of a core *in situ* STM target metalloprotein has prompted studies of other proteins and protein/DNA complexes, reviewed recently [176, 177, 202, 239]. We consider the electrochemical single-biomolecule system class, the blue copper redox *enzymes*, focused on the four-copper laccase (Lc) and the two-copper nitrite reductase (CuNiR), Figure 2.22.

A common observation is that enzymes show no voltammetry of their own, but once substrate, here dioxygen or nitrite is let in, strong electrocatalytic signals appear. CuNiR and laccase are "electrochemical" enzymes in the sense that electrons are let in at the blue type 1 center, and out towards dioxygen or nitrite reduction at the Type 2/3 or Type 2 center. This requires an efficient conductivity channel between the two centres. CuNiR (and laccase) offer single-molecule clues to substrate triggered enzyme voltammetry. A direct covalent link between the Cu-ligands connect the Type 1 and Type 2 centres, Figure 2.22. *In situ* STM of CuNiR on a cysteamine SAM displays single-molecule resolution but only when NO_2^- is present, as if NO_2^- triggers significant single-molecule conductivity enhancement from the resting to the active state. This is also the case for laccase, which shows high conductivity in potential ranges where O_2-substrate is reduced, but complete electronic transparency in ranges where no O_2 reduction occurs. The images show a variety of "shapes", i.e. conductivities. As expected for small molecular assemblies, some structures reflect the triangular structure expected for vertical CuNiR orientation, others multifarious other binding modes.

A recent AFM study offers other clues. Electrochemical AFM disclosed substantial CuNiR enzyme "swelling", from $\approx$ 4 nm to $\approx$ 5nm going from the resting to the electrochemically active enzyme state, Figure 2.24 [201]. Such conspicuous conformational changes would almost certainly be transmitted to the ET pathways between the two Cu-centres. A second clue is offered by the simpler two-heme cytochrome c_4 [242]. No intramolecular ET between the heme groups were detected in ET between cyt c_4 and Co-complexes in solution. On vertical cyt c_4 immobilization on a ω-thiol decanoic acid SAM covered Au(111), intriguing two-ET voltammetry, however, emerges which is only compatible with fast, sub-ms intramolecular ET. Conformational changes must therefore accompany immobilization. A computational study [172] showed in fact that intramolecular ET in the equilibrium conformation is blocked, but that even tiny conformational reorganization triggers fast intramolecular ET.

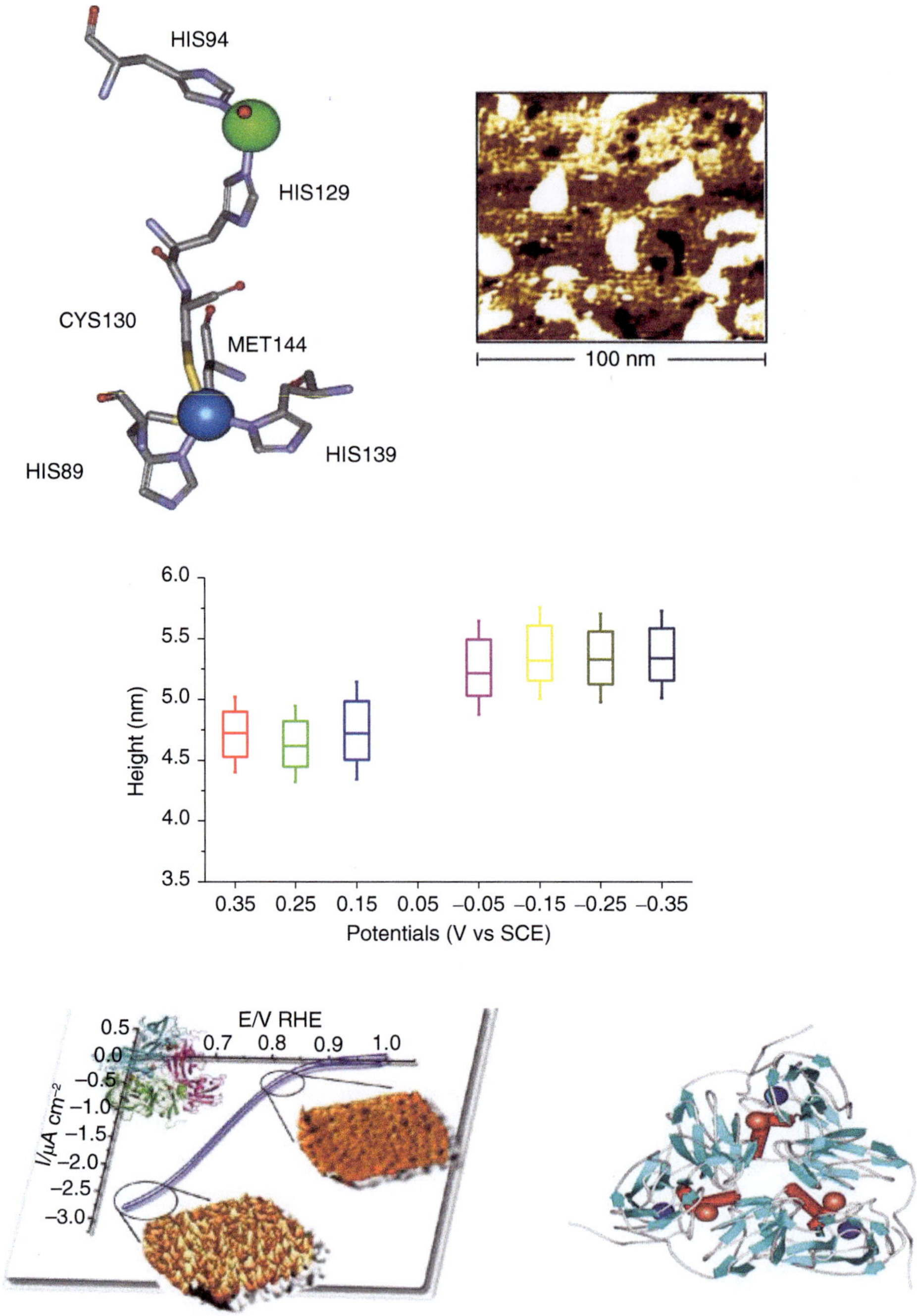

Figure 2.22 Top left: Intramolecular covalent link between the Type 1 (blue sphere) and the Type 2 centre (green sphere) in copper nitrite reductase. Top right: *In situ* STM image of CuNiR on Au(111) modified by a cysteamine SAM in the presence of 40 M KNO_2. Middle: Electrochemically controlled AFM of CuNiR on Au(111)/cysteamine electrode surface. Significant height increase accompanies transition from the resting (left) to the active enzyme state (right). Bottom left: Emerging *in situ* STM single-molecule laccase structures on O_2 substrate inlet. Bottom right: Top view of CuNiR trimer. Blue and orange spheres are the Type 1 and Type 2 centres, respectively. Adapted from [200, 201, 241].

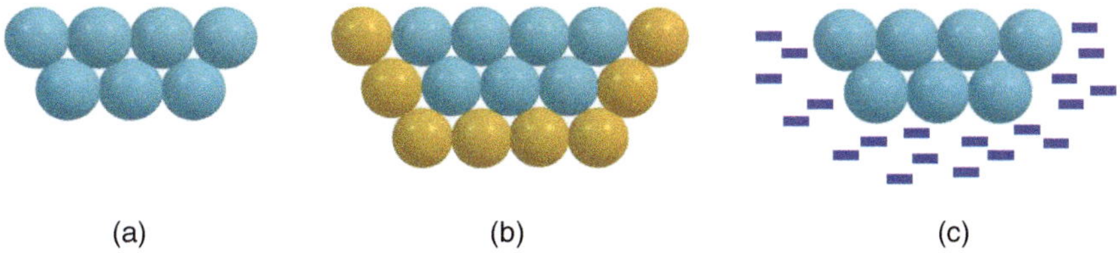

Figure 2.23 Three different cluster models of a metal surface. (a) Molecular cluster with free boundary conditions. (b) Embedded cluster model [245]. (c) Dipped cluster model [246].

2.6 Computational approaches to electrochemical surfaces and processes revisited

2.6.1 Theoretical methodologies and microscopic structure of electrochemical interfaces

The metal surface can be treated as an atomic cluster [243–246] (Figure 2.23). This approach is used to describe the electrode interaction with reactants and electrolytes, as well as to calculate the ET coupling matrix element. The cluster model is flexible and robust and can be combined with DFT and an arsenal of quantum chemical methods for calculations of the electronic structure of molecules. Clusters with several tens of atoms do not properly describe the bulk properties, but the local surface electronic structure (faces and defects) and metal–adsorbate chemical bonds can be addressed. Adsorption of charged species is, for example readily addressed. Cluster modeling of electrochemical interfaces and different ways to describe charged electrodes have been reviewed [244].

Another way to treat the metal surface is the periodical slab model [247] equally suitable for surface and bulk electrode properties. Periodical calculations are not faced with the main problem of the cluster approach, i.e. the effect of free boundary conditions, which makes computations broadly more reliable. Despite this advantage, the slab model has several shortcomings. First, there is a problem with calculations of charged systems [248, 249]. A unit cell must be electrostatically neutral, and both solvation of ionic species and counter ions must be taken into account. Without counter ions artificial charge separation occurs, which sometimes allows modeling the surface charge; for example, adsorbed solvated cations appear at negatively charged electrode surfaces. New approaches towards resolution of these issues have, however, been reported recently in the form of the effective screening method [251], and Lagrange multiplier and modified boundary condition approaches [252]. Secondly, periodical calculations are restricted basically to DFT (due to its computational efficiency), which strictly speaking is not an *ab initio* method. Pure *ab initio* methods e.g., the second order Møller-Plesset perturbation theory (MP2) and coupled cluster (CC) method are implemented in some "periodical" codes as well. Again, due to the semi-empirical nature of DFT there is even a "crisis" in this computational area caused by the diversity of exchange-correlation functionals [250]. Powerful *ab initio* schemes such as MP2, CC, CASSCF etc. are not fraught with this challenge. However, such calculations are more demanding in computer time and memory as compared with DFT, especially for large systems.

The solution side of the interface poses other challenges. The solvent is crucial and can be addressed both explicitly and implicitly. In the former case classical, *ab initio* or Car-Parrinello MD methods, as well as QM/MM schemes can be employed. The latter are based on the solvent as a continuum (polarized continuum model, effective screening medium [253] etc.) or on using the reference interaction site method (RISM) [253]. A second major issue is poor description of the asymptotic behaviour of the electrode wave functions (electronic density) at large distances. A still further important issue is excess charge on the electrode surface in contact with the electrolyte solution, which results in the electrochemical double layer. Flexible molecular modeling of the electrode surface charge (electrode potential) is either to apply an external electric field, or to exploit the concept of "a computational hydrogen electrode" [138, 254]. The latter is a special "trick" to introduce the electrode potential when calculating the energy of surface processes without directly including excess charge density or electric field.

2.6.2 The electrochemical process revisited

As noted, a key quantity in the distance dependent electronic transmission coefficient is the electronic coupling factor, which can be calculated for example by quantum mechanical perturbation theory [58, 60]. Non-perturbative models were introduced by Newton [255, 256], who employed *ab initio* SCF methods for electron exchange between transition metal complexes. The generalized Mulliken-Hush method [257], robust and suitable for large systems, is also frequently used. Several other methods such as Constrained DFT, Fragment Orbital DFT, Self-Consistent Charge Density Functional Tight-Binding, and Analytic Overlap Method were reviewed recently [258, 259].

The microscopic electronic transmission coefficient for interfacial ET, κ depends on the energy level of the electrode, ε most conveniently discussed for the weak-coupling diabatic limit, cf. section 2.2.2. We can recast the dimensionless ET rate constant k as

$$k = \frac{2\pi}{\hbar\omega_{eff}}\sqrt{\frac{\pi}{\lambda_R k_B T}}\int_{-U_0}^{0} \rho(\varepsilon)T_{AD}^2(\varepsilon)\exp\left\{-\frac{(\lambda_R-\varepsilon-e\eta)^2}{4\lambda_R k_B T}\right\}d\varepsilon \tag{2.59}$$

As usually, the Fermi-Dirac distribution is here replaced by a step-function, and the Fermi level taken as zero. U_0 is the bottom of the conduction band, $\rho(\varepsilon)$ is the electrode DOS and $T_{AD}(\varepsilon)$ the microscopic energy dependent coupling factor. At small overpotentials,

$$k \approx \frac{2\pi}{\hbar\omega_{eff}}\sqrt{\frac{\pi}{\lambda_R k_B T}}T_{AD}^2(\varepsilon_F)\rho(\varepsilon_F)k_B T\exp\left\{-\frac{(\lambda_R-e\eta)^2}{4\lambda_R k_B T}\right\} \tag{2.60}$$

The pre-exponential factor can be viewed as the effective transmission coefficient which is then a linear function of the electrode DOS at the Fermi level.

The dependence of κ on the DOS in the intermediate electronic coupling region is more complex and requires numerical approaches. The lattice Monte Carlo approach is one such way [260], Figure 2.24. The lattice is formed by the multitude of crossings of free energy surfaces of the reactants' and products' states and ET modeled as rambling through the lattice. The DOS effect was addressed by varying numbers of energy surfaces (n) in a given energy interval. Close to linear dependence of κ on $\rho(\varepsilon_F)$ emerges for small LZ factors $2\pi\gamma_e$ (0.01 and 0.001) (Eq. (2.61)), but step-like behavior (Figure 2.25) in the adiabatic limit, where κ

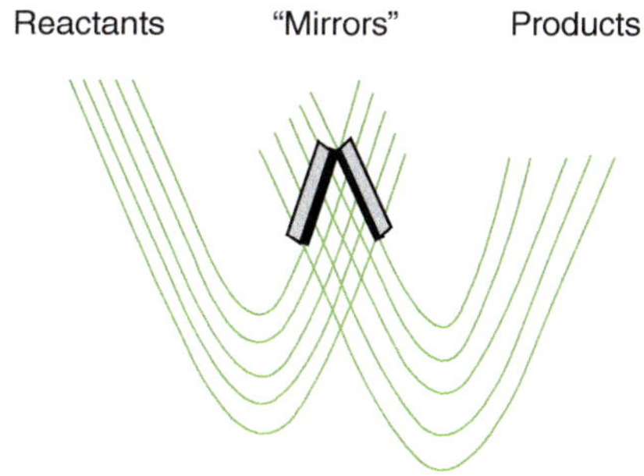

Figure 2.24 Two-dimensional lattice formed by crossing free energy surfaces along the reaction coordinate. Adapted from [260].

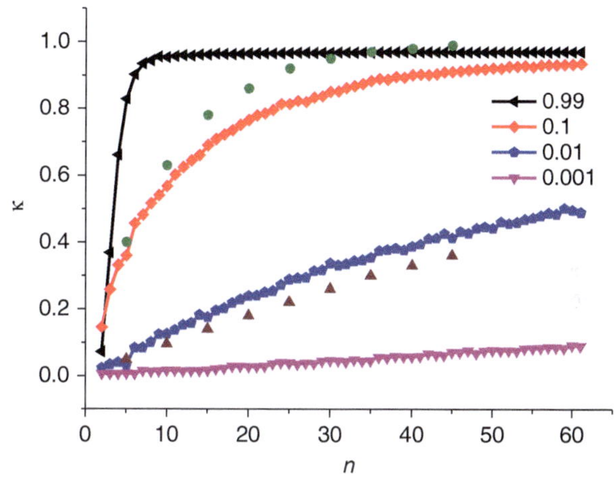

Figure 2.25 Dependence of κ on the number of reaction free energy surfaces (in a fixed energy interval) calculated for different LZ factors (0.99, 0.1, 0.01 and 0.001) at zero overpotential and room temperature obtained using the lattice MC simulations [260]. The results based on "multistate" LZ theory, Eq. (2.61) are also plotted: $2\pi\gamma_e = 0.1$ (green points), $2\pi\gamma_e = 0.01$ (brown triangles). Adapted from [260].

is practically independent of the DOS (i.e. the electrode material) [261], was also observed. The intermediate case ($2\pi\gamma_e = 0.1$), where κ displays significantly nonlinear behaviour is the most challenging. Recently Feldberg and Sutin [262] treated heterogeneous ET in the whole range from diabatic to adiabatic limit using a "multistate" LZ frame:

$$\kappa = 1 - \exp[-2\pi\gamma_e\rho(\varepsilon_F)k_BT] \tag{2.61}$$

Calculations of κ using Eq. (2.61) were compared with Monte Carlo simulations for two different LZ factor values ($2\pi\gamma_e = 0.01$ and 0.1) [259]. Figure 2.25 shows that Eq. (2.61) overestimates κ by 11% for $2\pi\gamma_e = 0.1$ but underestimates κ by 7 % for $2\pi\gamma_e = 0.01$.

2.7 Quantum and computational electrochemistry in retrospect and prospect

Since the electrochemical "renaissance" in the late 20th century, electrochemistry has evolved into highly sophisticated interdisciplinary condensed matter surface science. Strong theoretical framing is a core element, first by Levich, Dogonadze, Chizmadzhev, Kuznetsov, Hush, Gerischer, Marcus and their associates [24, 25, 36–41, 59–61], later on followed by other boundary traversing efforts by Schmickler and associates [78, 265], and by Nazmutdinov and associates [57, 223]. The largely classical theory of Marcus and, from the very beginning, fully quantum mechanical theories of Levich, Dogonadze and associates, and of Hush and by others is the theoretical basis of the new electrochemistry, which now extends to an impressive range of areas. These range from fundamentals of the electrochemical interface, in aqueous solution and other media, to fundamental electron and atom group transfer processes, to new approaches, and further to complex molecular systems such as transition metal complexes, and redox metalloproteins, as well as DNA-based molecules. A particular aspect is ultra-high resolution electrochemistry right down to the level of the single molecule. Theoretical approaches to the fundamental single-molecule electrochemical or *in situ* STM process are so far largely at the phenomenological level, but include a range of fluid systems and optical monitoring [171], in addition to single-molecule interfacial electrochemical processes. The latter extend to weak and strong molecular electronic coupling to the enclosing electrodes, solvent composition, and other properties beyond "mere" single-molecule imaging.

Our approach has been, first to offer a retrospective view of concepts and theoretical frames of interfacial electrochemical ET, including PT and PCET as well as processes involving bond fission such as the peroxodisulfate reduction [92, 93]. Here we pointed out the pioneering quantum mechanical ET theory founded by Dogonadze, Levich and their associates and in increasingly powerful form continued presently. Very much of present-day electrochemical charge transfer concepts and theory rest on the seminal results of these groups from the 1960s and 1970s. We have, secondly attempted to bridge established, largely analytical charge transfer theory with contemporary theoretical concepts and methodologies. These are illustrated by quantum mechanical and MD studies of both simple and complex electrochemical processes, and of novel type electrochemical interfaces such as single-crystal Au-surfaces modified by highly ordered alkanethiol and other SAMs. We have, third, addressed new electrochemical systems that could be in need of theoretical support in both analytical form and in terms of the contemporary computational arsenals. We summarize below other prospective challenges in theoretical electrochemical ET science, where we can conjecture progress on the way.

2.7.1 Prospective conceptual challenges in quantum and computational electrochemistry

Without offering an exhaustive list, *conceptual* challenges would, for example, be:

- Larger-scale fully quantum mechanical and quantum MD approaches to solvation dynamics. Carr-Parinello approaches to proton conductivity in chemistry and biology have, for example disclosed new PT phenomena [121, 125] and could transfer to electrochemistry.
- Solvation of molecules is a long-time electrochemical surface science notion. We refer to the three-volume monograph, *The Chemical Physics of Solvation* (1985–1988) [264]. Much of the material in this major opus is still of interest, but offers new challenges based on theoretical and computational methodology developed in the meantime.
- In spite of major progress in single-molecule electrochemistry, conceptual and theoretical issues keep popping up, relating to the nature of the "elementary" *in operando* STM process. Issues extend to coupling of the target molecule to the electrodes, the

stochastic nature of small molecular assemblies, and the molecular-scale solvent environment. Issues also relate to the "philosophical" question, of "what can be learned from single-molecule electrochemistry that cannot be learned from macroscopic electrochemistry?"

- The conceptual *correlations* between "ballistic" (Levich, Dogonadze, Marcus, Kuznetsov and associates) and "stochastic" (Kramers, Zusman, Wolynes, Sumi and others) approaches to ET processes in solution and at electrochemical interfaces are clarified in considerable detail e.g. by Kuznetsov and associates [143, 266], but still in need of applicable transparency.

2.7.2 Prospective interfacial electrochemical target phenomena

Intriguing, more specific electrochemical phenomena offer new theoretical and computational challenges. To this add experimental challenges:

2.7.2.1 Some conceptual, theoretical, and experimental notions and challenges

Among conceptual challenges in the new electrochemistry are still challenges relating to the fundamental adiabatic electrochemical ET process. Although strong electronic overlap appears to be immediately reflected solely in a lower activation free energy, concepts of the adiabatic limit will extend to other properties of the electrochemical interface. These will include for example magnetic properties, say in the dioxygen reduction, or transition metal complexes involving spin changes, as well as in other new areas listed below.

2.7.2.2 Non-traditional electrode surfaces and single-entity structure and function

What are now long-established notions of "clean", i.e. single-crystal metal electrode surfaces still pose computational challenges. Step edges, kinks, and other surface structural features are scattered over even the best surfaces [267, 268]. Recent computations of the electrochemical ET rate constants at model AuNPs has disclosed subtle dependence on the particular surface structural element [269]. Related categories are nanoporous metals as novel electrode materials with interesting properties [190, 191], in need of theoretical support.

Overviews of carbon materials and the "new wonder material" graphene are prodigiously available, e.g. [168, 186, 187]. Two recent, electrochemistry focused reviews are [169, 171]. Even perfect graphene is attractive for electrochemistry of simple redox couples due to unique electronic properties as noted, section 2.5.5. The complex ("crumbled") structure of 2D and 3D graphene-based materials offer obvious prospective theoretical challenges. N- and other graphene doping and surface functionalization add other exciting computational perspectives [214, 215] ("Will Any Crap We Put Into Graphene Increase Its Electrocatalytic Effect?" [269]).

Electrochemical ET covers a prodigious range of inorganic particle, wire, rod, "flower", layer etc. structures, structurally and functionally understood to different degrees. Electrochemistry of single *molecules* seems best understood, much due to *in situ* STM and AFM. Structural and functional mapping have reached high resolution, but we have noted unresolved challenges. Au-, Pt- and core-shell NPs [189] are used biomedically, in catalysis, and fuel cells. High-resolution structures, say of thiol-protected AuNPs, and understanding of the Au-S bond [192–195], however, only now offer firmer routes to AuNPs in real use. Nanowires, nanoflowers, etc. and single CNTs, offer other electrochemical challenges.

With single-entity electrochemistry on the move, entirely new electrochemical systems will appear. One new area covers electrical, optoelectronic, and magnetic properties of layered perovskites as "smarter" materials than carbon or graphene materials [186]. Surfaces of these 2D materials could be probed, using spintronics notions and magnetic STM/AFM. Another area is molecular magnets as engineered building blocks of strongly magnetic rare earth complexes, hierarchically tuned into magnetic materials [270]. Local electrochemical probing would here trigger new materials science. As in other single-entity electrochemistry, progress in these new areas offers interdisciplinary interplay between quantum chemical calculations and increasingly sophisticated experimental work.

2.7.2.3 Semiconductor and semimetal electrodes

From a prominent status created by Gerischer, Memming, and later on by Peter, Lewis, and others from the 1960s and 1970s [59–61, 271, 272], semiconductor electrochemistry long seemed to be lying "dormant", but conspicuously revitalized over the last couple of decades, stimulated by new photoelectrochemistry and solar energy conversion. The Gerischer symposium (2011) [273] testified strongly to new vibrant semiconductor electrochemistry and photoelectrochemistry, with challenges in theoretical and computational electrochemistry. Recently the Anderson-Newns formalism in the narrow band approximation was extended to describe adiabatic ET across a semiconductor electrode/solution interface [274]. These challenges extend to semimetal electrochemical surfaces with new carbon materials electrochemical surfaces, ranging from single-entity carbon nanotube surfaces over 2D and 3D graphene, to "hybrid" graphene/metallic nanoparticle surfaces [275]. Development of this whole area all the way from macroscopic to single-carbon entity electrochemistry is strongly established for example by McCreery and associates [186, 187]. Theoretical challenges, in interactions between the new surfaces and molecular target systems, are likely ahead.

2.7.2.4 Metal deposition and dissolution processes

Pure and applied aspects of metal deposition and dissolution processes are classical core areas. In combination with the impressive development in recent theoretical electrochemistry, these areas of complex processes offer other prodigious challenges for novel theoretical and computational efforts. A challenge would be elucidation at the molecular level, of why the deposition of some metal cations from aqueous solutions (e.g., Ag^+, Pb^{2+}) is very fast, while the deposition of others (e.g., Fe^{3+}, Ni^{2+}) is slow. The corresponding exchange current densities vary in wide ranges, from 10^2 to 10^{-5} A m^{-2}. A combination of quantum mechanical theory, DFT calculations and MD simulations [275] would here offer promising results. For example, it was shown [276] that the Ag^+ aqua complex can approach very closely to the silver surface without any significant distortion of its hydration sheath. Exactly the latter is what mainly facilitates the close approach of the cation, with strong Ag^+ – metal orbital overlap decreasing the ET energy barrier.

2.7.2.5 Chiral surfaces and ET processes of chiral molecules

Electrochemical surfaces are inherently chiral, reflected in both molecular adsorption and electrochemical kinetics of chiral or racemic adsorbates [194, 196, 277]. Higher-level chirality is based on chiral catalyst molecules, where metalloenzymes could be in focus, as these are naturally chiral being composed entirely of L-amino acids. The catalase catalyzed oxidation of chiral alcohols is such a case [278].

Early reports by Bernauer and associates [279] addressed ET reactions between chiral transition metal complexes and ET proteins such as cyt *c* and plastocyanin. ET between cyt *c* and the chiral enantiomers of $[Co(Ox)_3]^{3-}$ (Ox^{2-} = oxalate) was explored in a recent study using a combination of kinetic data and DFT [87]. Both binding constant and ET rate constant showed a small but significant preference for the Λ-form with chirality ratios of $\approx$1.1–1.2, assigned to intermolecular work terms and the electronic transmission coefficient. Extension of such efforts to electrochemical adsorption and ET processes is warranted. As noted in section 2.5.2, intriguing surface structural chirality was disclosed in studies of butanethiol enantiomers [194]. Notably, not only molecules but also whole adsorbate domains can exhibit chirality. Chiral domains of achiral molecules can emerge, and vice-versa.

2.7.2.6 ET reactions involving hot electrons (femto-electrochemistry)

Prompted by the evolving areas of nanoplasmonic [280], electrochemistry that involves "hot" electrons is, finally a prospective new ground for theoretical and computational efforts. "Hot" electrons is a metallic electron gas with high-energy electrons equilibrated at formal temperatures much higher than the physical temperature of the metallic phonons or solution [281]. Hot electrons are produced by ultra-fast (sub-picosecond) laser pulses, collisions of vibrationally excited molecules with the electrode, in metal-insulator junctions, and in other ways. Hot electron processes in ultra-high vacuum are long known, but efforts on hot electrons in electrochemistry are fewer, although stimulated by the new areas of nanoparticle plasmonics [280].

Otto [282], Benderskij [283] and their associates were the first to study electrochemical reactions involving hot electrons, particularly hydronium discharge at mercury electrodes. "Ultra-fast" ET and "small" (0.1–0.2) transfer coefficients indicative of activationless behaviour were immediate outcomes. Attractive formal, close to analytical frames were reported by Benderskij and associates [283] (non-adiabatic limit) and by Nazmutdinov and associates [281] (adiabatic limit). Molecular ET theory, with formally high temperatures of the electron gas, was the starting point. Both diabatic and adiabatic limits were addressed using the Newns-Anderson formalism. Temperature dependent free energy surfaces, activation free energy/overpotential correlations, and transfer coefficients were outcomes. The transparent analytical approaches reported [281] could offer starting points for new theoretical guidance in rapidly evolving metallic nanoplasmonics in photoelectrochemistry and photocatalysis.

2.8 A few concluding remarks

Retrospective overviews of theoretical and computational approaches to the electrochemical interface, with future prospects is a prodigious undertaking. The electrochemical interface is a complex, non-uniform, and anisotropic structure, further complex when ions, molecules, and biomolecules adsorb. We have chosen, first, to focus on electrochemical *processes*. Electronic and molecular structural properties have been comprehensively addressed elsewhere over the last decades. A comparison between the reviews of Schmickler [264] (1996) and of Magnussen and Gross [284] (2019) illuminates the development in the understanding of the structure of the electrochemical interface. We did address, however, what could be denoted as "non-traditional" electrochemical interfaces. Our choice was focused on thiol adsorption, metallic nanoparticles and nanoporous metals, and novel carbon-based materials, with perspectives in new generation electrochemical (bio)sensors and (bio)fuel cells [169]. All of these materials warrant theoretical efforts based on state-of-the-art computational methodologies.

In our retrospective/prospective we followed two major lines. We addressed, first, some experimentally well characterized, "classical" systems. These were outer sphere ET of transition metal complexes and, as a core case for electrochemical ET that involves all the theoretical ET parameters, reorganization energy, double-layer effects, bond fission etc., the electrochemical peroxodisufate reduction. Such studies have or will lead to understanding of the fundamental elementary processes in novel detail. Our other line was novel systems in the new (bio)electrochemistry, where we focused on single-entity molecular and biomolecular electrochemistry. These areas offer still other prospects together with "non-traditional" electrochemical materials science. Due to obvious limitations we were not concerned in this chapter with charge transfer in solids, which is of crucial importance in battery research. For such systems ion transfer in intercalation kinetics plays the most important role [285]. Altogether we can perhaps conclude that a new world of challenges for theoretical and computational electrochemistry, combining molecular electrochemistry and new electrochemical materials science, is ahead of us.

Acknowledgement

Financial support from the Russian Science Foundation (project no. 17-13-01274), the Russian Foundation for Basic Research (project no. 17-03-00619a), the Russian Ministry of Science and Higher Education and DTU Chemistry is acknowledged. We wish to thank Dr. Tamara T. Zinkicheva for help with the figures and for useful discussions.

References

1 Furtak, T.E., Kliewer, K.L., Lynch, D.W., Eds., Non-traditional Approaches to the Study of the Solid-electrolyte Interface. *Surf. Sci.,* 1980, *101*, 1–681.

2 Wieckowski, A., Ed. *Interfacial Electrochemistry. Theory, Experiment and Applications*, M. Dekker, New York, 1999.

3 Kolb, D.M. Electrochemical Surface Science. *Angew. Chem. Int. Ed.* 2000. 40, 1162–1181

4 Zhang, J., Albrecht, T., Chi, Q., Kuznetsov, A.M., Ulstrup, J. Charge Transfer and Interfacial Bioelectrochemistry at the Nanoscale and Single-molecule Levels. In: Hammerich, O., Ulstrup, J, Eds. *Bioinorganic Electrochemistry*, Springer, Dordrecht, 2008, pp. 249–302.

5 Carnie, S.L. and Torrie, G.M. The Statistical Mechanics of the Electrical Double-Layer. *Adv. Chem. Phys.*, 1984, 56, 141–253.

6 Henderson, D., Blum, L. A Simple Theory of the Electric Double-layer Including Solvent Effects. *J Electroanal. Chem.,* 1982, 132, 1–13.

7 Badiali, J.P., Rosinberg, M.R., Goodisman. Effect of Solvent on Properties of the Liquid-Metal Surface. *J. Electroanal. Chem.,* 1981, 130, 31–45.

8 Schmickler, W., Henderson, D. New Models for the Structure of the Electrochemical Interface. *Progr. Surf. Sci.* 1986, 22, 323–420.

9 Kornyshev, A.A. Solvation of a Metal Surface. In: Dogonadze, R.R., Kálmán, E., Kornyshev, A.A. and Ulstrup, Eds., *The Chemical Physics of Solvation. Part C. Solvation Phenomena in Specific Physical, Chemical and Biological Systems*, Elsevier, Amsterdam, 1988, pp. 355–400.

10 Smith, J.R. Self-consistent Many-electron Theory of Electron Work Function and Surface Potential Characteristic for Selected Metals. *Phys. Rev.* 1969, 181, 522–529.

11 Kornyshev, A.A., Ulstrup, J. Solvent Structural Effect on the Deviation from Linear Parsons-Zobel Plots with Increasing Electrolyte Concentration. *J. Electroanal. Chem.* 1985, 183, 87–89.

12 *a*. Schmickler, W. Adsorption at the Metal/Solution Interface. In: Guidelli, R., Ed., *Electrified Interfaces in Physics, Chemistry and Biology*, Nato ACI Series A: Mathematical and Physical Series, Vol. 355, pp. 399–425. *b*. Schmickler, W., Guidelli, R. The Patrial Charge Transfer. *Electrochim. Acta* 2014, 129, 489–505.

13 Edel, J.B., Kornyshev, A.A., Kucernak, A.R., Urbakh, M. Fundamentals and Applications of Self-assembled Plasmonic Nanoparticles. *Chem. Soc. Rev.* 2014, 45, 1581–1596.

14 Dzhavakhidze, P.G., Kornyshev, A.A., Liebsch, A, Urbakh, M. Theory of 2nd-Harmonic Generation at the Metal-Electrolyte Interface. *Phys. Rev. B* 1992, 45, 9339–9346.

15 Hubbard, A.T. Electrochemistry of Well-defined Surfaces. *Acc. Chem. Res.*, 1980, 13, 177–184.

16 a. Hamelin, A. Cyclic Voltammetry at Gold Single-crystal Surfaces. Part 1. Behaviour at Low-index Faces. *J. Electroanal. Chem.* 1996, 407, 1–11. b. Hamelin, A., Martins, A.M. Cyclic Voltammetry at Gold Single-crystal Surfaces. Part 2. Behaviour at High-index Faces. *J. Electroanal. Chem.*, 1996, *407*, 13–21.

17 Wiechers, J., Twomey, T., Kolb, D.M., Behm, R.J. *An In Situ Scanning* Tunneling Microscopy Study of Au(111) with Atomic Scale Resolution. *J. Electroanal. Chem.*, 1988, 248, 451–460.

18 Kolb. DM., Nichols, R.J., Behm, R.J. The Application of Scanning Tunnelling Microscopy to Electrochemistry. In: Guidelli, R., Ed., *Electrified Interfaces in Physics, Chemistry and Biology*, NATO Adv. Sci. Inst. Series C, Mathematical and Physical Sciences 1992, 355, pp. 275–292.

19 Gewirth, A.A, Niece, B.K., Electrochemical Applications of In Situ Scanning Probe Microscopy. *Chem. Rev.*, 1997, 97, 1129–1162.

20 Chen, C.J. *Introduction to Scanning Tunneling Microscopy*, 2nd Ed. Oxford Science Publications, Oxford 2007.

21 Yoshimoto, S., Itaya, K. Adsorption and Assembly of Ions and Organic Molecules at Electrochemical Interfaces: Nanoscale Aspects. *Ann. Rev. Anal. Chem.* 2013, 6, 13–35.

22 Dretschkow, T., Dakkouri, A.S., Wandlowski, T. In-Situ Scanning Tunneling Microscopy Study of Uracil on Au(111) and Au(100). *Langmuir* 1997, 13, 2843–2856.

23 Zhang, J., Kuznetsov, A.M., Igor G. Medvedev, I.G., Chi, Q., Albrecht, T., Palle S. Jensen, P.S., Jens Ulstrup, J. Single-molecule Electron Transfer in Electrochemical Environments. *Chem. Rev.* 2008, 108, 2737–2791.

24 Kuznetsov, A.M. *Charge Transfer in Physics, Chemistry and Biology*, Gordon & Breach, Reading, 1995.

25 Kuznetsov, A.M., Ulstrup, J. *Electron Transfer in Chemistry and Biology: An Introduction to the Theory*, Wiley, Chichester, 1999.

26 Butt, J., Armstrong, F.A. Voltammetry of Adsorbed Redox Enzymes: Mechanisms in the Potential Dimension. In: Hammerich, O., Ulstrup, J, Eds. *Bioinorganic Electrochemistry*, Springer, Dordrecht, 2008, pp. 91–128.

27 Willner, B., Willner, I. Reconstituted Redox Proteins on Surfaces for Bioelectronic Applications. In: Hammerich, O., Ulstrup, J, Eds. *Bioinorganic Electrochemistry*, Springer, Dordrecht, 2008, pp. 37–90.

28 Genereux, J.C., Barton, J.K. Mechanisms for DNA Charge Transport. *Chem. Rev.* 2010, 110, 1642–1662.

29 Hansen, A.G., Salvatore, P., K.K. Karlsen, R.J. Nichols, J. Wengel and J. Ulstrup. Electrochemistry and In Situ Scanning Tunnelling Microscopy of Pure and Redox-marked DNA- and UNA-based Oligonucleotides on Au(111)-electrode Surfaces, *Phys Chem Chem Phys* 2013, 15, 776–786.

30 Beratan, D.N., Naaman, R., Waldeck, D.H. Charge and Spin Transport through Nucleic Acids. *Curr. Opin. Electrochemistry*, 2017, 4, 175–181.

31 Katz E., Willner, I. Integrated Nanoparticle/Biomolecule Protein Hybrid Systems: Synthesis, *Properties and Applications. Angew. Chem.* 2004, 45, 6041–6108.

32 Li, Y., Wang, H., Wang, Z., Qiao, Y., Ulstrup, J., Chen, H.-Y., Zhou, G., Tao, N. Transition from Stochastic Events to Deterministic Ensemble Average in Electron Transfer Reactions Revealed by Single-molecule conductance measurement. *Proc. Nat. Acad. Sci. USA* 2019, 116, 3407–3412.

33 a. Kuznetsov, A.M., Ulstrup, J. Dissipative Relaxation of a Low-energy Intermediate Electronic State in Three-Level Electron Transfer, *Chem. Phys.* 1991, 157, 25–33. b. Kuznetsov, A.M., Sommer-Larsen, P., Ulstrup, J., Resonance and Environmental Fluctuation Effects in STM Currents through Large Adsorbed Molecules, *Surf. Sci.* 1992, *273*, 52–64.

34 Kuznetsov, A.M., Ulstrup, J. Mechanisms of In Situ Scanning Tunnelling Microscopy of Organized Redox Molecular Assemblies. *J. Phys. Chem. A* 2000, *104*, 11531–11540. Errata: J. *Phys. Chem. A* 2001, 105, 7494–7494.

35 Schmickler, W, Widrig, C. The Investigation of Redox Reactions with a Scanning Tunneling Microscope – Experimental and Theoretical Aspects. *J. Electroanal. Chem.* 1992, 336, 213–222.

36 Marcus, R.A. On the Theory of Oxidation-Reduction Reactions Involving Electron Transfer, 1. *J. Chem. Phys.* 1956, 24, 966–978.

37 Marcus, R.A., Sutin, N. Electron Transfers in Chemistry and Biology. *Biochim. Biophys. Acta* 1987, 811, 265–322.

38 **a.** Hush, Adiabatic Rate Processes at Electrodes. 1. Energy-Charge Relationships. *J. Chem. Phys* F1958, *28*, 962–972. **b.** Adiabatic Theory of Outer Sphere Electron Transfer Reaction in Solution. *Trans. Faraday Soc.* 1961, *57*, 557–580.

39 Hush, N.S. Electron Transfer in Retrospect and Prospect. 1: Adiabatic Electrode Processes. *J. Electroanal. Chem.* 1999, 460, 5–29.

40 Levich, V.G., Dogonadze, R.R. Theory of Non-radiation Electron Transitions from Ion to Ion in Solutions. *Dokl. Akad. Nauk SSSR* 1959, 124, 123–126.

41 Levich, V.G., Dogonadze, R.R. Adiabatic Theory of Electronic Processes in Solution. *Collect Czech. Chem. Comm.* 1961, 26, 193–214.

42 Allen, G.C., Hush, N.S. Intervalence-Transfer Absorption Part I. Qualitative Evidence for Intervalence-Transfer Absorption in Inorganic Systems in Solution and in the Solid State. *Progr. Inorg. Chem.* 1967, 8, 357–389.

43 Kuznetsov, A.M., Medvedev, I.G. Does Really Born-Oppenheimer Approximation Break Down in Charge Transfer processes? *An Exactly Solvable Model. Chem. Phys.* 2006, 324, 148–159.

44 **a.** Kuznetsov, A.M., Medvedev, I.G. On the Born-Oppenheimer Approximation in the Theory of Non-adiabatic Charge Transfer Reactions. An Exactly Solvable Model. *Elektrokhimiya*. 2008, 44, 182–201. **b.** Kuznetsov, A.M., Sokolov, V.V., Ulstrup, J. A Semiclassical Theory of Electron Transfer Reactions in Condon approximation and Beyond, *J. Electroanal. Chem.* 2001, *502*, 36–46.

45 **a.** Kuznetsov, A.M. *Charge Transfer in Physics, Chemistry and Biology*, Gordon & Breach, Reading, 1995, p. 204 and 237. **b.** R.R. Dogonadze, A.M. Kuznetsov, Calculation of the Probability of the Elementary Act of the Charge Transfer Reaction in a Polar liquid in the Landau-Zener Approximation, *Theoretical and Experimental Chemistry* 1973, 6, 243–248.

46 Rose D.A., Benjamin I. Molecular Dynamics of Adiabatic and Non-adiabatic Electron Transfer on the Metal-water Interface. *J. Chem. Phys.* 1994, 100, 3545–3555.

47 Hartnig, C., Koper, M.T.M. Solvent Reorganization in Electron and Ion Transfer Reactions near a Smooth Electrified Surface: a Molecular Dynamics Study. *JACS.* 2003, 125, 9840–9845.

48 LeBard, D.N., Matyushov, D.V. Protein-water Electrostatics and Principles of Bioenergetics. *PCCP.* 2010, 12, 15335–15348.

49 Lawrence, J.E., Manolopoulos, D.E. Path Integral Methods for Reaction Rates in Complex Systems. *Faraday Discuss.* 2020, 221, 9.

50 Nikitina, V.A., Kislenko, S.A., Nazmutdinov, R.R. Understanding the Nature of Heterogeneous Electron Transfer in Molecular and Ionic Solvents: Experiment, Theory and Computations. *J. Phys. Chem. C.* 2019, 123, 14370–14381.

51 Marenich, A.V., Ho., J., Coote, M.L., Cramer, C.J., Truhlar, D.G. Computational Electrochemistry: Prediction of Liquid-phase Reduction Potentials. *Phys. Chem. Chem. Phys.* 2014, *16*, 15068–15106.

52 Born, M. *Natural Philosophy of Cause and Chance*. Dover Publications, New York, 1964.

53 Kjaer, A.M., Ulstrup, J. Solvent Band Width Dependence and Band Asymmetry Features of Charge-transfer Transitions in N-pyridinium Phenolates. *J. Am. Chem. Soc.* 1987, 109, 1934–1942.

54 Blumbergen, J. Recent Advances in the Theory and Molecular Simulation of Biological Electron Transfer Reactions. *Chem. Rev.* 2015, 115, 11191–11238.

55 Hammes-Schiffer, S., Soudackov, A.V. Proton Coupled Electron Transfer in Solution, Proteins, and Electrochemistry. *J. Phys. Chem. B* 2008, 112, 14108–14123.

56 Xiang, L.M., Palma, J.L., Bruot, C., Mujica, V., Ratner, M.A., Tao, N.J. Intermediate Tunneling-C hopping Regime in DNA Charge Transport. *Nature Chemistry* 2015, 7, 221–226. *Errata: Nature Chemistry* 2017, 9, 295–295.

57 Nazmutdinov, R.R., Bronshtein, M.D., Zinkicheva, T.T., Glukhov. D.V. Modeling of Electron Transfer Across Electrochemical Interfaces: State-of-the Art and Challenges for Quantum and Computational Chemistry. *Int. J. Quantum Chem.* 2016, 116, 189–201.

58 Gurney, R.W. The Quantum Mechanics of Electrolysis. *Proc. Roy. Soc. London, Ser. A* 1931, 134, 137–154.

59 **a.** Gerischer, H. Über den Ablauf von Redoxreaktionen an Metallen und an Halbleitern. *Z. Physik. Chem. NF* 1960, 26, 223–247. **b.** *Z. Physik. Chem. NF* 1960, *26*, 325–338.

60 Dogonadze, R.R. Theory of Molecular Electrode Reactions. In: *Reactions of Molecules at Electrodes.* Hush, N.S., Ed. Interscience, London, 1971, pp. 135–227.

61 Dogonadze, R.R., Kuznetsov, A.M. Theory of Charge Transfer Kinetics at Solid-Polar Liquid Interfaces. *Prog. Surf. Sci.* 1975, 6, 1–41.

62 Frauenfelder, H., Wolynes, P.G. Rate Theories and Puzzles of Hemeproteins. *Science* 1985, 229, 337–345.

63 Dogonadze, R.R., Kuznetsov, A.M., Vorotyntsev, M.A. The Kinetics of the Adiabatic and Nonadiabatic Reactions at the Metal and Semiconductor Electrodes. *Croat. Chim. Acta* 1972, 44,257–273.

64 Kuznetsov A.M., Nazmutdinov R.R., Schmickler W. Monte Carlo Simulations of the Electrochemical Electron Transfer Processes. *J. Electroanal. Chem.* 2002, 532, 171–180.

65 Kuznetsov, A.M., Ulstrup, J. Adiabatic and Nonadiabatic Electronic Transitions in Bridge-assisted Electrochemical Tunnelling Contacts. *Russian J. Electrochem.* 1995, 31, 221–226.

66 Frumkin, A.N. The Double Layer in Electrochemistry. *J. Electrochem. Soc.* 1960, 107, 461–472.

67 Frumkin, A.N. Points of Zero Charge in Equations of Electrochemical Kinetics. *J. Electroanal. Chem.* 1965,9, 173–183.

68 Finklea, H.O., Hanshew, D.D. Electron Transfer Kinetics in Organized Thiol Monolayers with Attached Pentaammine(pyridine)ruthenium Redox Centers. *J. Am. Chem. Soc.* 1992, 114, 3173–3181.

69 Chidsey, C.E.D. Free Energy and Temperature Dependence of Electron Transfer at the Metal-Electrolyte Interface. *Science* 1991, 251, 919–922.

70 Chi, Q., Zhang, J., Andersen, J.E.T., Ulstrup, J. Ordered Assembly and Controlled Electron Transfer of the Blue Copper Protein Azurin at Gold(111) Single-crystal Substrates, *J. Phys. Chem. B* 2001, 105, 4669–4679.

71 Kharkats, Yu. I., Ulstrup, J. Quantum Theories of Elementary Steps in Electrochemical Hydrogen Evolution. *J. Electroanal. Chem.* 1975, 65, 555–572.

72 Kuznetsov, A.M. Variation of the Charge of the Adsorbed Hydrogen Atom in the Process of the Activation in the Elementary Act of Hydrogen Ion Discharge. *J. Electroanal. Chem.* 1983, 159, 241–255.

73 Kornyshev, A.A., Kuznetsov, A.M., Ulstrup, J. Effect of Overpotential on the Electronic tunnel Factor in Diabatic Electrochemical Processes. *J. Phys. Chem.* 1994, *98*, 3832–3837. *Errata: J. Phys. Chem.* 1994, 98, 8596–8596.

74 Kuznetsov, A.M., A Unified Theory of Adiabatic Electrochemical Reactions of Electron Transfer on Metals, *J. Electroanal. Chem.* 1988, 241, 45-56

75 Kuznetsov, A.M., Medvedev, I.G. A Model of the Surface Molecule for Adiabatic Electrochemical Electron Transfer Including Electron Correlation Effects: An Exact Solution. *J. Electroanal. Chem.* 2001, 502, 15–35.

76 Anderson, P.W. Localized Magnetic States in Metals, *Phys. Rev.* 1961, 124, 41–52.

77 Newns, D.M. Self-consistent Model of Hydrogen Chemisorption. *Phys. Rev.* 1969, 178, 1123–1135.

78 Schmickler, W. A Theory of Adiabatic Electron Transfer Reactions. *J. Electroanal. Chem.* 1986, 204, 31–43.

79 Sato, N. *Electrochemistry at Metal and Semiconductor Electrodes*, 2nd Ed. Elsevier, Amsterdam, 2003.

80 Schmickler, W., Santos, E. *Interfacial Electrochemistry*, 2nd Ed. Springer, Berlin-Heidelberg, 2010.

81 Savéant J.M. Electron-transfer, Bond Breaking, and Bond Formation. *Acc. Chem. Res.* 1993, 26, 455–461.

82 Vargaftik, V.N., German, E.D., Dogonadze, R.R., Syrkin, Y.K. Kinetics and Mechanism of Replacement of Ligands in Octahedral Complexes of Divalent Nickel. *J. Struct. Chem.* 1974, 15, 925–932.

83 Jortner, J. Temperature dependent Activation Energy for Electron Transfer between Biological Molecules. *J. Chem. Phys.* 1976, 64, 4860–4867.

84 Kuznetsov, A.M., Sondergaard, N.C., Ulstrup, J. Low-temperature Electron Transfer in Bacterial Photosynthesis. *Chem. Phys.* 1978, 29, 383–390.

85 [85] Nazmutdinov, R.R., Roznyatovskaya, N.V., Glukhov, D.V., Manyurov, I.R., Mazin V.M., Tsirlina, G.A., Probst, M. Medium and Interfacial Effects in the Multistep Reduction of Binuclear Complexes with Robson-type Ligand. *Inorganic Chemistry.* 2008, 47, 6659–6673.

86 Nazmutdinov, R.R., Rusanova, M. Yu., Van der Porten, D., Tsirlina, G.A., Fawcett, W.R., Towar, C. The Reactivity Prediction: Outersphere Electroreduction of Transition Metal Ammine Complexes. *J. Phys. Chem. C* 2009, 113, 2881–2890.

87 Nazmutdinov, R.R., Bronshtein, M.D., Zinkicheva, T.T., Hansen, N.S., Zhang, J., Ulstrup, J. Chiral Selectivity in Inter-reactant Recognition and Electron Transfer of the Oxidation of Horse Heart Cytochrome *c* by *Tris*-oxalato-cobaltate(III). *Inorganic Chemistry* 2016, 55, 9335–9345.

88 Jónsson, E.Ö, Thygesen, K.S., J. Ulstrup, J., Jacobsen, K.W. Ab Initio Calculations of the Electronic Properties of Polypyridine Transition Metal Complexes and their Adsorption on Metal Surfaces in the Presence of Solvent and Counter Ions. *J. Phys. Chem. B* 2011, 115, 9410–9416.

89 German, E.D Kinetics and Mechanism of the Outer Sphere Electron Transfer Reactions between Complex Ions. *Rev. Inorg. Chem.*1983, 5, 123–184.

90 Albrecht, T., Guckian, A., Ulstrup, J., Vos J.G. Transistor-like Behaviour of Transition Metal Complexes, *Nanoletters* 2005, 5, 1451–1456.

91 Kharkats, Y.I., Nielsen, H., Ulstrup, J. The Effect of a Low-dielectric Constant Interlayer on the Current-overvoltage Relationship for Simple Electrode Processes. *J. Electroanal. Chem.* 1984, 169, 47–57.

92 Nazmutdinov, R.R., Glukhov, D.V., Petrii, O.A., Tsirlina, G.A., Botukhova, G.N. Contemporary Understanding of the Peroxodisulfate Reduction at a Mercury Electrode. *J. Electroanal. Chem.* 2003, 552, 261–278.

93 Nazmutdinov, R.R., Glukhov, D.V., Tsirlina, G.A., Petrii, O.A. Molecular Description of the Persulfate Ion Reduction on a Mercury Electrode. *Russian J. Electrochemistry* 2002, 38, 720–731.

94 Santos, E., Schmickler, W. Electrocatalysis of Hydrogen oxidation – Theoretical Foundation. *Angew. Chem. Int. Ed.* 2007, 46, 8262–8265.

95 Nazmutdinov R.R., Glukhov D.V., Tsirlina G.A., Petrii O.A. Exploring Molecular Features of the Cationic Catalysis Phenomenon: Peroxodisulphate Reduction at a Mercury Electrode. *J. Electroanal. Chem.* 2005, 582, 118–129.

96 Nazmutdinov, R.R., Santos, E., Schmickler, W. Spin Effects in Oxygen Electrocatalysis: A Discussion. *Electrochemistry Communications.* 2013, 33, 14–17.

97 Lynden-Bell, R.M. Can Marcus Theory Be Applied to Redox Processes in Ionic Liquids? A Comparative Simulation Study of Dimethylimidazolium Liquids and Acetonitrile. *J. Phys. Chem. B* 2007, 111, 10800–10806.

98 Shim, Y., Kim, H.J. Free Energy and Dynamics of Electron-Transfer Reactions in a RooTemperature Ionic Liquid. *J. Phys. Chem. B* 2007, 111, 4510–4519.

99 Shim, Y., Kim, H.J. Adiabatic Electron Transfer in a Room-Temperature Ionic Liquid: Reaction Dynamics and Kinetics. *J. Phys. Chem. B* 2009, 113, 12964–12972.

100 Nikitina, V.A., Kislenko, S.A., Nazmutdinov, R.R., Bronshtein, M.D., Tsirlina, G.A. Ferrocene/ferrocenium Redox Couple at Au(111)/ionic Liquid and Au(111)/acetonitrile Interfaces: A Molecular-level View at the Elementary Act. *J. Phys. Chem. C.* 118 (2014) 6151–6154.

101 Fedorov, M.V., Kornyshev, A.A. Ionic Liquids at Electrified Interfaces. *Chem. Rev.* 2014, 114, 2978–3036.

102 Chen, M., Goodwin, Z.A.H., Feng, G., Kornyshev, A.A. On the Temperature Dependence of the Double Layer Capacitance of Ionic Liquids. *J. Electroanal. Chem.* 2018, 819, 347–358.

103 Su, Y.-Z., Fu, Y.-C., Wei, Y.-M., Yan, J.-W., Mao, B.-W. The Electrode/Ionic Liquid Interface: Electric Double Layer and Metal Deposition. *Chem Phys Chem* 2010, 11, 2764–2778.

104 Yan, J.-W., Tian, Z.-Q., Mao, B.-W. Molecular Level Understanding of Electric Double Layer in Ionic Liquids. *Curr. Opin. Electrochem.* 2017, 4, 105–111.

105 Nichols, R.J., Higgins, S.J. Single Molecule Electrochemistry in Nanoscale Junctions. *Curr. Opin. Electrochem.* 2017, 4, 98–104.

106 **a.** Gray, H.B., Winkler, J. Electron Tunnelling through Iron and Copper Proteins. In: Hammerich, O., Ulstrup, J., Eds. *Bioinorganic Electrochemistry*, Springer, Dordrecht, 2008, pp. 1–23. **b.** Warren, J.J., Ener, M.E., Vlček Jr., A, Winkler, J., Gray, H.B. Electron Hopping through Proteins. *Coord. Chem. Rev.* 2012, *256*, 2478–2487.

107 Wikström, M. The Respiratory Enzyme as an Electrochemical Energy Transducer. In: Hammerich, O., Ulstrup, J, Eds. *Bioinorganic Electrochemistry*, Springer, Dordrecht, 2008, pp. 25–35.

108 Kohen, A., Limbach H.-H. Eds. *Isotope Effects in Chemistry and Biology*. Taylor & Francis, Boca Raton 2006.

109 Kuznetsov, A.M., Ulstrup, J. Proton and Hydrogen Atom Tunnelling in Hydrolytic and Redox Enzyme Catalysis., *Can. J. Chem.* 1999, 77, 1085–1096.

110 Kuznetsov, A.M., Ulstrup, J. An S_{N2} Model for Proton Transfer in Hydrogen-bonded Systems. *Russian J. Electrochemistry* 2004, 40, 1000–1009.

111 Lam, Y.-C., Soudackov, A.V., Goldsmith, Z.K, Hammes-Schiffer, S. Theory of Proton Discharge on Metal Electrodes: Electronically Adiabatic Model. *J. Phys. Chem. C* 2019, 123, 12335–12345.

112 Migliore, A., Polizzi, N.F., Therien, M.J., Beratan, D.N. Biochemistry and Theory of Proton-coupled Electron Transfer. *Chem. Rev.* 2014, 114, 3381–3465.

113 Levich, V.G., Dogonadze, R.R., Kuznetsov, A.M. Theory of Hydrogen-Ion Discharge on Metals: Case of High Overvoltages. *Electrochim. Acta* 1968, 13, 1025–1044.

114 Levich, V.G., Dogonadze, R.R., German, E.D., Kuznetsov, A.M., Kharkats, Yu. I. Theory of Homogeneous Reactions Involving Proton Transfer. *Electrochim. Acta* 1970, 15, 353–367.

115 Nagel, Z.D., Klinman, J.P. Update 1 of: Tunneling and Dynamics in Enzymatic Hydride Transfer. *Chem. Rev.* 2010, 110, PR41-PR67.

116 Siebrand, W., Wildman, T.A., Zgierski, M.G. Temperature Dependence of Hydrogen Tunneling Rate Constants. *Chem. Phys. Lett.* 1983, 98, 108–112.

117 Kohen, A., Klinman, J.P. Enzyme Catalysis: Beyond Classical Paradigms. *Acc. Chem. Res.* 1998, 31, 397–404.

118 Eikerling, M., Kornyshev, A.A., Kuznetsov, A.M., Ulstrup, J., Walbran, S. Mechanisms of Proton Conductance in Polymer Electrolyte Membranes. *J. Phys. Chem. B* 2001, 105, 3646–3662.

119 Bell, R.P. *The Tunnel Effect in Chemistry*. Chapman and Hall, London, 1980.

120 Kuznetsov, A.M., Ulstrup, J. Dynamics of Low-barrier Proton Transfer in Polar Solvents and Protein Media. *Chem. Phys.* 1994, 188, 131–141.

121 Xu, J.C., Sharpe, M.A., Qin, L., Ferguson-Miller, S., Voth, G.A. Storage of an Excess Proton in the Hydrogen-bonded Network of the D-pathway of Cytochrome *c* Oxidase: Identification of a Protonated Water Cluster. *J. Am. Chem. Soc.* 2007, 129, 2910–2913.

122 Kornyshev, A.A., Kuznetsov, A.M., Spohr, E., Ulstrup, J. Kinetics of Proton Transport in Water. *J. Phys. Chem. B* 2003, 107, 3351–3366.

123 Agmon, N. *The Grotthuss Mechanism. Chem. Phys. Lett.* 1995, 244, 456–462.

124 Kuznetsov, A.M., Ulstrup, J. Theory of Inter-related Electron and Proton Transfer Processes. *Russian J. Electrochemistry* 2003, 39, 9–15.

125 Taraphder, S., Maupin, C.M., Swanson, J.M.J., Voth, G.A. Coupling Protein Dynamics with Proton Transport in Human Carbonic Anhydrase II. *J. Phys. Chem. B* 2016, 120, 8389–8404.

126 Soegaard-Andersen, P., Ulstrup, J. Semiclassical Calculation of Activation Parameters for Elementary Steps in Catalytic Olefin Hydrogenation in Liquid Media. *Acta Chem. Scand. Phys. Inorg. Chemistry* 1983, 37, 585–593.

127 German, E.D. Theoretical Models for X-H Bond Breaking (X = C, O, and H) over Metal Surfaces: Used for Simulation of Catalytic Steam Reforming. *Russian J. Chem.* 2017, 53, 1222–1231.

128 Klinman, J.P., Offenbacher, A.R. Understanding Biological Hydrogen Transfer through the Lens of Temperature Dependent Isotope Effects. *Acc. Chem. Res.* 2018, 51, 1966–1974.

129 Kuznetsov, A.M., Medvedev, I.G., Ulstrup, J. Electrochemical Proton Relay at the Single-molecule Level *Electrochem. Comm.* 2009, 11, 1170–1173.

130 Iofa, S. The Overvoltage on Mercury Cathode in Concentrated Solutions of Acids in the Presence of Surface Active Electrolytes. *Acta Physicochim (URSS).* 1939, 10, 903–912.

131 Vetter, K.J. *Electrochemical Kinetics*, 2nd. Ed. Wiley, New York, 1967. p.574.

132 Krishtalik, L.I. *Charge Transfer Reactions in Electrochemical and Chemical Processes*, Plenum, NY, 1986.

133 Nazmutdinov,R.R., Bronshtein, M.D., Wilhelm, F., Kuznetsov, A.M. Challenge of the Discharge of Hydronium Ion at a Mercury Electrode: beyond the Tafel Plots. *J. Electroanalyt. Chem.*, 200, 175–183.

134 Krishtalik, L.I. Kinetic Isotope Effect in the Hydrogen Evolution Reaction. *Electrochim. Acta*, 2001, 46, 2949–2960.

135 Sakaushi, K., Lyalin, A., Taketsugu, T. Observations and Theories of Quantum Effects in Proton Transfer Electrode Processes. *Curr. Opin. Electrochemistry* 2020, 19, 96–105.

136 Wilhelm, F., Schmickler, W., Nazmutdinov, Spohr, E. A Model for Proton Transfer to Metal Electrodes. *J. Phys. Chem. C*, 2008. 112. p.10814–10826.

137 Quaino, P., Juarez, F., Santos, E., Schmickler, W. Volcano Plots in Hydrogen Electrocatalysis: Uses and Abuses. *Beilstein J. Nanotechnology* 2014, 5, 846–854.

138 Skúlason, E., Tripkovic, V., Björketun, M.E., Gudmundsdóttir, S., Karlberg, G.S, Rossmeisl, J., Bligaard, T., Jónsson, H., J.K. Nørskov, J.K. Modeling the Electrochemical Hydrogen Oxidation and Evolution Reactions on the Basis of Density Functional Theory Calculations. *J. Phys. Chem. C.*, 2010, *114*, 18182–18197.

139 Kramers, H.A. Brownian Motion in a Field of Force and the Diffusion Model of Chemical Reactions. *Physica*, 1940, 7, 284–304.

140 Einstein, A. On the Motion of Small Particles Suspended in Liquids at Rest Required by the Molecular-Kinetic Theory of Heat, *Ann. Phys.* 1905, 17, 549–560.

141 Hynes, J.T. Outer Sphere Electron Transfer Reactions and Frequency Dependent Friction. *J. Phys. Chem.* 1986, 90, 3701–3706.

142 Zusman, L.D. The Theory of Electron Transfer Reactions in Solvents with Two Characteristic Relaxation Times. *Chem. Phys.* 1988, 119, 51–61.

143 Kuznetsov, A.M. *Stochastic and dynamic views of chemical reaction kinetics in solutions.* Presses Polytechniques et Universitaires Romandes, Lausanne, 1999.

144 Sumi, H. Marcus, R.A. Dynamic Effects in Electron-Transfer Reactions. *J. Chem. Phys.* 1986, 84, 4894–4914.

145 Titova, N.V., V.K. Laurinavichute, Z.V. Kuzminova, G.A. Tsirlina, Electroreduction of Peroxodisulfate on Mercury in Mixed Water-Carbohydrate Media: The Interplay of Solvent Effects and Concentration Dependent Structure and Reaction Layer. *Chem. Phys.* 2008, 352, 345–352.

146 Zagrebin, P.A., Buchner, R., Nazmutdinov, R.R, Tsirlina, G.A. Dynamic Solvent Effects in Electrochemical Kinetics: Indications for a Switch of the Relevant Solvent Mode. *J. Phys. Chem B.* 2010, 114, 311–320.

147 Ismailova, A.S. Berezin, M. Probst, R.R. Nazmutdinov. Interfacial Bond-Breaking Electron Transfer in Mixed Water-Ethylene Glycol Solutions: Reorganization Energy and Interplay between Different Solvent Modes. *J. Phys. Chem. B.* 2013, 117, 8793–8801.

148 Sparpaglione, M., Mukamel, S., Dielectric Friction and the Transfer from Adiabatic to Nonadiabatic Electron Transfer in Condensed Phases. *II. Application to Non-Debye Solvents. J. Chem. Phys.* 1986, 88, 4300–4311.

149 Allen, M.P., Tildesley, D.J. *Computer Simulation of Liquids*. Clarendon Press, Oxford, 1987.

150 Taube, H., Myers, H. Evidence for a Bridged Activated Complex for Electron Transfer Reactions. *J. Am. Chem. Soc.* 1954, 76, 2103–2111.

151 Kramers, H.A. The Interaction between Magnetogenic Atoms in a Paramagnetic Crystal. *Physica* 1934, 1, 182–192.

152 McConnell, H. Intramolecular Charge Transfer in Aromatic Free Radicals. *J. Chem. Phys.* 1961, 35, 508–515.

153 Yeganeh, S., Ratner, M.A. Dynamics of Charge Transfer: Rate Processes Formulated with Nonequilibium Green's Functions. *J. Chem. Phys.* 2007, *126*, *Art.* No. 161103.

154 Schmickler, W. Adiabatic and Non-adiabatic Electrochemical Electron Transfer in Terms of Green's Function Theory. *Russian J. Electrochem.* 2017, 53, 1182–1188.

155 Li, Y., Buerkle, M., Li, G., Rostamian, A., Wang, H., Wang Z., Bowler, D.R., Miyazaki, T., Xiang, L., Asai, Y., Zhou, G., Tao, N. Gate Controlling of Quantum Interference and Direct Observation of Anti-resonances in Single Molecule Charge Transport. *Nature Mat.* 2019, *18*, 357–363. *Errata: Nature Mat.* 2020, 19, 127–127.

156 Naghibi, S, Ismael, A.K., Vezzoli, A., Al-Khaykanee, M.K., Mohsin K., Zheng, X.J., Grace, I.M., Bethell, D., Higgins, S.J. Lambert, C.J., Nichols, R.J. Synthetic Control of Quantum Interference by Regulating Charge on a Single Atom in Heteroaromatic Molecular Junctions, *J. Phys. Chem. Lett.* 2019, 10, 6419–6424.

157 Jortner, J., Bixon, M., Langenbacher, T., Michel-Beyerle M.E. Charge Transfer and Transport in DNA. *Proc. Nat. Acad. Sci. USA* 1998, 95, 12759–12765.

158 Berlin, Yu. A., Voityuk, A.A., Ratner, M.A. DNA Base Pair Stacks with High Electric Conductance: A Systematic Structural Search. *ACS Nano* 2012, 6, 8216–8225.

159 Xiao, X.Y., Nagahara, L.A., Rawlett, A.M., Tao, N.J. Electrochemical Gate-controlled Conductance of Single Oligo(phenylene ethynylene)s. *J. Am. Chem. Soc.* 2005, 127, 9235–9240.

160 Nitzan, A. The Relationship between Electron Transfer Rate and Molecular Conduction. 2. The Sequential Hopping Case. *Israel J. Chem.* 2002, 42, 163–166.

161 Kharkats, Yu.I., Madumarov, A.K., Vorotyntsev, M.A. Application of Density Matrix Method iCJ Quantum Mechanical Calculation of Bridge-assisted Electron Transfer Probability in Polar Media. *J. Chem. Soc. Faraday Trans. II* 1974, 70, 1578–1590.

162 Kharkats, Yu. I., Ulstrup, Resonance Effects in 3-Center Long-range Electron Transfer. *Chem. Phys. Lett.* 1991, 182, 81–87.

163 Kuznetsov, A.M., Ulstrup, J. *Electron Transfer in Chemistry and Biology: An Introduction to the Theory*, Wiley, Chichester, 1999. Chapter 12.

164 Kharkats, Yu.I., Kuznetsov, A.M., Ulstrup, J. Coherence, Friction, and Electric Field Effects in Primary Charge Separation in Bacterial Photosynthesis. *J. Phys. Chem.* 1995, 99, 13545–13554.

165 Zakaraya, M.G., Ulstrup, J. Dynamic Solvent Effects on the Raman Scattering Cross Section and Resonance Profiles of Polyatomic Molecules in Solution. *Chem. Phys.* 1989, 135. 49–73.

166 Moth-Poulsen, K., Ed. *Handbook of Single-molecule Electronics*. Pan Stanford Publishing Pte, Singapore, 2016.

167 Bâldea, I., Ed. *Molecular Electronics: An Experimental and Theoretical Approach*. Pan Stanford Publishing Pte, Singapore, 2016.

168 Ambrosi, A., Chua, C.K., Bonanni, A., Pumera, M. Electrochemistry of Graphene and Related Materials. *Chem. Rev.* 2014, 114, 7150–7188.

169 Tang, J., Yan, X., Engelbrekt, C., Ulstrup, J., Magner, E., Xiao, X., Zhang, J. Development of Graphene-based Enzymatic Biofuel Cells: A Minireview, *Bioelectrochemistry* (2020) In press.

170 **a.** Rees, N.V., Zhou, Y.-G., Compton, R.G. Making Contact: Charge Transfer During Particle-Electrode Collisions. *RSC Adv.* 2012, *2*, 379–384. **b.** Sokolov, S.V., Eloul, S., Kätelhön, Batchelor-McAuley, C., Compton, R.G. Electrode-Particle Impacts: A Users Guide. *PCCP* 2017, 19, 28–43.

171 Mathwig, K., Aartsma, T.J., Canters, G.W., Lemay, S.G. Nanoscale Methods for Single-molecule Electrochemistry. *Ann. Rev. Anal. Chem.* 2014, 7, 383–404.

172 Albrecht, T., Horswell, S., Allerston, L.K., Rees, N.V., Rodriguez, P. Electrochemical Processes at the Nanoscale. *Curr. Opin. Electrochem.* 2018, 7, 138–145.

173 Hromadová, M., Vavrek, F. Electrochemical Electron Transfer and its Relation to Charge Transport in Single-Molecule Junctions. *Curr. Opin. Electrochem.* 2020, 19, 63–70.

174 Fan, Y., Anderson, T.J., Zhang, B. Single-molecule Electrochemistry: From Redox Cycling to Single Redox Events. *Curr. Opin. Electrochemistry* 2018, 7, 81–86.

175 Tao, N. Probing Potential-tuned Resonant Tunneling through Redox Molecules with Scanning Tunneling Microscopy. *Phys. Rev. Lett.* 1996, 76, 4066–4069.

176 Zhang, J., Chi, Q., Hansen, A.G., Jensen, P.S., Salvatore, P., Ulstrup, J. Interfacial Electrochemical Electron Transfer in Biology –towards the Level of the Single Molecule. *FEBS Letters* 2012, 586, 526–535.

177 Salvatore, P., Zeng, D., Karlsen, K.K., Chi, Q., Wengel, J., J. Ulstrup J., Electrochemistry of Single Metalloprotein and DNA-based Molecules at Au(111)-electrode Surfaces, *Chem Phys Chem* 2013, 14, 2101–2111.

178 Nichols, R.J., Higgins, S.J. Single Molecule Nanoelectrochemistry in Electrical Junctions. *Acc. Chem. Res.* 2016, 49, 2640–2648.

179 Petrangolini, P., Alessandrini, A., Berti, L., Facci, P. An Electrochemical Tunneling Microscopy Study of 2-(6-Mercaptoalkyl)hydroquinone Molecules on Au(111). *J. Am. Chem. Soc.* 2010, 132, 7445–7453.

180 Alessandrini, A., Gerunda, M., Canters, G.W., Verbeet, M.P., Facci, P.: Electron Tunnelling through Azurin is Mediated by the Active Site Cu ion. *Chem. Phys. Lett.* 2003, 376, 625–630.

181 Della Pia, E. A, Chi, Q., Macdonald, J.E., Ulstrup, J., Jones, D.D., Elliott, M. Fast Electron Transfer through a Single Molecule Natively Structured Redox Protein. *Nanoscale* 2012, 4, 7106–7113.

182 Lopez-Martinez, M., Artés, J.M., Sarasso, V., Carminati, M, Diez-Perez, I, Sanz, F. , Gorostiza, P. Differential Electrochemical Conductance Imaging at the Nanoscale. Small 2017, *13, Art. No. UNSP* 1700958.

183 Salvatore, P., Karlsen, K.K., Hansen, A.G., Zhang, J., Nichols, R.J., Ulstrup, J. Polycation Induced Potential Dependent Structural Transitions of Oligonucleotide Monolayers on Au(111)-surfaces, *J. Am. Chem. Soc.* 2012, 134, 19092–19098.

184 Chen, S.W., Murray, R.W., Feldberg, Quantized Capacitance Charging of Monolayer-protected Gold Clusters. *J. Phys. Chem. B* 1998, 102, 9898–9907.

185 Albrecht, T., Mertens, S.F.L., Ulstrup, J. Intrinsic Multi-state Switching of Gold Clusters through Electrochemical Gating, *J. Am. Chem. Soc.* 2007, 129, 9162–9167.

186 McCreery, R.L. Advanced Carbon Materials for Molecular Electrochemistry. *Chem. Rev.* 2008, 108, 2646–2687.

187 McCreery, R.L. Effects of Electronic Coupling and Electrostatic Potential on Charge Transport in Carbon-based Molecular Electronic Junctions. *Beilstein J. Nanotechnol.* 2016, 7, 32–46.

188 Manser, J.S., Christians, J.A., Kamat, P.V. Intriguing Optoelectronic Properties of Metal Halide Perovskites. *Chem. Rev.* 2016, 116, 12956–13008.

189 Šešelj, N., Engelbrekt, C., Ding, Y., Hjuler, H.A., Jens Ulstrup, J., Zhang, J. Tailored Electron Transfer Pathways in Aucore/Ptshell-graphene Nanocatalysts for Fuel Cells. *Adv. Energy Mat.* 2018, Paper No. 1702609.

190 Xiao, X, Si, P., Magner, E. An Overview of Dealloyed Nanoporous Gold in Bioelectrochemistry. *Bioelectrochemistry* 2016, 109, 117–226.

191 Ding Y., Zhang, Z. *Nanoporous Metals for Advanced Energy Technologies.* Springer, Dordrecht, 2016.

192 Reimers, J.R., Ford, M.J., Halder, A., Ulstrup, J., Hush, N.S. Gold Surfaces and Nanoparticles are Protected by Au(0)-thiyl Species and Are Destroyed when Au(I)-thiolates Form, *Proc. Nat. Acad. Sci. USA* 2016, 113, E1424-E1433.

193 Reimers, J.R., Ford, M.J., S. Marcuccio, J. Ulstrup and N.S. Hush, Competition of van der Waals and Chemical Forces on Gold-sulfur Surfaces and Nanoparticles. *Nature Reviews Chemistry* 2017, *1* (2017) 0017.

194 Chi, Q., Ford, M.J., Halder, A., N.S. Hush, J.R. Reimers and J. Ulstrup, Sulfur Ligand Mediated Electrochemistry of Gold surfaces and Nanoparticles: What, How, and Why?. *Curr. Opin. Electrochemistry* 2017, 1, 7–15.

195 Zhang, J., Welinder, A.C., Q. Chi, Q., Ulstrup, J. Electrochemically Controlled Self-assembled Monolayers Characterized with Molecular and Sub-molecular Resolution. *Phys Chem. Chem. Phys.* 2011, 12, 9999–10011.

196 Engelbrekt, C., Nazmutdinov, R.R., Zinkicheva, T.T., Glukhov, D.V., Yan, J.W., Mao, B.W., Ulstrup, J., Zhang, J.D. Chemistry of Cysteine Assembly on Au(100): Electrochemistry, *In Situ* STM and Molecular Modeling. *Nanoscale* 2019, 11, 17235–17251.

197 Wang, Y., Chi, Q., Zhang, J., Hush, N.S., Reimers, J.R., Ulstrup, J. Chain-branching Control of the Atomic Structure of Alkanethiol-based Gold-sulfur Interfaces. *J. Am. Chem. Soc.* 2011, 133, 14856–14859.

198 Ouyang, R., Yan, J., Jensen, P.S., Ascic, E., Gan, S., Tanner D., Mao, B., Niu, L., Zhang, J., Tang, C., Hush, N.S., Reimers, J.R., Ulstrup, J. Intermixed Adatom and Surface Bound Adsorbates in Regular Self-assembled Monolayers of Racemic 2-Butanethiol on Au(111). *Chem Phys Chem* 2015, 16, 928–932.

199 Zhang, J., Demetriou, A., Welinder, A.C., Albrecht, T., Nichols, R.J., Ulstrup, J. Potential-Induced Structural Transitions of Homocysteine Monolayers on Au(111) electrode surfaces, *Chem. Phys.* 2005, 319, 210–221.

200 Climent, V., Zhang, J., Friis, E.P., Oestergaard, L.H., Ulstrup, J. Voltammetry and In Situ Scanning Tunnelling Microscopy of Laccases and Bilirubin Oxidase in Electrocatalytic Dioxygen Reduction on Au(111) Single-crystal Electrodes. *J. Phys. Chem. C* 2012, 116, 1232–1243.

201 Hao, X., J. Zhang, J., Christensen, H.E.M., Wang, H., Ulstrup, J. Electrochemical Single-molecule AFM of the Redox Metalloenzyme Copper Nitrite Reductase in Action. *Chem Phys Chem* 2012, 13, 2919–2924.

202 Vivancos, J.A.M., Hihath, J., Diéz-Pérez, I. Biomolecular Electronics. In: Bâldea, I., Ed. *Molecular Electronics: An Experimental and Theoretical Approach.* Pan Stanford Publishing Pte, Singapore, 2016, pp. 281–313.

203 Kuznetsov, A.M., Medvedev, I.G., Ulstrup, J. Electric Double Layer Effect on Observable Characteristics of the Tunnel Current through a Bridged Electrochemical Contact. *J. Chem. Phys.* 2007, *127*, Art. No. 104708.

204 Kuznetsov, A.M., Ulstrup, J. Mechanisms of *In Situ* Scanning Tunnelling Microscopy of Organized Redox Molecular assemblies. *J. Phys. Chem. A* 2000, *104*, 11531–11540. *Errata: J. Phys. Chem. A* 2001, 105, 7494–7494.

205 Medvedev, I.G. Analytical Expression for the Tunnel Current through the Redox-mediated Tunnel Contact in the Case of the Adiabatic Electron Transfer at One of the Working Electrodes and Any Possible Type of the Electron Transfer at the Other Electrode. *J. Chem. Phys.* 2017, *124*, Art. No. 194108.

206 Wang, H., Wang, Z.X., Wang, Y., Hihath, J., Chen, H.I., Li, Y.Q. Tao, N.J. Potential Dependence of Mechanical Stability and Electronic Coupling of Single S-Au Bonds. *J. Am. Chem. Soc.* 2019, 140, 18074–18081.

207 Albrecht, T., Moth-Poulsen, K., Christensen, J.B., Hjelm, J., Bjornholm, T., Ulstrup, J. Scanning tunneling Spectroscopy in an Ionic Liquid. *J. Am. Chem. Soc.* 2007, 128, 6574–6575.

208 Higgins, S.J., Nichols, R.J. Metal/molecule/metal Junction Studies of Organometallic and Coordination Complexes; What can Transition Metals do for Molecular Electronics? *Polyhedron* 2018, 140, 25–34.

209 Fan, F.-RF., Bard, A.J. Electrochemical Detection of Single Molecules, *Science* 1995, 267, 871–874.

210 Faraday Discussions. Single Entity *Electrochemistry.* 2016, 193.

211 Dumitrescu, I., Dudin, P.V., Edgeworth, J.P., Macpherson, J.V., Unwin, P.R. Electron Transfer Kinetics at Single-Walled Carbon Nanotube Electrodes using Scanning Electrochemical Microscopy. *J. Phys. Chem. C* 2010, 114, 2633–2639.

212 Li Z.J., Jeanmairet G., Méndez-Morales T., Burbano M., Haefele M., Salanne, M. Confinement Effects on an Electron Transfer Reaction in Nanoporous Carbon Electrodes. *J. Phys. Chem. Lett.* 2017, 8, 1925–1931.

213 Niimi, Y., Matsui, T., Kambara, H., Tagami, K., Tsukada, M., and Fukuyama, H. Scanning Tunneling Microscopy and Spectroscopy of the Electronic Local Density of States of Graphite Surfaces near Monoatomic Step Edges. *Physical Review B*, 2006, *73*, *Art.* No. 085421.

214 Zhong, J.-H., Zhang, J., Jin, X., Liu, J.-Y., Li, Q., Li, M.-H., Cai, W., Wu, D.-Y., Zhan, D., and Ren, B. Quantitative Correlation between Defect Density and Heterogeneous Electron Transfer Rate of Single Layer Graphene. *J. Am. Chem. Soc.* 2014, 136, 16609–16617.

215 Li, Y., Wang, J., Li, X., Geng, D., Banis, M.N., Li, R., and Sun, X.: Nitrogen-doped Graphen Nanosheets as Cathode Materials with Excellent Electrocatalytic Activity for High Capacity Lithium-oxygen Batteries. *Electrochem. Comm.* 2012, 18, 12–15.

216 Wang, S., Dong, S., Wang, J., Zhang, L., Han, P., Zhang, C., Wang, X., Zhang, K., Lan, Z., and Cui, G. O Oxygen-enriched Carbon Material for Catalyzing Oxygen Reduction towards Hybrid Electrolyte Li-air Battery. *Journal of Materials Chemistry* 2012, 22, 21051–21056.

217 Unwin, P.R., Güell, A.G., Zhang, G. Nanoscale Electrochemistry of sp^2-Carbon Materials: From Graphite and Graphene to Carbon Nanotubes. *Acc. Chem. Res.* 2016, 49, 2041–2048.

218 Hodson, H., Li, X., Batchelor-McAuley, C., Shao, L., Compton, R. Single Nanotube Voltammetry: Current Fluctuations are Due to Physical Motion of the Nanotube. *J. Phys. Chem.* 2016, 120, 6281–6286.

219 Larromore, L., Nad, S. Zhou, X., Abruña, H., McEuen, P.L: Probing Electrostatic Potentials in Solution with Carbon Nanotube Transistors. *Nano Lett.* 2006, 6, 1329–1333.

220 [220].Nazmutdinov, R.R., Bronshtein, M.D., Schmickler, W. Electron Transfer across a Conducting Nanowire (-Tube)/Electrolyte Solution Interface. *Electrochim. Acta.* 2009, 55, 68–77.

221 Nazmutdinov, R.R,; Bronshtein, M.D., Berezin, A.S., Soldano, G., Schmickler, W. Bond Breaking Electron Transfer across a Conducting Nanowire (Nanotube)/Electrolyte Solution Interface: the Role of Electrical Double Layer Effects. *J. Electroanal. Chem.* 2011, 660, 309–313.

222 **a.** Pavlov, S.V., Nazmutdinov, R.R., Fedorov, M.V., Kislenko, S.A. Role of Graphene Edges in the Electron Transfer Kinetics: Insight from Theory and Molecular Modeling *J. Phys. Chem.* C. 2019, *123*, 6627–6634. **b.** Nazmutdinov, R.R., Bronshtein, M.D., Santos, E. Electron Transfer Across the Graphene Electrode/solution Interface: Interplay between Different Kinetic Regimes. *J. Phys. Chem.* C. 2019, *123*, 12346–12354. **c.** Kislenko, V.A., Pavlov S.V., Kislenko. S.A. Influence of Defects of Graphene on Electron Transfer Kinetics: The Role of the Surface Electronic Structure. *Electrochim. Acta.* 2020, 341, Art. No. 136011.

223 Nazmutdinov, R.R., Zinkicheva, T.T., Shermukhamedov, S.A., Zhang, J., Ulstrup, J. Electrochemistry of Single Molecules and Biomolecules, Molecular Scale Nanostructures, and Low-dimensional Systems, *Curr. Opin. Electrochem.* 2018. 7, 179–187.

224 **a.** Seo, M., Chung, N.D. Nanoconfinement Effects in Electrochemical Reactions. *Curr. Opin. Electrochem.* 2019, *13*, 47–54. **b.** Miners, S.A., Rance, G.A., Khlobystov, A.N. Chemical Reactions Confined within Carbon Nanotubes. *Chemical Society Reviews,* 2016, 45, 4727–4746.

225 [225] Kleijn, S.E., Lai, S.C.S., Koper, M.T.M., Unwin, P. Electrochemistry of Nanoparticles. *Angew. Chem. Int. Ed.* 2014, *53*, 3558–3586.

226 Louis, C, Pluchery, O. *Gold Nanoparticles for Physics, Chemistry and Biology*. Imperial College Press, London, 2012.

227 Astruc, M.C., Astruc, D. Gold Nanoparticles: Assembly, Supramolecular Chemistry, Quemtum-Size Related Properties, and Applications towards Biology, Catalysis, and Nanotechnology. *Chem. Rev.* 2004, 104, 293–346.

228 Kissling, G.P., Miles, D.O., Fermin, D.J., Electrochemical Charge Transfer Mediated by Metal Nanoparticles and Quantum Dots. *Physical Chemistry C Physics* 2011, 13, 21175–21185.

229 Razzaq, H., Qureshi R., Schiffrin, D.J. Enhanced Rate of Electron Transfer across Gold.Anthraquinone Hybrids, *Electrochem. Comm* 2014, 39, 9–11.

230 Jensen, P.S., Chi, Q., Grumsen, F.B., Abad, J.M., Horsewell, A., Schiffrin, D.J., Ulstrup, J. Gold Nanoparticle Assisted Assembly of a Heme Protein for Enhancement of Long-range Interfacial Electron Transfer. *J. Phys. Chem. C* , 2007, 111, 6124–6132.

231 Jensen, P.S., Chi, Q., Zhang, J., Ulstrup, J. Long-range Electrochemical Electron Transfer of *Pseudomonas Aeruginosa* Azurin-gold Nanoparticle Hybrid Systems. *J. Phys. Chem. C* 2009, 113, 13993–14000.

232 Gutiérrez-Sánchez, C., Pita, M., Vaz-Domíguez, C., Shleev, S., De Lacey, A.L. Gold Nanoparticles as Electronic Bridges for Laccase-Based Biocathodes. *J. Am. Chem. Soc.* 2012, 134, 17212–17220.

233 Kizling, M., Dzwonek, M., Wieckowska, A., Bilewicz, R., Size Does Matter – Mediation of Gol Clusters in Bioelectrocatalysis. *ChemCatChem* 2018, 10, 1998–1992.

234 Andersen, J.E.T., Moller, P., Pedersen, M.V., Ulstrup, J. Cytochrome *c* Dynamics at Gold and Glassy Carbon Surfaces Monitored by *In Situ* Scanning Tunnel Microscopy. *Surf. Sci.* 1995, 325, 193–205.

235 Zhang, J., Chi, Q., Dong, S.J., Wang, E.: In situ Electrochemical Scanning Tunnelling Microscopy Investigation of Structure for Horseradish Peroxidase and its Electrocatalytic Property. *Bioeletrochem. Bioenerg.* 1996, 39, 267–274.

236 Friis, E.P., Andersen, J.E.T., Madsen, L.L., Moller, P., Ulstrup, J. *In Situ* STM and AFM of the Copper Protein *Pseudonomans Aeruginosa* Azurin. *J. Electroanal. Chem.* 1997, 431, 35–38.

237 Chi, Q., Zhang, J., J.U. Nielsen, Friis, E.P., Chorkendorff, I., Canters, G.W., Andersen, J.E.T., Ulstrup, J. Molecular Monolayers and Interfacial Electron Transfer of Pseudomonas Aeruginosa Azurin on Au(111), *J. Am. Chem. Soc.* 2000, 122, 4047–4055.

238 Chi, Q., Farver, O., Ulstrup, J. Long-range Protein Electron Transfer Observed at the Single-molecule Level: *In Situ* Mapping of Redox-gated Tunnelling Resonance. *Proc. Nat. Acad. Sci. USA* 2005, 102, 16203–16208.

239 Alessandrini, A., Facci, P. *ECS/STM of Redox Metalloproteins and Co-Factors.* In: Bâldea, I., Ed. *Molecular Electronics: An Experimental and Theoretical Approach.* Pan Stanford Publishing Pte, Singapore, 2016, pp. 325–352.

240 Farver, O., Pecht, I. Electron Transfer in Blue Copper Proteins. *Coord. Chem. Rev.* 2011, 255, 757–773.

241 Zhang, J., Welinder, A.C., Hansen, A.G., Christensen, H.E.M., Ulstrup, J. Catalytic Monolayer Voltammetry and *In Situ* Scanning Tunnelling Microscopy of Copper Nitrite Reductase on Cysteamine-modified Au(111)- electrodes, *J. Phys. Chem. B* 2003, 104, 12480–12484.

242 Nazmutdinov, R.R., Bronshtein, M.D., Zinkicheva, T.T., Chi, Q., Zhang, J., Ulstrup. J. Modeling and Computations of the Intramolecular Electron Transfer in the Two-Heme Protein Cytochrome c_4. *Phys. Chem. Chem. Phys.* 2012, 14 5953–5965.

243 Kunz A., Cluster Chemisorption. In: *Theory of Chemisorption.* Smith, J.R., Ed. Springer-Verlag, Berlin, 1980.

244 Nazmutdinov, R.R., Shapnik, M.S. Contemporary Quantum Chemical Modeling of Electrified Interfaces. *Electrochim. Acta.* 1996, 41, 2253–2265.

245 Whitten, J.L., Yang, H. Theory of Chemisorption and Reactions on Metal Surfaces. *Surf. Sci. Rep.* 1996, 24, 55–124.

246 Nakatsuji, H. Dipped Adcluster Model for Chemisorption and Catalytic Reactions, *Progress in Surf. Sci.* 1997, 54, 1–68.

247 Groβ, A. *Theoretical Surface Science. A Microscopic Perspective*, 2nd Ed. Springer, Berlin, 2009.

248 Makov, G., Payne, M.C. Periodic Boundary-conditions in *Ab-initio* Calculations. *Phys. Rev. B.* 1995, 51, 4014–4022.

249 Quaino, P., Luque, N.B., Soldano, G., Nazmutdinov, R., Santos, E., Roman, T., Lundine, A., Groβ, A., Schmickler, W. Solvated Protons in Density Functional Theory – a Few Examples. *Electrochim. Acta.* 2013, 105, 248– 253.

250 Mardirossian, N., Head-Gordon, M. Thirty Years of Density Functional Theory in Computational Chemistry. *Molecular Physics.* 2017, 115, 2315–2372.

251 Otani, M., Sugino, O. First Principles Calculations of Charged Surfaces and Interfaces: A Plane-wave Non-repeated Approach. *Phys. Rev.* B. 2006, *73*, *Art.* No. 115407.

252 Melander, M.M., Kuisma, M.J., Christensen, T.E.K., Honkala, K. Grand-canonical Approach to Density Functional Theory of Electrocatalytic Systems: Thermodynamics of Solid-liquid Interfaces at Constant Ion and Electrode Potentials. *J. Chem. Phys.* 2019, 150, Art No. 041706.

253 Kovalenko, A., Hirata, F. Three-dimensional Density Profiles of Water in Contact with a Solute of Arbitrary Shape: a RISM Approach. *Chem. Phys. Lett.* 1988, 290, 237–244.

254 Gross, A., Sakong, S. Modelling the Electric Double Layer at Electrode/Electrolyte Interfaces. *Current Opin. Electrochemistry.* 2019, 14, 1–6.

255 Newton, M.D. Formalisms for Electron-Exchange Kinetics in Aqueous Solution and the Role of Ab Initio Techniques in Their Implementation. *Int. J. Quant. Chem.: Quant. Chem. Symp.* 1980, 14, 363–391.

256 Newton, M.D. Electronic Structure Analysis of Electron Transfer Matrix Elements for Transition Metal Redox Pairs. *J. Phys. Chem.* 1988, 92, 3049–3056.

257 Cave, R.J. and M.D. Newton, Generalization of the Mulliken-Hush Treatment for the Calculation of Electron Transfer Matrix Elements. *Chemical Physics Letters* 1996. 249, 15–19.

258 Kubas, A., Hoffmann, F., Heck, A., Oberhofer, H., Elstner, M., Blumberger, J. Electronic Couplings for Molecular Charge Transfer: Benchmarking CDFT, FODFT and FODFTB Against High-Level ab initio Calculations. *J. Chem. Phys.* 2014, 140, 104105–104225.

259 Kubas, A., Gajdos, F., Heck, A., Oberhofer, H., Elstner, M., Blumberger, J. Electronic Couplings for Molecular Charge Transfer: Benchmarking CDFT, FODFT and FODFTB Against High-Level ab initio Calculations. *II. Phys. Chem. Chem. Phys.* 2015, 17, 14342–14354.

260 Berezin, A.S., Nazmutdinov, R.R. Monte Carlo Simulations of Heterogeneous E Electron Transfer: New Challenges. *Rus. J. Electrochemistry.* 2017, 53, 1232–1238.

261 Iwasita, T.I., Schmickler, W., Schultze, J.W. Influence of the Metal on the Kinetics of Outer Sphere Redox Reactions. *Ber.r Bunsengeselschaft/Physical Chemistry Chemical Physics*. 1985, 89, 138–142.

262 Feldberg, S.W., Sutin, N. Distance Dependence of Heterogeneous Electron Transfer through the Nonadiabatic and Adiabatic Regimes. *Chem. Phys.* 2006, 324, 216–225.

263 Schmickler, W. Electronic Effects in the Electric Double Layer. *Chem. Rev.* 1996, 96, 3177–3200.

264 Dogonadze, R.R., Kálmán, E., A.A. Kornyshev, A.A., Ulstrup, J., Eds., *The Chemical Physics of Solvation, Elsevier, Amsterdam; Part A 1985: Theory of Solvation. Part B 1986: Spectroscopy of Solvation. Part C 1988: Solvation Phenomena in Specific Physical, Chemical and Biological Systems.*

265 Guidelli, R. , Schmickler, W. Recent Developments in Models for the Interface between a Metal and an Aqueous Solution. *Electrochim. Acta* 2000, 45, 2317–2338.

266 Belousov, A.A., Kuznetsov, A.M., Ulstrup, J. Evolution of Stochastic Chemical Rate Theory from the Dynamics of Molecular Reactant Nuclear Coupled to a Continuous Manifold of Heat Bath Oscillators. *Chem. Phys.* 1988, 129, 311–333.

267 Li, Z.H., Liu, Y.Q., Mertens, S.F.L., Pobelov, I.V., Wandlowski, T. From Redox Gating to Quantized Charging. *J. Am. Chem. Soc.* 2010, 132, 8187–8193.

268 Shermukhamedov, S.A., Nazmutdinov, R.R., Zinkicheva, T.T., Bronshtein, M.D., Zhang, J., Mao, B. Tian, Z., Yan, J., Wu, D.Y., Ulstrup, J.J. *Am. Chem. Soc.* 2020, In press.

269 Wang, L., Sofer, Z., Pumera, M. Will Any Crap We Put Into Graphene Increase Its Catalytic Effect? *ACS Nano* 2020, 14, 21–25.

270 Pedersen, K.S., Bendix, J., Clerac, R. Single-molecule Magnet Engineering: Building-block Approaches. *Chem. Comm.* 2014, 50, 4396–4415.

271 **a.** Memming, R. Photo-elecrrochemical Solar Energy Conversion. *Top. Curr.* Chem. Ser. 1988 C*143*, 79–112. **b.** Nozik, A.J., Memming R. Physical Chemistry of Semiconductor-Liquid Interfaces. *J. Phys. Chem.* 1996, *100*, 13061–13078.

272 Xiang, C.X., Papadantonakis, K.M., Lewis, N.S. Principles and Implementations of Electrolysis Systems for Water Splitting. *Mat. Horizons* 2016, 3, 179–183.

273 Kolb, D.M., Ed. Photoelectrochemistry: From Fundamentals to Solar Applications. Proc. 5. Gerischer Symposium, Berlin, 2011.

274 Schmickler, W., Santos, E., Bronshtein, M.D., Nazmutdinov, R.R. Adiabatic Electron Transfer on Semiconducting Electrode. *ChemPhysChem* 2017, 18, 111–116.

275 **a.** Zhu, N., Han, S., Gan, S., Ulstrup, J., Chi, Q. Graphene Paper Doped with Chemically Compatible Prussian Blue Nanoparticles as Nanohybrid Electrocatalyst. *Adv. Funct. Mat.* 2013, *23*, 5297–5306. **b.** Zhu, N., Hao, X., Ulstrup, J., Chi, Q.Single-nanoparticle Resolved Biomimetic Long-range Electron Transfer and Electrocatalysis of Mixed-valence Nanoparticles. *ACS Catalysis* 2016, 6, 2728–2738.

276 Pinto, L.M.C., Spohr, E., Quaino, P., Santos, E., Schmickler, W. Why Silver Deposition is so Fast: Solving the Enigma of Metal Deposition. *Angew. Chem. Int. Ed.* 2013, 52, 7883–7885.

277 Mandler, D. Chiral Self-assembled Monolayers in Electrochemistry. *Curr. Opin. Electrochem.* 2018, 7, 42–47.

278 Magner, E., Klabanov, A.M. The Oxidation of Chiral Alcohols Catalyzed by Catalase in Organic Solvents. *Biotechnol. Bioeng.* 1995, 46, 175–179.

279 Bernauer, K., Ghizdavu, S., Verardo, S. Chiral Metal Complexes as Probes in Electron Transfer Reactions Involving Metalloproteins. *Coord. Chem. Rev.* 1999, 190, 357–369.

280 Odom, T.W., Schatz, G.C., *Eds. Chem. Rev.* 2011, 111, 3667–3994.

281 Nazmutdinov, R.R., Manyurov, I.R., Schmickler, W. The Effects of "Hot" Electrons on the Heterogeneous Adiabatic Charge Transfer Reactions. *Chem. Phys. Lett.* 2006, 429, 457–463. Schatz, G.C., Ed.

282 Janssen, H., Diesing, D., Otto, A. Enhanced Inverse Photoemission by Hot Electrons at Strongly Charged Silver Surfaces. *Surf. Sci.* 1995, 331, 1267–1271.

283 Krivenko, A.G., Benderskij, V.A., Kruger, Kautek, W. Gigantic Hydrogen Ion Discharge Currents Initiated by a Subpicosecond Laser. *Russian J. Electrochem.* 1998, 34, 1068–1075.

284 Magnussen, O.M., Gross, A. Toward an Atomic-Scale Understanding of Electrochemical Interface Structure and Dynamics. *J. Am. Chem. Soc.* 2019, 141, 4777–4790.

285 Nikitina, V.A., Vassiliev, C.Y., Stevenson, K.J. Metal-Ion Coupled Electron Transfer Kinetics in Intercalation-Based Transition Metal Oxides. *Adv. Energy Mater.* 2020, Article No. 1903933.

Part III

3

Continuum Embedding Models for Electrolyte Solutions in First-Principles Simulations of Electrochemistry

Oliviero Andreussi[1], Francesco Nattino[2], and Nicolas Georg Hörmann[2,3]

[1] Department of Physics, University of North Texas, Denton, TX 76207, USA
[2] Theory and Simulations of Materials (THEOS) and National Centre for Computational Design and Discovery of Novel Materials (MARVEL), École Polytechnique Fédérale de Lausanne, CH-1015 Lausanne, Switzerland
[3] Chair of Theoretical Chemistry, Technische Universität München, 85748 Garching, Germany

3.1 Introduction to continuum models for electrochemistry

It can be said that electrochemistry was a core science for the nineteenth century and, equally well, that it will be a foundational science for the twenty-first century. Electrochemical storage and conversion of clean and renewable energy sources is rapidly becoming the most important field of theoretical and applied scientific research, boosted by the need to overcome fossil-fuels technologies and the environmental, economic, and political impacts intrinsically intertwined with their use. Such a key societal advance can only be guaranteed by a new generation of materials or materials with novel properties, designed to harvest solar energy, store it in fuels or in batteries, distribute it efficiently over the power grid, and sustain our information and communication revolution. The pressing need for a material's revolution in energy storage and conversion is the main reason behind the exponential growth in the last decades of surface-science research, dealing with the physics and chemistry of interfaces and encompassing important phenomena, such as heterogeneous catalysis, charge and energy transfer reactions, adsorption, and surface reconstruction. In particular, the study of solid-vacuum interfaces has witnessed profound developments in the past years, also due to the fundamental role played by first-principles quantum-mechanical simulations.

As a result of past efforts, solid-vacuum interfaces can now be treated routinely and efficiently via first-principles simulations: high-throughput studies started to appear in the literature, which allowed simulations to show their predictive power in the design of electrode materials, e.g. as reported in the chapters of Section V of this volume. Nonetheless, it is fair to say that, in the study of electrochemical and electro-catalytic reactions occurring in wet devices, solid-vacuum interfaces can only represent a simplified model of the real systems, as they miss or largely approximate the crucial role played by the solvent, the electrolyte, and the applied electrode potential.

Experimental research is expanding its tools to deal with wet interfaces and an increasing number of experimental probes with atomic resolution at the solid-liquid interface are

Atomic-Scale Modelling of Electrochemical Systems, First Edition.
Edited by Marko M. Melander, Tomi T. Laurila, and Kari Laasonen.

being developed to work in situ, i.e. under operational conditions. Nonetheless, the number of probes available to solid-liquid interface is still small compared to the solid-vacuum interface and important scientific opportunities are available for theoretical studies.

While density functional theory (DFT) has become the tool of choice to model both molecules and crystalline substrates, dealing with liquids and amorphous phases using first-principles simulations can be challenging: the length-scales and time-scales that need to be simulated can easily make the problem unmanageable with available, and future, computational resources. As a result, first-principles simulations of electrochemical systems with a fully explicit description of each component of the device are usually beyond reach. Even when state-of-the-art full atomistic first-principles molecular dynamics approaches can be attempted (see Chapters 5–7 of this volume), these calculations do not lend themselves to systematic high-throughput studies, performed over several materials and for different reaction pathways.

An alternative to simplified solid-vacuum simulations and fully atomistic approaches is represented by the use of a multiscale strategy, where less advanced classical models are exploited to handle some of the components of the electrochemical system. Among the different multiscale approaches, two are the most relevant: either the solvent's degrees of freedom are kept, but treated at a classical level (QM/MM approach), or the statistical nature of the liquid system is exploited to integrate out the degrees of freedom of the solution and replace them with continuum medium (QM/Continuum), where only selected phenomenological forms of interaction between the solvent and the solute are kept.

The connection between a statistical description of the system and the phenomenological description with reduced (or no) degrees of freedom can be made more formal under the approach of joint density functional theory, as described by Sundararaman and Arias in Chapter 4. Similarly, an empirical approach to reduce the degrees of freedom of the solvent using solute-solvent interactions derived from classical force-fields is also at the core of the RISM approach, in particular of 3D-RISM, that has been recently extended to the simulation of electrochemical systems by Ogino and , as reviewed in Chapter 8. While somehow similar to pure continuum embedding approaches, all these different strategies have the common goal of providing a description of the embedding environment that is accurate, but does not rely on too many empirical ingredients, so as to provide a general and transferable solvation model. The different competing efforts in this field have led in recent years to the development of a variety of methods whose computational costs pose a relatively negligible overhead on standard DFT simulations and whose parameterization is fast and general. Even though the balance between costs, empiricism and accuracy may be slightly different between alternative approaches, it is without doubt that these combined directions will all contribute to a new and improved understanding of electrochemical interfaces in future years.

Continuum models of solvation are among the most popular solvation approaches in the literature, with the Polarizable Continuum Model (PCM) of Tomasi and coworkers being probably the most widespread and influential one[1, 2]. Several reviews have appeared on the topic, covering the origin and evolution of the different features and capabilities of these models in the quantum-chemistry literature[3–6]. These models were initially designed in combination with simulation packages based on localized basis sets, for the study of isolated molecules and clusters. A reformulation of continuum solvation approaches in condensed-matter simulation codes was proposed in 2002 by Fattebert and Gygi (FG) [7, 8], opening the field for the development of approaches ideally suited for the study of two-dimensional interfaces. The main ingredient of the FG model is the same as of PCM, namely a homogeneous polarizable dielectric medium embedding a quantum-mechanical system. At odds with PCM and its sharp two-dimensional (2D) interface, the FG approach features a smoothly varying dielectric function defined inside a three-dimensional (3D) simulation cell. Due to this different dimensionality, the problem at the core of the FG model is intrinsically more complex than the PCM one. Nonetheless, this choice allowed Fattebert and Gygi to provide an elegant and compact variational derivation of the continuum model, formulated in terms of a physically-sound free energy functional. Moreover, the choice was motivated by practical numerical reasons linked to the computational environment for which the model was developed, namely condensed-matter DFT calculations in periodic boundary conditions (PBC). The end result is a continuum dielectric model that is perfectly suited for periodic systems and allows the seamless coupling with first-principles molecular dynamics simulations.

Extensions of the work of Fattebert and Gygi have led to more advanced models, including accurate parameterization and tuning of electrostatic and non-electrostatic interactions [9, 10], testing and validation on charged systems [11], proper handling of PBC artifacts [12], characterization of solvation effects on the response properties of the system [13], etc. While most of these developments follow the path of similar progresses in the PCM and quantum-chemistry literature, more recent developments have started to explore and improve the capabilities of continuum models for electrochemical systems. In particular, some of the recent developments have involved a reparameterization of continuum models for noble metal surfaces [14], the use of continuum approaches to model the electrochemical diffuse layer [15], the possibility of using hybrid implicit/explicit approaches to improve the description of specific interaction on surfaces [16], the need to overcome artifacts of continuum models when dealing with complex multicomponent systems [17].

In the following we will review the main ingredients of continuum models of solvation, focusing on smooth-interfaces approaches developed in the field of condensed-matter simulation packages. Following the description of the main details of the models designed for neutral solutions, a review of the recent hierarchy of approaches for the description of the electrochemical diffuse layer will be provided. Finally, applications of continuum models of solvation to electrochemistry and electro-catalysis will be reviewed. In particular, the possibility of using multiscale approaches to estimate the range of validity and overcome the main limitation of the Computational Hydrogen Electrode will be reported.

3.2 Continuum models of liquid solutions

The basic assumption of continuum models of solvation is that solvent molecules, when statistically averaged, are homogeneous everywhere around the solvated system, as reported schematically in Fig. 3.1. Their effects on the properties of the solute can thus be modelled in a mean-field kind of picture as the interaction with a continuum medium. Due to the ergodic nature of most liquid systems, the continuum assumption holds when no

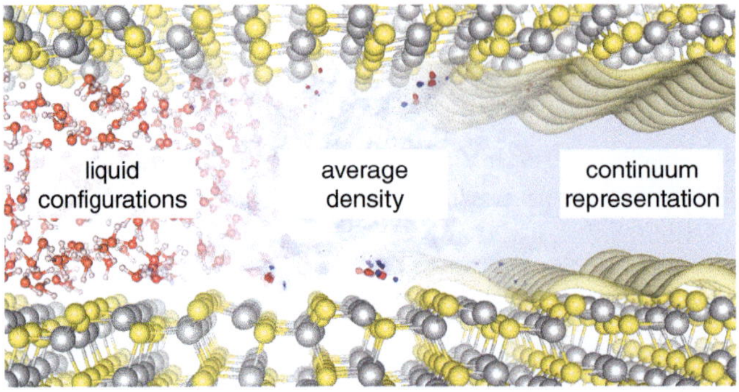

Figure 3.1 Illustration of the continuum approximation. On the left a single configurations of a liquid solution (consisting of explicit water molecules) near a QM system (CdS substrate) is reported. In the middle, the averaged distribution of solvent molecules is reported, with red/blue isosurfaces showing probability distributions higher/lower than the average. Even though these points are indications of specific long-lasting solvent-solute interactions, the continuum approximation (right part of the illustration) seems to be a reasonable assumption for most of the liquid environment in this system.

long-lasting interactions between the solute and the solvent exist or when the structure of the solvent at the interface with the solute is sufficiently mobile. Failures of the continuum assumption are usually the result of specific interactions, such as the formation of hydrogen bonds, ionic bonds, and covalent bonds between the solute and the solvent. The continuum assumption can also show limited validity in the presence of overstructured solvation layers, that may form at low temperature around charged systems as well as apolar molecules (hydrophobic solvation [18]). Similarly, solvent layers on top of solid substrates may present more rigid and/or long-lasting structures, due to the presence of specific interactions and a reduced conformational freedom at the interface. While these effects may hinder a successful application of continuum approaches, hybrid approaches that feature a reduced number of explicit solvent molecules embedded in otherwise implicit environments have been shown to provide accurate results [16, 19, 20] and represent the most promising strategy to model solvation effects in a general and unbiased fashion.

3.2.1 Continuum interfaces

A key component of any continuum embedding approach is the definition of the boundary between the embedded system and its surrounding environment. In principlesthis boundary should reflect the region of space occupied by the degrees of freedom of the environment, i.e. for liquid solutions it should correspond to the solvent-accessible volume of the simulation cell. However, since continuum models fully integrate out the degrees of freedom of the environment, the definition of the interface can only rely on the knowledge of the system's degrees of freedom. In general, embedding interfaces can be sharp or smooth, giving rise to 2D or 3D domains, respectively. Moreover, interfaces can be defined on the ionic degrees of freedom, on the electronic density of the embedded system, or on both.

Formally, the interface function can be expressed as

$$s(\mathbf{r}) = \begin{cases} 1 & \mathbf{r} \in \text{system} \\ 0 & \mathbf{r} \notin \text{system}. \end{cases} \tag{3.1}$$

Historically, continuum models have relied on interfaces defined on the positions of the atoms, $\{\mathbf{R}_a\}$, composing the quantum-mechanical system of interest [3, 4]:

$$s(\mathbf{r}) \equiv s(\mathbf{r}; \{\mathbf{R}_a\}). \tag{3.2}$$

In particular, PCM exploits a cavity built from atom-centered interlocking spheres [21–24], where each sphere has a radius proportional to the van der Waals radius of the elements, $\{R_a^{vdW}\}$, scaled by a homogenous factor, α. While PCM and similar approaches in quantum-chemistry adopt sharp interface functions, an analogous soft-sphere continuum solvation (SSCS) model was recently introduced by Fisicaro et al. [25] in terms of the product of interlocking smooth spherical functions:

$$s(\mathbf{r}) = 1 - \prod_i h_i(|\mathbf{r} - \mathbf{R}_i|), \tag{3.3}$$

where the spherical functions are expressed in terms of error functions as

$$h_i(\mathbf{r}) = \frac{1}{2}\left[1 + \text{erf}\left(\frac{r - \alpha R_i^{vdW}}{\Delta}\right)\right] \tag{3.4}$$

and Δ is the additional smoothing parameter.

The main advantage of using interface functions based on ionic degrees of freedom is the fact that the interface is uniquely defined at the beginning of the electronic optimization algorithm (e.g. the self-consistent field – SCF – cycle): the interface contribution to the total energy does not depend explicitly on the electronic density, so it does not contribute to the potential acting on the electrons (e.g. the Kohn–Sham potential) and it can usually be computed with a reduced computational effort. Nonetheless, the explicit dependence of the model on ionic positions implies that the interface contribution to the total energy will affect the calculation of inter-atomic forces. Thus, numerical inaccuracies in the calculation of these derivatives may result in inaccurate forces, which may pose serious problems when performing geometry optimization and, even more critical, may lead to poor energy conservation in micro-canonical molecular dynamics simulations.

An additional flexibility is present in models based on atom-centered spheres, in that the spheres' radii can be adjusted to improve the accuracy of the calculations. In particular, by appropriately scaling the solvation radius of individual elements, when present in specific functional groups, continuum models can somehow recover part of the errors associated to the continuum approximation and the neglect of specific interactions. For example, shrinking the radius of oxygen atoms in acidic carboxylic groups allows their solvation energy to be increased, thus accounting for the larger propensity of these atoms to form hydrogen bonds with protic solvents. This additional flexibility may be useful for modelling charged systems, where anions have been shown to require a continuum interface that lies much closer to the solvated system [11]. However, while improved accuracies are usually reported for this kind of approach [26], these improvements may come at the expense of transferability.

Moreover, when used to treat charged systems, enforcing smaller or larger solvation radii requires the *a priori* knowledge of the charge state of each individual components of the system, which may be far from trivial in complex electrochemical setups.

As an alternative to ionic interfaces, interface functions defined on the electronic density, ρ^{el}, of the embedded system have been proposed:

$$s(\mathbf{r}) \equiv s(\rho^{el}(\mathbf{r})). \tag{3.5}$$

In particular, sharp-interfaces defined on frozen or self-consistent isosurfaces of the electronic density have been proposed in the PCM framework [27], but found limited application, possibly due to the increased complexity in the numerical discretization of the resulting 2D domains. These limitations are not present in the context of smooth-interface continuum solvation approaches, where Fattebert and Gygi first proposed [7, 8] an interface function in terms of the electronic density as

$$s(\mathbf{r}) = \frac{1}{2}\left(1 + \frac{1 - (\rho^{el}(\mathbf{r})/\rho_0)^{2\beta}}{1 + (\rho^{el}(\mathbf{r})/\rho_0)^{2\beta}}\right), \tag{3.6}$$

where ρ_0 and β are the parameters of the model. With respect to ionic interfaces, electronic interfaces do not involve empirical solvation radii, but only rely on a limited number of parameters which are meant to identify an isosurface of the electronic density at the outskirts of the quantum-mechanical system. Thus, the use of electronic interfaces can significantly reduce the number of parameters required by the model, improving its generality and transferability.

Moreover, as the interface function follows the electronic density of the system, its shape gets optimized along the SCF cycle, thus readjusting automatically with the charge distribution in the system. However, for negatively charged systems, this automatic readjustment goes in the nonphysical direction of an expansion of the interface away from the solute, instead of a shrinking [11]. In addition, the flexibility of electronic interfaces comes at the additional computational cost of generating a new interface function and solving a new numerical problem at each SCF step. Contrary to ionic interfaces, the contribution of the interface energy to the electronic potential needs to be explicitly included in the simulation, while the contribution to the interatomic forces vanishes, making these approaches more suitable for molecular dynamics simulations [7, 8].

Different functional forms for the interface in terms of the electronic density can be adopted. The optimal definition should satisfy the following conditions:

1. The interface function should smoothly assume a uniform value of 1 inside the quantum mechanical system. Variations of the interface function inside the quantum-mechanical system may give rise to unphysical pockets of embedding environment in a region close to the ions and the chemical bonds and, thus, very sensitive to small perturbations.
2. The interface function should smoothly assume a value of 0 in the region far from the quantum mechanical system. While less stringent for the convergence of the methods, small variations of the interface function in the bulk of the environment are also non-physical and may interact and promote unphysical fluctuations of the electronic density far from the nuclei.
3. The transition between the two uniform regimes described above should be as smooth as possible, so that the typical resolution of the structured 3D grids used in condensed

matter simulations is fine enough to describe it correctly. This condition is particularly relevant as the interface function is defined in a region of space, at the outskirts of the quantum mechanical system, where the electronic density is known to decay exponentially fast.

Based on the above conditions, Andreussi and coworkers proposed a revised definition of the interface function [10], which exploits a step-wise expression and a switching function in terms of the logarithm of the electronic density:

$$s(\mathbf{r}) = \begin{cases} 1 & \rho^{el}(\mathbf{r}) > \rho_{max} \\ t(\ln(\rho^{el}(\mathbf{r}))) & \rho_{max} > \rho^{el}(\mathbf{r}) > \rho_{min} \\ 0 & \rho^{el}(\mathbf{r}) < \rho_{min} \end{cases} \tag{3.7}$$

where $t(x)$ is a smooth function. This contribution was later adopted by several other models, e.g. [28–31]. Similarly to the expression introduced by Fattebert and Gygi, this function depends on only two adjustable parameters, ρ_{min} and ρ_{max}, which correspond to the isosurfaces of the electronic density where the transition from the embedded region to the embedding one occurs. Higher values of the two thresholds correspond to the interface getting closer to the quantum mechanical system and, thus, to increasing the strength of solvation effects. In addition, thresholds of similar values correspond to sharper transitions, which in turn increase the interaction with the embedding environment.

Calculations on a comprehensive set of neutral organic molecules show that variations in the two parameters are strongly correlated, so that different combinations of the two can provide similar agreement with other simulations and/or the experiments [10]. One can thus tune a single global parameter to control the shape of the interface, as compared to the extensive set of solvation radii required in ionic interfaces. The definition and parametrization of the above function represent the key ingredients of the self-consistent continuum solvation (SCCS) approach [10, 11, 17]. As discussed above, while a single parametrization provides consistent results for neutral species (see Fig. 3.2), in order to

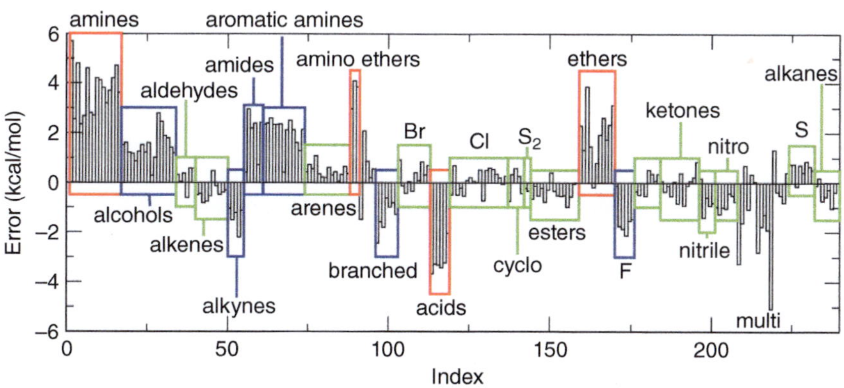

Figure 3.2 Errors in aqueous solvation free energies for a set of 240 neutral organic molecules, as computed with the SCCS model and compared to experimental values. Several compounds (reported in the green boxes) show errors lower than 0.5 kcal/mol, well within chemical accuracy. Instead, larger errors are observed for compounds (red boxes), such as acids and bases, that are known to form strong hydrogen-bond interactions with water molecules.

Table 3.1 Cavity parameterizations proposed for the SCCS and the SSCS interfaces, with the corresponding non-electrostatic parameters. The parametrizations labeled with a (*) refer to the SCCS non-electrostatic re-parametrizations proposed for an interface function as in Eq. 3.7, while the original SCCS cavity [10] was defined as a linear function of $\epsilon(\mathbf{r})$. Note that two parametrizations have been proposed for neutral isolated systems for each interface type, including and excluding the non-electrostatic parameter β.

Training set	Interface	Electrostatic parameters	$\alpha + \gamma$ (dyn cm^{-1})	β (GPa)	Ref.
Neutral	SCCS	$\rho_{max} = 0.00500$ a.u., $\rho_{min} = 0.0001$ a.u.	50	−0.350	[10]
Neutral	SCCS	$\rho_{max} = 0.00500$ a.u., $\rho_{min} = 0.0001$ a.u.	11.5	–	[10]
Neutral	SCCS(*)	$\rho_{max} = 0.00500$ a.u., $\rho_{min} = 0.0001$ a.u.	47.9	−0.360	[17]
Neutral	SCCS(*)	$\rho_{max} = 0.00500$ a.u., $\rho_{min} = 0.0001$ a.u.	12.4	–	[17]
Neutral	SSCS	$\alpha = 1.12$, $\Delta = 0.5$ Bohr	50	−0.350	[25]
Neutral	SSCS	$\alpha = 1.16$, $\Delta = 0.5$ Bohr	11.5	–	[25]
Cations	SCCS	$\rho_{max} = 0.00350$ a.u., $\rho_{min} = 0.0002$ a.u.	5	0.125	[11]
Cations	SSCS	$\alpha = 1.10$, $\Delta = 0.5$ Bohr	50	−0.350	[25]
Anions	SCCS	$\rho_{max} = 0.01550$ a.u., $\rho_{min} = 0.0024$ a.u.	0	0.450	[11]
Anions	SSCS	$\alpha = 0.98$, $\Delta = 0.5$ Bohr	50	−0.350	[25]
Metal surfs.	SCCS	$\rho_{max} = 0.01025$ a.u., $\rho_{min} = 0.0013$ a.u.	–	–	[32]

increase the accuracy of the method on systems with strong specific interactions and/or in the presence of overstructured solvation layers, a reparametrization of the interface function can be required [32]. In Table 3.1 we report the main parameterizations proposed in the literature for the SCCS and SSCS models. In Fig. 3.3 we report an example of the two interfaces for the case of a platinum slab with an absorbed hydroxy group.

In particular, Dupont et al. showed that, while cations are well reproduced by the parameters optimized for the neutral systems, anions show severe discrepancies and require a separate parametrization [11]. Similarly, when studying electrochemical systems with continuum solvation, Hörmann et al. showed that an alternative set of parameters for the interface function is able to recover a reasonable agreement with experimental values for the potential of zero charge of noble metal surfaces [32]. The parameters proposed by Hörmann et al. are also able to reproduce the correct magnitude of the differential capacitance of silver surfaces [15], thus further supporting its accuracy for the study of electrochemical interfaces. However, the lack of transferability of the interface thresholds can lead to unreliable simulations, especially when multiple components with different parametrizations need to be treated at the same time. As the main cause of the problem is linked to overstructured solvent molecules or solvation layers, hybrid approaches hold the potential of providing a unifying and accurate approach to complex systems [16]. However, as the computational cost of these methods grows exponentially fast with the number of molecules included in the quantum-mechanical system, alternative definitions of the interface function and/or of the interaction energy can be devised, so as to increase the applicability of purely continuum approaches.

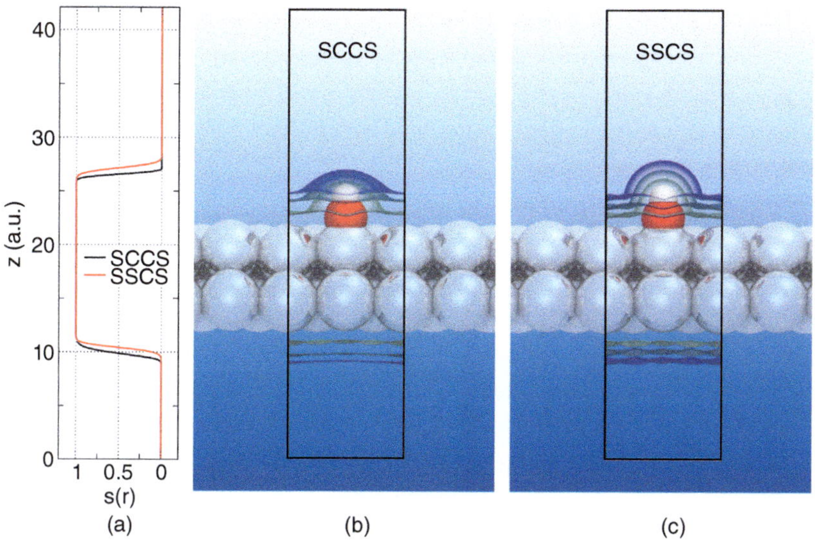

Figure 3.3 Examples of electronic (SCCS, in b) and ionic (SSCS, in c) continuum interfaces embedding a Pt (111) slab with an absorbed hydroxy group. Isosurfaces of the two interface functions corresponding to values of 0, 0.5, and 1 are visualized in panels b) and c). In a) the behaviors of the two interface functions are reported along the line passing through the hydroxy group and as a function of the position in the direction perpendicular to the slab.

3.2.2 Beyond local interfaces

As described above, different definitions of continuum interfaces share similar advantages and limitations. In general, when properly parametrized on the same set of data, similar accuracies have been observed for different formulations [25]. However, both ionic and electronic interfaces fail to correctly reproduce the effects of changes in the charge state of the system. This has a profound implication for complex systems composed by different components, where the electrons can transfer from one part of the system to another. Moreover, when performing electrochemical simulations, the charge state of the system represents a key external variable, as it is connected by a Legendre transformation to the applied electrochemical potential. Thus, being able to correctly reproduce changes in solvation effects with applied potential is crucial for the accuracy of continuum models. In order to overcome the problem, non-local interfaces that depend on both the ionic and electronic degrees of freedom have been proposed [33, 34]. In particular, a scaling factor proportional to the local net charge embedded by the interface can be introduced. Such a net charge can be estimated by exploiting Gauss's law in terms of the normal component of the electrostatic field $\mathbf{E}$ of the system at the electronic interface [33]

$$E_n(\mathbf{r}) = \mathbf{E}(\mathbf{r}) \cdot \nabla s(\mathbf{r}) \tag{3.8}$$

or from its fluxes through the spherical functions composing the ionic interface [34]

$$\Phi^a = \int \mathbf{E}(\mathbf{r}) \cdot \nabla h_a(|\mathbf{r} - \mathbf{R}_a|) \prod_{b \neq a} h_b(|\mathbf{r} - \mathbf{R}_b|) d\mathbf{r}. \tag{3.9}$$

Since the electrostatic field in vacuum is a non-local functional of the total charge density of the system,

$$\mathbf{E}(\mathbf{r}) = \int \mathbf{K}(\mathbf{r} - \mathbf{r}')\rho^{sys}(\mathbf{r})d\mathbf{r}', \tag{3.10}$$

where $\mathbf{K}$ is a generic electrostatic kernel that depends on the boundary conditions,

$$\rho^{sys}(\mathbf{r}) = \left(\rho^{el}(\mathbf{r}) + \sum_i z_i \delta(|\mathbf{r} - \mathbf{R}_i|)\right), \tag{3.11}$$

and z_i are the ionic charges, these field-aware interfaces non-trivially depend on both the electronic density and the ionic positions of the solute. The asymmetry in solvation of positive and negative compounds is recovered by using an asymmetric scaling function that depends on the sign of the field normal or field flux. This strategy has shown to provide accurate results for neutral species, cations and anions, while relying on a single parametrization. However, further improvements in accuracy may be obtained if not only the shape of the interface, but also the interaction energy between the solute and the continuum is defined in terms of the local polarity or charge state of the system, as proposed by Chipman and coworkers [35] or in the approach of Bardhan and coworkers [36–38].

A second limitation of straightforward definitions of ionic and electronic interfaces is that small regions of space not occupied by the embedded system may be assigned to the continuum environment, despite being physically inaccessible to solvent molecules [17]. While this is usually not a problem for small and compact systems, most local definitions of interfaces present some limitations when complex systems are considered. A typical example is the one of molecular complexes and absorption, where small interstices are present between different molecules or between molecules and substrates. An even more common occurrence of the problem is found in condensed-matter simulations, when studying substrates with open or porous structures, such as nanotubes, semiconductors, zeolites, etc.

While simplified approaches, e.g. the introduction of rigid exclusion regions, allow the continuum embedding from entering these holes to be avoided, these do not usually provide smooth derivatives. Thus, geometry optimizations and molecular dynamics simulations may suffer from artifacts at the interface, due to the presence of external constraints. Similarly, the concepts of solvent-accessible and solvent-excluded interfaces have been proposed in the quantum-chemistry literature [22, 39–41], but the algorithms adopted may present numerical instabilities and are not straightforward to extend to smooth interfaces.

Recently the use of non-local definitions of the interface, which exploits convolutions with spherical functions [33, 42] to sample larger regions of space surrounding the point where the interface is computed, were shown to handle the problem effectively. The use of convolutions, especially in simulations that already make large use of fast Fourier transforms, allows the computational cost of the algorithm to be kept under control, while maintaining continuous derivatives with respect to the underlying degrees of freedom (atomic or electronic) adopted in the construction. Following this strategy, Andreussi and coworkers proposed a general solvent-aware [17] definition $\hat{s}$ of continuum interfaces by complementing a local interface s with a smooth switching function of the filled fraction, i.e. the fraction of volume occupied by the continuum within a spherical region of space:

$$\hat{s}(\mathbf{r}) = s(\mathbf{r}) + (1 - s(\mathbf{r}))t(f^{ff}(\mathbf{r})). \tag{3.12}$$

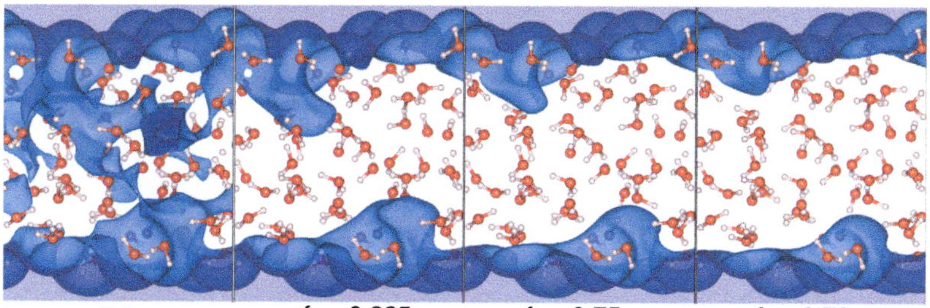

Figure 3.4 Solvent-aware continuum interfaces in contact with a slab of liquid water. By increasing the effects of the non-local correction (from left to right, i.e. by decreasing the threshold on the filled fraction f_0) spurious regions of continuum environment, smaller than the solvent-radius, are effectively removed from the simulation.

The filled fraction, f^{ff}, is obtained by a convolution of the local interface with a spherical probe $u_{R_{solv}}$ whose size is proportional to the solvent radius

$$f^{ff}[s(\mathbf{r}')](\mathbf{r}) \equiv \int s(\mathbf{r}')u_{R_{solv}}(|\mathbf{r}-\mathbf{r}'|)d\mathbf{r}'. \tag{3.13}$$

The approach provides analytic and smoothly varying derivatives, allowing its use in accurate geometry optimizations and molecular dynamics simulations. Even though the solvent-awareness introduces a number of new parameters, none of them requires the use of empirical fitting of experimental data. While simple geometrical assumptions allow a model to be defined that is able to overcome the presence of small continuum artifacts, the main model parameter can be converged exploiting bulk calculations. Simulations on realistic and complex systems, such as liquid water (as reported in Fig. 3.4) and semiconductor slabs in contact with continuum water, showed the effectiveness of the approach and allowed a more general value of the threshold parameter for condensed-matter applications to be proposed, featuring open and liquid/amorphous components [17].

3.2.3 Electrostatic interaction: polarizable dielectric embedding

In addition to the definition of the interface, the main physical ingredient of continuum approaches is represented by the way the embedded system interacts with the environment. A full QM picture of this interaction should be based on the electrostatic interaction between the electrons and ions of the two components. When the environment degrees of freedom are integrated into a continuum, approximations must be made on the form of these interactions. These approximations usually provide a definition of the energy of interaction with the environment as a functional of the continuum interface. As a result, the total energy of the quantum-mechanical system can usually be expressed as the sum of two terms, the first only related to the system's degrees of freedom and with an expression equivalent to the energy of the system in vacuum, and the second that instead explicitly depends on the interface function:

$$E^{tot}[\rho^{el}, \{\mathbf{R}_a\}] = E^{system}[\rho^{el}, \{\mathbf{R}_a\}] + E^{interface}[s(\rho^{el}, \{\mathbf{R}_a\})]. \tag{3.14}$$

The strength of the interaction terms is controlled by tunable parameters, which may be related to a physically meaningful and well-characterized property of the continuum (e.g., its dielectric permittivity or surface tension) or can be fitted to improve the accuracy of more complex model.

Electrostatic interactions play a crucial role in most embedding environments: as the embedded system interacts with the solvent molecules, the flexible and mobile environment will readjust and polarize, so as to stabilize the charge distribution of the system. In classical electrostatics, the ability of the solvent molecules to screen embedded charges is described in a compact way in terms of the static dielectric permittivity of the solvent, ϵ_0. The picture of a classical polarizable dielectric medium is at the core of most of the continuum models of solvation, starting from the original approach of Onsager [43], where electrostatic solvation effects are described in terms of a reaction field, induced by the embedded system on the environment and acting on the embedded system degrees of freedom. The presence of this additional field reduces the electrostatic energy of the system, but the work spent to generate the field needs to be accounted for in computing the total free energy of solvation. The same physical principles behind the model of Onsager apply to more complex dielectric embedding schemes [3]. In most of the proposed approaches, as the work done to polarize the medium is removed from the electrostatic energy of the system, the total quantum mechanical energy of the system embedded in a dielectric medium is labeled as an electrostatic free energy [3]. It is also worth pointing out that, in the context of continuum calculations, both Helmholtz F and Gibbs G free energies are used, owing to the fact that the continuum may represent both a constant volume or constant pressure environment.

In particular, Fattebert and Gygi [7, 8] recast this electrostatic picture in terms of three-dimensional fields, potentials and charge distributions, with the total electrostatic free energy of the embedded system expressed as

$$F[\rho^{el}, \{\mathbf{R}_i\}] = \int \rho^{sys}(\mathbf{r})\phi(\mathbf{r})d\mathbf{r} - \int \frac{1}{8\pi}\epsilon(\mathbf{r})|\nabla\phi(\mathbf{r})|^2 d\mathbf{r}, \tag{3.15}$$

where $\phi(\mathbf{r})$ is the electrostatic potential, and the dielectric permittivity of the medium $\epsilon(\mathbf{r})$ reflects the system/continuum separation and is thus a function of the continuum interface, e.g.

$$\epsilon(\mathbf{r}) \equiv \epsilon(s(\mathbf{r})) = 1 + (\epsilon_0 - 1)(1 - s(\mathbf{r})), \tag{3.16}$$

with ϵ_0 the bulk dielectric permittivity of the environment.

By taking the functional derivative of the above expression with respect to the electrostatic potential, the generalized Poisson equation (GPE) in a dielectric medium is obtained

$$\nabla \cdot \epsilon(\mathbf{r})\nabla\phi(\mathbf{r}) = -4\pi\rho^{sys}(\mathbf{r}). \tag{3.17}$$

The difference between the electrostatic potential obtained from Eq. ((3.17)) and the solution of the Poisson equation (PE) in vacuum,

$$\nabla^2\phi^0(\mathbf{r}) = -4\pi\rho^{sys}(\mathbf{r}) \tag{3.18}$$

provides the polarization potential

$$\phi^{pol}(\mathbf{r}) = \phi(\mathbf{r}) - \phi^0(\mathbf{r}). \tag{3.19}$$

Alternatively, the effects of the dielectric embedding can be rationalized in terms of an auxiliary polarization density, spread at the continuum interface and expressed as [10]

$$\rho^{pol}(\mathbf{r}) = \frac{1}{4\pi}\nabla \ln \epsilon(\mathbf{r}) \cdot \nabla\phi(\mathbf{r}) - \frac{\epsilon(\mathbf{r}) - 1}{\epsilon(\mathbf{r})}\rho^{sys}(\mathbf{r}). \tag{3.20}$$

As this auxiliary density depends on the gradient of the logarithm of the dielectric function, Andreussi et al. proposed an alternative formulation of the dielectric in terms of the continuum interface [10], in order to provide a more robust expression for the polarization density, namely

$$\epsilon(\mathbf{r}) = e^{\log \epsilon_0[1-s(\mathbf{r})]}. \tag{3.21}$$

In order to compute the polarization potential, or alternatively, the polarization charge, a numerical procedure to solve the GPE needs to be devised. While the original approach of Fattebert and Gigy relied on a multigrid solver [7], iterative schemes based on the use fast Fourier transforms have been proposed [10, 44], which provide fast and robust solutions to the partial differential equation, with a negligible overhead with respect to simulations in vacuum. When optimizing the degrees of freedom of the embedded system, the functional derivatives of the energy with respect to the electronic density, $V_{KS} \equiv \delta F[\rho^{el}, \{\mathbf{R}_i\}]/\delta\rho^{el}$, or the partial derivatives of the energy with respect to atomic positions, $\mathbf{f}_i \equiv \partial F[\rho^{el}, \{\mathbf{R}_i\}]/\partial\mathbf{R}_i$, need to be computed. Depending on whether the continuum interface is defined in terms of the electronic or atomic degrees of freedom, explicit contributions to either the Kohn–Sham potential V^{el}_{KS} or to the inter-atomic forces $\mathbf{f}_i$ will be present. By introducing the functional derivative of the free energy with respect to the interface function, $\delta F[\rho^{el}, \{\mathbf{R}_i\}]/\delta s$, both contributions can be expressed in a unified framework, namely

$$V^{interface}_{KS}(\mathbf{r}) = \int \frac{\delta s(\mathbf{r}')}{\delta\rho^{el}(\mathbf{r})}\frac{\delta F[s]}{\delta s(\mathbf{r}')}\mathbf{d}\mathbf{r}' \tag{3.22}$$

$$\mathbf{f}^{interface}_i = -\int \frac{\partial s(\mathbf{r})}{\partial \mathbf{R}_a}\frac{\delta F[s]}{\delta s(\mathbf{r})}\mathbf{d}\mathbf{r}, \tag{3.23}$$

where the last ingredients required, namely the functional derivative of $s(\mathbf{r})$ with respect to the electronic density and its partial derivative with respect to the ionic positions, will depend on the chosen form of the interface function.

3.2.4 Beyond electrostatic interactions

While electrostatic solute-solvent interactions are well characterized by the empirical model of a polarizable dielectric medium, other interactions are more ambiguous and their definition is bound to rely on arbitrary partitions. The most widespread approach is the one of Ben-Naim [45, 46] where the global solvation free energy of a solute is decomposed as

$$\Delta G^{sol} = \Delta G^{el} + G^{cav} + G^{rep} + G^{dis} + P\Delta V + \Delta G^{tm}. \tag{3.24}$$

The electrostatic contribution ΔG^{el} can be computed as the electrostatic free energy difference between the system relaxed in the polarizable dielectric environment and the system optimized in vacuum. The non-electrostatic terms, instead, comprise interaction energies that are of more statistical mechanics or quantum-mechanical nature. In particular, the cavitation energy G^{cav} is associated to the energy required to form a cavity in the embedding medium to host the solute. Repulsion, G^{rep}, and dispersion, G^{dis}, interactions are instead related to quantum-mechanical phenomena linked to the electronic repulsion due to Pauli principle or the electronic dispersion due to non-local correlations effects. Changes in thermal motion, ΔG^{tm} due to the embedding environment and the effect of the external pressure, $P\Delta V$, are usually regarded as less important for solvation of molecules in standard conditions [3].

However, volume changes can play an important role in specific applications, such as the characterization of finite clusters under the effect of an external pressurizing medium. For these applications, Cococcioni et al. developed a continuum medium approach [47] where a volume dependent term is included in the total energy to define a quantum-mechanical enthalpy functional

$$H = E + PV_q \tag{3.25}$$

and the quantum-volume V_q of the system is expressed as a functional of the embedding interface as

$$V_q \equiv V[s(\mathbf{r})] = \int s(\mathbf{r})d\mathbf{r}. \tag{3.26}$$

Statistical mechanics approaches, such as the scaled particle theory [3, 48], are usually exploited to account for cavitation contributions. Alternatively, Scherlis et al. proposed an approach based on a microscopic extension of the surface tension model [9], where, following the strategy of Cococcioni et al. [47], the free energy cost to create a molecular sized cavity is expressed as proportional to the solute's surface

$$G^{cav} = \gamma S_q, \tag{3.27}$$

where the quantum-surface S_q of the system can also be expressed as a function of the embedding interface as

$$S_q \equiv S_q[s(\mathbf{r})] = \int |\nabla s(\mathbf{r})|d\mathbf{r}. \tag{3.28}$$

In the context of continuum solvation in quantum-chemistry, Amovilli and Mennucci proposed a strategy based on the theory of McWeeny and on dynamic polarizabilities to model dispersion effect [4, 49, 50]. Repulsion effects were instead proposed by the same authors to be proportional to the amount of electronic density that lies outside of the sharp PCM interface [4, 49]. Alternatively, approaches based on fragment contributions proportional to the surface of the continuum interface belonging to each atomic type have been adopted [5, 51].

Possible ways to describe these purely QM effects are to follow the same strategies that are used to model intra-molecular van der Waals interactions, but extending these schemes to homogeneous continuum media. In the context of smooth-interface models based on the electronic density, the escaped charge can be considered to be proportional to the exposed surface of the embedded system. Thus, the repulsion contribution can be defined to have the same form as the surface tension one [10]. Starting from this assumption, Andreussi et al. proposed a compact model for aqueous solutions based on the SCCS cavity and involving two non-electrostatic contributions to the total solvation free energy, expressed through two global parameters that multiply the quantum-surface and the quantum-volume of the system, namely

$$\Delta G^{sol} = \Delta G^{el} + (\alpha + \gamma)S_q + \beta V_q. \tag{3.29}$$

For interfaces, the slab geometry typical of condensed-matter simulations introduces an approximation that reduces the validity of the volume contribution, thus a single non-electrostatic functional proportional to the surface is exploited (see e.g. Table 3.1). The same strategy was later extended to ionic interfaces in the SSCS model [25].

Still in the context of smooth-interface solvation models, Sundararaman et al. extended the empirical approach of Grimme to include vdW effects into DFT, to include the interaction with a continuum distribution of electrons in the embedding environment [52]. Jinnouchi and Anderson proposed an approach [53] based on surface fragments similar to the one exploited by the Minnesota family of solvation models.

Despite their crucial role to provide a reliable comparison with the experiments, in most condensed-matter applications, non-electrostatic effects are usually neglected, due to their less general definition and parametrization. Efforts to extend the parameterization to a wider range of solvents are currently ongoing and may lead to more general definitions.

3.3 Continuum diffuse-layer models

3.3.1 Continuum models of electrolytes

The implicit solvation framework described so far can be extended to account for the presence of ions in the embedding solution. In particular, a continuum description of the electrolyte can be obtained by introducing smooth functions $\{c_i(\mathbf{r})\}$ that define the local concentrations of the various solvated ionic species. These concentration functions are expected to approach the corresponding bulk electrolyte concentrations $\{c_i^0\}$ deep in the bulk solution. While the bulk electrolyte solution is expected to be charge neutral (i.e. $\sum_i z_i c_i^0 = 0$, where z_i is the charge of the i-th electrolyte) in proximity of the solute region the interaction with the quantum-mechanical system can drive the accumulation or the depletion of some of the ionic species. Thus, this region is characterized by a local unbalance between positive and negative ions, which can be described in terms of a classical charge distribution $\rho_{\text{ions}}(\mathbf{r}) = \sum_i z_i c_i(\mathbf{r})$.

A common element to all the hybrid atomistic-continuum models of electrolyte solutions is the interface that defines the volume of space that can be occupied by the solvated ions. One can formally introduce this boundary using an ion-exclusion function $\gamma(\mathbf{r})$ that smoothly switches between the values of 0 and 1 in the electrolyte-free and electrolyte-accessible region, respectively. In particular, the electrolyte domain can be defined using the interface function $s(\mathbf{r})$, analogous to the one introduced in Section 3.2.1:

$$\gamma(\mathbf{r}) = 1 - s(\mathbf{r}). \tag{3.30}$$

Continuum models of electrolytes are particularly relevant in the context of electrochemical interfaces. Solvated ions play a dominant role in screening the electrode charge, and the structure of the charge distribution $\rho_{\text{ions}}(\mathbf{r})$ that emerges in the liquid solution, also known as the diffuse layer, affects many interfacial properties. In addition to very simple double layer models consisting of a countercharge plane positioned at a fixed distance from the model electrode surface [54–56], other popular electrolyte models have been based on the solution of some form of the Poisson–Boltzmann equation (PBE) [57, 58]. This second class of models can be formally introduced within a grand-canonical framework [57, 59], where the system constituted by the simulation cell is assumed to be in thermodynamic equilibrium with a reservoir of electrolyte species, i.e. the bulk solution. In this picture, for given values of the chemical potentials of the ions, one looks for the equilibrium electrolyte concentrations that minimize the system grand potential function. This quantity is typically expressed as a functional of the electrostatic potential of the embedded system $\phi(\mathbf{r})$ so that ultimately accounting for the presence of electrolytes in the embedding solution becomes equivalent to solving a modified electrostatic problem.

In this section, we will describe the most popular diffuse-layer models that have been implemented in periodic DFT codes [58, 60–69]. These models essentially differ in the type of interactions that are considered in the free-energy functional of the embedded system. We will review how these models can be formally derived and we will show how they perform for a representative observable that is sensitive to the diffuse-layer structure, i.e. the differential capacitance of single-crystal metal surfaces.

3.3.2 Helmholtz double-layer model

A charged electrode in an electrolyte solution electrostatically attracts counter-ions from the bulk solution. To first approximation, one can assume that the solvated ions would accumulate in a plane parallel to the electrode surface, giving rise to the so-called double-layer structure. Under this approximation, the diffuse layer can be represented as a planar charge distribution, which is generally known as the outer Helmholtz plane [70]. Such Helmholtz model of the double layer assumes the electrolyte screening of the surface charge to be independent on the bulk ionic concentration and to take place in an arbitrarily narrow volume of the liquid solution.

The planar counter-charge model has been employed for a long time in combination with electronic structure calculations [54–56]. In typical computational implementations, the counter-charge plane is broadened along the surface normal direction (here chosen to lie along the x axis) to have a finite width. For instance, the Helmholtz layer can be modelled as a planar Gaussian distribution:

$$\rho_{ions}(x) = \frac{q}{A\sqrt{\pi}\Delta} \exp\left(-\frac{(x - x_0)^2}{\Delta^2}\right). \tag{3.31}$$

Here q is the integrated charge distribution, A is the cell area in the surface plane and x_0 and Δ are the two model parameters that define the distribution's center and width, respectively. Among these parameters, only the former significantly affects interface properties such as the differential capacitance, while the latter can be treated as a numerical parameter with the purpose of guaranteeing an accurate representation of the counter-charge distribution on the real-space grid[15]. The inclusion of $\rho_{ions}(x)$ to the charge distributions of the system modifies the electrostatic energy of the system. The overall system's free-energy can be computed as:

$$F[\rho(\mathbf{r}), \phi(\mathbf{r})] = \int \left[-\frac{\varepsilon(\mathbf{r})}{8\pi} |\nabla \phi(\mathbf{r})|^2 + \rho(\mathbf{r})\phi(\mathbf{r}) + \rho^{ions}(\mathbf{r})\phi(\mathbf{r}) \right] d\mathbf{r}, \tag{3.32}$$

where the electrostatic potential $\phi(\mathbf{r})$ is the overall electrostatic potential of the system, and it thus includes the Helmholtz-plane contribution. Note that once the counter-charge plane is added to the charge distributions of the system, one can solve the GPE without additional iterative steps for the electrolyte.

3.3.3 Poisson–Boltzmann model

The Helmholtz double-layer model represents a rather simplified approximation of the diffuse layer. A more realistic diffuse-layer model can be derived from a functional that includes terms related to the bulk electrolyte reservoir [57, 63, 71]:

$$\begin{aligned} F[\rho(\mathbf{r}), \phi(\mathbf{r}), \{c_i(\mathbf{r})\}] = \int \Bigg[& -\frac{\varepsilon(\mathbf{r})}{8\pi} |\nabla \phi(\mathbf{r})|^2 + \rho(\mathbf{r})\phi(\mathbf{r}) + \rho^{ions}(\mathbf{r})\phi(\mathbf{r}) + \\ & - \sum_i \mu_i (c_i(\mathbf{r}) - c_i^0) - T(s(\{c_i(\mathbf{r})\}) - s(\{c_i^0\})) \Bigg] d\mathbf{r}, \end{aligned} \tag{3.33}$$

where T is the temperature, μ_i is the chemical potential of the i-th ionic species, and $s(\{c_i\})$ is the solution entropy density. Note that we have referred the chemical-potential and the entropy density terms to the corresponding values in the bulk solution. Also note that we have now made explicit the dependence of F on the electrolyte concentration functions $\{c_i(\mathbf{r})\}$, with respect to which F needs to be minimized to find the corresponding equilibrium concentrations.

The entropy density of a point-charge electrolyte, assuming ideal mixing, can be computed as:

$$s(\{c_i(\mathbf{r})\}) = -k_B \sum_i c_i(\mathbf{r}) \ln \frac{c_i(\mathbf{r})}{\gamma(\mathbf{r})}, \tag{3.34}$$

where k_B is the Boltzmann constant. This entropy density expression, together with the following expression for the chemical potential of the ions:

$$\mu_i = k_B T (\ln c_i^0 + 1), \tag{3.35}$$

allows the well-known Poisson–Boltzmann equation (PBE) to be derived, which is obtained by minimizing the free-energy functional in Eq. 3.33 first with respect to the local electrolyte concentration functions, and then with respect to the electrostatic potential:

$$\nabla \cdot \varepsilon(\mathbf{r}) \nabla \phi(\mathbf{r}) = -4\pi(\rho(\mathbf{r}) + \underbrace{\sum_i z_i c_i(\mathbf{r})}_{\rho_{ions}(\mathbf{r})}). \tag{3.36}$$

The PBE represents the modified electrostatic problem that needs to be solved instead of the Poisson equation for standard vacuum calculations. The local electrolyte concentration function takes the following form:

$$c_i(\mathbf{r}) = \gamma(\mathbf{r}) c_i^0 e^{-\frac{z_i \phi(\mathbf{r})}{k_B T}}. \tag{3.37}$$

In the context of electrochemical interfaces, the solution of the PBE leads to an electrolyte concentration profile also known as the Gouy[72]–Chapman[73] diffuse layer. Note that the presence of the ion-exclusion function $\gamma(\mathbf{r})$ ensures a zero electrolyte concentration inside the solute region. The solute region also gives a zero contribution to the solution entropy (see Eq. 3.34), consistently with the fact that this volume should be considered as external from the electrolyte solution.

Whenever the electrostatic interaction is small with respect to the thermal energy, the Boltzmann factor in Eq. 3.37 can be approximated with a linear function. Such approximation leads to the following expression for the electrolyte concentrations:

$$c_i(\mathbf{r}) \approx \gamma(\mathbf{r})c_i^0 \left(1 - \frac{z_i\phi(\mathbf{r})}{k_B T}\right). \tag{3.38}$$

Substitution in the Eq. 3.36 leads to a linearized-version of the PBE (LPBE), which can be written as:

$$\nabla \cdot \varepsilon(\mathbf{r})\nabla\phi(\mathbf{r}) = -4\pi\rho(\mathbf{r}) + k^2\gamma(\mathbf{r})\phi(\mathbf{r}), \tag{3.39}$$

where $k^2 = 4\pi\frac{\sum_{i=1}^{p} z_i^2 c_i^0}{k_B T}$. It is difficult to determine *a priori* the range of validity of the linear-regime approximation of the Poisson–Boltzmann model, which is expected to be very much dependent on the observable of interest. For instance, quantities related to electrostatic potential differences, such as the differential capacitance, are expected to be very sensitive to the diffuse-layer profile. The linearization of the PBE is thus expected to be valid for a very narrow range of potentials around the potential of zero charge (PZC) of the electrode surface. Integrated quantities, such as interface or solvation energies, are instead expected to be less affected by the linearization procedure. The advantage introduced by the linearization of the electrostatic problem lies in the fact that it allows for a simpler numerical solution [15, 44, 64, 66, 67, 71]. More elaborate iterative procedures have been recently implemented to solve the full non-linear PBE within periodic electronic structure codes [15, 44, 67, 71].

It is worth mentioning that in the case of electrochemical interfaces, one can take advantage of the reduced dimensionality of the problem, as long as the interface between the quantum-mechanical region and the electrolyte solution can be approximated as homogeneous in two dimensions [58, 62, 63]. In particular, for a charge-symmetric electrolyte one can exploit the solution of the one-dimensional (L)PBE along the surface normal, which can be obtained analytically once the boundary conditions of the problem are specified. Similarly to the Helmholtz double-layer model, also the planar-averaged (linearized) Poisson–Boltzmann model does not require any iterative procedure to determine the diffuse-layer contribution to the overall electrostatic potential, since it makes use of its analytic solution beyond a certain distance from the quantum-mechanical system [15]. This approach also bypasses the need of large simulation cells, which are instead necessary with numerical approaches. The latter class of methods, in fact, requires the whole diffuse-layer charge density to be modeled, which can extend for various nanometers, in order to obtain the correct potential drop across the interface.

3.3.4 Size-modified Poisson–Boltzmann model

The point-charge approximation that is part of the standard Poisson–Boltzmann model typically leads to an overestimate of the electrolyte concentration at the solute-solvent interface. A more physical description of the electrolyte particles accounts instead for their finite size and for the steric hindrance that limits the ionic concentrations to a maximum finite value [60, 61, 71, 74]. Such size-modified Poisson–Boltzmann model can be formally derived using a modified expression for the entropy density of the electrolyte solution, which is derived from a lattice-gas model [57, 59]. Note that models coupled to electronic-structure methods should ensure that the volume assigned to the quantum-mechanical region gives a zero contribution to the entropy of the solution. One way to enforce this limit is to make use of the ion-exclusion function to introduce a space dependence for the maximum local concentration [62, 63]. The entropy density of the solution, expressed as the sum of a ionic and a solvent contribution, can thus be written as [57, 59, 62, 63]:

$$
\begin{aligned}
s(\{c_i(\mathbf{r})\}) = & -k_B \sum_i c_i(\mathbf{r}) \ln \frac{c_i(\mathbf{r})}{c_{max}\gamma(\mathbf{r})} + \\
& -k_B \left(c_{max}\gamma(\mathbf{r}) - \sum_i c_i(\mathbf{r}) \right) \ln \left(1 - \sum_i \frac{c_i(\mathbf{r})}{c_{max}\gamma(\mathbf{r})} \right),
\end{aligned}
\tag{3.40}
$$

where the maximum local concentration c_{max} is related to the minimum volume that can be occupied by a solvent or electrolyte particle a^3 through $c_{max} = 1/a^3$. We assume here an equal size for all electrolyte particles (i.e. anions and cations) [57, 59]. Models that account for the disparity of ionic sizes have been proposed, but they do not lead to an analytical expression for the ion concentration functions [75, 76]. Assuming a random close packing for the electrolyte particles (i.e. a packing efficiency $P = 0.64$), the parameter c_{max} can also be related to the radius of the solvated ions: $c_{max} = 3P/4\pi r_i^3$.

The minimization of the free-energy functional in Eq. 3.33, using the expression of Eq. 3.40 for the entropy density and the following expression for the electrolyte chemical potentials:

$$
\mu_i = k_B T \ln \left(\frac{c_i^0}{c_{max} - \sum_i c_i^0} \right),
\tag{3.41}
$$

leads to the size-modified version of the PBE (MPBE), which differs from the standard PBE (Eq. 3.36) only in the expression of the local electrolyte concentrations:

$$
c_i(\mathbf{r}) = \frac{\gamma(\mathbf{r}) c_i^0 e^{-\frac{z_i \phi(\mathbf{r})}{k_B T}}}{1 - \sum_i \frac{c_i^0}{c_{max}} \left(1 - e^{-\frac{z_i \phi(\mathbf{r})}{k_B T}} \right)}.
\tag{3.42}
$$

Within this model the concentration of all the ionic species remain bound even for diverging values of the electrostatic potential. The expressions for the entropy density, the chemical potential, and the ionic concentration in the size-modified Poisson–Boltzmann model (Eqs. 3.40, 3.41, and 3.42) reduce to the corresponding expressions in the standard model

(Eqs. 3.35, 3.35, and 3.37) by taking their point-charge limit (i.e. $c_{max} \to \infty$ or $r_i \to 0$). Similarly to the standard PBE for general interface geometries, the solution of the non-linear MPBE is typically carried out with numerical iterative algorithms.

We note in passing that an alternative derivation of the size-modified Poisson–Boltzmann assumes that the solute cavity forms from effective repulsive interactions between the solute and the electrolyte [61]. Such repulsive interaction can also be recasted in terms of an ion-exclusion function, leading, however, to a slightly different formulation of the MPBE [71, 77].

3.3.5 Stern layer and additional interactions

In addition to marking the limit of the electrolyte continuum domain, the ion-exclusion function $\gamma(\mathbf{r})$ can be employed to displace the electrolyte charge density from a portion of the solvent region at the boundary with the solute. By using a scaled solvent cavity as electrolyte cavity function, one can indeed set an effective finite spacing between the solute and the electrolytes [63, 68, 71, 74]. This solvent-accessible but electrolyte-free region, also known as the Stern layer [78], is expected to surround solvated systems since tightly bound water molecules impede direct contact between the solute and the ionic species. As mentioned in Section 3.3.4, the electrolyte cavity can be equivalently introduced by means of *ad-hoc* repulsive interactions, which can be tuned to leave an ion-free region at the boundary with the solute [61]. Both approaches predict a zero-entropy density for the Stern layer. This is consistent with the absence of diffusing ions and for the significant conformational restriction that is expected for the solvent molecules in the first solvation shells of the solute.

A third possible approach is based on an electrolyte cavity that coincides with the solvent cavity and whose aim is to separate the quantum-mechanical region from the continuum solution. One then resorts to effective repulsive interactions to displace the continuum electrolyte from the Stern-layer region [15]. A similar formalism can be exploited to account for the selective adsorption of ions at electrochemical interfaces: Baskin and Prendergast have proposed using an effective binding interaction potential to account for mean-field electrostatic effects that follow the preferential adsorption of anionic species on electrode surfaces [79]. In particular, they have made use of a Morse-like potential to mimic the interaction of the negatively charged species with a model surface.

3.3.6 Performance of the diffuse-layer models

In order to illustrate the performance of the various electrolyte models, we present here differential capacitance curves computed for few representative model systems. The differential capacitance is an observable that is often studied in the context of electrochemical interface simulations because in addition of being experimentally accessible it is also highly sensitive on the details of the model employed for the electrolyte solution.

Differential capacitance curves have been computed by taking the numerical derivative of the charge-potential relations that we calculate through an approach that is micro-canonical with respect to the electrons of the system: we set a given surface charge by adding or removing (fraction of) electrons, and we determine the corresponding absolute potential Φ (with respect to vacuum) from the difference between the electrostatic potential in the bulk solution and the computed Fermi energy. The calculations have been performed with the Quantum ESPRESSO distribution [80, 81] using the open-source ENVIRON module for the continuum embedding [82], where all diffuse-layer model illustrated here have been implemented [15].

Four surfaces have been considered: Cu(111), Ag(111), Au(111), and Pt(111). The metal surfaces have been modelled by means of symmetric 9-layer slabs, exposing two identical interfaces to the continuum electrolyte solution. The metal slabs have been relaxed only in vacuum. Computational details include a 24x24x1 Γ-centered k-point grid to sample the first Brillouin zone of the metals, the Perdew-Burke-Ernzerhof exchange-correlation functional [83, 84], and pseudo-potentials and cutoffs for the plane-wave expansion of the wave-function from the standard solid-state pseudo-potential library (SSSP efficiency set, version 1.0). Considering that the goal here is to illustrate the performance of the various diffuse-layer models described, without aiming at producing quantitative predictions for the differential capacitance, we make use of a simplified planar cavity with an error-function profile along the surface normal for both the solvent and the electrolyte. For close-packed metal surfaces, such as the considered (111) surfaces of metals with a face-centered cubic structure, the rigid interface described is expected to very closely resemble the SSCS cavity, which is based on atom-centered interlocking spheres (see Eq. 3.3 in Section 3.2.1) [25]. We thus assume a certain degree of transferability with the SSCS parametrization, and set the interface of the solute-solvent region at a surface distance equal to $d = \alpha \cdot r_{vdW}^{UFF}$, where r_{vdW}^{UFF} is the van der Waals radius of a given atom type in the universal force-field (UFF) parametrization [85] and $\alpha = 1.16$ is a scaling parameter fitted to solvation energies of neutral isolated systems (see also Table 3.1) [25]. Also following the original SSCS parametrization [25], we set the width parameter for the error-function profile to $\Delta = 0.5$ Bohr. We choose here to use similar electrolyte cavities for the numerical models and for the models that make use of the planar-averaged approximation, so that we can more directly compare results obtained with the two approaches. We thus set the boundary of the diffuse layer at a fixed distance from the metal surface, introducing a 5 Å spacing between the outermost metal layer and the onset of the electrolyte charge. A small width parameter ($\Delta = 0.001$ Å) has been employed to broaden the ion-exclusion function in the calculations with the numerical solvers, to mimic the sharp diffuse-layer onset in the planar-averaged analytical methods.

Differential capacitance curves computed with the Helmholtz double-layer model are presented in Fig. 3.5. As mentioned above, the counter-charge plane that characterizes this model screens any surface charge and there is no dependence on the bulk electrolyte concentration. The electron-density tail extends less and less towards the bulk solution for increasing potential values (i.e. for increasing surface charge), effectively increasing the spacing between the surface and the continuum media, which lie at a fixed distance from the surface. This phenomenon explains the observed capacitance decrease as a function of the applied potential. This trend is readily understood from a simple capacitor model, for which the capacity is inversely proportional to the distance between its plates.

Similar capacitance trends are observed if the linearized Poisson–Boltzmann model is employed instead, as illustrated in Fig. 3.6. This model predicts a decreasing capacitance for decreasing electrolyte concentrations, in agreement with experimental trends. Note that the computed capacitance curves approach the corresponding Helmholtz-model

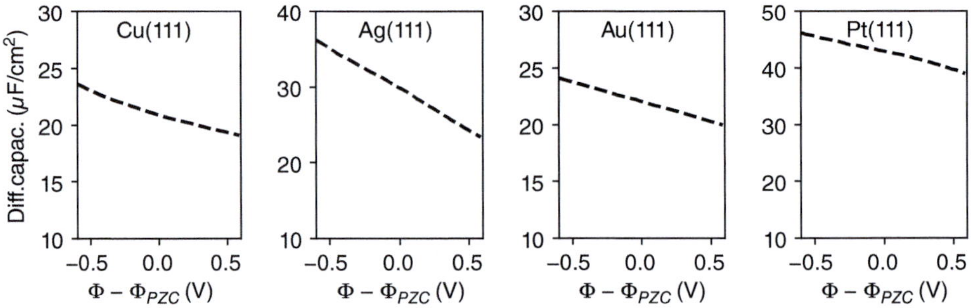

Figure 3.5 Differential capacitance computed for four clean metal slabs as a function of the applied potential. The Helmholtz double-layer model has been employed, using a planar counter-charge plane located 5 Å away from the outermost metal layer. A spread parameter $\Delta = 0.25$ Å has been employed.

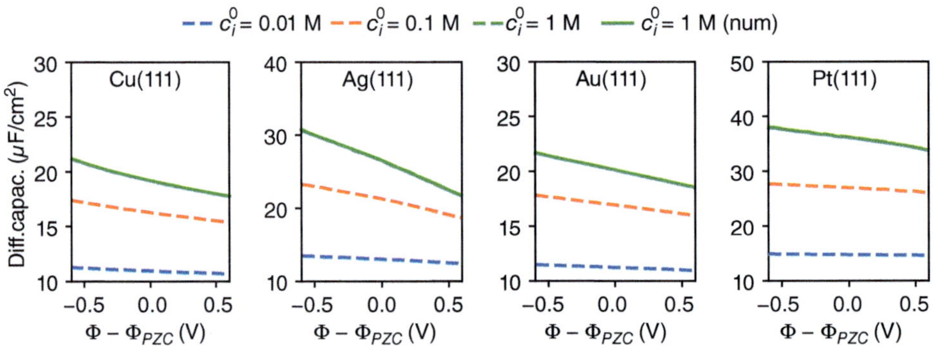

Figure 3.6 Same as Fig. 3.5, but the linearized Poisson–Boltzmann model has been employed. Three bulk ionic concentrations have been simulated: 0.01 M (blue), 0.1 M (orange), 1M (green). Calculations have been performed with the planar-averaged method, exploiting the analytical solution of the LPBE (dashed lines) and with the corresponding numerical method for general geometries (solid line).

curves for increasing concentrations (cf. Fig. 3.5). The Helmholtz model, in fact, with its counter-charge-plane screening, represents an infinite ionic-strength limit for the electrolyte solution. Monotonic capacitance curves are computed for all the metal surfaces, failing to reproduce the typical capacitance minimum at the PZC (0 V on the scale employed in Fig. 3.6). Similarly to what observed for the Helmholtz model, the observed capacitance decrease for increasing applied potential is consistent with the use of rigid solvent and electrolyte cavities, which keep the onset of the embedding charges at a fixed distance from the metal atoms. Note that an opposite trend (i.e. a capacitance decrease for decreasing potential) has been observed with density-dependent cavities [15, 64], which follows from the fact that in this case the embedding continuum moves further away from the QM system as the system's charge becomes more and more negative, increasing the effective separation between the surface and the solution. Fig. 3.6 also illustrates the comparison of results obtained through the numerical solution of the LPBE and results obtained exploiting the analytical solution of the LPBE along the surface normal. The

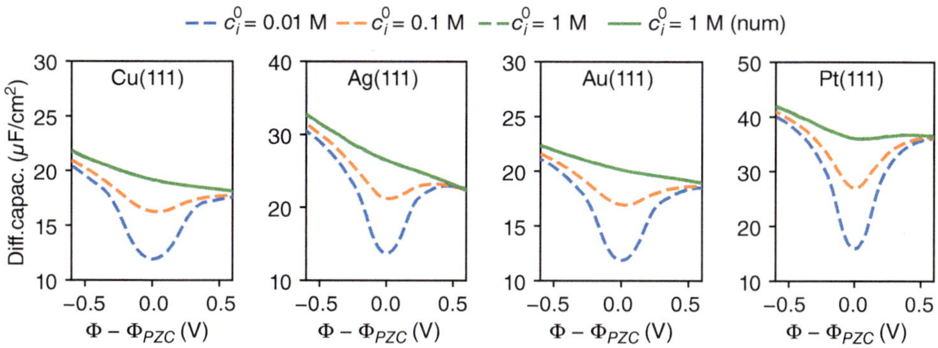

Figure 3.7 Same as Fig. 3.6, but the full non-linear Poisson–Boltzmann model has been employed here. Also here, calculations have been performed with the planar-averaged method, exploiting the analytical solution of the PBE (dashed lines) and with the corresponding numerical method for general geometries (solid line).

perfect match between the two approaches validate the planar-average approximation that is part of the latter, as expected for close-packed transition metal surfaces.

Fig. 3.7 illustrates the performance of the full non-linear Poisson–Boltzmann model. This model correctly predicts a capacitance minimum at the PZC, where the predicted capacitance values also agree well with the estimates from the linear-regime model (cf. Fig. 3.6). The overall potential dependence of the computed capacitance can be explained on the basis of a well-known model that assumes two series-capacitor contributions [58]: a diffuse-layer contribution and a Helmholtz-layer contribution. The first contribution dominates close to the PZC, where the diffuse-layer capacitance exhibits a minimum. The Poisson–Boltzmann model predicts a monotonously increasing capacitance [15, 58], so that the Helmholtz contribution is expected to become dominant at larger applied potentials, where the surface charge is rapidly screened by the electrolyte. And in fact, at larger potential values all the capacitance curves for the systems investigates approach the same overall-decreasing trend, which is similar to the one observed for the counter-charge plane double-layer model (Fig. 3.5). As also observed for the linear-regime model, also for the full non-linear Poisson–Boltzmann model the capacitance curve obtained with the planar-averaged analytic approach is in very good agreement with the capacitance obtained with the numerical solver for general geometries.

Finally, the differential capacitance curves obtained with the size-modified Poisson–Boltzmann model are presented in Fig. 3.8, where we compare results obtained with different values of the maximum-concentration parameter c_{max}. Similar capacitance values to the ones obtained with the standard Poisson–Boltzmann model are obtained for potentials around the PZC, for which the electrolyte concentration at the interface is still significantly smaller than c_{max} due to the weak electrostatic attraction. For larger applied potentials, the steric hindrance between ions opposes to the electrostatic attraction towards the electrode, increasing the effective distance between the charged surface and the electrolyte charge density. We thus observe a capacitance decrease for large values of the applied potential, in agreement with experimental observations. Also in agreement with previous results from purely-continuum models [79, 86, 87], we thus find that the

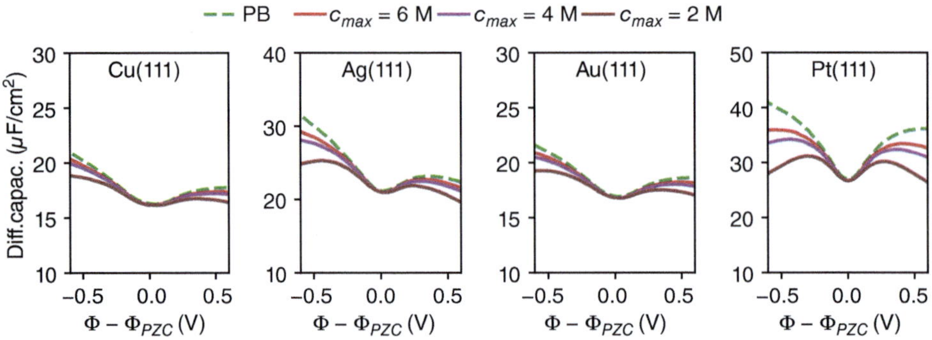

Figure 3.8 Same as Fig. 3.7, but the results of the standard Poisson–Boltzmann model (green, $c_{max} = \infty$ M) are compared to results obtained with the size-modified version of the Poisson–Boltzmann model. A bulk ionic concentration of 0.1 M has been simulated, together with a maximum concentration parameter equal to: 2 M (red), 4 M (purple), and 6 M (brown).

size-modified version of the Poisson–Boltzmann model is able to qualitative reproduce all the experimentally-observed features of the differential capacitance for single-crystal metal surfaces: the capacitance minimum at the PZC, an initial capacitance increase for increasing potentials, and a final capacitance decrease after reaching two local maxima. This is achieved with a linear dielectric model, without accounting for non-linear solvent effects [67–69]. While the focus here is on the qualitative trends that the most wide-spread diffuse-layer models are able to provide, it is important to mention that the choice of the dielectric interface function strongly affects the absolute values of the computed capacitance [15, 32, 67, 68]. A suitable parametrization of such cavity is thus highly relevant in order to aim at quantitative description of experimental data.

3.4 Grand canonical simulations of electrochemical systems

Having established a computational model that can treat the most relevant interfacial energy contributions of electrified solid-liquid interfaces appropriately, we present in this section how the grand canonical energetics of electrochemical interfaces can be derived in practice from QM/Continuum simulations.

This is in particular important, as the precise atomistic structure and composition of electrochemical interfaces under operation conditions are rarely known though of high interest. It is here where an ab-initio-based thermodynamics approach can help to predict stable interface structures and configurations, which is the basis for a better understanding of electrochemical processes as well as for the interpretation of the results of advanced experimental techniques e.g. in-situ spectroscopies (Raman, STM, NEXAFS, NMR, ...), which very often necessitate a comparison to electronic-structure simulations.

Until recently, most ab-initio studies on electrochemical interfaces have used methods developed originally for the thermodynamic description of solid-gas interfaces which represent, as shall be shown, a significant approximation to the correct grand canonical description of electrified interfaces. In order to understand better the approximations of this so-called Computational Hydrogen Electrode (CHE) approach [88–91], we will review shortly how the thermodynamics of solid-gas interfaces can be derived from ab-initio simulations and formulate subsequently a grand canonical (GC) description of electrochemical interfaces within QM/Continuum models, from which the CHE approximation follows automatically as a 0th order approximation. We will outline how to simulate such interfaces in practice and present a selection of recent results.

3.4.1 Thermodynamics of interfaces

Interfaces occur between two different phases. In some contexts they are also referred to as *interphases*, which highlights the fact that they describe a specific region in space of *finite extent* located in between two homogeneous bulk *phases*, which has specific properties different from the adjacent bulk phases. Different interface types are depicted schematically in Fig. 3.9, where the potentially large extents of electrochemical interfaces are highlighted specifically.

The thermodynamic description of interfacial systems is based on two central pillars:

1. Inclusion of all energy contributions within the *interphase*.
2. Correct statistical mechanics treatment of the relevant degrees of freedom.

Given the geometry of Fig. 3.9, the most straightforward approach to simulate interfaces based on ab-initio methods uses the slab geometry and periodic density functional theory, which allows the in-plane degrees of freedom of the interface to be treated using appropriate periodic boundary conditions while keeping artificial periodic image interactions in the out-of-plane direction perpendicular to the interface under good control e.g. by the use of sufficiently large slab separations and appropriate correction methods [12].

Typically, quantum-mechanical simulations allow the 0 K energetics of systems to be accessed at fixed volume and composition. In contrast, most experimental results are obtained while controlling the temperature T, the pressure p and/or the chemical potential μ of species. Therefore, in order to derive the correct thermodynamic description based on ab-initio simulations, it is necessary to augment them with thermodynamic sampling or complement them with approximate correction methods.

For solid, bulk systems very accurate results for the Gibbs free energy $G(p, T, N)$ can be obtained by including the vibrational degrees of freedom via the harmonic or

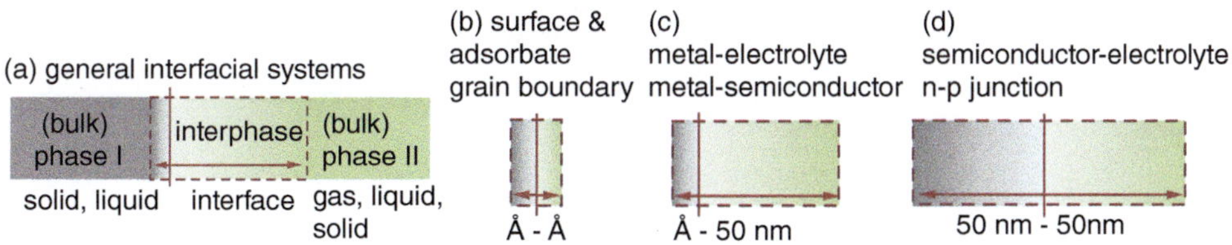

Figure 3.9 Relevant parts of interfacial systems: a) the energetics is defined by all energy contributions from the *interphase*, the 3 dimensional region between two homogeneous bulk phases. Depending on the system under study, the spatial extent can vary between b) a few Angstrom and c), d) several tens of nanometers e.g. due to long-range space charge layers as relevant for electrochemical interfaces.

quasi-harmonic approximation [92, 93] and potentially including also contributions due to additional degrees of freedom (e.g. configurational (alloys), magnetic (ferromagnets), polaronic (battery systems) [94, 95]). The discussed derivation of approximate free energies can be applied equivalently to interfacial systems, as e.g. studied in a slab approach; in particular, for describing the thermodynamics of solid-gas interfaces, where the interfacial regions extend typically only over a few atomic layers below the substrate surface (~Å) and include potential adsorbates, such that all relevant energy contributions can be captured in relatively small quantum mechanical cells (see Fig. 3.9 b).

In general, *absolute* values of free energies are not well defined.[1] In fact, it is only free energy differences that can be formulated and determined unambiguously. Such energy differences are typically called *excess* free energies; in interfacial systems, they are conveniently normalized to the surface area yielding surface (*excess*) energies γ. The *excess* free energy is defined as the difference of the Gibbs free energy of the system under study and the energy cost of removing its constituents from certain thermodynamic reservoirs it is assumed to be coupled to. Alternative terms used in the literature are Gibbs excess energy, Gibbs free energy change, or simply formation energy.

As an example, the surface excess energy γ for a specific surface, given a unary material of specie s, can be calculated in a symmetric slab approach by determining the Gibbs free energy of a sufficiently thick slab G^{slab}, consisting of N_s atoms and subtracting $\mu_s N_s$ - the energy cost for removing those N_s atoms from a reservoir characterized by μ_s. In the present case, the bulk material itself is assumed to correspond to the reservoir, which is why $\mu_s N_s$ is equivalent to the Gibbs free energy of an equally sized bulk system with $\mu_s N_s = G^{\text{bulk}}$ such that:

$$\gamma = \frac{1}{2A}(G^{\text{slab}} - \mu_s N_s) = \frac{1}{2A}(G^{\text{slab}} - G^{\text{bulk}}) \tag{3.43}$$

The factor $\frac{1}{2A}$ derives from the symmetric slab setup, with two equivalent surfaces. Here and in the following we neglect the thermodynamic variables T and p for clarity. In addition, it is worth noting that for solid systems under ambient conditions (and also solid slabs) the Helmholtz free energy $F(V, T, N)$ is basically equivalent to $G(p, T, N)$ due to the negligible volume variations for ambient pressures p. Therefore, although G is used in the equations, in practice, all simulations are performed at constant volume (surface area), with lattice constants fixed at the bulk equilibrium values.[2] For sufficiently large slab thicknesses, the numerical value of Eq. 3.43 eventually converges, e.g. when the free energy contribution of an additional atomic layer in the center of the slab coincides with the corresponding bulk value. In case of systems that are composed of several species i, e.g. substrate species s and adsorbates a with $(\{N_i\}) = (\{N_s\}, \{N_a\})$, the corresponding additional reservoirs have to be included, yielding:

$$\hat{J}(\{N_i\}) = G^{\text{slab}}(\{N_i\}) - \sum_i \mu_i N_i. \tag{3.44}$$

1 This can be easily understood from the fact that reported internal energies from force fields or from different DFT codes typically differ, depending on e.g. how an arbitrary 0 was chosen or how many electrons are treated (cf. pseudo-potential approximation).

2 Note also that typical accuracy thresholds for the pressure e.g. for variable cell simulations are much larger than the ambient pressure.

We used here the symbol $\hat{J}$ for the Gibbs excess energy as the minimum of $\hat{J}$ in the space of variables $\{N_i\}$ at fixed system size is equivalent with the grand potential J, the Legendre transform of G. This is easily seen by formulating:

$$\min_{\{N_i\}}(\hat{J}) \leftrightarrow \left.\frac{\partial \hat{J}}{\partial N_i}\right|_{N_i^0} = 0, \forall i \leftrightarrow \left.\frac{\partial G}{\partial N_i}\right|_{N_i^0} = \mu_i, \forall i. \tag{3.45}$$

The last relation yields the equilibrium composition $\{N_i^0(\mu_i)\}$ as a function of the chemical potentials $\{\mu_i\}$, which, when inserted in $\hat{J}$, corresponds to the traditional formulation of the Legendre transform of G and thus the grand potential J. $\hat{J}$ is conveniently interpreted as a grand canonical free energy landscape in the variables $\{N_i\}$.

A similar and more detailed description of the outlined approach to determine the energetics of solid-gas interfaces is given in [96], including also a discussion of how to derive μ_i for gas phase species from ab-initio simulations. Examples for real-world applications are presented in [91, 96–105].

3.4.2 Ab-initio based thermodynamics of electrochemical interfaces

The thermodynamics of electrochemical interfaces can be formulated in perfect correspondence to the above discussion [32, 106]. Only two minor but important differences need to be clarified, that follow directly from the definition of electrochemical systems [107]:

> *Electrochemistry deals with the processes at the interface between an electron conductor and an ion conductor.*

Hence, in addition to charge-neutral species, electrochemical interfaces are also composed of charged species, namely electrons and ions which derive from two different, decoupled thermodynamic reservoirs: the electron conducting phase (the electrode) and the ion conducting phase (the electrolyte solution). These reservoirs are characterized by *electro*chemical potentials, which we denote by $\tilde{\mu}$. This indicates that a thermodynamic treatment as in the case of solid-gas interfaces based on systems with different composition might be hampered by the fact that charged cells will have to be considered, which poses a problem in periodic calculations where total energies are not well defined for charged cells (cf. homogeneous background charge). On the other hand, as both, the electron and ion conducting phase, exhibit mobile charge carriers all electrostatic fields will be eventually screened at large enough distances from the interface. This means that large enough cells exhibit no net charge when they incorporate the whole interphasial region including in particular the diffuse layer of electrolyte counter charges. As illustrated in Fig. 3.9 c) and d), however, the double layer can easily extend over several tens of nanometers, depending on the electrolyte concentration. As a result, it can be hard to simulate such systems purely based on atomistic models, considering also that the liquid nature of the solution phase necessitates thermodynamic sampling e.g. using molecular dynamics.

As discussed previously, most of these difficulties are alleviated in quantum-mechanical models coupled to a continuum solvent model, that includes mean field free energy contributions from the solution while including in particular also appropriate electrolyte counter charges that will keep the simulation cell charge neutral automatically. In addition, as shown in the previous sections, treating the long range part of the diffuse layer with the analytic solution to the Poisson–Boltzmann problem allows rather small cells of similar size to vacuum simulations to be used, without altering any of the mentioned properties. In particular, a flat electrostatic potential V^{im} outside the double layer region is always established, also when varying the total electron number of the system; V^{im} can therefore serve as a universal reference for the electrostatic potential, which is important as the electrochemical potentials of charged species (electrons and ions) vary with the electrostatic potential and need to be defined on a consistent potential scale.

The Gibbs excess energy of a slab in implicit solvent composed of $\{N_s\}$ slab atoms, $\{N_a\}$ potentially charged adsorbates and N_e^{abs} electrons can be formulated in perfect analogy to the considerations for non-electrochemical interfaces:

$$\hat{J}(\{N_s\},\ \{N_a\},\ N_e^{\text{abs}}) = G^{\text{slab}}(\{N_s\},\ \{N_a\},\ N_e^{\text{abs}}) - \sum_s \mu_s N_s - \sum_a \tilde{\mu}_a N_\alpha^a + e\Phi_E N_e^{\text{abs}} \tag{3.46}$$

Note that the applied electrode potential Φ_E corresponds to the electrochemical potential of electrons $\tilde{\mu}_e$ with [32]:

$$\tilde{\mu}_e = -e\Phi_E. \tag{3.47}$$

Φ_E is measured according to the electrochemistry convention, namely with increasing values going downwards. The experimental value for the standard hydrogen electrode (SHE) is e.g. located at +4.44 V on the absolute scale [108]. Note that the electrode potential Φ_E times e on the absolute scale corresponds to the work function of a system immersed in the solution; in implicit models the work function is naturally determined as the difference of the Fermi level and the flat electrostatic potential in the center of the implicit region V^{im}.

As shown recently [15, 32] for the SCCS implicit model and as mentioned already before, the so-determined work functions (potentials of zero charge, PZC) and interfacial capacitances of metallic slabs in QM/Continuum simulations yield good agreement with experiment for optimized parameters of the implicit model. In particular, the PZC of Pt surfaces is in perfect agreement with the experimental and theoretical PZC on the absolute scale for the parameters determined in [32]. However, a consistent offset of approx. 0.4 eV was found for the other coinage metals Au, Ag, and Cu, which is potentially related to the non-negligible, material-dependent contributions to the potential drop at metal-water interfaces that derive from interfacial charge transfer between surfaces and the first water layer(s) [109–111], which is not included in implicit-model-derived electrostatics. In addition, more recent studies on semiconductor band alignment in implicit models [16] suggest that a universal offset of 0.33 V with respect to the absolute potential scale is to be included to align the implicit reference potential of the SCCS model with the absolute scale. This is probably due to the absence of the potential drop at the water-vacuum interface in the implicit model (see discussions in [16, 112] for more details). We are currently assessing these issues and note here only that the work function in implicit simulations might be shifted with respect to the absolute potential scale. Such a shift needs to be paid attention to whenever comparing to experimental values. In the subsequent equations we leave it away for clarity, and note only that in all cases the electrode potential Φ_E is to be measured consistently with the in-simulation work function, i.e. against the flat electrostatic potential in the implicit region.

Furthermore, it is important to clarify, that the absolute electron number N_e^{abs} in the excess free energy expression Eq. 3.46 counts only the electrons beyond the charge neutral, clean system, which means:

$$N_e^{\text{abs}} = N_e^{\text{net}} + \frac{q_a}{e} N_a. \tag{3.48}$$

N_a is the number of adsorbed ions of charge q_a and $\frac{q_a}{e} N_a$ thus the number of electrons that would compensate identically the ionic charges; $-eN_e^{\text{net}}$ is the charge of the quantum mechanical region beyond charge neutrality, which is equivalent in absolute size with the neutralizing electrolyte counter charges (in the diffuse layer or Helmholtz plane, depending on electrolyte representation). Other electron counting rules – e.g. using the actual total number of electrons – can be applied as long as all constituents of the systems are considered correctly (no double counting!) [113]. The present choice is convenient as N_e^{abs} corresponds to the number of electrons that are provided from the external circuit measured with respect to the number of electrons of the charge-neutral, pristine slab, yielding interfacial charges consistent with the Lippmann equation [114] (see derivation in [32]) and as it allows a direct analysis of the Computational Hydrogen Electrode approach [88, 89], as shown below.

The electrochemical potentials of ions can be determined by combining DFT simulations of the involved species with experimental values. As an example, for the proton in water, the electrochemical potential $\tilde{\mu}_{\text{H}}$ can be defined by the absolute potential of the standard hydrogen electrode (SHE @ 4.44 V) [108] and the equilibrium of following reaction:

$$H_2(g) \rightleftharpoons 2H^+ + 2e^- \tag{3.49}$$

$$\Rightarrow \tilde{\mu}_{\text{H}} = \frac{\mu(H_2(g))}{2} - k_B T \ln(10)\text{pH} + 4.44\ \text{eV}, \tag{3.50}$$

with $\mu(H_2(g))$ evaluated at 1 bar and 298 K, which can be calculated from the ideal gas partition function [96]. Other ions can be treated accordingly [32, 91].

As in the case of solid-gas interfaces, the thermodynamically most stable interface configurations minimize $\hat{J}$ at given system size with respect to $(\{N_s\}, \{N_a\}, N_e^{\text{abs}})$. As mentioned before, in most applications, the slab is assumed in equilibrium with the bulk material which means $\sum_s \mu_s N_s = G^{\text{bulk}}$, such that only adsorbate and electron degrees of freedom, as well as different structural configurations (e.g. adsorbates at different sites) need to be considered in the minimization. Excess energies at applied potential with equilibrated electron charge, will be denoted with $\tilde{J}$ and are obtained by partial minimization of Eq. 3.46 w.r.t. the number of electrons.

3.4.3 Grand canonical simulations and the CHE approximation

In contrast to the presented description of the thermodynamics of electrochemical interfaces, the widely used Computational Hydrogen Electrode (CHE) approach approximates the grand canonical (GC) energetics of electrochemical interfaces only from charge-neutral DFT simulations. Such an approximation removes the discussed problem of charging the interface in periodic calculations without appropriate electrolyte counter charges,

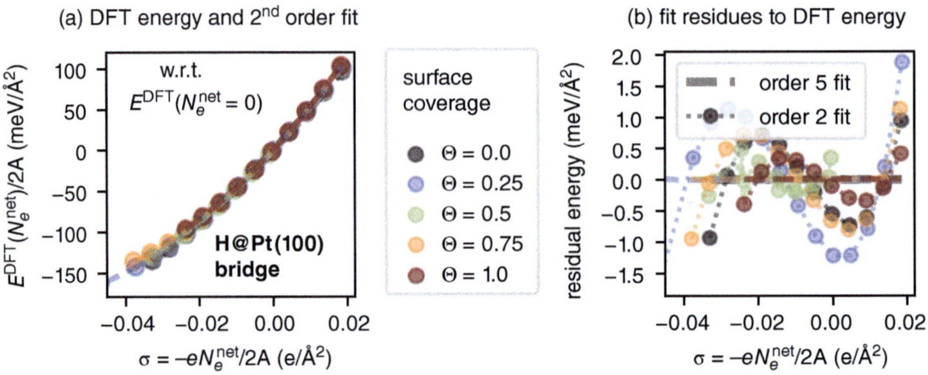

Figure 3.10 a) Total energy $E^{\mathrm{DFT,\,slab}}/2A$ of clean and hydrogen-covered Pt(100) slabs [H on the bridge site] as a function of surface charge $\sigma = -eN_e^{\mathrm{net}}/2A$. b) Fit residues for a 2nd and 5th order fit of the energy-charge dependence $E^{\mathrm{DFT,\,slab}}(N_e^{\mathrm{net}})/2A$ from panel a). The computational parameters as well as more details on these calculations can be found in [32, 106]. Source: From G. Hörmann, O. Andreussi, and N. Marzari, The Journal of Chemical Physics 150,041730 (2019).

however, it neglects all energetic contributions that arise from the double layer. In particular, considering only a single, specific charge state with $N_e^{\mathrm{net}} = 0$ corresponds to neglecting the necessary minimization of the excess energy $\hat{J}$ with respect to the number of electrons and thus is just an upper boundary estimate for the correct, charge-equilibrated energy $\tilde{J}(\Phi_E)$ [32].

As recently shown [106], an insightful connection between GC simulations and the CHE approach can be made by developing G^{slab} (cf. Eq. 3.46) up to second order in the net electron charges N_e^{net} around $N_e^{\mathrm{net}} = 0$ given a certain interface configuration and fixed atomic composition. As commonly done, we approximate G^{slab} and $\sum_s \mu_s N_s = G^{\mathrm{bulk}}$ with the 0 K total energies of the relaxed ground state structures of the slab $E^{\mathrm{DFT,\,slab}}$ and of the bulk material $E_0^{\mathrm{DFT,\,bulk}}$ plus a correction ΔE^{corr} due to vibrational and/or configurational degrees of freedom of the adsorbates, which is largely independent of the charge state N_e^{abs}:

$$
\begin{aligned}
&G^{\mathrm{slab}}(\{N_s\},\{N_a\},N_e^{\mathrm{abs}}) - G^{\mathrm{bulk}}(\{N_s\}) \approx \\
&\quad \approx E^{\mathrm{DFT,\,slab}}(\{N_s\},\{N_a\},N_e^{\mathrm{abs}}) - E_0^{\mathrm{DFT,\,bulk}} + \Delta E^{\mathrm{corr}}.
\end{aligned}
\tag{3.51}
$$

Note that this approximation is typically very accurate, as significant error cancellation occurs for the substrate degrees of freedom. Furthermore, $E^{\mathrm{DFT,\,slab}}$ includes the free energy contributions of the electrolyte solution as represented by the implicit model and elaborated before (section 3.3). Fig. 3.10 plots the change of $E^{\mathrm{DFT,\,slab}}$ for pristine as well as hydrogen-covered Pt(100) slabs in implicit solvent as a function of the net surface charges, which indicates that following second order expansion is a rather accurate approximation:

$$
E^{\mathrm{DFT,\,slab}}(N_e^{\mathrm{abs}}) = E_0^{\mathrm{DFT,\,slab}} + \underbrace{\left.\frac{\partial E^{\mathrm{DFT,slab}}}{\partial N_e^{\mathrm{net}}}\right|_0}_{-e\Phi_0} N_e^{\mathrm{net}} + \frac{1}{2}\underbrace{\left.\frac{\partial^2 E^{\mathrm{DFT,slab}}}{\partial (N_e^{\mathrm{net}})^2}\right|_0}_{\frac{e^2}{2AC}} (N_e^{\mathrm{net}})^2. \tag{3.52}
$$

$E_0^{\text{DFT, slab}}$ is the total energy, Φ_0 the work function, and C the differential interfacial capacitance, all evaluated at zero net surface charge; the factor $2A$ arises from the assumed symmetric slab setup. Note that N_e^{abs} and N_e^{net} coincide up to a constant shift for fixed surface compositions, which is why the derivatives in both variables are interchangeable in the present analysis. Within the 2nd-order approximation, the partial minimization of $\hat{J}$ with respect to electron number can be performed analytically with:

$$\frac{\partial \hat{J}}{\partial N_e^{\text{abs}}} = 0 \leftrightarrow \underbrace{\frac{\partial E^{\text{DFT, slab}}}{\partial N_e^{\text{net}}}}_{-e\Phi} = -e\Phi_E \tag{3.53}$$

$$N_e^{\text{net}}(\Phi_E) = \frac{2AC}{e} \cdot (\Phi_0 - \Phi_E). \tag{3.54}$$

Eq. 3.53 indicates that the charge equilibrated system that minimizes $\hat{J}$, corresponds to the system where the work function Φ is equivalent to the externally applied (target) potential Φ_E, which is equivalent to the results obtained from more direct GC approaches [113, 115, 116]. The charge-equilibrated, excess free energy $\tilde{J}(\Phi_E)$ within the 2nd-order model is obtained by evaluation of $\hat{J}$ (eqs. 3.46–3.52) at the equilibrium charge (Eq. 3.54), which allows the discussed connection between the CHE and the GC approach to be clarified:

$$\tilde{J}(\Phi_E) \approx \underbrace{\Delta E_0^{\text{DFT, interface}} + (q_a\Phi_E - \tilde{\mu}_a)N_a}_{E^{\text{CHE}}\text{(zero net charge approach)}} \underbrace{-AC(\Phi_0 - \Phi_E)^2}_{|E^{\text{DL}}\text{(double layer effects)}} \tag{3.55}$$

$$\text{with } \Delta E_0^{\text{DFT, interface}} = E_0^{\text{DFT, slab}} - E_0^{\text{DFT, bulk}} + \Delta E^{\text{corr}}. \tag{3.56}$$

The first two terms in Eq. 3.55 correspond exactly to the excess free energy expression E^{CHE} of the CHE approach, where the GC energetics is approximated from zero-net-charge calculations. The third term corresponds to additional double-layer-related energy contributions E^{DL}, which can be sizable, whenever $\Phi_0 - \Phi_E \neq 0$, so for applied potentials Φ_E different than the potential of zero charge Φ_0 of the slab system; in the second order model, this energy contribution is proportional to the interfacial capacitance C and the square of the potential drop $\Phi_0 - \Phi_E$ across the double layer.

To see more clearly that E^{CHE} corresponds identically to the CHE approximation, consider the case of electrosorbed protons, for which the following relations hold [32, 56]:

$$\Phi_E = U_{\text{RHE}} + 4.44\text{V} - \tfrac{k_B T}{e}\ln(10)\text{pH} \tag{3.57}$$

$$e\Phi_E - \tilde{\mu}_{\text{H}} \overset{\text{eqs. 3.50 and 3.57}}{\widehat{=}} eU_{\text{RHE}} - \frac{\mu(H_2(g))}{2}. \tag{3.58}$$

The term 4.44 V corresponds to the experimental value for the SHE potential on the absolute scale [108]. Insertion of Eq. 3.58 into the second term in E^{CHE} (Eq. 3.55) shows that this indeed corresponds to the known CHE expression, with the adsorption-energy-like term $\Delta E_0^{\text{DFT, interface}} - \frac{1}{2}\mu(H_2(g))N_{\text{H}}$ plus an additional term $+eU_{\text{RHE}}N_{\text{H}}$, that scales linearly with the potential U_{RHE} on the reversible hydrogen electrode (RHE) scale and the surface coverage N_{H}. Considering Eq. 3.58, however, such a behaviour is only correct up to 0th order within a GC description. In particular, the CHE approximation cannot describe following potential-induced effects, which are directly understandable from the second order equation 3.55 (see [106] for a more detailed analysis of the second order model):

- A potential-induced change of the most stable adsorption site: Within the CHE approximation the relative stability of sites is invariant as the term $U_{RHE}N_H$ is the same for systems with equal coverage. In a GC approach, different work function shifts and interfacial capacitances due to adsorbates on different sites can induce however relative stability shifts (cf. E^{DL} in Eq. 3.55).
- A potential-induced lifting or creation of a surface reconstruction: This is an analogue problem to the previous case, with the additional condition $N_H = 0$, creating no potential dependence in interface energy at all within the CHE approximation.
- Electrocatalytic reactions, where intermediates are not only formed via proton-coupled electron transfer steps, and ionic adsorbates which drag a non-integer number of electrons onto the electrode (non-integer *electrosorption valencies* [117]): Such phenomena are excluded by definition from the CHE model, can however be understood and simulated within a GC approach, where the total electron number N_e^{abs} flowing onto the electrode is only determined by the requirement to equilibrate the system work function with the applied potential (Eq. 3.53), which does not have to be exactly compensating the adsorbate charge.

Note that potential-induced effects correspond to pH effects if we measure on the RHE scale as a change in pH at fixed U_{RHE} is equivalent with a change of the absolute potential Φ_E according to Eq. 3.57.

Some of these effects have been studied using the described methodology in three recent publications [32, 106, 118], which we summarize briefly in the following section.

3.5 Selected applications

In [32] rather simple systems that are however of fundamental interest were studied using the described GC methodology, namely the pristine low index surfaces of Pt, Au, Ag and Cu as well as H- and Cl-covered Pt electrodes. All results were obtained using the SCCS cavity as implemented in ENVIRON [82] and with a Helmholtz model for the electrolyte, with the planar counter charge layers in a distance of ~ 4Å with respect to the outermost atomic layer (corresponding to high electrolyte concentrations, see the original paper for more details). It was demonstrated that the discussed GC methodology is capable of predicting a) the lifting of the quasi-hexagonal surface reconstruction of Au(100) by application of positive potentials [see Fig. 3.11 a)] and b) the experimentally observed, non-trivial shifts with pH on the RHE scale of the pseudocapacitance peak induced by proton electrosorption [see Fig. 3.11 b)].

To understand better how these results were obtained practically we summarize briefly the major steps of the simulations involved:

1. Construct slab structures of the pristine surfaces as well as with the adsorbates on all possible adsorption sites for all considered coverages.

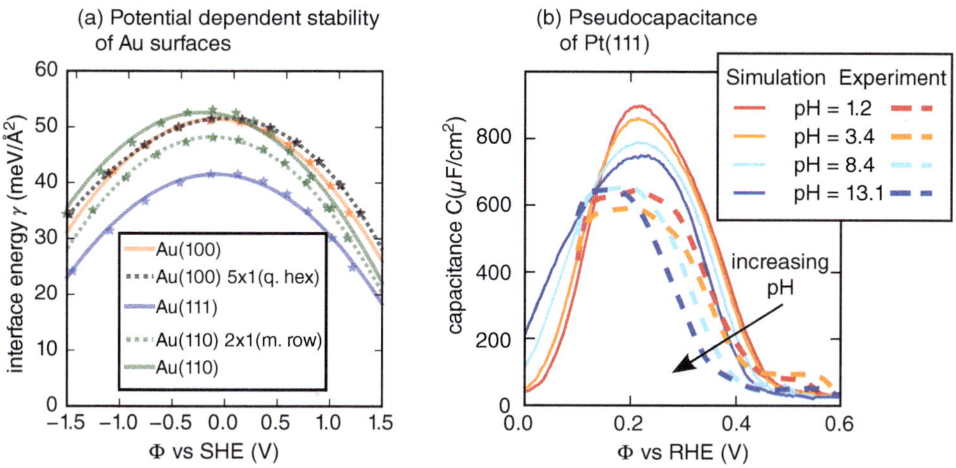

Figure 3.11 a) GC interface energies of pristine Au surfaces at applied potential. The stars represent the actual simulations and the lines the interpolated interface energies. In agreement with experiment [119], the (in vacuum) more stable quasi hexagonal, 5x1 reconstructed, (100) surface termination (dotted, grey) is becoming destabilized against the regular (100) surface (solid, orange) by application of a positive potential. b) Simulated and experimental pseudocapacitance of Pt(111) and their dependence on pH. While good agreement between experiment and theory is observed for the overall position and shape of the capacitance peak due to proton electrosorption, very good agreement is found for the relative shifts with pH (see [32] for more details). Sources: Based on N.G. Hörmann, O. Andreussi, and N. Marzari, The Journal of Chemical Physics 150, 041730 (2019); N.G. Hörmann and N. Marzari, The effects of pH and potential at electrochemical interfaces," (2019).

2. Simulate the systems for different net surface charges $\{N_e^{\text{net}}\}$ (10–15 different values) within a certain range around 0. An estimated differential interfacial capacitance, e.g. approx. 30 μF/cm^2 for the studied systems, can be used to estimate the possible potential range of work functions that these net charges will cover.
3. Eq. 3.53 indicates that the constant charge calculations at $\{N_e^{\text{net}}\}$ give access to the charge-equilibrated, GC interface excess energy $\tilde{J}$ for electrode potentials $\{\Phi_E\}$ which coincide with the observed system's work functions. Considering that the variation of the GC interface excess energy with Φ_E is determined by the PZC and the interfacial capacitance (see Eq. 3.55 and the discussion in [32]), the variation of $\tilde{J}$ with Φ_E can be approximated accurately by an interpolation through these points $\{\Phi_E\}$ using a low order polynomial.
4. In the case that adsorbates are involved the equilibrium interface compositions and GC interface energies $\gamma(\Phi_E)$ need to be determined by minimizing across electronic and ionic degrees of freedom. For each studied composition N_a, charge-equilibrated excess energies $\tilde{\gamma}(N_a, \Phi_E)$ were determined according to step iii) and subsequently the whole set of terminations $\{N_a\}$ was considered for the minimization with respect to surface composition and to adsorption site. In order to be able to simulate continuous observables, e.g. the variation of the absolute surface charge N_e^{abs} with applied potential, or the pseudocapacitance [32] $\frac{-e}{2A}\frac{\mathrm{d}N_e^{\text{abs}}}{\mathrm{d}\Phi_E} = -\frac{\mathrm{d}^2\gamma}{\mathrm{d}\Phi_E^2}$, it is advisable to interpolate the charge-equilibrated interface energies $\tilde{\gamma}(N_a, \Phi_E)$ as well with respect to surface coverages N_a and perform

the minimization w.r.t. N_a on the interpolated energy landscape. Note however, that the interpolation with coverage is only advisable for similar adsorbate configurations (e.g. adsorbates on the same adsorption site), as the energy landscape is rather smooth only in this case.

The same methodology was also applied to study the surface composition and stability of Cu nanoparticles under applied potential conditions relevant to CO_2 reduction [118]. Having access to well defined GC interface energies, it was possible to relate the experimentally observed degradation of Cu nanocubes under reactive conditions to negative values of the interface energy, found for applied electrode potentials smaller than approx −0.5 vs RHE; negative interface energies mean that an increase of interface area leads to a reduction of the total system energy, which represents a thermodynamic driving force towards nanoclustering leading to nanoparticles with large surface to volume ratios. These results were obtained from studying H, CO, and mixed H and CO covered surfaces for which strong potential dependencies were observed e.g. in the interfacial compositions and configurations as evidenced by the surface Pourbaix diagrams in Fig. 3.12. Note how adsorbed CO molecules sit at different adsorption sites on Cu(100) depending on the pH for a coverage of 0.25 ML which is proof for the strong impact of the explicit application of a potential in GC simulations. Similar strong potential dependencies have been reported recently in [116] also for other adsorbates on Cu. While all technical details of these simulations are reported in the Supporting Information file of [118], we wish to note here, that the Pourbaix diagrams were constructed by fixing the chemical potential of the CO_2 molecule (gas, 1 mbar, T = 298 K) and determining the electrochemical potential of CO at the surface assuming it derives

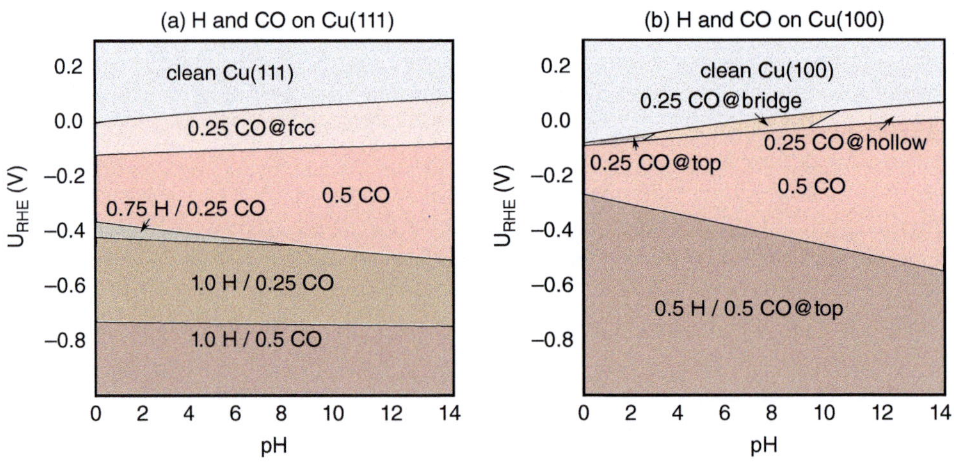

Figure 3.12 GC surface Pourbaix diagrams for Cu(111) and Cu(100) in aqueous solution with H and CO as possible adsorbates. The numbers indicate the surface coverage (in ML) for the respective termination. In case the atoms sit on well identified adsorption sites they are reported. For mixed coverages, multiple or ill-defined adsorption sites are typically occupied. As explained in the text, non-GC simulations that use the CHE approximation can show no pH dependence on the RHE scale, i.e. only horizontal lines would be visible in CHE-derived Pourbaix diagrams. Note how the CO adsorption site switches with pH for the 0.25 ML coverage on Cu(100). Similar potential dependencies are also observed for other adsorbates on Cu (see e.g. [116]). Source: Based on H. Zhang, W. A. Goddard, Q. Lu, and M.-J. Cheng, Phys. Chem. Chem. Phys. 20, 2549 (2018).

from CO_2 which has reacted with 2 protons to form a water molecule:

$$CO_2(g) + 2H^+ + 2e^- \rightarrow CO + H_2O(l) \tag{3.59}$$

Thus, the chemical potential of CO $\tilde{\mu}(CO)$ can be expressed as a function of the (electro-) chemical potentials of H_2O, CO_2, H^+ and e^- giving it additional dependencies on Φ_E and pH.

These selected computational results show how GC simulations within QM/Continuum models are able to overcome the limitations of the CHE approximation, at decent computational overhead. However, there are still several open questions that have to be addressed in future work. Among these are the accurate absolute potential alignment in implicit models, the accuracy of total energies including solvation energy contributions in slab geometries as well as of DFT energies (advanced functionals and empirical correction methods, e.g. by inclusion of experimental references) and the possibly necessary explicit treatment of interfacial water layers [16].

Acknowledgments

This project has received funding from the European Union's Horizon 2020 research and innovation programme under grant agreements No. 798532. The methods and software described in this chapter have been developed over the years thanks to the support of the Swiss Platform for Advanced Scientific Computing (PASC), MARVEL National Centre of Competence in Research of the Swiss National Science Foundation, and the University of North Texas. O.A. acknoledge support from ORAU for the organization of hackathon events, which have allowed some of the presented developments. We are grateful to Nicola Marzari for his fundamental role in pushing the research on continuum models for electrochemistry. We also acknowledge Davide Ceresoli, Michele Pavanello, Matthew Truscott, and Ismaila Dabo for their contributions to the presented work and for useful discussions.

References

1 S. Miertuš, E. Scrocco, and J. Tomasi. Electrostatic interaction of a solute with a continuum. A direct utilizaion of AB initio molecular potentials for the prevision of solvent effects. *Chemical Physics*, 55(1):117–129, 2 1981.

2 B Mennucci, E. Cancès, and J Tomasi. Evaluation of solvent effects in isotropic and anisotropic dielectrics and in ionic solutions with a unified integral equation method: theoretical bases, computational implementation, and numerical applications. *The Journal of Physical Chemistry B*, 101(49):10506–10517, 12 1997.

3 J Tomasi and M Persico. Molecular-interactions in solution – an overview of methods based on continuous distributions of the solvent. *Chemical Reviews*, 94(7):2027–2094, 1994.

4 J Tomasi, B Mennucci, and R Cammi. Quantum mechanical continuum solvation models. *Chemical Reviews*, 105(8):2999–3093, 2005.

5 C J Cramer and D G Truhlar. Implicit solvation models: Equilibria, structure, spectra, and dynamics. *Chemical Reviews*, 99(8):2161–2200, 1999.

6 Wei Zhao, Frédéric Leroy, Berit Heggen, Stefan Zahn, Barbara Kirchner, Sundaram Balasubramanian, and Florian Müller-Plathe. Are there stable ion-pairs in room-temperature ionic liquids? Molecular dynamics simulations of 1-n-Butyl-3-methylimidazolium hexafluorophosphate. *J. Am. Chem. Soc.*, 131(43):15825–15833, 2009.

7 J L Fattebert and F Gygi. Density functional theory for efficient ab initio molecular dynamics simulations in solution. *Journal of Computational Chemistry*, 23(6):662–666, 2002.

8 Jean-Luc Fattebert and François Gygi. First-principles molecular dynamics simulations in a continuum solvent. *International Journal of Quantum Chemistry*, 93(2):139–147, 2003.

9 Damián A Scherlis, Jean-Luc Fattebert, François Gygi, Matteo Cococcioni, and Nicola Marzari. A unified electrostatic and cavitation model for first-principles molecular dynamics in solution. *The Journal of Chemical Physics*, 124(7):074103, 2 2006.

10 Oliviero Andreussi, Ismaila Dabo, and Nicola Marzari. Revised self-consistent continuum solvation in electronic-structure calculations. *The Journal of Chemical Physics*, 136(6):064102, Feb 2012.

11 C. Dupont, O. Andreussi, and N. Marzari. Self-consistent continuum solvation (SCCS): The case of charged systems. *The Journal of Chemical Physics*, 139(21):214110, Dec 2013.

12 Oliviero Andreussi and Nicola Marzari. Electrostatics of solvated systems in periodic boundary conditions. *Physical Review B*, 90(24):245101, 12 2014.

13 Iurii Timrov, Oliviero Andreussi, Alessandro Biancardi, Nicola Marzari, and Stefano Baroni. Self-consistent continuum solvation for optical absorption of complex molecular systems in solution. *The Journal of Chemical Physics*, 142(3):034111, 1 2015.

14 Nicolas G. Hörmann, Oliviero Andreussi, and Nicola Marzari. Grand canonical simulations of electrochemical interfaces in implicit solvation models. *The Journal of Chemical Physics*, 150(4):041730, 1 2019.

15 Francesco Nattino, Matthew Truscott, Nicola Marzari, and Oliviero Andreussi. Continuum models of the electrochemical diffuse layer in electronic-structure calculations. *The Journal of Chemical Physics*, 150(4):041722, Jan 2019.

16 Nicolas Georg Hörmann, Zhendong Guo, Francesco Ambrosio, Oliviero Andreusi, Alfredo Pasquarello, and Nicola Marzari. Absolute band alignment at semiconductor-water interfaces using explicit and implicit descriptions for liquid water. submitted, 2019.

17 Oliviero Andreussi, Nicolas Georg Hörmann, Francesco Nattino, Giuseppe Fisicaro, Stefan Goedecker, and Nicola Marzari. Solvent-aware interfaces in continuum solvation. *Journal of Chemical Theory and Computation*, 15(3):1996–2009, 3 2019.

18 David Chandler. Interfaces and the driving force of hydrophobic assembly. *Nature* 2005 437:7059, 9 2005.

19 Casey P Kelly, Christopher J Cramer, and Donald G Truhlar. Adding explicit solvent molecules to continuum solvent calculations for the calculation of aqueous acid dissociation constants. *The journal of physical chemistry. A*, 110(7):2493–9, 2 2006.

20 Lars Blumenthal, Juhan Matthias Kahk, Ravishankar Sundararaman, Paul Tangney, and Johannes Lischner. Energy level alignment at semiconductor–water interfaces from atomistic and continuum solvation models. *RSC Adv.*, 7(69):43660–43670, 2017.

21 Juan Luis Pascual-Ahuir and Estanislao Silla. GEPOL: An improved description of molecular surfaces. I. Building the spherical surface set. *Journal of Computational Chemistry*, 11(9):1047–1060, 10 1990.

22 J L Pascualahuir, E Silla, and I Tunon. Gepol - an improved description of molecular-surfaces. 3. A new algorithm for the computation of a solvent-excluding surface. *Journal of Computational Chemistry*, 15(10):1127–1138, 1994.

23 G Scalmani and M J Frisch. Continuous surface charge polarizable continuum models of solvation. I. General formalism. *Journal of Chemical Physics*, 132(11):114110, 2010.

24 Maurizio Cossi, Giovanni Scalmani, Nadia Rega, and Vincenzo Barone. New developments in the polarizable continuum model for quantum mechanical and classical calculations on molecules in solution. *Journal of Chemical Physics*, 117(1):43–54, 2002.

25 Giuseppe Fisicaro, Luigi Genovese, Oliviero Andreussi, Sagarmoy Mandal, Nisanth N. Nair, Nicola Marzari, and Stefan Goedecker. Soft-sphere continuum solvation in electronic-structure calculations. *Journal of Chemical Theory and Computation*, 13(8):3829–3845, Jul 2017.

26 Andreas Klamt, Benedetta Mennucci, Jacopo Tomasi, Vincenzo Barone, Carles Curutchet, Modesto Orozco, and F. Javier Luque. On the performance of continuum solvation methods. a comment on "universal approaches to solvation modeling". *Accounts of Chemical Research*, 42(4):489–492, 4 2009.

27 J B Foresman, T A Keith, K B Wiberg, J Snoonian, and M J Frisch. Solvent effects. 5. Influence of cavity shape, truncation of electrostatics, and electron correlation ab initio reaction field calculations. *Journal of Physical Chemistry*, 100(40):16098–16104, 1996.

28 Alexander Held and Michael Walter. Simplified continuum solvent model with a smooth cavity based on volumetric data. *The Journal of Chemical Physics*, 141(17):174108, 11 2014.

29 Kiran Mathew, Ravishankar Sundararaman, Kendra Letchworth-Weaver, T. A. Arias, and Richard G. Hennig. Implicit solvation model for density-functional study of nanocrystal surfaces and reaction pathways. *The Journal of Chemical Physics*, 140(8):084106, 2 2014.

30 Stefan Ringe, Harald Oberhofer, Christoph Hille, Sebastian Matera, and Karsten Reuter. Function-space-based solution scheme for the size-modified Poisson–Boltzmann equation in full-potential DFT. *Journal of Chemical Theory and Computation*, 12(8):4052–4066, 8 2016.

31 Wen-Jin Yin, Matthias Krack, Xibo Li, Li-Zhen Chen, and Li-Min Liu. Periodic continuum solvation model integrated with first-principles calculations for solid surfaces. *Progress in Natural Science: Materials International*, 27(2):283–288, 4 2017.

32 Nicolas G. Hörmann, Oliviero Andreussi, and Nicola Marzari. Grand canonical simulations of electrochemical interfaces in implicit solvation models. *The Journal of Chemical Physics*, 150(4):041730, Jan 2019.

33 Ravishankar Sundararaman and William A. Goddard. The charge-asymmetric nonlocally determined local-electric (CANDLE) solvation model. *The Journal of Chemical Physics*, 142(6):064107, 2 2015.

34 Matthew Truscott and Oliviero Andreussi. Field-aware interfaces in continuum solvation. *The Journal of Physical Chemistry B*, 123(16):3513–3524, 4 2019.

35 Anna Pomogaeva and Daniel M. Chipman. Field-extremum model for short-range contributions to hydration free energy. *Journal of Chemical Theory and Computation*, 7(12):3952–3960, 12 2011.

36 Jaydeep P. Bardhan, Pavel Jungwirth, and Lee Makowski. Affine-response model of molecular solvation of ions: Accurate predictions of asymmetric charging free energies. *The Journal of Chemical Physics*, 137(12):124101, 9 2012.

37 Jaydeep P. Bardhan and Matthew G. Knepley. Communication: Modeling charge-sign asymmetric solvation free energies with nonlinear boundary conditions. *The Journal of Chemical Physics*, 141(13):131103, 10 2014.

38 Amirhossein Molavi Tabrizi, Spencer Goossens, Ali Mehdizadeh Rahimi, Matthew Knepley, and Jaydeep P. Bardhan. Predicting solvation free energies and thermodynamics in polar solvents and mixtures using a solvation-layer interface condition. *The Journal of Chemical Physics*, 146(9):094103, 3 2017.

39 Chaoyu Quan and Benjamin Stamm. Mathematical analysis and calculation of molecular surfaces. *Journal of Computational Physics*, 322:760–782, 10 2016.

40 Chaoyu Quan and Benjamin Stamm. Meshing molecular surfaces based on analytical implicit representation. *Journal of Molecular Graphics and Modelling*, 71:200–210, 1 2017.

41 Chaoyu Quan, Benjamin Stamm, and Yvon Maday. A domain decomposition method for the polarizable continuum model based on the solvent excluded surface. *Mathematical Models and Methods in Applied Sciences*, 28(07):1233–1266, 6 2018.

42 Ravishankar Sundararaman, Kathleen A. Schwarz, Kendra Letchworth-Weaver, and T. A. Arias. Spicing up continuum solvation models with SaLSA: The spherically averaged liquid susceptibility ansatz. *The Journal of Chemical Physics*, 142(5):054102, 2 2015.

43 Lars Onsager. Electric moments of molecules in liquids. *Journal of the American Chemical Society*, 58(8):1486–1493, 1936.

44 G. Fisicaro, L. Genovese, O. Andreussi, N. Marzari, and S. Goedecker. A generalized poisson and poisson-boltzmann solver for electrostatic environments. *The Journal of Chemical Physics*, 144(1):014103, Jan 2016.

45 A Ben-Naim. *Solvation Thermodynamics.* Plenum Press:New York, 1987.

46 A Ben-Naim. *Molecular Theory of Solutions.* Oxford University Press, 2006.

47 Matteo Cococcioni, Francesco Mauri, Gerbrand Ceder, and Nicola Marzari. Electronic-enthalpy functional for finite systems under pressure. *Physical Review Letters*, 94(14):145501, 4 2005.

48 H. Reiss, H. L. Frisch, and J. L. Lebowitz. Statistical mechanics of rigid spheres. *The Journal of Chemical Physics*, 31(2):369–380, 8 1959.

49 Claudio Amovilli and Benedetta Mennucci. Self-consistent-field calculation of pauli repulsion and dispersion contributions to the solvation free energy in the polarizable continuum model. *The Journal of Physical Chemistry B*, 101(6):1051–1057, 2 1997.

50 Ville Weijo, Benedetta Mennucci, and Luca Frediani. Toward a general formulation of dispersion effects for solvation continuum models. *Journal of Chemical Theory and ...*, (1):3358–3364, 2010.

51 Christopher J Cramer and Donald G Truhlar. A universal approach to solvation modeling. *Accounts of Chemical Research*, 41(6):760–768, 6 2008.

52 Ravishankar Sundararaman, Deniz Gunceler, and T a Arias. Weighted-density functionals for cavity formation and dispersion energies in continuum solvation models. *The Journal of chemical physics*, 141(13):134105, 10 2014.

53 Ryosuke Jinnouchi and Alfred B. Anderson. Aqueous and surface redox potentials from self-consistently determined Gibbs energies. *The Journal of Physical Chemistry C*, 112(24):8747–8750, 6 2008.

54 C. L. Fu and K. M. Ho. External-charge-induced surface reconstruction on ag(110). *Physical Review Letters*, 63(15):1617–1620, Oct 1989.

55 Alexander Y. Lozovoi and Ali Alavi. Reconstruction of charged surfaces: General trends and a case study of pt(110) and au(110). *Physical Review B*, 68(24), Dec 2003.

56 Nicéphore Bonnet and Nicola Marzari. First-principles prediction of the equilibrium shape of nanoparticles under realistic electrochemical conditions. *Physical Review Letters*, 110(8), Feb 2013.

57 Itamar Borukhov, David Andelman, and Henri Orland. Steric effects in electrolytes: A modified Poisson-Boltzmann equation. *Physical Review Letters*, 79(3):435–438, Jul 1997.

58 Ismaila Dabo, Yanli Li, Nicéphore Bonnet, and Nicola Marzari. Ab initio electrochemical properties of electrode surfaces. In *Fuel Cell Science*, pages 415–431. John Wiley & Sons, Inc., Oct 2010.

59 I. Borukhov, D. Andelman, and H. Orland. Adsorption of large ions from an electrolyte solution: a modified poisson–boltzmann equation. *Electrochimica Acta*, 46(2-3):221–229, Nov 2000.

60 M. Otani and O. Sugino. First-principles calculations of charged surfaces and interfaces: A plane-wave nonrepeated slab approach. *Physical Review B*, 73(11), Mar 2006.

61 Ryosuke Jinnouchi and Alfred B. Anderson. Electronic structure calculations of liquid-solid interfaces: Combination of density functional theory and modified poisson-boltzmann theory. *Physical Review B*, 77(24), Jun 2008.

62 Ismaila Dabo. *Towards first-principles electrochemistry*. PhD thesis, Massachusetts Institute of Technology, 2008.

63 I Dabo, E Cancès, Y L Li, and N Marzari. Towards First-principles Electrochemistry. *arXiv*, page 0901.0096, 12 2008.

64 Kendra Letchworth-Weaver and T. A. Arias. Joint density functional theory of the electrode-electrolyte interface: Application to fixed electrode potentials, interfacial capacitances, and potentials of zero charge. *Physical Review B*, 86(7), Aug 2012.

65 Deniz Gunceler, Kendra Letchworth-Weaver, Ravishankar Sundararaman, Kathleen A Schwarz, and T A Arias. The importance of nonlinear fluid response in joint density-functional theory studies of battery systems. *Modelling and Simulation in Materials Science and Engineering*, 21(7):074005, Oct 2013.

66 Kiran Mathew and Richard G. Hennig. Implicit self-consistent description of electrolyte in plane-wave density-functional theory. *arXiv*, page 1601.03346, 1 2016.

67 Ravishankar Sundararaman and Kathleen Schwarz. Evaluating continuum solvation models for the electrode-electrolyte interface: Challenges and strategies for improvement. *The Journal of Chemical Physics*, 146(8):084111, Feb 2017.

68 Ravishankar Sundararaman, Kendra Letchworth-Weaver, and Kathleen A. Schwarz. Improving accuracy of electrochemical capacitance and solvation energetics in first-principles calculations. *The Journal of Chemical Physics*, 148(14):144105, Apr 2018.

69 Marko M. Melander, Mikael J. Kuisma, Thorbjørn Erik Køppen Christensen, and Karoliina Honkala. Grand-canonical approach to density functional theory of electrocatalytic systems: Thermodynamics of solid-liquid interfaces at constant ion and electrode potentials. *The Journal of Chemical Physics*, 150(4):041706, Jan 2019.

70 H. Helmholtz. Ueber einige gesetze der vertheilung elektrischer ströme in körperlichen leitern mit anwendung auf die thierisch-elektrischen versuche. *Annalen der Physik und Chemie*, 165(6):211–233, 1853.

71 Stefan Ringe, Harald Oberhofer, Christoph Hille, Sebastian Matera, and Karsten Reuter. Function-space-based solution scheme for the size-modified poisson–boltzmann equation in full-potential DFT. *Journal of Chemical Theory and Computation*, 12(8):4052–4066, Jul 2016.

72 M. Gouy. Sur la constitution de la charge électrique à la surface d'un électrolyte. *Journal de Physique Théorique et Appliquée*, 9(1):457–468, 1910.

73 David Leonard Chapman. LI. a contribution to the theory of electrocapillarity. *The London, Edinburgh, and Dublin Philosophical Magazine and Journal of Science*, 25(148):475–481, Apr 1913.

74 Robert C. Harris, Alexander H. Boschitsch, and Marcia O. Fenley. Sensitivities to parameterization in the size-modified poisson-boltzmann equation. *The Journal of Chemical Physics*, 140(7):075102, Feb 2014.

75 Alexander H. Boschitsch and Pavel V. Danilov. Formulation of a new and simple nonuniform size-modified poisson-boltzmann description. *Journal of Computational Chemistry*, 33(11):1152–1164, Feb 2012.

76 Artem Baskin and David Prendergast. Exploring chemical speciation at electrified interfaces using detailed continuum models. *The Journal of Chemical Physics*, 150(4): 041725, Jan 2019.

77 Stefan Ringe, Harald Oberhofer, and Karsten Reuter. Transferable ionic parameters for first-principles poisson-boltzmann solvation calculations: Neutral solutes in aqueous monovalent salt solutions. *The Journal of Chemical Physics*, 146(13):134103, Apr 2017.

78 Otto Stern. Zur Theorie der Elektrolytischen Doppelschicht. *Z. Elektrochem.*, 30:508, 1924.

79 Artem Baskin and David Prendergast. Improving continuum models to define practical limits for molecular models of electrified interfaces. *Journal of The Electrochemical Society*, 164(11):E3438–E3447, 2017.

80 Paolo Giannozzi, Stefano Baroni, Nicola Bonini, Matteo Calandra, Roberto Car, Carlo Cavazzoni, Davide Ceresoli, Guido L Chiarotti, Matteo Cococcioni, Ismaila Dabo, Andrea Dal Corso, Stefano de Gironcoli, Stefano Fabris, Guido Fratesi, Ralph Gebauer, Uwe Gerstmann, Christos Gougoussis, Anton Kokalj, Michele Lazzeri, Layla Martin-Samos, Nicola Marzari, Francesco Mauri, Riccardo Mazzarello, Stefano Paolini, Alfredo Pasquarello, Lorenzo Paulatto, Carlo Sbraccia, Sandro Scandolo, Gabriele Sclauzero, Ari P Seitsonen, Alexander Smogunov, Paolo Umari, and Renata M Wentzcovitch. QUANTUM ESPRESSO: a modular and open-source software project for quantum simulations of materials. *Journal of Physics: Condensed Matter*, 21(39):395502, Sep 2009.

81 P Giannozzi, O Andreussi, T Brumme, O Bunau, M Buongiorno Nardelli, M Calandra, R Car, C Cavazzoni, D Ceresoli, M Cococcioni, N Colonna, I Carnimeo, A Dal Corso, S de Gironcoli, P Delugas, R A DiStasio, A Ferretti, A Floris, G Fratesi, G Fugallo, R Gebauer, U Gerstmann, F Giustino, T Gorni, J Jia, M Kawamura, H-Y Ko, A Kokalj, E Küçükbenli, M Lazzeri, M Marsili, N Marzari, F Mauri, N L Nguyen, H-V Nguyen, A Otero de-la Roza, L Paulatto, S Poncé, D Rocca, R Sabatini, B Santra, M Schlipf, A P Seitsonen, A Smogunov, I Timrov, T Thonhauser, P Umari, N Vast, X Wu, and S Baroni. Advanced capabilities for materials modelling with quantum ESPRESSO. *Journal of Physics: Condensed Matter*, 29(46):465901, Oct 2017.

82 O Andreussi, F Nattino, I Dabo, I Timrov, G Fisicaro, Goedecker S., and N Marzari. www.quantum-environment.org, 2018.

83 John P. Perdew, Kieron Burke, and Matthias Ernzerhof. Generalized gradient approximation made simple. *Physical Review Letters*, 77(18):3865–3868, Oct 1996.

84 John P. Perdew, Kieron Burke, and Matthias Ernzerhof. Generalized gradient approximation made simple [phys. rev. lett. 77, 3865 (1996)]. *Physical Review Letters*, 78(7):1396–1396, Feb 1997.

85 A. K. Rappe, C. J. Casewit, K. S. Colwell,W. A. Goddard, and W. M. Skiff. UFF, a full periodic table force field for molecular mechanics and molecular dynamics simulations. *Journal of the American Chemical Society*, 114(25):10024–10035, Dec 1992.

86 Martin Z. Bazant, Mustafa Sabri Kilic, Brian D. Storey, and Armand Ajdari. Towards an understanding of induced-charge electrokinetics at large applied voltages in concentrated solutions. *Advances in Colloid and Interface Science*, 152(1-2):48–88, Nov 2009.

87 Yasuya Nakayama and David Andelman. Differential capacitance of the electric double layer: The interplay between ion finite size and dielectric decrement. *The Journal of Chemical Physics*, 142(4):044706, Jan 2015.

88 J. K. Nørskov, J. Rossmeisl, A. Logadottir, L. Lindqvist, J. R. Kitchin, T. Bligaard, and H. Jonsson. Origin of the Overpotential for Oxygen Reduction at a Fuel-Cell Cathode. *J. Phys. Chem. B*, 108(46):17886–17892, 2004.

89 Jens Kehlet Nørskov, Thomas Bligaard, Ashildur Logadottir, J.R. Kitchin, J.G. Chen, S. Pandelov, and U. Stimming. Trends in the exchange current for hydrogen evolution. *J. Electrochem. Soc.*, 152(2):J23–J26, 2005.

90 Jan Rossmeisl, Jens K. Nørskov, Christopher D. Taylor, Michael J. Janik, and Matthew Neurock. Calculated phase diagrams for the electrochemical oxidation and reduction of water over pt(111). *J. Phys. Chem. B*, 110(43):21833–21839, 2006.

91 Nicolas Hörmann, Markus Jäckle, Florian Gossenberger, Tanglaw Roman, Katrin Forster-Tonigold, Maryam Naderian, Sung Sakong, and Axel Groß. Some challenges in the first-principles modeling of structures and processes in electrochemical energy storage and transfer. *J. Power Sources*, 275(0):531–538, 2015.

92 M. T. Dove. *Introduction to Lattice Dynamics*. Cambridge, UK: Cambridge University Press, November 1993.

93 Atsushi Togo and Isao Tanaka. First principles phonon calculations in materials science. *Scripta Materialia*, 108:1–5, 2015.

94 Fritz Körmann, Abed Al Hasan Breidi, Sergei L. Dudarev, Nathalie Dupin, Gautam Ghosh, Tilmann Hickel, Pavel Korzhavyi, Jorge A. Muñoz, and Ikuo Ohnuma. Lambda transitions in materials science: Recent advances in calphad and first-principles modelling. *physica status solidi (b)*, 251(1):53–80, 2014.

95 Nicolas G. Hörmann and A. Groß. Phase field parameters for battery compounds from first-principles calculations. *Phys. Rev. Materials*, 3:055401, May 2019.

96 J Rogal and K. Reuter. Ab initio atomistic thermodynamics for surfaces: A primer. In *Experiment, Modeling and Simulation of Gas- Surface Interactions for Reactive Flows in Hypersonic Flights*, pages (pp. 2–1 –2–18). Educational Notes RTO–EN–AVT–142, Paper 2. Neuilly–sur–Seine, 2007.

97 E. Kaxiras, Y. Bar-Yam, J. D. Joannopoulos, and K. C. Pandey. Ab initio theory of polar semiconductor surfaces. i. methodology and the (22) reconstructions of gaas(111). *Phys. Rev. B*, 35:9625–9635, Jun 1987.

98 Mathias Scheffler. Thermodynamic aspects of bulk and surface defects – first-principles calculations. In J. Koukal, editor, *Physics of Solid Surfaces*. Elsevier, Amsterdam, 1988.

99 Guo-Xin Qian, Richard M. Martin, and D. J. Chadi. First-principles study of the atomic reconstructions and energies of ga- and as-stabilized gaas(100) surfaces. *Phys. Rev. B*, 38:7649–7663, Oct 1988.

100 Karsten Reuter and Matthias Scheffler. First-principles atomistic thermodynamics for oxidation catalysis: Surface phase diagrams and catalytically interesting regions. *Phys. Rev. Lett.*, 90:046103, Jan 2003.

101 W. Zhang, J. R. Smith, and X.-G. Wang. Thermodynamics from ab initio computations. *Phys. Rev. B*,70:024103, Jul 2004.

102 Jutta Rogal, Karsten Reuter, and Matthias Scheffler. CO oxidation at Pd(100): A first-principles constrained thermodynamics study. *Phys. Rev. B*, 75:205433, May 2007.

103 Jutta Rogal, Karsten Reuter, and Matthias Scheffler. First-principles statistical mechanics study of the stability of a subnanometer thin surface oxide in reactive environments: Co oxidation at pd(100). *Phys. Rev. Lett.*, 98:046101, Jan 2007.

104 N.G. Hörmann and A. Groß. Stability, composition and properties of li2fesio4 surfaces studied by dft. *Journal of Solid State Electrochemistry*, 18(5):1401–1413, May 2014.

105 N. G. Hörmann, A. Groß, J. Rohrer, and P. Kaghazchi. Stabilization of the γ-Sn phase in tin nanoparticles and nanowires. *Appl. Phys. Lett.*, 107(12):123101, 2015.

106 Nicolas Georg Hörmann and Nicola Marzari. The effects of pH and potential at electrochemical interfaces. submitted, 2019.

107 Wolfgang Schmickler and Elizabeth Santos. *Interfacial Electrochemistry* 2nd edition. Springer Verlag Berlin Heidelberg, Berlin, Heidelberg, 2nd edition edition, 2010.

108 Sergio Trasatti. The "absolute" electrode potential – the end of the story. *Electrochim. Acta*, 35(1):269–271, 1990.

109 Luisa D'Urso, Cristina Satriano, Giuseppe Forte, Giuseppe Compagnini, and Orazio Puglisi. Water structure and charge transfer phenomena at the liquid–graphene interface. *Phys. Chem. Chem. Phys.*, 14:14605–14610, 2012.

110 Jiabo Le, Marcella Iannuzzi, Angel Cuesta, and Jun Cheng. Determining potentials of zero charge of metal electrodes versus the standard hydrogen electrode from density-functional-theory-based molecular dynamics. *Phys. Rev. Lett.*, 119:016801, Jul 2017.

111 Remi Khatib, Ashwinee Kumar, Stefano Sanvito, Marialore Sulpizi, and Clotilde S. Cucinotta. The nanoscale structure of the Pt-water double layer under bias revealed. *arXiv*, May 2019.

112 Lars Blumenthal, Juhan Matthias Kahk, Ravishankar Sundararaman, Paul Tangney, and Johannes Lischner. Energy level alignment at semiconductor-water interfaces from atomistic and continuum solvation models. *RSC Adv.*,7:43660–43670, 2017.

113 Ravishankar Sundararaman, William A. Goddard, and Tomas A. Arias. Grand canonical electronic density-functional theory: Algorithms and applications to electrochemistry. *J. Chem. Phys.*, 146(11):114104, 2017.

114 A. Frumkin, O. Petry, and B. Damaskin. The notion of the electrode charge and the lippmann equation. *J. Electroanal. Chem. Interfacial Electrochem.*, 27(1):81 –100, 1970.

115 Jan Rossmeisl, Karen Chan, Rizwan Ahmed, Vladimir Tripkovic, and Marten E. Björketun. ph in atomic scale simulations of electrochemical interfaces. *Phys. Chem. Chem. Phys.*, 15:10321–10325, 2013.

116 Haochen Zhang, William A. Goddard, Qi Lu, and Mu-Jeng Cheng. The importance of grand-canonical quantum mechanical methods to describe the effect of electrode potential on the stability of intermediates involved in both electrochemical CO_2 reduction and hydrogen evolution. *Phys. Chem. Chem. Phys.*, 20:2549–2557, 2018.

117 K. J. Vetter and J. W. Schultze. Stromflußbei Elektrosorptionsprozessen und Elektrosorptionswertigkeit γ. *Berichte der Bunsengesellschaft für physikalische Chemie*, 76(9):927–933, 1972.

118 Jianfeng Huang, Nicolas Hörmann, Emad Oveisi, Anna Loiudice, Gian Luca De Gregorio, Oliviero Andreussi, Nicola Marzari, and Raffaella Buonsanti. Potential-induced nanoclustering of metallic catalysts during electrochemical CO_2 reduction. *Nat. Commun.*, 9(1):3117, August 2018.

119 D.M. Kolb and J. Schneider. Surface reconstruction in electrochemistry: Au(100-(5 × 20), au(111)-(1 23) and au(110)-(1 2). *Electrochim. Acta*, 31(8):929–936, 1986.

4

Joint and grand-canonical density-functional theory

Ravishankar Sundararaman and Tomás A. Arias

4.1 Introduction

This chapter introduces and describes a novel approach to first principles computational electrochemistry that occupies a unique intermediate niche among available approaches, offering the advantages of the alternate approaches without the consequent disadvantages. While, certainly, a first principles approach is the right path for the development of a truly predictive and accurate computational electrochemistry, great care is needed to avoid the pitfalls of reductionism. For example, knowledge of the full many-body wave function of electronic and nuclear coordinates, even if somehow obtainable, would be far too detailed to be useful or informative. Moreover, such a description itself would not be truly 'first principles', as it would not consider processes within the nuclei and likely would ignore various relativistic processes. As with any scientific pursuit, one must describe nature in terms of appropriately chosen entities. If the resulting description is to be quantitatively predictive, so too must be the laws which the chosen entities follow.

The preceding chapter of this work (Chapter 3) presents an approach which treats the reactive components of an electrochemical system in terms of electrons and nuclei and the laws of quantum mechanics, but which treats the electrolyte as an effective continuum material that responds to and acts upon the active components of the system (Figure 4.1(c)). This approach gains a great computational benefit by recognizing that the precise details of the locations and arrangements of the vast majority of the atoms in the electrolyte are largely irrelevant to electrochemical processes. By radically reducing computational costs, this approach has the scientific benefit of allowing for the study of much more complex systems and the exploration of far more varied systems than could be studied through an approach tracking the positions of the atoms making up the electrolyte. The continuum approach, however, does sacrifice some accuracy and predictive power because the explicit arrangements of electrolyte atoms ('shell structure') near an active electrochemical process can be quite important and also may be of direct interest in themselves to electrochemistry. Complementing the continuum approach, the two following chapters (Chapters 5 & 6) present molecular dynamics approaches contending directly with the full statistics of the detailed arrangements of all of the electrolyte atoms (Figure 4.1(a)). This latter approach deals

Atomic-Scale Modelling of Electrochemical Systems, First Edition.
Edited by Marko M. Melander, Tomi T. Laurila, and Kari Laasonen.

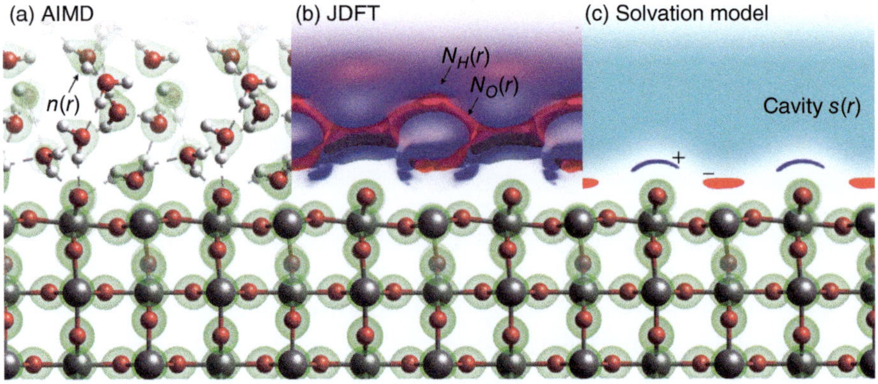

Figure 4.1 Theoretical representations of fluid interacting with a quantum system: (a) *ab initio* molecular dynamics (AIMD) with full atomistic detail, (b) joint density-functional theory (JDFT) employing a continuum description that retains atomic-scale liquid structure, and (c) continuum solvation model without atomic-scale information. Results shown above for an aqueous rutile TiO_2(110) surface include electron density $n(r)$ (green contours), solvent hydrogen and oxygen density distributions $N_H(r)$, $N_O(r)$ (blue and red contours), continuum solvent cavity shape function (light blue field) and bound charge density (± contours in blue/red respectively).

with more of the relevant underlying degrees of freedom, in particular shell structure, and, correspondingly, is more accurate and predictive. However, the increase in accuracy does come at the cost of *several orders of magnitude* in needed computational power. This limits not only the complexity of systems that can be studied, but also the breadth of searches that can be carried out in the search for new or improved electrochemical materials.

The focus of the present chapter is a new, complementary approach which maintains both computational efficiency and quantitative predictive power by focusing most closely upon the relevant microscopic entities for electrochemical processes. As with the two approaches described so far, the reactive components of the electrochemical system are treated fully quantum mechanically. Recognizing that contending with the full statistical distribution of all possible arrangements of the electrolyte atoms is not relevant, this new approach deals instead with the *average* density of each atomic species within the electrolyte environment. By not simply replacing the electrolyte with an effective continuum material, this new approach recognizes the atomic nature of the electrolyte and allows direct description of explicit arrangements of atoms near active electrochemical processes (Figure 4.1(b)). The resulting *joint density-functional theory* (JDFT) thus recovers the computational efficiency of continuum approaches while maintaining much of the accuracy of molecular dynamics approaches, thereby occupying an important niche in the suite of methods available to computational electrochemists.

That the statistical distribution of relevant arrangements of the electrolyte atoms can be replaced by the average density of each atomic species is guaranteed by a theorem first proved in 2005 [1] and refined in 2007 [2]. This theorem rigorously *joins* a classical density-functional theory to describe the electrolyte (an approach pursued in Chapter 8) with standard electronic density-functional theory to describe the reactive parts of the electrochemical system. The short-hand statement of this theorem (described in more detail below) is that the total free energy A associated with the equilibrium between a quantum-mechanical system and an electrolyte environment can be obtained *exactly* through the variational principle

$$A_0 = \min_{n,\{N_\alpha\}} \left(A_{\text{HK}}[n, V] + F_{\text{lq}}[N_\alpha] + \Delta A[n, V; \{N_\alpha\}]\right). \tag{4.1}$$

Here $n(\mathbf{r})$ is the quantum-statistical average electron density (green contours in Figure 4.1(b)) of the quantum system alone, $\{N_\alpha(\mathbf{r})\}$ are the statistically averaged densities of each atomic species α in the electrolyte (red and blue contours for α=O,H respectively in Figure 4.1(b)), and $V(\mathbf{r})$ is the electrostatic potential from the nuclei which make up the explicit, electrochemically active system.

The free energy minimized in the above principle (Eq. 4.1) includes three terms. The first term represents the free energy of the quantum mechanical system $A_{\text{HK}}[n, V]$ [3] were it in isolation by itself, as expressed in terms of its electron density and the electrostatic potential $V(\mathbf{r})$ from its nuclei. The second term is the *internal* free energy of the electrolyte system $F_{\text{lq}}[\{N_\alpha\}]$ [4, 5] were it in isolation by itself, while the final term $\Delta A[n, V(\mathbf{r}); \{N_\alpha\}]$ couples the two sub-systems.[1] This important theorem establishes not only that minimization of Equation 4.1 yields the temperature-dependent equilibrium free energy of the joint system, but also that the value of the $\{N_\alpha(\mathbf{r})\}$ at that minimum are indeed the equilibrium averages of the corresponding atomic site densities in the electrolyte.

There are tremendous advantages in having direct access to the *free* energy from a calculation of a single configuration of the quantum mechanically described system. First, equilibrium phases can now be determined directly by comparison of the free energy, without the need to extract the free energy as the potential of mean force, which generally requires extensive sampling to obtain accurate thermodynamic integration. Likewise, reaction energies are directly accessible as free energy differences, again without the need for thermodynamic integration. Finally, activation free energies ΔA, which control reaction rates, can be computed directly as the difference in free energy between just two calculations, one for the original state and one for the activated state.

Treatment of electrode potentials now also becomes straightforward. This arises because taking appropriate derivatives of the free energy yields direct access to all relevant thermodynamic average quantities, without the need for computationally expensive averaging over multiple configurations. For example, when treating an electrode within a canonical ensemble of a fixed total number of electrons N_{el}, the electrode potential comes directly from the thermodynamic derivative $\mu = \partial A/\partial N_{\text{el}}$. Note that, unlike molecular dynamics calculations which may report different 'electrode potentials' for different arrangements of atoms in the electrolyte, the actual electrode potential is a single thermodynamic value *averaged* over all atomic arrangements across the electrode surface in space and over time. Thus, rather than needing to average the potential over many repeated electrode potential calculations, joint density-functional theory provides the electrode potential directly from a single calculation. Also, as in Section 4.5, by a straight-forward Legendre transformation, Eq. 4.1 can be generalized to the grand potential to allow for direct calculation at fixed electrode potentials.

As a final application, the nuclei of the explicit quantum-mechanical calculation are generally held at fixed locations $\mathbf{R}_1, \ldots, \mathbf{R}_N$ during the minimization of Eq. 4.1, so that the free

1 Here and throughout, this chapter uses standard functional notation in which functions which take as input other functions and output real numbers are denoted with [ ]'s.

energy becomes a function of these coordinates, $A(\mathbf{R}_1, \ldots, \mathbf{R}_N)$. The corresponding derivatives then give the mean forces on the nuclear coordinates $f_i = -\partial A/\partial \mathbf{R}_i$, which can then be used to carry out molecular dynamics in the potential of mean force, yielding an accelerated molecular dynamics approach.

4.2 JDFT variational theorem and framework

4.2.1 Variational principle and underlying theorem

Joint density-functional theory describes the thermodynamic equilibrium between a system described at the quantum mechanical level and a liquid electrolyte environment. Hohenberg and Kohn's original rigorous formulation of density-functional theory [6], however, contemplates only the quantum-mechanical ground-state (zero temperature) properties of a system of interacting electrons moving in an external potential. This theory describes the electronic system entirely in terms of its average density and yields the ground-state energy and density as a result of minimization of the sum of a universal functional of the electron density and the coupling between the external potential and this density. Several extensions to this original theory are needed to provide a rigorous foundation for joint-density functional theory.

First, classical liquid electrolytes are inherently statistical in nature, existing only at non-zero temperatures and with properties which are critically dependent upon that temperature, whereas Hohenberg-Kohn theory only considers zero temperatures. Joint density-functional theory thus requires extension of Hohenberg-Kohn theory to yield free energies at non-zero temperatures. This extension was developed shortly after the original Hohenberg-Kohn theory by Mermin in 1965 [7].

Second, as illustrated in Figure 4.1(b), joint density-functional theory seeks to represent the state of the overall system in terms of not only the average density of electrons but also the average density of each atomic species making up the environment. The first to introduce such a multi-component density functional theory were Capitani *et al.*, who extended density functional theory to include not only electronic, but also nuclear degrees of freedom. This new density-functional theory framework indeed describes the ground state of a combined system of electrons and nuclei moving in an external potentials in terms of only the average electron density and the average nuclear densities, all within a unified variational framework [8].

A straightforward combination of the above unified density-functional theory framework with Mermin's non-zero temperature framework leads to the following, *exact* variational principle for the total thermodynamic free energy of an electron-nuclear system in a fixed external electrostatic potential $V(\mathbf{r})$,

$$A_0 = \min_{n_t(\mathbf{r}),\{N_\alpha(\mathbf{r})\}} \left\{ F_U[n_t(\mathbf{r}), \{N_\alpha(\mathbf{r})\}] + \int d^3r\, V(\mathbf{r}) \left(\sum_\alpha Z_\alpha N_\alpha(\mathbf{r}) - n_t(\mathbf{r}) \right) \right\}, \quad (4.2)$$

where $n_t(\mathbf{r})$ is the thermally and quantum mechanically averaged total number density of electrons, $N_\alpha(\mathbf{r})$ is the likewise averaged density of the nuclear species α (of atomic number Z_α), and F_U is a new internal free-energy functional for the *unified* electron-nuclear system that is directly analogous to the traditional Hohenberg-Kohn energy functional.[2] Note that the results here generalize naturally to various grand-canonical thermodynamic ensembles allowing for exchange of electrons and/or electrolyte species with external reservoirs to directly treat the impact of, for example, working at fixed electrode potentials or solution pH values.

The universality properties of the unified functional F_U may be seen directly from its construction within Levy's constrained search procedure [9],

$$F_U[n_t(\mathbf{r}), \{N_\alpha(\mathbf{r})\}] \equiv \min_{\hat{\rho}\to[n_t(\mathbf{r}),\{N_\alpha(\mathbf{r})\}]} \mathrm{Tr}\left(\hat{\rho}\hat{H} + k_B T\hat{\rho}\ln\hat{\rho}\right). \tag{4.3}$$

Here $k_B T$ is the thermal energy, $\hat{H}$ represents all interactions among and the kinetic energy of the electrons and nuclei, and $\hat{\rho}$ is the full quantum-mechanical density matrix for the electronic and nuclear degrees of freedom. The minimization is carried out over only those $\hat{\rho}$ which lead to the given densities $n_t(\mathbf{r})$ and $\{N_\alpha(\mathbf{r})\}$. From this construction it is clear that F_U, like $\hat{H}$ from which it derives, is independent of the external potential $V(\mathbf{r})$ and depends only upon the identities of the nuclear species α (and, implicitly, upon the temperature T), as in Capitani *et al.*'s original $T = 0$ formulation.

Applying the variational principle Eq. 4.2 in the joint density-functional theory context requires careful interpretation of the quantities involved. Joint density-functional theory takes the nuclear species α to be those comprising the electrolyte environment, so that the densities $\{N_\alpha(\mathbf{r})\}$ correspond precisely to the atomic densities in joint density-functional theory (Eq. 4.1), provided the position of each nucleus identifies the location of the corresponding atom. Finally, within joint density-functional theory, the electrostatic potential $V(\mathbf{r})$ in Eq. 4.2 is that which arises from the nuclei of the quantum mechanically treated part of the system. This gives

$$V(\mathbf{r}) \equiv \sum_A Z_A/|\mathbf{r} - \mathbf{R}_A|, \tag{4.4}$$

where these nuclei remain at *fixed* locations $\mathbf{R}_A$ and have atomic numbers Z_A. Note that, although the nuclei of the explicit system are at fixed locations within this framework, the nuclear species of the environment in Eq. 4.3 are quantum mechanical and fully free without any such restriction. Thus, Eqs. (4.2 and 4.3) account for all zero-point motion effects associated with lighter nuclear species in the electrolyte, such as may be associated with the protons in liquid water. Finally, note that $n_t(\mathbf{r})$ represents the *total* electron density of the entire system under consideration, including not only the electrons in the quantum mechanically treated part of the system, but also the large number of electrons in the electrolyte environment. Much of the potential computational advantage of the joint density-functional framework would be lost to computing the behavior of electrolyte electrons, unless there were a way to remove these electrons from explicit consideration.

The final development needed for a rigorous, workable theory is thus to effectively 'integrate out' the electrons associated with the electrolyte. To accomplish this, joint density-functional theory writes $n_t(\mathbf{r}) = n(\mathbf{r}) + n_e(\mathbf{r})$, where $n_e(\mathbf{r})$ is the electron density

2 Here and throughout, this chapter employs atomic units, in which the reduced Planck constant, electron mass and electron charge all have value unity, $\hbar = m_e = e = 1$.

associated with the electrolyte and $n(\mathbf{r})$ is the electron density associated with the quantum mechanically treated portion of the system (following the usage in Eq. 4.1). The next step is to first minimize over all allowable $n_e(\mathbf{r})$, and finally to minimize over all allowable $n(\mathbf{r})$. For this purpose, $n_t(\mathbf{r})$, $n_e(\mathbf{r})$ and $n(\mathbf{r})$ are all taken among the sets of N-representable functions satisfying the criteria of Gilbert[10] and integrating to the corresponding number of electrons for the respective subsystem. Because all thus defined $n_t(\mathbf{r})$ can be constructed as the sum of some allowable $n_e(\mathbf{r})$ and some allowable $n(\mathbf{r})$, and because all such allowable $n_e(\mathbf{r})$ and $n(\mathbf{r})$ sum to an allowable $n_t(\mathbf{r})$, this procedure is guaranteed to recover the final free energy in Eq. 4.2. Thus,

$$A_0 = \min_{n(\mathbf{r}),\{N_i(\mathbf{r})\}} \left\{ F_{\text{JDFT}}[n(\mathbf{r}), \{N_i(\mathbf{r})\}, V(\mathbf{r})] - \int d^3r \; V(\mathbf{r})n(\mathbf{r}) \right\} \tag{4.5}$$

where $V(\mathbf{r})$ is defined above in Eq. 4.4. Here

$$F_{\text{JDFT}}[n(\mathbf{r}), \{N_i(\mathbf{r})\}, V(\mathbf{r})] \equiv \min_{n_e(\mathbf{r})} \left\{ F_U[n(\mathbf{r}) + n_e(\mathbf{r}), \{N_\alpha(\mathbf{r})\}] + \int d^3r \; V(\mathbf{r}) \left(\sum_\alpha Z_\alpha N_\alpha(\mathbf{r}) - n_e(\mathbf{r}) \right) \right\} \tag{4.6}$$

is a new free energy functional that is universal in the sense that its functional form, like F_U from which it derives, depends only on the nature of the electrolyte and, implicitly, the temperature. Its dependence on the explicit electrochemically active system is only through the electrostatic potential of the explicit nuclei contributing to $V(\mathbf{r})$. The choice to separate the interaction $-\int d^3r \; V(\mathbf{r})n(\mathbf{r})$ in Eq. 4.5 from the definition of F_{JDFT} limits the interactions which the unknown functional F_{JDFT} must describe, easing the task of finding good approximations. Note, for example, that this choice leads to the coupling of $V(\mathbf{r})$ in Eq. 4.6 with a completely neutral charge distribution, thereby limiting to the greatest extent possible the dependence of F_{JDFT} on $V(\mathbf{r})$.

Eq. 4.5 gives the exact free energy and exact configuration of the solvent $\{N_\alpha(\mathbf{r})\}$. However, care must be taken in the interpretation of the $n(\mathbf{r})$ which yields the minimum value. The indistinguishability of electrons implies that there can be no fundamentally meaningful assignment of electrons as belonging either to the environment or to the system, and thus no exact formulation can give a unique result for $n(\mathbf{r})$ without some additional prescription. Indeed, for the exact $\{N_\alpha(\mathbf{r})\}$ and any $n(\mathbf{r})$ which integrates to the correct number of electrons and is everywhere less than the exact solution $n_t(\mathbf{r})$ so that $n_e(\mathbf{r}) = n_t(\mathbf{r}) - n(\mathbf{r})$ is allowable in the above sense, the minimization in Eq. 4.6 will find $n_e(\mathbf{r}) = n_t(\mathbf{r}) - n(\mathbf{r})$ and thus ultimately produce the same exact value for F_{JDFT} (Figure 4.2(a,b)). Only in cases where $n(r)$ exceeds $n_t(r)$ so that no physically valid minimizing $n_e(r)$ can be found will the functional F_{JDFT} be forced to take larger values (Figure 4.2(c)). There is thus a large set of $n(\mathbf{r})$ which yield the same minimum value in Eq. 4.6 (Figure 4.2(d)), and the variational principle embodied in Eqs. (4.5 and 4.6) satisfies the fundamental condition of not enforcing any particular, arbitrary decomposition of the total electron density into solvent and environment contributions.

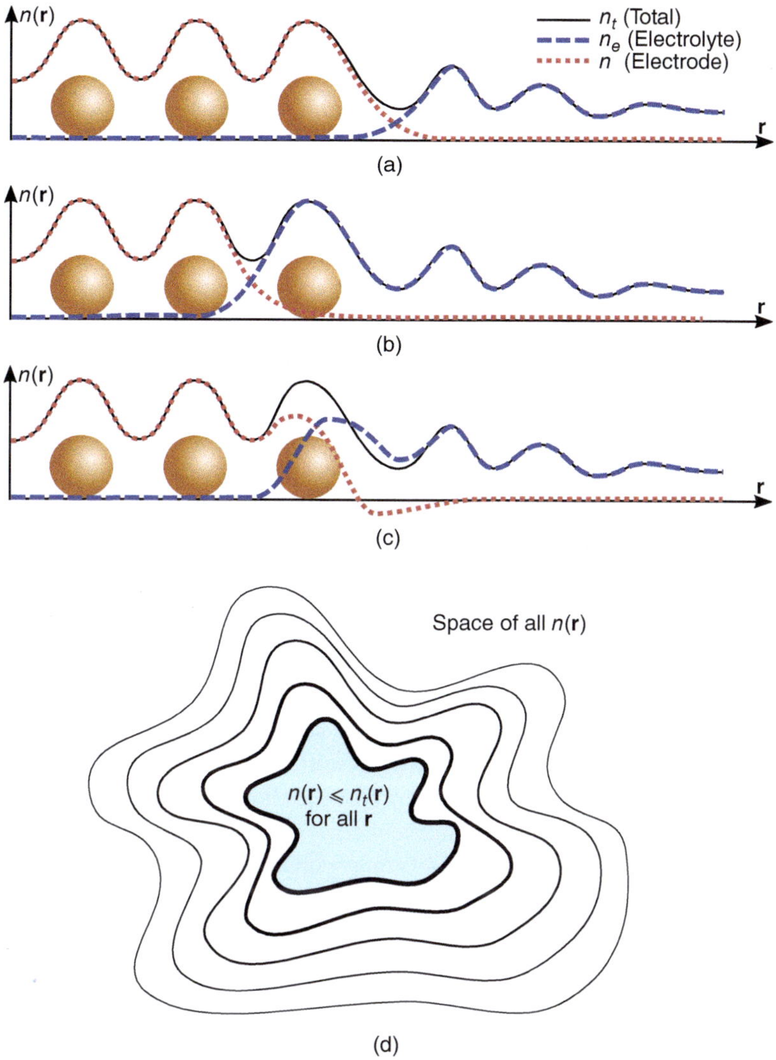

Figure 4.2 (a, b) Different allowed partitions of $n_t(\mathbf{r})$ to electrolyte $n_e(\mathbf{r})$ and explicit electrochemically active system $n(\mathbf{r})$, and (c) a disallowed partition where $n(\mathbf{r})$ exceeds $n_t(\mathbf{r})$. (d) Conceptual diagram of isosurfaces of the functional F_{JDFT} from Eq. 4.6 as a function of $n(r)$. Subset of all densities $n(r)$ nowhere exceeding the optimal total density $n_t(r)$ (shaded region), within which the functional is constant.

In practice, however, approximations to Eq. 4.6 break the above degeneracy and pick out a unique solution. The standard pseudopotential method, which replaces the effects of the nuclei and (relatively) inert core electrons of a solid or molecule with an effective or "pseudo"-potential [11], is in fact an example of an approximation which tracks only a portion of the total electron density. As is well known, this method provides results approaching chemical accuracy while suffering no pathologies related to the underlying degeneracy of an apportionment of electrons between two subsystems.

4.2.2 Separation of effects and regrouping of terms

With the functional dependence of $F_{\rm JDFT}$ established in Eq. 4.5, joint density-functional theory next separates out known components and leaves an unknown part to be approximated,

$$\begin{aligned} F_{\rm JDFT}[n(\mathbf{r}), \{N_i(\mathbf{r})\}, V(\mathbf{r})] \equiv F_{\rm KS}[n] \\ + F_{\rm lq}[\{N_\alpha(\mathbf{r})\}] + \int d^3r \; (V(\mathbf{r}) - \phi(\mathbf{r})) \sum_\alpha Q_\alpha N_\alpha(\mathbf{r}) \\ + U[n(\mathbf{r}), \{N_\alpha(\mathbf{r})\}; V(\mathbf{r})]. \end{aligned} \tag{4.7}$$

Above, $F_{\rm KS}[n(\mathbf{r})]$ is the standard, universal non-zero temperature quantum mechanical Kohn-Sham electron-density functional for the electrochemically active component of the system, $F_{\rm lq}[\{N_\alpha(\mathbf{r})\}]$ is the corresponding "classical" internal free-energy density-functional for the liquid electrolyte environment, $\phi(\mathbf{r}) \equiv -\int d^3\mathbf{r}' n(\mathbf{r}')/|\mathbf{r}-\mathbf{r}'|$ is the mean field electrostatic potential from the quantum mechanically treated part of the system, and Q_α are (approximate) atomic site densities for the atoms of the electrolyte. The new functional $U[n(\mathbf{r}), \{N_\alpha(\mathbf{r})\}; V(\mathbf{r})]$ describes the coupling between the two systems, and is defined formally and exactly as the difference between the exact functional $F_{\rm JDFT}$ and the remaining terms on the right-hand side of Eq. 4.7. Consequently, it has the same universality properties as the functional $F_{\rm JDFT}$ from which it derives. With the internal energies of the electrochemically active system and of the electrolyte subtracted out, U includes only coupling effects between the two systems. Moreover, because the second line of Eq. 4.7 already includes the mean electrostatic coupling between the total potential from the electrochemically active subsystem and the liquid through an atomic site-charge model, U primarily involves interactions beyond mean-field electrostatics, such as Thomas-Fermi repulsion and van der Waals dispersion effects. Since the *fixed* potential $V(\mathbf{r})$ of the explicit nuclei is already treated exactly at the mean-field level, U is only very weakly dependent on $V(\mathbf{r})$.

Combining the separation of effects (Eq. 4.7) with the variational principle (Eq. 4.5) and rearranging terms yields the standard form for the JDFT variational principle

$$\begin{aligned} A_0 = \min_{n(\mathbf{r}),\{N_i(\mathbf{r})\}} \Bigg\{ & F_{\rm KS}[n] - \int d^3r \; V(\mathbf{r})n(\mathbf{r}) \\ & + F_{\rm lq}[\{N_\alpha(\mathbf{r})\}] \\ & + \int d^3r \; (V(\mathbf{r}) - \phi(\mathbf{r})) \sum_\alpha Q_\alpha N_\alpha(\mathbf{r}) + U[n(\mathbf{r}), \{N_\alpha(\mathbf{r})\}, V(\mathbf{r})] \Bigg\}, \end{aligned} \tag{4.8}$$

where the terms on each line of the right hand side correspond, respectively, to the three terms $A_{\rm HK}[n, V]$, $F_{\rm lq}[N_\alpha]$, and $\Delta A[n, V; \{N_\alpha\}]$ as defined in Eq. 4.1.

4.2.3 Practical functionals and universal form for coupling

Practical calculations require explicit approximations to the functionals required in Eq. 4.8. For the solute energy functional $F_{\mathrm{HK}}[n]$, there are many established, efficacious density functionals [12] and even quantum-chemical methods [13] available. For the remaining two functionals (F_{lq} and U), two general approaches are available. First, one may combine these two terms into a single approximation for the liquid response, yielding a hierarchy of continuum solvation models, as described in Section 4.4 below and in [1] and [14]. Although such JDFT-based approximations have computed successfully the solvation free energies of molecules and ions [15, 16], much like the previous continuum methods they closely resemble, these approximations ignore atomic-scale liquid shell structure. Restoring proper treatment of such effects requires realistic, atomically-detailed classical density functionals F_{lq} that accurately capture effects pertinent to solvation, including the phase diagram, surface tension, and nonlinear dielectric response. Section 4.3 below discusses approaches for such functionals [4, 5]. This leaves the coupling functional U as the last functional requiring discussion.

As described above, with mean-field electrostatic interactions already separated out from the coupling functional $U[n, \{N_\alpha\}; V]$, to a good approximation, the functional can be taken as completely independent of the nuclear potential V. It should then be possible to construct a universal, highly accurate approximation $\tilde{U}[n, \{N_\alpha\}]$ without any dependence on V that is applicable *universally* to *any* quantum-mechanical system in *any* electrolyte environment. Because these remaining effects are all electronic in nature without any reference to an external potential, the most natural way to describe this coupling is the subtraction-based approach often used in embedding theory, [17–21],

$$\tilde{U}[n, \{N_\alpha\}] \equiv \mathcal{F}[n_t] - \mathcal{F}[n] - \mathcal{F}[n_e]. \tag{4.9}$$

Here $n_t \equiv n + n_e$ represents the electron density of the entire system, n_e represents the electron density of the electrolyte environment, and $\mathcal{F}$ is a suitable approximate functional for the kinetic and exchange-correlation energy of the electron gas.

Two final ingredients thus are needed to give an explicit description of Eq. 4.9 suited for calculations. First, the functional $\tilde{U}$ does not accept as input the electron density of the electrolyte n_e, but rather the atomic site densities $\{N_\alpha\}$. A model is thus needed for the electrolyte electron density in terms of the atomic site densities. For this, a convolution of atomic-site electron densities $n_\alpha(\mathbf{r})$

$$n_e(\mathbf{r}) = \sum_\alpha \int d^3R\, n_\alpha(\mathbf{r} - \mathbf{R}) N_\alpha(\mathbf{r}), \tag{4.10}$$

proves to give accurate results[22, 23]. Finally, a computationally efficacious form for $\mathcal{F}$ must be determined. To capture kinetic energy and exchange-correlation effects, the simple local Thomas-Fermi kinetic energy approximation plus the standard local-density approximation to the exchange-correlation functional $E_{xc}[n]$ of standard density-functional theory proves to describe the kinetic energy and local correlation effects well[22, 23]. Long-range dispersion interactions, however, are quite important in the determination of solvation free energies[16, 24], and recent work shows that inclusion of such effects does indeed lead to a very accurate and efficient description for $\tilde{U}$ [23].

4.3 Classical DFT with atomic-scale structure

The full power of the general framework of joint density-functional theory described above lies in the possibility of capturing atomic-scale liquid structure in a computationally expedient first-principles calculation. This critically requires that the classical DFT functional, $F_{\text{lq}}[\{N_\alpha(\mathbf{r})\}]$, reproduces the free energy and atomic density profiles $\{N_\alpha(\mathbf{r})\}$ of the liquid upon minimization. This section outlines the key requirements for a classical DFT functional to be suitable for JDFT, and recipes for developing functionals that satisfy them.

The first step in developing practical density-functional approximations is capturing a large fraction of the energy or free energy exactly using known terms, leaving behind the smallest possible unknown part to approximate. For example, Kohn-Sham electronic density functional theory [25] separates the total electronic energy functional as

$$F_{\text{KS}}[n] \equiv T[\Psi_0[n]] + \frac{1}{2}\int d\mathbf{r}d\mathbf{r}'\frac{n(\mathbf{r})n(\mathbf{r}')}{|\mathbf{r}-\mathbf{r}'|} + E_{\text{xc}}[n]. \tag{4.11}$$

The first term captures the energy of a non-interacting electron gas of density $n(r)$, which consists solely of kinetic energy, by first finding $\Psi_0[n]$, the lowest energy *non-interacting* wave function corresponding to that density, and then evaluating the kinetic energy of this wave function. This term is often abbreviated $T_s[n] \equiv T[\Psi_0[n]]$. The second term, the 'Hartree' energy, captures the mean-field Coulomb repulsion among the electrons, and the final term, the exchange-correlation functional, captures the remainder. The kinetic and Hartree terms capture a large fraction of the total energy, leaving behind a smaller piece that is relatively more manageable to approximate.

Similarly, classical density-functional theory of simple fluids decomposes the total free energy of the fluid as

$$F_{\text{lq}}[N] \equiv F_{\text{id}}[N] + F_{\text{ex}}[n], \tag{4.12}$$

$$\text{where } F_{\text{id}}[N] = \int d\mathbf{r}N(\mathbf{r})\left[k_B T\left(\log\frac{N(\mathbf{r})}{N_{\text{ref}}} - 1\right) + V(\mathbf{r}) - \mu\right] \tag{4.13}$$

is the exact *grand* free-energy of the inhomogeneous ideal gas in an external potential $V(\mathbf{r})$ and chemical potential μ. (N_{ref} is an arbitrary reference density that effectively controls the zero of the chemical potential.) The excess functional $F_{\text{ex}}[N]$ captures the remainder, analogous to the exchange-correlation functional in electronic DFT, and must be approximated.

For a classical DFT to be suitable for JDFT, $F_{\text{lq}}[\{N_\alpha(\mathbf{r})\}]$ should describe the free energy of a fluid in terms of the densities $N_\alpha(\mathbf{r})$ of each atom type α in the fluid. A real electrolyte with a molecular fluid and ions will therefore always involve more than one type of atomic site density. It is true that Eq. 4.13 can straightforwardly be summed over atomic species to describe a mixture of atomic ideal gases. However, in describing a molecular fluid such as water, this would imply that the starting point is a mixture of oxygen and hydrogen atoms, requiring F_{ex} to capture the intramolecular bonds to compose water molecules. This is a tall order, and we need to find a better starting point. The subsections below describe strategies to efficiently capture molecular geometry within the non-interacting (ideal gas) functional and then proceed to approximations of the intermolecular interactions described by F_{ex}.

4.3.1 Ideal gas functionals with molecular geometry

A better starting point for atomically-detailed classical DFT of electrolytes is a mixture of ideal *molecular* gases, such that F_{id} captures all primary *intramolecular* bonding, while F_{ex} is only responsible for relatively weaker intermolecular interactions. Specifically, development begins with an ideal gas of rigid molecules, where each molecule has atoms at positions $\mathbf{R}_{\alpha k}$ in a reference orientation while centered at the origin. Here, α identifies the atomic species and each molecule may have multiple atoms of type α, labeled by index k. Deviations of the molecular geometry from this rigid configuration, such as vibrations, may then be treated perturbatively from this starting point.

The ideal gas of rigid molecules poses a new challenge: the constraint that atoms must be bonded together as molecules implies that the atomic densities $\{N_\alpha(\mathbf{r})\}$ are no longer independent. Minimization of the free energy while satisfying the molecular geometry constraints becomes extremely challenging, requiring the introduction of auxiliary variables such as effective ideal gas atomic potentials $\psi_\alpha(\mathbf{r})$ [26, 27]. This, in turn, requires computationally cumbersome mapping from atomic densities to corresponding effective potentials and back, termed the 'inversion problem'. Three successive approaches have been pursued to address this challenge: use of effective ideal gas potentials as the *independent* variables, explicit integration over molecular orientations, and introduction of auxiliary effective fields.

4.3.1.1 Effective ideal gas potentials

Lischner et al. proposed to address the molecular constraint issue by switching to the ideal gas effective potentials $\psi_\alpha(\mathbf{r})$ as the independent variables instead [28], analogous to Kohn-Sham electronic DFT, which used non-interacting orbitals instead the of electron density n as the independent variables for the kinetic energy term in Eq. 4.11. Within Lischner et al.'s approach, the exact energy of the ideal molecular gas in external atomic potentials V_α and chemical potentials μ_α is

$$F_{\text{id}}[\{\psi_\alpha(\mathbf{r})\}] = \Omega^{(ni)}[\{\psi_\alpha\}] + \sum_\alpha \int d\mathbf{r} N_\alpha(\mathbf{r})(V_\alpha(\mathbf{r}) - \mu_\alpha - \psi_\alpha(\mathbf{r})) \tag{4.14}$$

$$\text{with } \Omega^{(ni)} \equiv -N_{\text{ref}} k_B T \int \prod_{\alpha,k} e^{-\psi_\alpha(\mathbf{r}_{\alpha k})/k_B T} d\mathbf{r}_{\alpha k} s(\{\mathbf{r}_{\alpha k}\}). \tag{4.15}$$

Here, the constraint function $s(\{\mathbf{r}_{\alpha k}\})$ picks out configurations $\{\mathbf{r}_{\alpha k}\}$ which satisfy the rigid molecule geometry (i.e. equivalent to the reference geometry $\{\mathbf{R}_{\alpha k}\}$ under rotations and translations), and the atomic densities are dependent variables given by the functional derivative, $N_\alpha = \delta\Omega^{(ni)}/\delta\psi_\alpha$ [28]. This approach eliminated the inversion problem; however, dealing with constraints in evaluating $\Omega^{(ni)}$ and its functional derivatives remained computationally cumbersome, especially with increasing complexity of the molecular geometry [29].

4.3.1.2 Integration over molecular orientations

The constraints above can be removed by sampling only sets of $\{\mathbf{r}_{\alpha k}\}$ that satisfy the molecular geometry. Specifically, one can achieve this by generating $\{\mathbf{r}_{\alpha k}\}$ from all possible translations $\mathbf{r}$ and rotations $\omega \in SO(3)$. This transforms (4.15) to

$$\Omega^{(ni)} = -N_{\text{ref}} k_B T \int \frac{\text{d}\mathbf{r}\text{d}\omega}{8\pi^2} \prod_{\alpha,k} e^{-\psi_\alpha(\mathbf{r}+\omega\circ\mathbf{R}_{\alpha k})/k_B T} \tag{4.16}$$

where $\omega\circ\mathbf{R}$ denotes rotating vector $\mathbf{R}$ by ω. Further, defining

$$p_\omega(\mathbf{r}) = N_{\text{ref}} \prod_{\alpha,k} e^{-\psi_\alpha(\mathbf{r}+\omega\circ\mathbf{R}_{\alpha k})/T}, \tag{4.17}$$

simplifies the above expressions for the ideal gas free energy and atomic densities to

$$\Phi_{\text{id}} = T \int \frac{\text{d}\mathbf{r}\text{d}\omega}{8\pi^2} p_\omega(\mathbf{r}) \left(\log \frac{p_\omega(\mathbf{r})}{N_{\text{ref}}} - 1 \right) + \sum_\alpha \int \text{d}\mathbf{r} N_\alpha(\mathbf{r})(V_\alpha(\mathbf{r}) - \mu_\alpha) \tag{4.18}$$

$$\text{and } N_\alpha(\mathbf{r}) = \sum_k \int \frac{\text{d}\omega}{8\pi^2} p_\omega \left(\mathbf{r} - \omega\circ\mathbf{R}_{\alpha k} \right). \tag{4.19}$$

Note that $p_\omega(\mathbf{r})$ is simply a product of Boltzmann factors for each atom in the molecule, given that $\psi_\alpha(\mathbf{r})$ are ideal-gas effective potentials that equal $V_\alpha(\mathbf{r}) - \mu_\alpha$ when Φ_{id} is minimized. In this framework, $p_\omega(\mathbf{r})$ is thus exactly the probability of finding a molecule centered at location $\mathbf{r}$ with orientation ω in the ideal gas.

Importantly, there are no constraints on $p_\omega(\mathbf{r})$, which makes it a convenient choice for the independent variables for free energy minimization in the classical DFT of molecular fluids [4]. Both the ideal gas free energy and the atomic densities can be evaluated readily from $p_\omega(\mathbf{r})$ using Eqs. 4.18 and 4.19, and the excess free energy can then be evaluated from the site densities. Indeed, [4] shows that direct free energy minimization over $p_\omega(\mathbf{r})$ exhibits superior iterative convergence compared to the ideal gas effective potential ($\{\psi_\alpha\}$) minimization.

In order to switch to minimization over $p_\omega(\mathbf{r})$, it is important that this approach produces the same result as minimizing the functional of atomic densities, carried out in terms of ideal gas effective potentials as the independent variables. A constrained search procedure [9] easily proves these two approaches to be equivalent [4]. Specifically, the liquid free energy in equilibrium is

$$\begin{aligned} F_{\text{lq}} &= \min_{p_\omega(\mathbf{r})} \left(F_{\text{id}}[p_\omega(\mathbf{r})] + F_{\text{ex}}[\{N_\alpha\}] \right) \\ &= \min_{p_\omega(\mathbf{r})} \left(\underbrace{T \int \frac{\text{d}\mathbf{r}\text{d}\omega}{8\pi^2} p_\omega(\mathbf{r}) \log \frac{p_\omega(\mathbf{r})}{N_{\text{ref}}}}_{-S_{\text{id}}[p_\omega(\mathbf{r})]} + E_{\text{id}}[\{N_\alpha\}] + F_{\text{id}}[\{N_\alpha\}] \right). \end{aligned} \tag{4.20}$$

The second line separates out the ideal gas entropy, S_{id}, which depends explicitly on p_ω from the ideal gas internal energy, E_{id}, which only depends on N_α. Going now to a constrained search splits the minimization over all $p_\omega(\mathbf{r})$ into an inner minimization over $p_\omega(\mathbf{r})$ corresponding to atomic densities $\{N_\alpha(\mathbf{r})\}$, followed by an outer minimization over the atomic densities,

$$F_{\text{lq}} = \min_{\{N_\alpha(\mathbf{r})\}} \left(\min_{p_\omega(\mathbf{r}) \mapsto \{N_\alpha(\mathbf{r})\}} T \int \frac{\text{d}\mathbf{r}\text{d}\omega}{8\pi^2} p_\omega(\mathbf{r}) \log \frac{p_\omega(\mathbf{r})}{N_{\text{ref}}} + E_{\text{id}}[\{N_\alpha\}] + F_{\text{id}}[\{N_\alpha\}] \right). \tag{4.21}$$

Finally, performing the inner, constrained minimization over $p_\omega(\mathbf{r})$ by introducing Lagrange multipliers $\psi_\alpha(\mathbf{r})$ for each $N_\alpha(\mathbf{r})$ constraint, results in Euler-Lagrange equations for the inner minimization, (obtained by taking $\delta/\delta p_\omega$) yielding precisely Eq. 4.17. The result of free energy minimization over $p_\omega(\mathbf{r})$ is therefore exactly the same as the previous ideal-gas effective potential methods of [26] and [28].

4.3.1.3 Auxiliary fields

The primary disadvantage of direct minimization over $p_\omega(r)$ is the computational time and storage requirements of processing several ($\sim$ 100) orientations for each point in space. The previous approach of optimizing over effective potentials dramatically reduced this computational complexity to computing with a limited number of effective potentials $\{\psi_\alpha(\mathbf{r})\}$ at each point. This latter approach provides an effective compression of the information contained within $p_\omega(r)$. There are, however, other possible ways to compress this information that prove computationally advantageous.

To understand the $\{\psi_\alpha(\mathbf{r})\}$ as just one possible effective compression of the information in $p_\omega(r)$ requires a broadened perspective. The exact equivalence between minimization over $p_\omega(\mathbf{r})$ and minimization over $\{\psi_\alpha(\mathbf{r})\}$ requires that all terms but the ideal gas entropy depend only on the atomic densities. This was indeed true above because the external potential took the form of atomic potentials, $V_\alpha(\mathbf{r})$. To gain a more general perspective, consider the case of a more general potential $V_\omega(\mathbf{r})$ that explicitly depends on the orientation orientation ω. In this orientation-dependent case, minimization over $\{\psi_\alpha(\mathbf{r})\}$ effectively maximizes the molecular ideal gas entropy $S_{\text{id}}[p_\omega(\mathbf{r})]$ over the subspace of $p_\omega(\mathbf{r})$ corresponding to a given set of atomic densities $\{N_\alpha(\mathbf{r})\}$. However, since the interaction no longer depends solely on $\{N_\alpha(\mathbf{r})\}$, the resulting free energy will in general be higher than direct unconstrained minimization over $p_\omega(\mathbf{r})$. The $\{\psi_\alpha(\mathbf{r})\}$ thus represent an approximate, *compressed representation* of the full $p_\omega(\mathbf{r})$ [4], in close analogy with lossy compression algorithms used for images and videos. Extracting $p_\omega(\mathbf{r})$ from the $\{\psi_\alpha(\mathbf{r})\}$, corresponding to "decompression", then proceeds by maximizing the entropy with the atomic density constraints implemented using $\{\psi_\alpha(\mathbf{r})\}$ as Lagrange multipliers.

[4] uses this perspective to define such representations in general, and then identifies particularly useful representations for classical DFT in the presence of strong electric fields, which is the most important case for solvation with JDFT. Following [4], one defines a set of variables $\{X_i = \hat{X}_i[p_\omega(\mathbf{r})]\}$ as a maximum-entropy compressed representation of $p_\omega(\mathbf{r})$, such that $p_\omega(\mathbf{r})[\chi_i]$ is the result of the constrained maximization

$$\frac{\delta}{\delta p_\omega(\mathbf{r})}\left(TS_{\text{id}}[p_\omega(\mathbf{r})] + \sum_i (\hat{X}_i[p_\omega(\mathbf{r})] - X_i)\chi_i\right) = 0. \tag{4.22}$$

using Lagrange multipliers χ_i for the X_i constraints. These Lagrange multipliers then become the independent variables for free energy minimization,

$$F_{\text{lq}} = \min_{\{\chi_i\}} \left(F_{\text{id}}\left[p_\omega(\mathbf{r})[\chi_i]\right] + F_{\text{ex}}\left[\{N_\alpha[p_\omega(\mathbf{r})[\chi_i]]\}\right], \right) \tag{4.23}$$

where $\Phi_{\text{id}}[p_\omega(\mathbf{r})]$ and $N_\alpha[p_\omega(\mathbf{r})]$ are given by (4.18) and (4.19) respectively. Within this general form, picking $X_i = N_\alpha(\mathbf{r})$ yields the ideal-gas effective potential representation, while picking $X_i = p_\omega$ itself yields p_ω itself as the representation (no compression).

In the context of solvation within JDFT, the external potential on the liquid is due primarily to the electrostatic potential from the electronic system. Because the electronic system is spatially separated from the liquid, the potential experienced by a molecule or ion can be expanded in a multipole series, with lower order multipoles being more important than higher order ones. Consequently, as developed in [4], multipole probability densities give a particularly effective compressed representation,

$$\hat{M}^j_{m_1 m_2}(\mathbf{r})[p_\omega(\mathbf{r})] = \frac{2j+1}{8\pi^2} \int d\omega p_\omega(\mathbf{r}) D^j_{m_1 m_2}(\omega), \tag{4.24}$$

where $D^j_{m_1 m_2}(\omega)$ are the Wigner D-matrices, which are the irreducible matrix representations of the $SO(3)$ rotation group [30].

As discussed above, the independent variables for minimization are then the Lagrange multipliers $\mu^j_{m_1 m_2}(\mathbf{r})$ involved in constraining the multipole densities $\hat{M}^j_{m_1 m_2}(\mathbf{r})$, and the orientation probability depends on these independent variables as

$$p_\omega(\mathbf{r})[\mu^j_{m_1 m_2}(\mathbf{r})] = N_{\mathrm{ref}} \prod_j \prod_{m_1,m_2=-j}^{+j} \exp\left(-\frac{\mu^j_{m_1 m_2}(\mathbf{r}) D^j_{m_1 m_2}(\omega)}{T}\right). \tag{4.25}$$

While this approach is exact when including all terms $j \to \infty$, practical calculations must truncate to $j \leq j_{\max}$ for some $j_{\max}$. In particular, $j_{\max} = 1$ accounts for the dominant monopole and dipole densities, with the corresponding independent variables being the ideal-gas effective local chemical potential and local electric fields (instead of local atomic potentials). [4] shows robust iterative minimization of classical DFT functionals in this representation that is simultaneously accurate and computationally practical in both time and memory requirements.

4.3.2 Minimal excess functionals for molecular fluids

Having captured exactly the first and largest contribution to the liquid free energy functional $F_{\mathrm{lq}}[\{N_\alpha\}]$, namely the intramolecular bonds that convert an ideal gas of atoms to an ideal gas of molecules, the remaining task is to account for all intermolecular interactions, as captured by $F_{\mathrm{ex}}[\{N_\alpha\}]$ in Eq. 4.12. Analogous to the development of exchange-correlation functionals in electronic DFT, there are several complementary approaches to classical DFT excess functional development ranging from the simple and non-empirical to the highly-parametrized and computationally expensive. The discussion below focuses on a specific, minimally-empirical approach developed in [4] and [5] that is suitable for JDFT solvation with atomic-scale structure.

Two model fluids inform the classical DFT treatment of physical liquids. First, virtually all classical DFT functionals begin with the hard sphere fluid, which consists of particles that perfectly repel one another within a given cutoff distance but do not interact otherwise. For this system, the free energy of the fluid is purely entropic, resulting from a reduction of configuration space from overlapping spheres. A family of excess functionals known as Fundamental Measure Theory (FMT) provides highly accurate approximations to the entropy of such fluids, including the case of mixtures of spheres of different sizes. Briefly, the FMT class of excess functionals for the 3D hard sphere fluid take the form

$$F_{\mathrm{HS}}[N] = T \int d\mathbf{r} \left(n_0 \log \frac{1}{1-n_3} + \frac{n_1 n_2 - \mathbf{n}_{v1} \cdot \mathbf{n}_{v2}}{1-n_3} + \cdots \right), \tag{4.26}$$

with variants differing in the final term(s) [31–33]. The quantities n_3, n_2 etc. are weighted densities that measure the number of spheres that overlap in volume, at the surface and so on. For example, $n_3(\mathbf{r}) = \int d\mathbf{r}'\Theta(r_{HS} - |\mathbf{r} - \mathbf{r}'|)N(\mathbf{r}')$ counts the number of particles within the hard sphere radius r_{HS}. See [34] for a comprehensive and accessible review of these FMT hard sphere entropy functionals.

The second fluid informing classical DFTs is the Lennard-Jones fluid, which interacts with the standard 6-12 potential,

$$U_{LJ} = 4\epsilon\left[\left(\frac{\sigma}{r}\right)^{12} - \left(\frac{\sigma}{r}\right)^{6}\right]. \tag{4.27}$$

This interaction features repulsion below separation $r = \sigma$ at larger distances of an energy scale set by ϵ, which eventually decays to zero at larger distances. Classical DFT treatment of such fluids begins by decomposing the potential into repulsive and attractive parts [35] as

$$U_R(r) = \begin{cases} \epsilon + 4\epsilon\left[\left(\frac{\sigma}{r}\right)^{12} - \left(\frac{\sigma}{r}\right)^{6}\right], & r < 2^{1/6}\sigma \\ 0, & r \geq 2^{1/6}\sigma \end{cases} \tag{4.28}$$

$$U_A(r) = \begin{cases} -\epsilon, & r < 2^{1/6}\sigma \\ 4\epsilon\left[\left(\frac{\sigma}{r}\right)^{12} - \left(\frac{\sigma}{r}\right)^{6}\right], & r \geq 2^{1/6}\sigma. \end{cases} \tag{4.29}$$

The free energy functional for this fluid can be approximated by treating the fluid interacting with $U_R(r)$ alone using fundamental measure theory, typically with a hard sphere radius $R_{HS} = \sigma/2$, and then accounting for the effects of $U_A(r)$ perturbatively. Mean-field perturbation, in particular, leads to the excess functional

$$F_{ex}^{(MF)}[N(\mathbf{r})] \approx F_{HS}[N] + \frac{1}{2}\int d\mathbf{r}\int d\mathbf{r}' N(\mathbf{r})U_A(|\mathbf{r} - \mathbf{r}'|)N(\mathbf{r}'), \tag{4.30}$$

where the second term simply evaluates the attractive potential integrated over the average liquid densities.

However, such approaches do not capture correlation effects in the density, and do not reproduce the equation of state of the liquid, especially at high densities. In contrast, in electronic density-functional theory, great care is taken to ensure that the energy of the uniform electron gas is always recovered exactly, even with the simplest local density approximation (LDA) to the exchange correlation functional. The analogous strategy in classical DFT fluids, where nonlocality is critical and LDA does not suffice, is to employ the weighted density approximation (WDA) [36], which applies the equation of state constraint to a weighted (convolved) liquid density instead of simply the local density. For Lennard-Jones fluids in particular, the modified WDA functional [37] recasts the mean-field term into a nonlinear weighted-density form

$$F_{ex}^{(MWF)}[N(\mathbf{r})] \approx F_{HS}[N] + \int d\mathbf{r} N(\mathbf{r}) A_{att}^{LJ}(w_A * N), \tag{4.31}$$

with the mean-field weight function $w_A(r)$ set to the normalized perturbation potential $w_A(r) = U_A(r)/\int 4\pi r'^2 dr' U_A(r')$. Here, $A_{att}^{LJ}(N)$ is the attractive component of the uniform fluid free energy, defined as the difference $A_{LJ}(N) - A_{HS}(N)$ of the free energies of uniform

Lennard-Jones and hard sphere fluids at the same density. This form is now explicitly constrained to reproduce the uniform fluid limit (and the equation of state) exactly, and reduces to the mean-field form in the inhomogeneous low density limit.

[4] and [5] build on the above modified WDA approach [37] to additionally capture the key quantities relevant for JDFT solvation:

1. *Nonlinear dielectric response*: F_{lq} must correctly capture the electric interaction energy with the electronic system even at high electric fields $\sim 10^{10}$ V/m that are ubiquitous at the atomic scale.
2. *Cavitation free energy*: F_{lq} must correctly account for the free energy resulting from displacing the fluid in the region of space occupied by the electronic system.
3. *Atomic-scale liquid structure*: upon minimizing F_{lq}, the resulting N_α should correspond to the thermodynamically averaged structure of the liquid down to the atomic scale.

An implicit design choice for these functionals was simplicity and stability: the functionals should primarily require only bulk experimental/molecular dynamics data for the fluid equation of state and microscopic molecular properties derivable from electronic DFT calculations, and should remain stable in the quite strong short-range atomic potentials involved within JDFT. Previous complex functionals for real liquids such as water require structural data such as pair correlations from neutron scattering, were highly complex and computationally intensive, or were not stable when subject to strongly attractive atomic potentials [27–29, 38–40].

A general recipe for free energy functionals for molecular fluids [5] that satisfies the above requirements takes the overall form

$$F_{\text{lq}}[N_\alpha] = F_{\text{id}}[p_\omega] + F_{\text{HS}}[N_0] + \int d\mathbf{r}\bar{N}A_{\text{att}}(w_A * \bar{N}) + F_\epsilon[p_\omega]. \tag{4.32}$$

Here, the first two terms form a reference molecular hard-sphere fluid with $N_0(\mathbf{r}) = \int d\omega p_\omega(\mathbf{r})$ being the density of the molecule geometric center. The third term captures the attractive interactions similar to the modified WDA for the Lennard-Jones fluid, applied to an average liquid density, $\bar{N}(\mathbf{r}) = \sum_\alpha N_\alpha(\mathbf{r})\chi_\alpha / \sum_\alpha \chi_\alpha$, weighted by the dipole polarizabilities χ_α of the atoms. The free energy function $A_{\text{att}}(N)$ is constrained to experimental equations of state for the fluids, such as the Jeffrey-Austin equation of state for liquid water [41] or the Tao-Mason equation of state [42] for less polar fluids.

The weight function w_A in the third term of Eq. 4.32 is set to the normalized Lennard-Jones attractive part as discussed above. The Lennard-Jones range parameter σ is not yet determined, but strongly affects the cohesive energy of the liquid and its interfacial tension. Fortunately, selecting σ to reproduce the bulk interfacial tension of the liquid also captures the free energy for forming cavities with sizes down to the atomic scale, as the discussion of Figure 4.3 below will show.

The final term of Eq. 4.32 captures all long-range Coulomb interactions in the fluid, which in competition with the ideal gas entropy, critically determines the nonlinear dielectric response of the fluid. Specifically, [5] proposes the *ansatz*

$$F = \sum_\alpha \int d\mathbf{r} \frac{N_\alpha \mathcal{P}_\alpha^2}{2C_{\text{pol}}\chi_\alpha} + \frac{1}{2}\int d\mathbf{r} \int d\mathbf{r}' \frac{\rho_{\text{MF}}(\mathbf{r})\rho_{\text{MF}}(\mathbf{r}')}{|\mathbf{r}-\mathbf{r}'|} + \frac{C_{\text{rot}}^{-1} - 1}{N_{\text{bulk}} p_{\text{mol}}^2/3T} \int d\mathbf{r}\bar{\mathcal{P}}_{\text{rot}}^2, \tag{4.33}$$

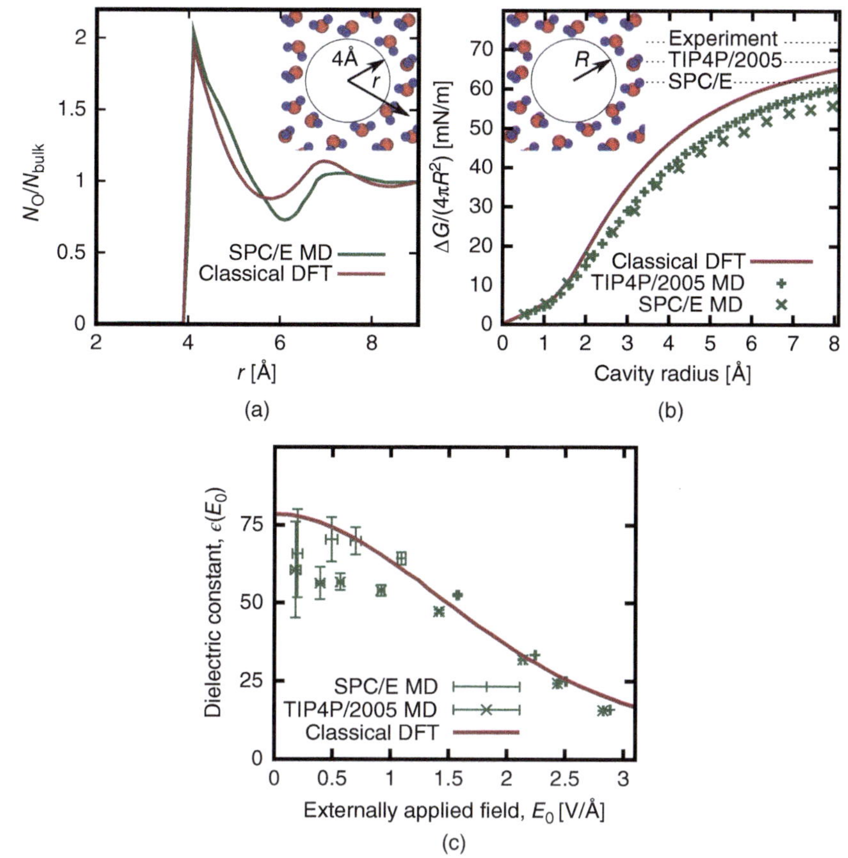

Figure 4.3 Classical DFT suitable for JDFT solvation must accurately capture (a) atomic-scale inhomogeneous liquid structure, (b) interfacial free energy from atomic to macroscopic scales, and (c) non-linear dielectric response for characteristic field strengths at the atomic scale. Results shown compare the liquid water functional in [5] against SPC/E [43] and TIP4P/2005 [44] molecular dynamics simulations. (Parts (b–c) adapted from [5] with permission from AIP Publishing.)

where the first term captures the internal energy of polarization due to vibrational and electronic response of the fluid, the second term captures mean-field interactions between all induced charge densities (due to molecular rotation and polarization), and the final term accounts for correlations in the rotational response in a local polarization-density approximation on the rotational polarization. The first term is required because, although the free energy for molecular rotation is implicitly captured by the ideal gas entropy discussed above, the internal energy for molecular polarization is not included in an ideal gas of *rigid* molecules.

The polarization response above introduces an additional independent variable $\mathcal{P}_\alpha(\mathbf{r})$, which is the dipole induced per atom of type α at location $\mathbf{r}$. The first term of Eq. 4.33 captures the internal energy cost of inducing this dipole, either by vibration or electronic polarization, totaling to a dipole polarizability of χ_α on each atom. The net charge density from the rotational and translational response captured by $N_\alpha(\mathbf{r})$ and this polarization response,

$$\rho_{\mathrm{MF}} = \sum_{\alpha} Z_\alpha w_{\mathrm{MF}} * N_\alpha(\mathbf{r}) - \nabla \cdot \sum_{\alpha} w_{\mathrm{MF}} * N_\alpha \boldsymbol{P}_\alpha, \tag{4.34}$$

where Z_α are the effective atomic charges in the fluid, then contributes a mean-field interaction in the second term of Eq. 4.33, which is exactly like the Hartree term. However, exactly as in electronic DFT, this mean-field approximation introduces self-interaction errors and misses correlations. This functional minimizes self-interaction errors by using a weight function w_{MF} to limit the Coulomb interaction at short ranges [5]. Additionally, missing correlations result in incorrect bulk dielectric response. The functional introduces local polarization density corrections $\propto \int d\mathbf{r}|P(\mathbf{r})|^2$ to capture the correct bulk static and optical dielectric constants. These corrections effectively scale up the local field that interacts with the polarization (analogous to the Clausius-Mossoti relation) by factors C_{pol} and C_{rot} respectively. While C_{pol} integrates with the polarization internal energy term (first term of Eq. 4.33), the rotational contribution stays distinct as the third term of Eq. 4.33 in terms of the weighted rotational polarization-density,

$$\bar{\boldsymbol{P}}_{\mathrm{rot}} = w_{\mathrm{MF}} * \int \frac{d\omega}{8\pi^2} p_\omega(\mathbf{r}) \omega \circ \mathbf{p}_{\mathrm{mol}}, \tag{4.35}$$

where $\omega \circ \mathbf{p}_{\mathrm{mol}}$ is the dipole moment of the molecule at orientation ω. Analytically evaluating the polarization and rotational dielectric response of the bulk fluid introduced by Eq. 4.33 then connects C_{pol} and C_{rot} to the experimentally known bulk dielectric constants,

$$C_{\mathrm{pol}} = \frac{\epsilon_\infty - 1}{4\pi N_{\mathrm{bulk}} \sum_\alpha \chi_\alpha} \quad \text{and} \quad C_{\mathrm{rot}} = \frac{\epsilon_b - \epsilon_\infty}{4\pi N_{\mathrm{bulk}} p_{\mathrm{mol}}^2 / 3T}. \tag{4.36}$$

thereby fully constraining all unknowns in the functional.[3]

The development above is fully general for any liquid containing relatively small molecules or ions that can be treated from a starting point of rigid molecules. It also easily generalizes to mixtures, as required for treatment of electrolytes. The rigid-molecule ideal gas pieces add together for the various constituents, while the hard sphere entropy functional accounts for mixtures of hard sphere fluids by adding together the weighted density contributions from each [34]. The long-range parts straightforwardly include the mean-field interactions between all constituents, adding together the charge densities from polarization and rotational response of all species in solution. The modified WDA remains in place to enforce the equation of state of the solvent, but needs to be augmented with functionals $F_{\mathrm{mix}}[N_{\mathrm{solv}}, N_{\mathrm{ion}}]$ to account for the contributions to the cohesive energy from the electrolyte ions. This can be done using mean-field perturbation theory at low concentrations, or using another modified WDA piece at higher concentrations, provided the equation of state of the electrolyte is available from experiment or simulations. For simplicity, this section discusses classical DFT results for pure solvents (water), and Section 4.4 below will discuss electrolytes in more detail in the context of JDFT-derived continuum solvation models.

Figure 4.3 shows that the resulting functional F_{lq} captures the three characteristics necessary for accurate JDFT solvation. First, Figure 4.3(a) shows that the classical DFT predicts

3 All atomic charges, dipole polarizabilities, molecular geometries etc. are obtained directly from electronic DFT calculations [5].

the atomic-scale shell structure around a 4 Å cavity (within which O atoms are excluded by a repulsive potential) in excellent agreement with classical molecular dynamics simulations. Next, with varying cavity size, the free energy per unit surface area for introducing this cavity into the liquid (Figure 4.3(b)) is also in excellent agreement between classical DFT and molecular dynamics. The free energy is proportional to the volume of small cavities and transitions to an interfacial tension (proportional to the area) on the scale of a few nanometers. In fact, the molecular dynamics simulations with different pair potentials (SPC/E [43] and TIP4P/2005 [44]) underestimate the bulk interfacial tension, while the classical DFT captures it by construction. Similarly, Figure 4.3(c) shows that the classical DFT captures the dielectric response at fields exceeding 10^{10} V/Å, accounting for the saturation of the rotational response, in excellent agreement with molecular dynamics simulations. Once again, at low fields, the classical DFT reproduces the experimental bulk dielectric constant exactly by construction, unlike the molecular dynamics models.

Underscoring the computational efficiency of the classical DFT approach, the calculations presented in Figure 4.3 require only a single free energy minimization to predict the equilibrium density, free energy, and polarization response. In contrast, the molecular dynamics simulations require statistical averaging over several nanoseconds (10^5–10^6 configurations), and an additional coupling constant integration to calculate the free energies [5, 45]. This is particularly important when the classical DFT is coupled to electronic DFT in JDFT because, in contrast with *ab initio* molecular dynamics, this reduces the number of required electronic DFT calculations to *one*.

4.4 Continuum solvation models from JDFT

Joint density-functional theory (JDFT) provides a rigorous framework for capturing liquid and electrolyte environment effects in terms of atomic densities, and classical DFT liquid functionals are available to realize the full potential of JDFT to deliver atomic scale liquid structure. Conventional continuum solvation models may be thought of as a special, highly-simplified approximation to the liquid and coupling functionals, where the JDFT theorem simply gives license to construct such approximations. Much more importantly, 'full' JDFT with classical DFT liquids provides a powerful starting point for the systematic construction of improved continuum models.

Whereas conventional continuum solvation models generally describe the dielectric fluid response as *linear* and/or *local*, from JDFT it is apparent that reality is significantly more complex. As discussed in Section 4.3, to accurately describe the environment, the classical DFT component of JDFT must capture the *nonlinear* response of the fluid, including saturation of the rotations of the molecular dipoles. Additionally, the atomic scale structuring of the liquid is intrinsically a *non-local* response, wherein molecules respond to atomic potentials or electric fields on characteristic molecular dimensions and then produce atomic and charge-density response on a similar scale. The following sections show that it indeed is possible to capture these effects at a continuum theory level by constructing varying levels of approximation to JDFT, stepping down in complexity from nonlinear and nonlocal classical DFT, through continuum models that are either nonlocal or nonlinear, all the way down to local and linear models.

4.4.1 JDFT linear response: nonlocal 'SaLSA' solvation

The central concept in a continuum solvation model is the 'cavity', a region of space occupied by the electronic system from which the solvent is excluded. In classical DFT, the atomic densities $N_\alpha(\mathbf{r})$ are fully optimized to minimize the free energy and thereby automatically avoid the region occupied by the electronic system in order to reduce the Thomas-Fermi repulsion inherent in the coupling functional. The resulting liquid density profile indeed corresponds to a cavity and is relatively simple to model in terms of the molecule center density $N_0(\mathbf{r}) = N_{\text{bulk}}s(\mathbf{r})$ through a cavity shape function $s(\mathbf{r})$ normalized so that its value transitions from zero inside the cavity to one outside the cavity. Taking as a reference point a fluid conforming to a model cavity shape $s(r)$ but with unpolarized molecules with a uniform distribution of angles, one can construct a continuum model by focusing on how the fluid responds to minimize its free energy. The result of this analysis is the Spherically-averaged Liquid Susceptibility *Ansatz* (SaLSA) solvation model [16], which retains the nonlocality of JDFT in a continuum solvation model.

The SaLSA analysis begins with full JDFT (Eq. 4.8) and the classical DFT from Section 4.3.2. The first step is to separate all terms that depend on the molecule center density from those depending on molecular orientation and polarization [16],

$$\begin{aligned}F_{\text{JDFT}} &= F_{\text{HK}}[n] + F_0[N_0] + T\int d\mathbf{r}\int \frac{d\omega}{8\pi^2}p_\omega(\mathbf{r})\left[\ln\frac{p_\omega(\mathbf{r})}{N_0(\mathbf{r})} - 1\right] \\ &+ \sum_\alpha \int d\mathbf{r}\frac{N_\alpha(\mathbf{r})\mathcal{P}_\alpha(\mathbf{r})^2}{2C_{\text{pol}}\chi_\alpha} + \frac{C_{\text{rot}}^{-1} - 1}{N_{\text{bulk}}p_{\text{mol}}^2/3T}\int d\mathbf{r}\left(\int \frac{d\omega}{8\pi^2}p_\omega(\mathbf{r})\omega\circ\mathbf{p}_{\text{mol}}\right)^2 \\ &+ \int d\mathbf{r}\int d\mathbf{r}'\left(\rho_{\text{el}}(\mathbf{r}) + \frac{\rho_{\text{lq}}(\mathbf{r}) - \rho_{\text{lq}}^0(\mathbf{r})}{2}\right)\frac{1}{|\mathbf{r}-\mathbf{r}'|}(\rho_{\text{lq}}(\mathbf{r}') - \rho_{\text{lq}}^0(\mathbf{r}')).\end{aligned} \tag{4.37}$$

Here, only the ideal gas (rotational) entropy and long-range parts from the classical DFT parts appear explicitly, while all remaining terms are grouped into $F_0[N_0]$. The final mean-field interaction term now also includes the interaction between the solute charge density $\rho_{\text{el}}(\mathbf{r})$ and the induced charge density in the liquid $\rho_{\text{lq}}(\mathbf{r}) - \rho_{\text{lq}}^0(\mathbf{r})$ (where ρ_{lq}^0 is the charge in the initial isotropic distribution), which emerges from the coupling functional rather than the classical DFT. The remaining contributions in the coupling functional such as the Thomas-Fermi repulsion are also lumped in with $F_0[N_0]$, which is now related to the energy of forming the cavity.

The key idea of SaLSA is to treat the changes which occur as the fluid relaxes from the unpolarized and uniformly oriented reference state *perturbatively*. Specifically, SaLSA expands the orientation density at linear order as $p_\omega(\mathbf{r}) = N_0(\mathbf{r})(1 + \sum_{lmm'}x_{mm'}^l(\mathbf{r})D_{mm'}^l(\omega))$ by using the Wigner D-matrices as a representation of the rotation group [30]. Then, writing out the Euler-Lagrange equations for minimizing the total free energy with respect to the coefficients $x_{mm'}^l$, substituting back into Eq. 4.37 and retaining terms up to second order in F_{JDFT}, gives

$$F_{\text{SaLSA}} = F_{\text{HK}}[n] + F_0[N_0] + \frac{1}{2}\int d\mathbf{r}\rho_{\text{el}}(\mathbf{r})\left[(\hat{K}^{-1} - \hat{\chi})^{-1} - \hat{K}\right]\rho_{\text{el}}(\mathbf{r}), \tag{4.38}$$

where $\hat{K}(r, r')$ is the free-space Coulomb potential operator $1/|r - r'|$, and $\hat{\chi}(r, r')$ is the non-local solvent susceptibility operator. (See [16]) for details.) In this result, the final term effectively captures the energy due to liquid response as the difference between the electrostatic self energy of the solute charge density $\rho_{\text{el}}(\mathbf{r})$ with the solvent-screened Coulomb interaction $(\hat{K}^{-1} - \hat{\chi})^{-1}$ from that resulting from the bare Coulomb interaction $\hat{K}$. This is a fully general form that describes all common continuum solvation models, with differences restricted to the prescription of $\hat{\chi}$.

As derived above, in its full form, the susceptibility operator in reciprocal space is

$$\hat{\chi}(\mathbf{G}, \mathbf{G}') \equiv -\sum_{\alpha} \tilde{N}_{\alpha}^{0}(\mathbf{G} - \mathbf{G}')C_{\text{pol}}\chi_{\alpha}\mathbf{G} \cdot \mathbf{G}'\tilde{w}_{\alpha}(G)\tilde{w}_{\alpha}^{*}(G')$$
$$- \tilde{N}_{0}(\mathbf{G} - \mathbf{G}')\sum_{lm} \frac{C_{\text{rot}}^{l}}{k_B T}\frac{P_l(\hat{G} \cdot \hat{G}')}{4\pi}\tilde{\rho}_{\text{mol}}^{lm}(G)\tilde{\rho}_{\text{mol}}^{lm*}(G'), \tag{4.39}$$

where the first term represents the polarizability response of the liquid distributed according to the atomic densities $N_{\alpha}(\mathbf{r})$ (Fourier transformed to $\tilde{N}_{\alpha}^{0}(\mathbf{G} - \mathbf{G}')$) and the second term represents the rotational response distributed according to the molecular density $N_0(\mathbf{r})$ (see [16].)

Each of the two terms above represent physically correct generalizations of effects commonly included on a *post hoc* basis in standard continuum theories. For the first term, the weight functions w_{α} capture the range of the polarization response in the classical DFT [5]. Forcing the response to be local by setting $w_{\alpha}(r) \to \delta(r)$ (so that $\tilde{w}(G) = 1$) removes the weight function terms above. The remaining expression is then proportional to $\tilde{N}(\mathbf{G} - \mathbf{G}')\mathbf{G} \cdot \mathbf{G}'$, which corresponds precisely to the operator $-\nabla \cdot (N(\mathbf{r})\nabla(\cdot))$ in real space, exactly the form of the susceptibility due to a continuum dielectric! In the more realistic SaLSA theory, the polarizability weight functions give an explicit prescription for precisely the correct nonlocal generalization of the strictly local dielectric response commonly used in common continuum theories. The second term from Eq. 4.39 likewise represents nonlocal rotational responses, but now in the form of an expansion in spherical harmonics. The $l = 1$ response from this term has precisely the same mathematical form as the polarization response and thus likewise gives an explicit prescription for the nonlocal generalization of the rotational contribution to the dielectric response. Beyond this, the $l = 0$ monopole response is negligible for neutral molecules, but dominant for charged species like ions. In real space, this term in the local limit yields simply a multiplicative constant to the local potential proportional to the ion density $N_{\text{ion}}(\mathbf{r})$, which exactly converts the Poisson equation to a linearized Poisson-Boltzmann equation $(-\nabla \cdot (\epsilon\nabla\phi) + \kappa^2\phi = 4\pi\rho)$! Therefore SaLSA produces an explicit prescription for a nonlocal (but linear) solvation model directly from JDFT perturbation theory, which recovers exactly conventional continuum solvation models in the (somewhat unphysical) local limit.

Two additional prescriptions are necessary for arriving at a complete solvation model. The first is determination of the reference $N_0(\mathbf{r})$, or equivalently the cavity shape function $s(\mathbf{r})$, which forms the basis for the above perturbation theory. The second is specification of an approximation to the 'cavity energy' $F_0(N_0)$, which is independent of the orientation/polarization response of the liquid. [16] introduces a nonlocal model for $s(\mathbf{r})$ in terms of the overlap between solute electron density $n(\mathbf{r})$ and a model solvent electron density $n_{\text{lq}}(\mathbf{r})$, such that $s = 0$ when a test molecule's electron density would overlap strongly with $n(\mathbf{r})$ and

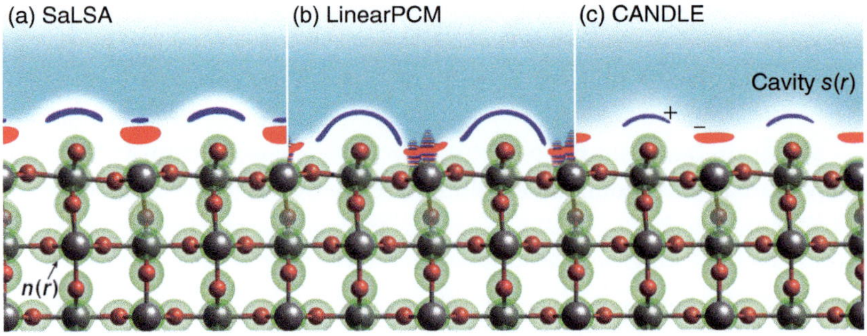

Figure 4.4 Comparison of cavities and charge response of (a) the nonlocal SaLSA solvation model [16], (b) local solvation models [47–49], and (c) the hybrid CANDLE solvation model [24], shown for an aqueous rutile $TiO_2(110)$ surface (as in Figure 4.1).

$s = 1$ when it overlaps negligibly. It calibrates this model to the distance of nearest approach of all pairs of atoms in the periodic table, without any reference to solvation, such that $s(\mathbf{r})$ is defined non-empirically with respect to solvation. The cavity energy is decomposed to a cavitation free energy that emerges from the classical DFT and a dispersion energy that emerges from the coupling, and SaLSA adopts nonlocal models [46] for both of these pieces that fully capture their atomic-scale size dependence (Figure 4.3(b)).

Figure 4.4(a) shows the cavity shape function and induced charge density of the SaLSA solvation model for an aqueous rutile $TiO_2(110)$ surface. Figure 4.4(a) shows that the SaLSA charge response *extends past the cavity surface*, a direct manifestation of its inherent nonlocality. This nonlocality allows the SaLSA model to give the proper location of the charge response while simultaneously maintaining the cavity surface at the distance of closest approach of molecular centers. In contrast, conventional local solvation models must artificially bring the cavity boundary closer to the solute to reproduce the correct charge distribution and free energy, as in Figure 4.4(b). As a result, the parameters in such models cannot be determined from the underlying microscopic physics and must be fit empirically, lessening the transferability of such approaches.

4.4.2 JDFT local limit: nonlinear continuum solvation

As shown above, the linear response limit of JDFT solvation with classical DFT results in the nonlocal SaLSA solvation model. Further taking the local limit then reduces the SaLSA response to the Poisson equation or linearized Poisson-Boltzmann equation for pure solvents or electrolytes, respectively.[4] Proceeding in the opposite order of first taking the local limit of the JDFT response, while maintaining nonlinearity, results in a second class of useful, physical solvation models.

Taking the local limit effectively reduces the sizes of the molecules and atoms to zero, preserving only their dominant multipole moment, thereby reducing neutral solvent molecules to dipoles and charged ions to point charges. Starting from Eq. 4.37, this leaves

4 As described above, however, the further reduction to local response requires changing the cavity definition to preserve the charge response and solvation free energy, introducing unavoidable empiricism and lack of transferability into the solvation model.

the cavity energy and mean-field electrostatic interactions unchanged, but the remaining terms for solvent molecules and ions reduce to separate internal dielectric and ionic free energy terms, A_ϵ and A_κ, respectively. The resulting 'NonlinearPCM' model [47] takes the form

$$F_{\text{NPCM}} = F_{\text{HK}}[n] + F_0[N_0] + A_\epsilon + A_\kappa + \int d\mathbf{r} \int d\mathbf{r}' \frac{\rho_{\text{lq}}(\mathbf{r}')}{|\mathbf{r} - \mathbf{r}'|} \left(\rho_{\text{el}}(\mathbf{r}) + \frac{\rho_{\text{lq}}(\mathbf{r})}{2} \right). \tag{4.40}$$

Here, as before, ρ_{lq} will capture the charge density response from both the solvent and the electrolyte ions.[5]

First consider the simpler case of electrolyte ions: the orientation density $p_\omega(\mathbf{r})$ no longer matters because the ions have been reduced to point charges with no orientational degree of freedom. This reduces the ideal gas entropy functional in the $p_\omega \log p_\omega$ form to the simpler $N \log N$ form for the mono-atomic ideal gas, as in Eq. 4.13. Additionally, the ionic response involves translation which changes the average density N_0, leading to a contribution from the hard-sphere entropy functional in the general nonlinear case. Putting these together gives the internal free energy of several ionic species with charge Z_i and bulk concentrations N_i as [47]

$$A_\kappa[\{\eta_i(\mathbf{r})\}] = k_B T \sum_i \int d\mathbf{r} N_i s(\mathbf{r}) \left[(\eta_i(\log \eta_i - 1) + 1) + \frac{(x(\mathbf{r}) - x_0)^2}{x_0(1 - x_0)^2(1 - x(\mathbf{r}))^2} \right], \tag{4.41}$$

where $\eta_i(\mathbf{r})$ is the (dimensionless) local enhancement of the ionic densities relative to their (cavity-scaled) bulk values. The ions then contribute a charge density $\sum_i Z_i N_i s(\mathbf{r}) \eta_i(\mathbf{r})$ to the total liquid charge density. The second term in Eq. 4.41 is a simplified local form of the hard sphere functional subtracted from its bulk value, which enforces an upper bound or packing limit on the ion concentration. Here $x(\mathbf{r})$ is the local packing fraction, equal to the local limit of n_3 from Fundamental Measure Theory [34], and x_0 is its bulk value.

Similarly, for the solvent molecules now reduced to dipoles, the orientation density simplifies from a general $p_\omega(\mathbf{r})$ where ω is a rotation in SO(3), to $p_{\hat{e}}(\mathbf{r})$ where $\hat{e}$ is a unit vector along the dipole direction. (One Euler angle for rotating the dipole about its own axis is now irrelevant.) This reduces the internal free energy for the dielectric to

$$A_\epsilon[\mathbf{P}_{\text{pol}}, p_{\hat{e}}] = \int d\mathbf{r} N_{\text{mol}} s(\mathbf{r}) \frac{\chi_{\text{mol}} \mathbf{P}_{\text{pol}}^2}{2} + \int d\mathbf{r} T N_{\text{bulk}} s(\mathbf{r}) \left[\int \frac{d\hat{e}}{4\pi} p_{\hat{e}} \log p_{\hat{e}} - \frac{\alpha \mathbf{P}_{\text{rot}}^2}{2} \right], \tag{4.42}$$

where $\mathbf{P}_{\text{pol}}(\mathbf{r})$ and χ_{mol} are the induced dipole moment and dipole polarizability per molecule (summed over the atomic contributions from the classical DFT case), $\mathbf{P}_{\text{rot}}(\mathbf{r})$ is the local rotational polarization, and α is the prefactor to the rotational polarization correlation functional. The solvent polarization then contributes total charge density

5 Note that $\rho_{\text{lq}}^0 = 0$ in the local limit for the unpolarized net-neutral liquid, and so it does not appear in Eq. 4.40, as opposed to the more general Eq. 4.37.

$-\nabla \cdot (N_{mol} s(\mathbf{r})(\mathbf{P}_{pol}(\mathbf{r}) + \mathbf{P}_{pol}(\mathbf{r})))$ due to polarization and rotations. This completes the formulation of the nonlinear solvation model. Further simplification is possible by evaluating the Euler-Lagrange equation for $p_{\hat{e}}(\mathbf{r})$. Specifically, this shows that $p_{\hat{e}}(\mathbf{r}) \propto \exp(\hat{e} \cdot \boldsymbol{\varepsilon}(\mathbf{r}))$ for some vector field $\boldsymbol{\varepsilon}(\mathbf{r})$. Using this, [47] recasts the total dielectric internal energy as

$$A_{\epsilon}[\boldsymbol{\varepsilon}(\mathbf{r})] = \int d\mathbf{r} k_B T N_{mol} s(\mathbf{r}) \times \left[\varepsilon^2 \left(f(\varepsilon) - \frac{\alpha}{2} f^2(\varepsilon) + \frac{\chi_{mol} k_B T}{2 p_{mol}^2} (1 - \alpha f(\varepsilon))^2 \right) - \log \frac{\sinh \varepsilon}{\varepsilon} \right], \tag{4.43}$$

in terms of the single independent variable $\varepsilon(\mathbf{r})$. Note that this is a generalization of the internal energy function that produces the Langevin saturation response for dipole rotation, except that it includes the polarization response and accounts for *correlations in dipole rotations* in order to reproduce the bulk dielectric response.

Finally, note the linear limit of the fully specified nonlinear model above results in the conventional local linear solvation models, just as before but from a different sequence of limits. In the linear limit of the nonlinear model, the Euler-Lagrange equations can be used to eliminate η_i in terms of the total electrostatic potential ϕ and the vector field $\boldsymbol{\varepsilon}(\mathbf{r})$ in terms of $\nabla \phi$, reducing the internal energy functionals to

$$A_{\epsilon} + A_{\kappa} = \frac{1}{4\pi} \int d\mathbf{r} s(\mathbf{r}) \left[(\epsilon_b - 1) \frac{|\nabla \phi|^2}{2} + \kappa^2 \frac{\phi^2}{2} \right]. \tag{4.44}$$

Here ϵ_b is the bulk dielectric constant and $\kappa^2 = 8\pi N_{ion} Z_{ion}^2 / (k_B T)$ is the inverse Debye-length squared (scaled by ϵ_b). Optimizing this functional with respect to ϕ, in turn, results in the usual linearized Poisson-Boltzmann equation,

$$-\nabla \cdot \left[1 + (\epsilon_b - 1) s(n(\mathbf{r})) \right] \nabla \phi(\mathbf{r})) + \kappa^2 s(n(\mathbf{r})) \phi(\mathbf{r}) = 4\pi \rho_{el}(\mathbf{r}). \tag{4.45}$$

As was the case for the SaLSA solvation model, the cavity shape function $s(\mathbf{r})$ and the cavity energy $F_0[N_0]$ remain to be specified. As discussed above, the cavity can no longer appear at the physical location and must be chosen empirically. Following previous developments in iso-density continuum solvation models (see Chapter 3), the cavity is chosen as a function of the local solute electron density $s(\mathbf{r}) = s(n(\mathbf{r}))$. The function $s(n)$ transitions smoothly from one to zero as n crosses some threshold n_c [1, 14, 47, 49] (or in a range n_{min} to n_{max} [48, 50]). Here, n_c (or equivalently, $n_{min\,/\,max}$) are empirical parameters that are adjusted to reproduce the solvation free energies of small organic molecules in the solvent. The cavity energies are also typically treated empirically, such as an effective surface tension multiplying the cavity area calculated as $\int d\mathbf{r} |\nabla s(\mathbf{r})|$. See [47] for a detailed overview of the numerical implementation and comparison between linear and nonlinear response models.

Figure 4.4(b) shows the cavity and induced charge of the linear solvation model from [47] at an aqueous rutile TiO_2 surface. Unlike the nonlocal solvation model, the induced charge appears at the surface of the cavity which requires the cavity surface to be pulled closer to the solute. Notice that in this case, the cavity gets too close to the positively-charged titanium atoms at the surface, near which the electron density is very low. This, in turn, creates numerical issues for the solvation model resulting in the oscillating charge density artifacts (red and blue stripes). Similar issues are commonplace for widely-available iso-density local solvation models, such as the present LinearPCM model [47], the closely-related VASPsol model [49] and the SCCS model [48], when they are applied to ionic solids that are highly different from the molecules used in their parametrization. Additionally, in many cases, these models can create 'cavity bubbles' in the interstices of the solid where the electron density is low enough, and these must be specifically removed. Consequently, applying these versatile and computationally efficient solvation models to new material systems should be done with care to ensure the cavity parametrization is adequately transferable.

4.4.3 Hybrid semi-empirical approaches: 'CANDLE' solvation

The above discussion shows two paths from JDFT to conventional local linear solvation models, one through the nonlocal linear SaLSA model, and the other through the nonlinear local NonlinearPCM model. The nonlocality in SaLSA allows for a *non-empirical* stable cavity determination, but the nonlocal response involves an angular momentum expansion that can be relatively computationally expensive and difficult to implement compared to the conventional solvation models. The usual linear local solvation models are easy to implement, but require an empirical parametrization that can create issues with transferability as demonstrated above.

The CANDLE solvation model [24] is a hybrid between the nonlocal and local models that takes the non-empirical nonlocal cavity determination and cavity free energy terms directly from SaLSA. These terms are relatively simple and computationally efficient compared to the nonlocal response, and on par with the traditional iso-density models, only requiring a few extra convolutions. However, as discussed above, with a local charge response, the same cavity describing the centers of the solvent molecules cannot be used to determine accurate locations for the charge response through solution of the Poisson equation. The CANDLE model addresses this issue by smoothing out the cavity by an empirical 'electrostatic radius' parameter η fit to solvation free energies. This allows the induced charge density to approach closer to the solute by an amount proportional to η.

Figure 4.4(c) (and Figure 4.1(c)) show the cavity and bound charge of the CANDLE solvation model for the same aqueous rutile TiO_2 example as before. Because of the cavity smoothing, the cavity transition is centered on the physical location of the molecule centers, while the charge response is closer *despite the local response*, exactly as in the nonlocal SaLSA solvation model but at a fraction of the computational expense. Even with η empirically determined, the cavity functions are far more transferable being based on the physically-meaningful density overlap $s(\mathbf{r})$ from SaLSA, rather than thresholding on the solute electron density alone as in all iso-density models. Consequently, the charge response does not show any of the numerical issues exhibited by the traditional solvation model in Figure 4.4(b).

The discussion above specifies a nonlocally-determined local-electric (NDLE) solvation model. The CANDLE solvation model additionally captures charge asymmetry (CA-) in the response of the solvent by adding a solute electric-field dependence to the SaLSA cavity prescription. This is critical because solvents like water do not respond equally to positive and negative charges as must do fluids of point dipoles because of symmetry. For example, water solvates negative ions of the same effective radius more strongly than positive ions

Table 4.1 Mean absolute errors (MAEs) of solvation energy predictions by several solvation models for a standard set of neutral solutes, cations and anions in liquid water. (Adapted from [24] with permission from AIP Publishing.)

			MAE [kcal/mol]			
Category	**Model/variant**		**Neutral**	**Cations**	**Anions**	**All**
Linear	LinearPCM [47] = VASPsol [49]		1.27	2.10	15.1	3.59
	SCCS [48, 51]	Neutrals	1.20	2.55	17.4	3.97
		Cations	–	2.26	–	–
		Anions	–	–	5.54	–
	CANDLE [24]		1.27	2.62	3.46	**1.81**
Nonlinear	NonlinearPCM [47]		1.28	16.1	27.0	7.55
Nonlocal	SaLSA [16]		1.36	3.20	19.7	4.55

because the H atoms forming the positive termini of water can get closer to a negative charge than can the bulkier O atom forming the negative terminus.

Table 4.1 compares the accuracy of several solvation models for the solvation of the same set of neutral molecules, cations and anions in liquid water. Most models consistently underestimate the solvation energy of anions when parametrized to neutral molecules or cations, sometimes addressed by developing different parameter sets for differently-charged solutes, as in the charged SCCS models [51]. Building in charge asymmetry (empirically) makes CANDLE uniformly accurate for solutes of all charges within a *single parametrization*, while retaining the stability and transferability of the nonlocal SaLSA cavity based on solute-solvent electron density overlap.

4.5 Grand-canonical DFT

In periodic DFT calculations used for calculating properties of solids and surfaces relevant for electrochemistry, a well-known limitation is the requirement that the unit cell of the calculation must remain overall electrically neutral. In Kohn-Sham DFT, this issue appears in all of the long-range terms, specifically the nucleus-electron interaction, the nuclear repulsion (Ewald term) and the electron repulsion (Hartree term), and must be managed carefully. Additionally, the electric potential and hence all electron energy levels are indeterminate up to an overall offset. These issues originate in the fact that the electrostatic potential $\phi(\mathbf{r})$ is the solution of

$$-\nabla^2\phi(\mathbf{r}) = 4\pi\rho_{\text{el}}(\mathbf{r}), \tag{4.46}$$

where ρ_{el} is the total charge density of the electronic system including electrons and nuclei. The electrical neutrality issue becomes apparent upon integrating Eq. 4.46. The integrand on the left hand side is a total derivative and integrates to zero in periodic boundary conditions. The equation can therefore be consistent only if the right hand side

$4\pi \int d\mathbf{r}\rho_{el}(\mathbf{r}) = 4\pi Q_{el}$ is zero, *i.e.* the electronic system has a net charge $Q_{el} = 0$. Second, the indeterminacy of ϕ up to an additive constant is evident in the fact that changing $\phi(\mathbf{r}) \to \phi(\mathbf{r}) + \phi_0$ for a constant ϕ_0 leaves the equation and the (periodic) boundary conditions unchanged.

The situation changes entirely in the presence of an electrolyte, even in the simplest continuum description of the electrolyte as satisfying the linearized Poisson-Boltzmann equation

$$-\nabla \cdot \epsilon(\mathbf{r})\nabla\phi(\mathbf{r})) + \kappa^2(\mathbf{r})\phi(\mathbf{r}) = 4\pi\rho_{el}(\mathbf{r}). \tag{4.47}$$

Now, the second term need not integrate to zero in a unit cell, allowing $Q_{el} \neq 0$, and the second term changes when $\phi(\mathbf{r}) \to \phi(\mathbf{r}) + \phi_0$ so that the overall offset of $\phi(\mathbf{r})$ is also well-determined [14]. Further, the charge density induced in the electrolyte ions is precisely $\hat{\chi}_{ion}\phi(\mathbf{r}) = -\kappa^2(\mathbf{r})/(4\pi)\phi(\mathbf{r})$. Therefore, the total induced charge in the electrolyte exactly balances the net charge of the electronic system, thereby neutralizing the unit cell [14, 47].

The simple observation for the linear local model that the absolute potential is well-determined is actually completely general for *any continuum model of the electrolyte.* For a general nonlinear model, the charge neutrality is not automatic however, and must be implemented using a Lagrange multiplier constraint. When this is done, the Lagrange multiplier that maintains neutrality turns out to precisely determine the constant offset of the potential [47]. Regardless of the details, the final result is that the electrostatic potential and electronic energy levels have meaningful absolute scales when using any continuum electrolyte model that maintains charge neutrality of the unit cell. Specifically, this absolute scale references the potential to zero far from the electronic system in the bulk of the continuum liquid [14].

The absolute scale of the electron energies prove vitally useful for computational electrochemistry. Conventional Kohn-Sham DFT calculations in the canonical ensemble evaluate the electron density from Fermi occupations, $f_i = 1/\left(1 + \exp\frac{\varepsilon_i - \mu}{T}\right)$, evaluated at the Kohn-Sham orbital energies ε_i, where the electron chemical potential μ is a Legendre multiplier that enforces a constraint on the total number of electrons. Because an electrolyte environment now places the ε_i on a well-defined absolute scale, so now too is the electron chemical potential, μ. Moreover, because this scale references the deep interior of the electrolyte, it relates directly to the electrode potential in an electrochemical system through the fixed potential offset between the reference electrode and locations deep in the electrolyte. This fixed offset can be determined experimentally, or calibrated within a solvation model by comparing experimental potentials of zero charge (PZC) with predicted neutral chemical potentials [14].

First-principles electrochemistry calculations take advantage of the direct relation between the now meaningful electron chemical potential and electrode potentials on standard electrochemical scales to study reaction mechanisms on experimentally-relevant potential scales. However, conventional DFT methods require the specification of a fixed number of electrons in the calculations, and the resulting electron chemical potential will vary vary as the configuration of the species on the electrode surface changes during a reaction. This necessitates a cumbersome manual scan over electron numbers to target a specific chemical potential (electrode potential) of interest, or an expensive 'outer loop' over calculations to automatically adjust the electron number [52].

Grand-canonical DFT [53] addresses this issue by directly switching Kohn-Sham DFT to the grand canonical ensemble, such that the input for the calculation is no longer the electron number, but instead the electron chemical potential. [53] develops two complementary algorithms for directly converging the electronic system in this ensemble. First, the self-consistent field (SCF) iteration solves Kohn-Sham equations in the potential from a given density, computes a new density, and iterates until the input and output densities become equal, within a set tolerance. Within this approach, the Fermi occupations are determined from the Kohn-Sham eigenvalues in each iteration, and these can be evaluated at a fixed μ instead of imposing an electron number constraint. While trivial in concept, this approach introduces significant numerical instabilities as charge can oscillate back and forth between the electron reservoir representing the electrode and the electronic system as the minimization proceeds. [53] develops preconditioners to stabilize the Pulay mixing [54] to reduce these oscillations in the SCF iteration in the grand canonical ensemble. As a second approach, [53] also develops a direct variational minimization of the grand free energy expressed as a functional of the Kohn-Sham orbitals and an auxiliary subspace Hamiltonian [55] that generates the Fermi occupations. This direct minimization exhibits superior stability to the SCF technique for converging Kohn-Sham DFT in the grand-canonical ensemble and enables fixed-μ calculations at the same computational cost of a *single* conventional fixed electron number calculation. [53] provides a detailed description of the underlying algorithms.

Calculation of the charging curve of electrodes provides an important and illustrative example of the use of this grand canonical approach. Figure 4.5 shows the charge density versus electrode potential of a TiN(001) in an aqueous electrolyte using grand-canonical DFT within the SaLSA model, along with visualizations of the induced charge in the adjacent solvent and electrolyte at selected potentials. TiN (titanium nitride) is a metallic transition-metal nitride in the rock-salt structure, which has gained recent attention as a high-temperature and chemically stable metal for plasmonic devices [56] and plasmon-enhanced photochemistry [57].

Such a calculation begins by computing the neutral surface in the open-source JDFTx software [58] using the input file below. (See http://jdftx.org for tutorials, input documentation and to download the code).

```
lattice Tetragonal 5.686 30      #Lattice geometry [Bohrs]
kpoint-folding 6 6 1             #Brilluoin zone sampling
ion-species GBRV/$ID_pbe.uspp   #Pseudopotential set
elec-cutoff 20 1a00              #Plane-wave cutoffs [Hartrees]
elec-smearing Fermi 0.01         #Fermi smearing [Hartrees]
ion Ti 0.5 0.5 -0.134  0         #Atom positions [fractional]
ion Ti 0.0 0.0  0.000  0
ion Ti 0.5 0.5  0.134  0
ion  N 0.0 0.0 -0.134  0
ion  N 0.5 0.5  0.000  0
```

```
ion    N 0.0 0.0    0.134    0
fluid SaLSA                #or LinearPCM, ClassicalDFT etc.
fluid-solvent H2O          #Solvent
fluid-cation Na+ 1.        #1M cations
fluid-anion     F- 1.      #1M anions
dump End BoundCharge #Output induced charge density
```

For brevity of the code listing, this example employs a three-layer slab of material, which is easily expanded with very little computational cost. Fluid calculations within JDFTx, at minimum require specification of only the model, solvent, and electrolyte ions if any. This example considers a 1M solution of NaF, a non-adsorbing electrolyte, because solvation models do not include explicit adsorption; any explicitly-adsorbing species must be included in the electronic DFT directly. The final line of the input file requests the induced charge density in the fluid as an optional output. The neutral calculation correctly yields $N_0 = 51$ valence electrons and finds a neutral chemical potential of $\mu_0 = -0.08366$ Hartrees. This corresponds to the potential of zero charge (PZC), which converts to the experimental scale using the calibration of the standard hydrogen electrode for the SaLSA solvation model ($\mu_{\text{SHE}} = -4.55$ eV [16, 24]), yielding $(\mu_0 - \mu_{\text{SHE}})/(-e) \approx -2.3$ V vs SHE for the PZC of TiN(001).

Next, performing a grand-canonical DFT calculation requires the addition of just one simple command to the input file,

```
target-mu -0.050    #electron chemical potential in Hartrees
```

which sets the electron chemical potential to the specified value. Repeating the calculations for various uniformly-spaced values of μ ranging from -0.050 Hartrees to -0.120 Hartrees, spanning a range of approximately 1 V above and below the neutral value, then yields the raw data needed to determine the charging curve. (Each calculation takes approximately 1 minute on a Tesla K80 GPU.) Finally, the charge density on the surface of the electrode is given by $\sigma = -e(N - N_0)/(2A)$, where N is the converged number of electrons in each charged calculation, and A is the cross-sectional area of the unit cell (the factor of two accounts for the top and bottom surfaces in the calculation).

As Figure 4.5 shows, unlike the surface of an elemental metal, the transition metal nitride exhibits significant charge polarization in the solvent even for the neutral surface, due to the distribution of charge between positively-charged Ti and negatively-charged N atoms on the surface. As the surface charges positive or negative, opposing charges appear in the bound solvent charges to precisely compensate. Also note that the solvent bound charge approaches closer to the surface for positive surface charges as the surface electron density reduces, thereby reducing the overlap with solvent molecule electron density. Finally, SaLSA shows a minimal shell structure, with a smaller positive counter charge appearing above the negative induced charge, reflecting its capability of describing nonlocal atomic scale structure.

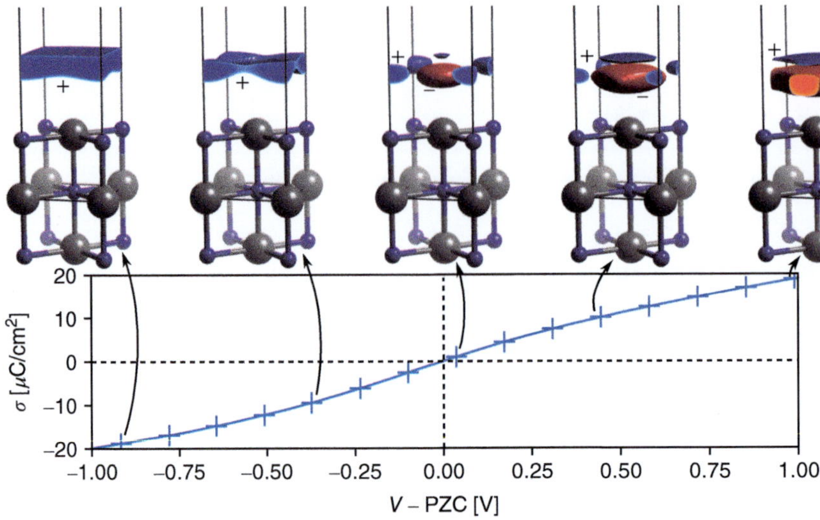

Figure 4.5 Electrochemical charging curve of a metallic TiN(001) surface in 1M non-adsorbing aqueous NaF electrolyte calculated using the SaLSA solvation model [16] and grand-canonical DFT [53] in the open-source JDFTx software [58]. Top panels show the induced charge density in the electrolyte for selected charge states.

4.6 Conclusions

This chapter shows that joint density-functional theory (JDFT) provides a variational principle that puts continuum solvation on a firm footing, and provides a framework for accurate solvation in electrochemical systems with *atomic-scale* liquid structure. Liquid free energy-functionals within the framework of classical DFT realize this potential for atomistic detail in continuum solvation for real molecular fluids, including liquid water.

Simultaneously, JDFT and classical DFT provide a pathway for developing an entire hierarchy of microscopically derived continuum solvation models capturing nonlocality or nonlinearity of the solvent response, all the way down to the simplest linear and local solvation models. Importantly, this paves the way for the SaLSA and CANDLE solvation models with a nonlocal cavity determination based on physically-motivated solvent-solute electron density overlap that is transferable to a wide class of solutes.

Continuum solvation including electrolytes at any level of this hierarchy additionally make the electron chemical potential meaningful on an absolute scale that can be calibrated against the experimental scale for electrode potentials. With grand-canonical DFT algorithms that converge DFT at a fixed electron potential, this makes it possible to directly study the impact of experimental electrochemical conditions from within first-principles calculations.

References

1 S. Petrosyan, A. A. Rigos, and T. A. Arias. Joint density-functional theory: Ab initio study of Cr_2O_3 surface chemistry in solution. *J. Phys. Chem B.*, 109:15436–15444, 2005.

2 S. A. Petrosyan, J. F. Briere, D. Roundy, and T. A. Arias. Joint density-functional theory for electronic structure of solvated systems. *Physical Review B*, 75, 2007.

3 W. Kohn. Nobel lecture: Electronic structure of matter, wave functions, and density functionals. *Rev. Mod. Phys.*, 71:1253–1266, 1999.

4 R. Sundararaman and T. A. Arias. Efficient classical density-functional theories of rigid-molecular fluids and a simplified free energy functional for liquid water. *Comp. Phys. Comm.*, 185:818, 2014.

5 R. Sundararaman, K. Letchworth-Weaver, and T. A. Arias. A recipe for free energy functionals of polarizable molecular fluids. *Journal of Chemical Physics*, 140:144504, 2014.

6 P. Hohenberg and W. Kohn. Inhomogeneous electron gas. *Phys. Rev.*, 136:B864, 1964.

7 N. D. Mermin. Thermal properties of the inhomogeneous electron gas. *Phys. Rev.*, 137:A1441, 1965.

8 J. F. Capitani, R. F. Nalewajski, and R. G. Parr. Non-Born-Oppenheimer density functional theory of molecular systems. *J. Chem. Phys.*, 76:568, 1982.

9 M. Levy. Universal variational functionals of electron densities, first–order density matrices, and natural spin–orbitals and solution of the V–representability problem. *Proc. Natl. Acad. Sci.*, 76:6062–6065, 1979.

10 T. L. Gilbert. Hohenberg-Kohn theorem for nonlocal external potentials. *Phys. Rev. B*, 12:2111–2120, 1975.

11 M. C. Payne, M. P. Teter, D. C. Allan, T. A. Arias, and J. D. Joannopoulos. Iterative minimization techniques for ab initio total energy calculations: molecular dynamics and conjugate gradients. *Reviews of Modern Physics*, 64:1045–1097, 1992.

12 Ann Mattson. Density functional theory: In pursuit of the 'divine' functional. *Science*, 298:759–760, 2002.

13 K. A. Schwarz, R. Sundararaman, K. Letchworth-Weaver, T. A. Arias, and R. G. Hennig. Framework for solvation in quantum monte carlo. *Phys. Rev. B., Rapid Comm.*, 85:201102, 2012.

14 K. Letchworth-Weaver and T. A. Arias. Joint density functional theory of the electrode-electrolyte interface: Application to fixed electrode potentials, interfacial capacitances, and potentials of zero charge. *Physical Review B*, 86:075140, 2012.

15 R. Sundararaman, D. Gunceler, and T. A. Arias. Weighted-density functionals for cavity formation and dispersion energies in continuum solvation models. *The Journal of Chemical Physics*, 141:134105, 2014.

16 R. Sundararaman, K. A. Schwarz, K. Letchworth-Weaver, and T. A. Arias. Spicing up continuum solvation models with SaLSA: The spherically averaged liquid susceptibility ansatz. *J. Chem. Phys.*, 142:054102, 2015.

17 P. Cortona. Self-consistently determined properties of solids without band-structure calculations. *Phys. Rev. B*, 44:8454–8458, 1991.

18 Chen Huang and Emily A. Carter. Potential-functional embedding theory for molecules and materials. *J. Chem. Phys.*, 135:194104, 2011.

19 Tomasz Adam Wesolowski and Arieh Warshel. Frozen density functional approach for ab initio calculations of solvated molecules. *The Journal of Physical Chemistry*, 97(30):8050–8053, 1993.

20 T. Wesolowski and A. Warshel. Ab initio free energy perturbation calculations of solvation free energy using the frozen density functional approach. *The Journal of Physical Chemistry*, 98(20):5183–5187, 1994.

21 Andrey Laktionov, Emilie Chemineau-Chalaye, and Tomasz A. Wesolowski. Frozen-density embedding theory with average solvent charge densities from explicit atomistic simulations. *Phys. Chem. Chem. Phys.*, 18:21069–21078, 2016.

22 Manuel Plaza, Xin Huang, J. Y. Peter Ko, Mei Shen, Burton H. Simpson, Joaquín Rodríguez-López, Nicole L. Ritzert, Kendra Letchworth-Weaver, Deniz Gunceler, Darrell G. Schlom, Tomás A. Arias, Joel D. Brock, and Héctor D. Abruña. Structure of the photo-catalytically active surface of $SrTiO_3$. *J. Am. Chem. Soc.*, 138(25):7816–7819, 2016.

23 K. Letchworth-Weaver, R. Sundararaman, and T. A. Arias. First principles free-energy theory of solvation with atomic scale liquid structure. preprint arXiv:1710.05005, 2017.

24 R. Sundararaman and W. A. Goddard III. The charge-asymmetric nonlocally-determined local-electric (CANDLE) solvation model. *J. Chem. Phys*, 142:064107, 2015.

25 W. Kohn and L. Sham. Self-consistent equations including exchange and correlation effects. *Phys. Rev.*, 140:A1133, 1965.

26 D. Chandler, J. McCoy, and S. Singer. Density functional theory of nonuniform polyatomic systems. ii. rational closures for integral equations. *J. Chem. Phys.*, 85:5978, 1986.

27 K. Ding, D. Chandler, S. J. Smithline, and A. D. J. Haymet. Density-functional theory for the freezing of water. *Phys. Rev. Lett.*, 59:1698, 1987.

28 J. Lischner and T. A. Arias. Kohn-sham-like approach toward a classical density-functional theory of inhomogeneous polar molecular liquids: An application to liquid hydrogen chloride. *Phys. Rev. Lett.*, 101:216401, 2008.

29 J. Lischner and T. A. Arias. Classical density-functional theory of inhomogeneous water including explicit molecular structure and nonlinear dielectric response. *J. Phys. Chem. B*, 114:1946, 2010.

30 E. P. Wigner. *Group theory and its application to the quantum mechanics of atomic spectra*. Academic Press, New York, 1959.

31 Y. Rosenfeld. Free-energy model for the inhomogeneous hard-sphere fluid mixture and density-functional theory of freezing. *Phys. Rev. Lett.*, 63:980, 1989.

32 P. Tarazona. Density functional for hard sphere crystals: A fundamental measure approach. *Phys. Rev. Lett.*, 84:694, 2000.

33 H. Hansen-Goos and R. Roth. Density functional theory for hard-sphere mixtures: the white bear version mark ii. *J. Phys.: Cond. Matt.*, 18:8413, 2006.

34 R. Roth. Fundamental measure theory for hard-sphere mixtures: a review. *J. Phys.: Cond. Matt.*, 22:063102, 2010.

35 J. D. Weeks, D. Chandler, and H. C. Andersen. Role of repulsive forces in determining the equilibrium structure of simple liquids. *J. Chem. Phys.*, 54:5237, 1971.

36 W. A. Curtin and N. W. Ashcroft. Weighted-density-functional theory of inhomogeneous liquids and the freezing transition. *Phys. Rev. A*, 32:2909, 1985.

37 B. Peng and Y.-X. Yu. A density functional theory with a mean-field weight function: Applications to surface tension, adsorption, and phase transition of a Lennard-Jones fluid in a slit-like pore. *J. Phys. Chem. B*, 112:15407, 2008.

38 M. S. Wertheim. Thermodynamic perturbation theory of polymerization. *J. Chem. Phys.*, 87:7323, 1987.

39 G. N. I. Clark, A. J. Haslam, A. Galindo, and G. Jackson. Developing optimal wertheim-like models of water for use in statistical associating fluid theory (saft) and related approaches. *Mol. Phys.*, 104:3561, 2006.

40 R. Sundararaman, K. Letchworth-Weaver, and T. A. Arias. A computationally efficacious free-energy functional for studies of inhomogeneous liquid water. *J. Chem. Phys.*, 137:044107, 2012.

41 C. A. Jeffery and P. H. Austin. A new analytic equation of state for liquid water. *J. Chem. Phys*, 110:484, 1999.

42 F.-M. Tao and E. A. Mason. Statistical-mechanical equation of state for nonpolar fluids: Prediction of phase boundaries. *J. Chem. Phys*, 100:9075, 1994.

43 H. J. C. Berendsen, J. R. Grigera, and T. P. Straatsma. The missing term in effective pair potentials. *J. Phys. Chem.*, 91:6269, 1987.

44 J. L. F. Abascal and C. Vega. A general purpose model for the condensed phases of water: TIP4P/2005. *J. Chem. Phys.*, 123:234505, 2005.

45 D. M. Huang, P. L. Geissler, and D. Chandler. Scaling of hydrophobic solvation free energies. *J. Phys. Chem. B*, 105:6704, 2001.

46 R. Sundararaman, D. Gunceler, and T .A. Arias. Weighted-density functionals for cavity formation and dispersion energies in continuum solvation models. *J. Chem. Phys.*, 141:134105, 2014.

47 D. Gunceler, K. Letchworth-Weaver, R. Sundararaman, K. A. Schwarz, and T. A. Arias. The importance of nonlinear fluid response in joint density-functional theory studies of battery systems. *Modelling Simul. Mater. Sci. Eng.*, 21:074005, 2013.

48 O. Andreussi, I. Dabo, and N. Marzari. Revised self-consistent continuum solvation in electronic-structure calculations. *J. Chem. Phys*, 136:064102, 2012.

49 K. Mathew, R. Sundararaman, K. Letchworth-Weaver, T. A. Arias, and R. G. Hennig. Implicit solvation model for density-functional study of nanocrystal surfaces and reaction pathways. *J. Chem. Phys.*, 140:084106, 2014.

50 J.-L. Fattebert and F. Gygi. First-principles molecular dynamics simulations in a continuum solvent. *Int. J. Quant. Chem.*, 93:139, 2003.

51 C. Dupont, O. Andreussi, and N. Marzari. Self-consistent continuum solvation (sccs): the case of charged systems. *J. Chem. Phys*, 139:214110, 2013.

52 J. D. Goodpaster, A. T. Bell, and M. Head-Gordon. Identification of possible pathways for C-C bond formation during electrochemical reduction of CO_2: New theoretical insights from an improved electrochemical model. *J. Phys. Chem. Lett.*, 7:1471, 2016.

53 R. Sundararaman, W. A. Goddard III, and T. A. Arias. Grand canonical electronic density-functional theory: Algorithms and applications to electrochemistry. *J. Chem. Phys.*, 146:114104, 2017.

54 P. Pulay. Ab initio calculation of force constants and equilibrium geometries in polyatomic molecules. *Mol. Phys.*, 17:197, 1969.

55 C. Freysoldt, S. Boeck, and Jörg Neugebauer. Direct minimization technique for metals in density functional theory. *Phys. Rev. B*, 79:241103(R), 2009.

56 Urcan Guler, Alexandra Boltasseva, and Vladimir M. Shalaev. Refractory plasmonics. *Science*, 344:263, 2014.

57 Adela Habib, Fred Florio, and Ravishankar Sundararaman. Hot carrier dynamics in plasmonic transition metal nitrides. *J. Opt.*, 20:064001, 2018.

58 R. Sundararaman, K. Letchworth-Weaver, K. Schwarz, D. Gunceler, Y. Ozhabes, and T. A. Arias. JDFTx: Software for joint density-functional theory. *SoftwareX*, 6:278–284, 2017.

5

Ab initio modeling of electrochemical interfaces and determination of electrode potentials

Jia-Bo Le, Xiao-Hui Yang, Yong-Bing Zhuang, Feng Wang, and Jun Cheng

State Key Laboratory of Physical Chemistry of Solid Surfaces, iChEM, College of Chemistry and Chemical Engineering, Xiamen University, Xiamen 361005, China

5.1 Introduction

Electrochemical interfaces, e.g. metal/water interfaces, widely exist in a variety of energy devices such as fuel cells, supercapacitors and batteries [1–5]. The interfacial processes involved in electrochemical energy conversion and storage are of paramount importance for designing electrode materials and improving the efficiency in these devices and thus has attracted a great deal of research interests. A prerequisite would be elucidating the structures of the interfaces at an atomic level. However, buried between electrodes and electrolyte solutions, it is extremely challenging to probe the electrochemical interfaces that are only 5∼20 Å [6] thick. Enormous efforts have been dedicated to development of experimental techniques to study electrochemical interfaces. For example, electrochemical scanning tunneling microscopy (EC-STM) has been used to observe the microscopic structures of the electrode surfaces under electrochemical conditions; it becomes possible to study the surface reconstructions, and the patterns of the chemically adsorbed species on surfaces [7, 8]. While we still lack techniques for directly observing the species which are not strongly adsorbed on electrodes, like water molecules and non-specifically adsorbed ions at interfaces, which limits further understanding of charge transfer and reaction mechanisms at interfaces. Generally, the interfacical structures are more often probed by spectroscopic methods, e.g. Infrared [9–11], Raman [12–14], sum frequency generation (SFG) [15–17], X-ray scattering [18, 19], and X-ray absorption [20, 21]. These techniques are sensitive to the intramolecular vibration, intermolecular hydrogen bonding or atomic radial distribution, and have demonstrated their power for unraveling the interfacial structures in the past decades. However, due to difficulties in interpreting complex spectra, there is no general consensus yet on the microscopic structures of electrochemical interfaces.

Computer modeling offers a complementary approach to these experimental techniques for studying the interfacial structures. In recent years, there is more and more computational work simulating electrochemical interfaces, and new methods have been implemented for representing electrochemical interfaces. In many cases, electrode surfaces are often modelled at the level of electronic structure theories, typically density functional

Atomic-Scale Modelling of Electrochemical Systems, First Edition.
Edited by Marko M. Melander, Tomi T. Laurila, and Kari Laasonen.

theory (DFT), because electronic effects such as chemisorption can only be reliably treated by electronic structure methods. The choice for representation of the electrolyte phase can vary from continuum models [22–26], molecular mechanics (MM) to *ab initio* molecular dynamics (AIMD) method [27–33]. The continuum models focus on solvation effects of the surfaces and adsorbates [34], often omitting chemical interaction among electrode surfaces, solute ions, and solvent molecules. The latter, however, often plays an important role in electrochemical processes including electrocatalysis. AIMD simulations, though expensive, faithfully account for these non-trivial effects. Recently, increase of computational power has encouraged applications of AIMD to electrochemical interfaces, for example calculating interfacial structures [14, 27, 35–41], potentials of zero charge (PZC) of metal/water interfaces [31, 33, 42], surface acidities [43–45], and band alignment [32, 46–49] of semiconductor/water interfaces, thermochemistry of surface reactions [50, 51], etc.

A combination of experiment and theory has been proven to be an effective approach in many research areas, but an important question to be asked is how realistic the computational models used are in comparison to experiment. In single crystal electrochemistry, single crystal electrodes such as atomically flat metal surfaces, have been widely used to obtain fundamental understanding of electrochemical interfaces with well-defined structures, and can serve as model systems for computation and bridge between experiment and theory [14, 52]. A fundamental variable for electrochemical interfaces is the electrode potential, which is key in any electrochemical measurements and should be explicitly specified in computation when attempting to compare with experiment. A special reference condition is the so-called PZC for metal electrodes at which there is no net surface charge and thus electric double layers are absent. Therefore, one of the main aspects in computational electrochemistry is developing methods for determining the electrode potentials of interfaces.

A key issue in determining electrode potentials is how to choose a common potential reference. The vacuum level is widely used as reference in experiment, for example, measurement of the ionization potential, i.e. valence band maximum (VBM) of liquid water with photoemission spectroscopy [53]. However, the vacuum level is not conveniently accessible in electrochemical experiment. Instead, various reference electrodes, such as standard hydrogen electrode (SHE) and Ag/AgCl electrode, have been devised for satisfying various electrochemical working conditions. The relative potential difference between two reference electrodes can be easily measured by setting up an electrochemical cell, while determining the absolute potentials of reference electrodes, i.e. referencing to vacuum, is a non-trivial task. Even for the SHE, the reported absolute potential still has a significant uncertainty, i.e. 4.28 to 4.85 V [54–62], 63, [63]. Two types of computational methods based on difference choices of potential references have been developed for determining electrode potentials. The first is the work function method using vacuum potential as a reference. It is somewhat easy to implement by explicitly including a vacuum region in computational models, and then the obtained absolute potential is shifted by the experimental estimate of absolute SHE potential to be converted into the SHE scale [42, 64–66]. The other is the computational reference electrode method, which is more compatible to electrochemical measurements [31, 46, 67, 68], and avoids the uncertainty in the absolute potential of the reference electrode.

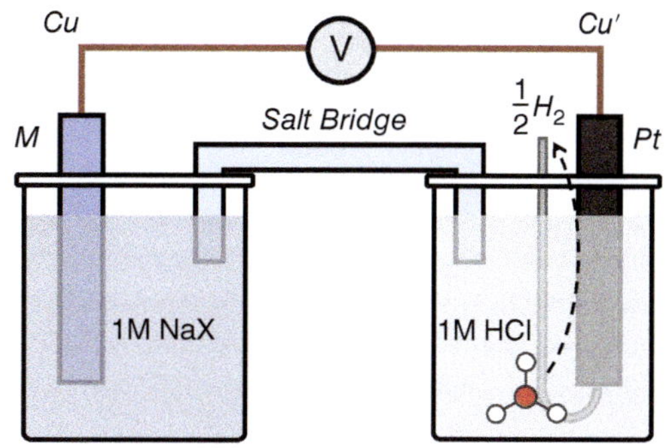

Figure 5.1 Illustration of a typical electrochemical cell connecting a work electrode and reference electrode. The left side represents the a metal electrode immersed in an aqueous solution, and the right side shows the working principle of the standard hydrogen electrode.

This chapter starts by introducing related theoretical background of electrochemistry in the next section, including the concept of the electrode potential and reference electrode. In section 5.3, we summarize the computational methods for modeling electrochemical interfaces. In section 5.4, we review the two main methods for computing the electrode potentials of the solid/liquid interfaces, i.e. work function based and reference electrode based methods. Finally, some applications of these methods for simulations of interfaces and determination of electrode potentials are mentioned in section 5.5.

5.2 Theoretical background of electrochemistry

Figure 5.1 shows a typical electrochemical cell at open circuit potential condition, it is composed by two single electrodes (working electrode and reference electrode), which are connected via a salt bridge. The reference electrode normally has a well defined potential and can keep constant during electrochemical processes, serving as a stable reference. There are many kinds of reference electrodes which have been developed in experiments for different electrochemical applications. For instance, the SHE is one of the most important reference electrodes for experiments in aqueous solution. It is defined as the potential energy by reducing a solvated $H^+(aq)$ to $\frac{1}{2}H_2(g)$ at $T = 298.15$ K in the standard condition ($p = 1$ bar and pH = 0). In addition, the reversible hydrogen electrode (RHE) is more popular for removing the pH effect for electrochemical reactions involving $H^+(aq)$, and saturated calomel electrode (SCE) and Ag/AgCl electrode are also widely used in aqueous solution. In the case of non-aqueous electrochemical cells, reference electrodes like the Ag/AgCl electrode are often used, including some so-called quasi- and pseudo-reference electrodes.

5.2.1 Definition of electrode potential

In this section, we will show the electrochemical definition of electrode potential at a metal electrode/aqueous solution interface. As illustrated in Figure 5.1, the electrochemical potential of the working electrode U with respect to the reference electrode can be written as

$$U = \frac{\tilde{\mu}_e^{Cu} - \tilde{\mu}_e^{Cu'}}{-e_0}, \tag{5.1}$$

where $\tilde{\mu}_e^{Cu}$ and $\tilde{\mu}_e^{Cu'}$ denote the electrochemical potential of electron in Cu and Cu' wire connecting to the working electrode and reference electrode, respectively, and $-e_0$ is the charge of an electron. The electrochemical potential of the electron in Cu wire is composed by two parts,

$$\tilde{\mu}_e^{Cu} = \mu_e^{Cu} - e_0\phi^{Cu}, \tag{5.2}$$

the terms μ_e^{Cu} and ϕ^{Cu} represent the chemical and electrostatic contributions, respectively. Note that the working electrode and the reference electrode are terminated with the same metal (Cu). Therefore, the working electrode potential of the cell can be expressed as

$$U = \phi^{Cu} - \phi^{Cu'}. \tag{5.3}$$

We can expand the Eq. (5.3) in series by adding the inner (Galvani) potentials of the components in the cell,

$$U = (\phi^{Cu} - \phi^{M}) + (\phi^{M} - \phi^{S}) + (\phi^{S} - \phi^{Pt}) + \left(\phi^{Pt} - \phi^{Cu'}\right), \tag{5.4}$$

where ϕ^{Pt}, ϕ^{S} and ϕ^{M} denote the inner potential of Pt in the SHE, electrolyte solution and working electrode M, respectively. Note that the metal M is bound to Cu in the cell, indicating that they are in electrochemical equilibrium,

$$\tilde{\mu}_e^{M} = \tilde{\mu}_e^{Cu}, \tag{5.5}$$

and therefore the first term in the right side of Eq. (5.4) can be rewritten as

$$\phi^{Cu} - \phi^{M} = \frac{1}{e_0}\left(\mu_e^{Cu} - \mu_e^{M}\right), \tag{5.6}$$

and similarly, the last term in Eq. (5.4) is expressed as

$$\phi^{Pt} - \phi^{Cu'} = \frac{1}{e_0}\left(\mu_e^{Pt} - \mu_e^{Cu'}\right). \tag{5.7}$$

Moreover, the right-hand side of the cell in Figure 5.1 is the reference electrode SHE, and the Pt electrode potential should be in electrochemical equilibrium with the half reaction ($H^+(aq) + e^-(vac) \rightarrow \frac{1}{2}H_2(g)$). Thus,

$$\phi^{S} - \phi^{Pt} = \frac{1}{e_0}\left(\frac{1}{2}\mu_{H_2}^{g,o} - \mu_{H^+}^{S,o} - \mu_e^{Pt}\right), \tag{5.8}$$

where $\mu_{H^+}^{S,o}$ and $\mu_{H_2}^{g,o}$ terms in Eq. (5.8) denote the standard chemical potential of a solvated proton and hydrogen gas, respectively. Then, combining Eq. (5.4), (5.6), (5.7), and (5.8) we obtain

$$U = \frac{1}{e_0}\left(\frac{1}{2}\mu_{H_2}^{g,o} - \mu_{H^+}^{S,o} - \mu_e^{M}\right) + (\phi^{M} - \phi^{S}). \tag{5.9}$$

The Eq. (5.9) is a general expression of the electrode potential on the scale of SHE. To better understand the electrode potential, we will discuss two specific conditions of metal electrode/aqueous solution interface. First, in the case that the working electrode is also in electrochemical equilibrium with a redox couple ($X^\bullet/X^-$) in the solution, the inner potential difference of the solid and liquid can be rewritten as the combination of the chemical potential of species,

$$\phi^{M} - \phi^{S} = \frac{1}{e_0}\left(\mu_{X^\bullet}^{S,o} - \mu_{X^-}^{S,o} + \mu_e^{M}\right). \tag{5.10}$$

The terms $\mu_{X^\bullet}^{S,o}$ and $\mu_{X^-}^{S,o}$ denote the standard chemical potential of $X^\bullet$ and X^- in the solution, respectively. After substituting Eq. (5.9) into Eq. (5.10), we then obtain

$$U = \frac{1}{e_0}\left[\left(\mu_{X^\bullet}^{S,o} - \mu_{X^-}^{S,o}\right) - \left(\mu_{H^+}^{S,o} - \frac{1}{2}\mu_{H_2}^{g,o}\right)\right]. \tag{5.11}$$

It is interesting to notice that the electrode potential is independent of the material of the working electrode at this condition, and only determined by the potential difference of two redox couples.

The second special condition is the PZC of a metal/water interface, defined as the potential at which there is no excess charge on the electrode surface. Unlike the former case, the working electrode is not in electrochemical equilibrium with the solution, and the term $\phi^M - \phi^S$ in Eq. (5.9) can not be expanded with the way shown in Eq. (5.10). Instead, we can express the PZC on the scale of SHE (U_{PZC}^{SHE}) by taking the difference between the absolute potential between two electrodes in Figure 5.1,

$$U_{PZC}^{SHE} = U_{PZC}^{abs} - \frac{\Phi_{SHE}}{e_0}. \tag{5.12}$$

The terms U_{PZC}^{abs} and Φ_{SHE} in Eq. (5.12) denote the absolute electrode potential of the working electrode and the absolute potential energy of SHE, respectively.

As shown in Figure 5.2, the U_{PZC}^{abs} is closely related to the work function of metal surface (W_e^M); they correspond to the energy cost for moving an electron from the Fermi level of metal to the vacuum just outside of the surface of electrolyte solution and the surface of metal, respectively. Therefore, the U_{PZC}^{abs} can be further written as

$$U_{PZC}^{abs} = \frac{W_e^M}{e_0} + \left(\psi^M - \psi^S\right). \tag{5.13}$$

Then, by combining Eq. (5.12) and (5.13), the expression of the electrode potential at PZC is transformed to

$$U_{PZC}^{SHE} = \frac{1}{e_0}\left(W_e^M - \Phi_{SHE}\right) + \left(\psi^M - \psi^S\right). \tag{5.14}$$

The term of outer potential difference $\psi^M - \psi^S$ in Eq. (5.14) coincides to the energy difference of W_e^M and U_{PZC}^{abs}, and it is caused by the interaction between solution and electrode at the interface. By reviewing related experimental papers for work function and PZC measurements, it has been reported that the term $\psi^M - \psi^S$ is almost a constant ~ -0.3 V for

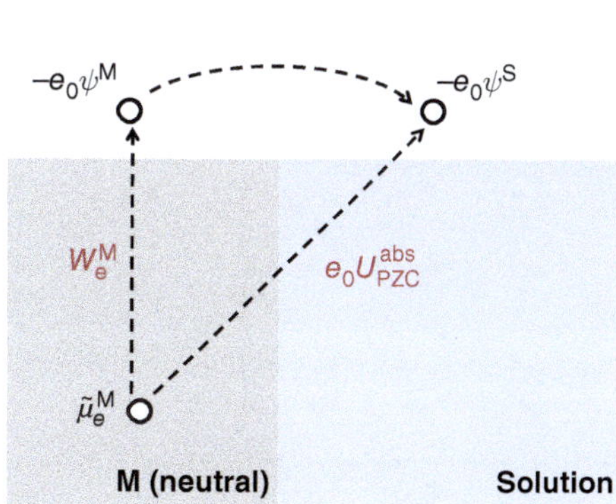

Figure 5.2 The relation between work function (W_e^M) and absolute potential of PZC (U_{PZC}^{abs}) for a metal/solution interface. The ψ^M and ψ^S denote the outer potential of metal surface and solution surface, respectively. Source: Jia-Bo Le, Jun Cheng. Modeling electrochemical interfaces from ab initio molecular dynamics: water adsorption on metal surfaces at potential of zero charge. Curr. Opin. Electrochem., 19:129, 2020.

a series of *sp* metal, such as Cd, In, and Hg [6, 69, 70]. However, the transition metals, for example, Ag, Pt and Pd, normally deviate from this relation. In the following section of this chapter, we will further discuss this difference between transition metal and *sp* metal.

5.2.2 Absolute potential energy of SHE

As shown in Eq. (5.15), the energy conversion between U_{PZC}^{abs} and U_{PZC}^{SHE} requires the value of Φ_{SHE}. This relation is indeed very important for many applications, and therefore calibration of the energy level of Φ_{SHE} has been an important issue in the past few decades. So far, several methods have been proposed to estimate Φ_{SHE}. First of all, based on Eq. (5.14) the measurements of the terms W_e^M, U_{PZC}^{SHE} and $\psi^M - \psi^S$ separately can lead to the determination of Φ_{SHE}. Using this approach, Trasatti [62] proposed the value of Φ_{SHE} is 4.44 V, which has been accepted by IUPAC as the recommendation value. Since the surface condition of the metal in ultrahigh vacuum (UHV) may differ from that in electrochemical environment, this inconsistency in the measurement of the W_e^M, U_{PZC}^{SHE}, and $\psi^M - \psi^S$ may cause error in the estimation of Φ_{SHE}.

Alternatively, in surface science, researchers prefer to access Φ_{SHE} via directly measuring U_{PZC}^{abs} in vacuum, dosing the metal surface with water until the work function tends to be stable [71, 72], the value of which is regarded as U_{PZC}^{abs}. Then, subtracting the U_{PZC}^{SHE} measured from electrochemical experiment can lead us to Φ_{SHE} as indicated in Eq. (5.15). However, some factors may give unexpected deviations in this method. For example, the structure of interfacial water is very likely sensitive to the temperature: the work function measurements are carried out in the condition of ultrahigh vacuum (UHV) and low temperature (150~200K), while the electrochemical experiment is usually performed at room temperature. To overcome this drawback, Hansen and Kolb proposed another method called emersed electrode [58]. Briefly, it measures the work function of the metal electrode just emersed from a solution, the structure of the dragged solvent layer on which is considered to be still controlled by the electrode potential in short time. Based on this method, the proposed value of Φ_{SHE} is 4.7 V.

A more direct approach to estimate the Φ_{SHE} is via the Born-Haber cycle, as shown in Figure 5.3. The Φ_{SHE} can be expressed with

$$\Phi_{SHE} = \mu_{H^+}^{S,o} - \frac{1}{2}\mu_{H_2}^{g,o} + e_0\chi^S = -W_{H^+} + \Delta_f G_{H^+}^{g,o}, \tag{5.15}$$

where the χ^S is the surface potential of a water/air interface and the W_{H^+} denotes the work function by moving a proton from aqueous solution to gas phase. The $\Delta_f G_{H^+}^{g,o}$ is the formation free energy of a H^+ from H_2 in gas, which has been well measured.

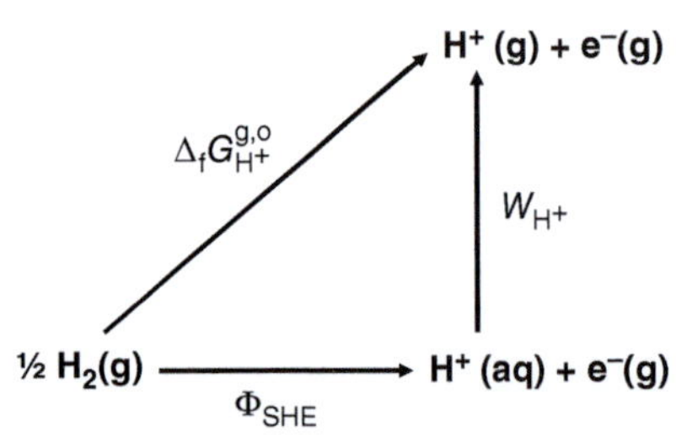

Figure 5.3 The Born-Haber cycle of the standard hydrogen electrode. The $\Delta_f G_{H^+}^{g,o}$ denotes the formation free energy of a H^+ in the gas phase, and W_{H^+} is the work function for moving a H^+ from solution to gas phase.

Recently, Isse and Gennaro [59] reported a reasonable estimate of Φ_{SHE} 4.28 eV through the measurement of the solvation free energy of a proton $\Delta_s G^o_{H^+}$, while it should be noted that the term $\Delta_s G^o_{H^+}$ differs from W_{H^+} by a surface potential of water χ^S,

$$\Delta_s G^o_{H^+} = -W_{H^+} - e_0\chi^S. \tag{5.16}$$

That may be the reason why the estimated Φ_{SHE} by Isse and Gennaro [59] is slightly smaller than the recommended value 4.44 eV. Another estimate of Φ_{SHE} is reported by Fawcett [55] from measuring the real potential of the proton $\alpha^{S,o}_{H^+}$. The $\alpha^{S,o}_{H^+}$ is defined as

$$\alpha^{S,o}_{H^+} = -W_{H^+} + \Delta_f G^{g,o}_{H^+}, \tag{5.17}$$

and the obtained value with this method is 4.42 eV, close to the value (4.44 eV) recommended by Trasatti.

Overall, the values of the Φ_{SHE} estimated from different approaches vary in a wide potential region, from 4.28 to 4.85 V [54–63, 71, 72]. It is still under debate which value is most reliable, and researchers generally have their own preference for the value of Φ_{SHE}.

5.3 Short survey of computational methods for modeling electrochemical interfaces

Theoretical modeling of electrochemical interface started a few decades ago [73–75]. In the early stage, classical molecular dynamics (MD) based on force fields was used for simulating fully atomistic electrochemical interfaces, in which atoms are treated as point charges, and pairwise interaction is parameterized by fitting experimental data or *ab initio* calculations [3, 75–83]. These simulations are computationally efficient, and can be used for studying systems with very large size ($\sim 10^8$ atoms), and processes at large time scale ($\sim 10^2$ nanoseconds), for example, porous carbon/ionic liquid interfaces [3] and proton transfer at interfaces [83]. The challenges of the classical MD for simulating electrochemical interfaces lie in the difficulties in finding accurate force fields for describing polarization effects of electrodes and ions, and accounting for chemical adsorption and faradaic processes.

The electronic structure theory, such as DFT, has unique advantage for correctly describing the chemical interaction between electrodes and electrolytes. *Ab initio* molecular dynamics (AIMD), independent of the pre-set parameters, treats electrodes and the electrolytes at the same level of DFT [84–88]. It also allows for statistical sampling of configurations of dynamic liquid solutions, and thus is an excellent tool for simulating electrochemical interfaces. Compared to classical MD, the downside of AIMD is computationally much more expensive. The state-of-the-art algorithms and computing power enable us to model interface models consisting of up to $10^3 \sim 10^4$ atoms and $10^1 \sim 10^2$ ps.

One of the main sources of high computational cost of AIMD comes from the necessity for statistical sampling of the liquid phase, and thus one way of reducing the cost is simplifying the representation of the liquid solution of the interface. A popular approach is combining DFT treatment of electrodes and various levels of continuum models for electrolyte solutions (DFT-CM) [22, 23, 89–95], such as the reference interaction site model (RISM) [94], joint density-functional theory (JDFT) [95], and dielectric continuum models [24–26]. The empirical parameters in these models are usually fitted against *ab initio* calculations or experimental data. Another approach is based on quantum mechanics/-molecular mechanics (QM/MM) [96–98], in which the electrode is still computed at the quantum mechanical level, and the liquid (explicit solvent molecules and counter ions) is modeled with MM, and hence the computational cost can be decreased. Both DFT-CM and QM/MM method have been often applied to studying electrocatalysis, such as methanol oxidation reaction [92], CO_2 reduction reaction [99], water oxidation reaction [23], and formic acid oxidation reaction [100]. Note that embedding DFT with CM or MM may not correctly represent the chemical interaction between electrodes and electrolytes, and thus special care is required when dealing with systems that strong chemisorption is present.

5.4 *Ab initio* determination of electrode potentials of electrochemical interfaces

Knowing the electrode potentials of the constructed interface models is a prerequisite to compare simulation results with electrochemical experiment. To date, there are two main approaches to adjust the electrode potential of interfaces in *ab initio* computational electrochemistry. One is called the constant potential method (constant μ_e) [33, 101–106], and just as its name implies, the electrode potential in this method is a pre-set variable, and can be kept constant during the electrochemical processes, for example, Volmer reaction [33]. Generally the constant potential is reached with the employment of the grand canonical ensemble, in which the number of electrons in the interface model varies in exchange with an external electron reservoir with a constant potential. In the standard DFT implementation, the electron number of the simulation model is kept constant (constant N_e), and the electrode potential can be adjusted by varying the surface charge density [107–113]. In this section we will concentrate on how the electrode potential can be computed for given models of electrochemical interfaces in the framework of constant N_e.

5.4.1 Work function based methods

5.4.1.1 Vacuum reference

An interface is constructed with an electrode/water/vacuum model (see Figure 5.4) in work function based methods in which vacuum potential can be directly used as reference. As illustrated in Figure 5.2, the electrode potential on the absolute potential scale can be obtained by referencing the Fermi level (or electrochemical potential) of the interface to the potential of vacuum just outside of the liquid phase, which is the same as the definition of work function and can be expressed as

$$e_0 U^{\mathrm{abs}} = -\epsilon_{\mathrm{F}} - e_0 V_{\mathrm{vac}}, \tag{5.18}$$

where ϵ_F and V_{vac} represent the computed Fermi level and vacuum potential in the interface model, respectively.

Note that two sides of the electrode model shown in Figure 5.4 are often not symmetric in surface normal direction, which can result in a dipole in the cell, particularly when only one surface of the electrode slab is charged. The dipole-dipole interaction between the model and its periodic images will lead to unphysical structures, and moreover the

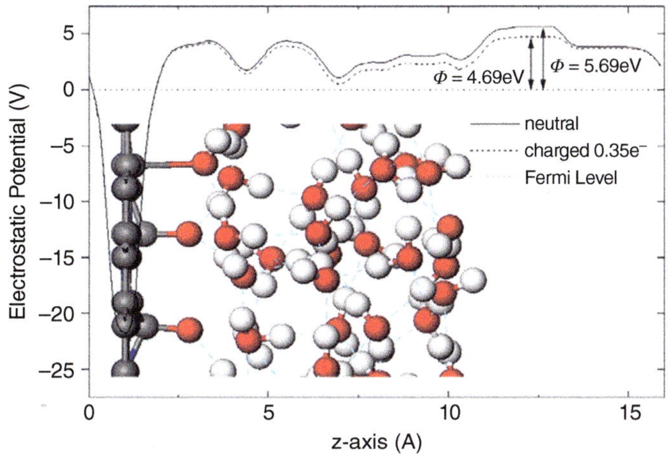

Figure 5.4 The averaged electrostatic potential profiles of a nitrogen-doped graphene/water interface along the surface normal direction [64]. The absolute electrode potential at different surface charge conditions are calculated by referencing the Fermi level of the interface to the vacuum potential. A dipole correction scheme is used to cancel the dipole-dipole interaction between the interface slab and its periodic images [64].

vacuum region will show a shifting potential profile due to the presence of an electric field. To overcome this problem, a dipole correction scheme has to be used by inserting a dipolar layer in the vacuum region to compensate the electric field [114], and hence the potential of one side of the vacuum can be still taken as the reference as shown in Figure 5.4.

In order to compare with electrochemical experiment, the absolute potential of a reference electrode (e.g. SHE) should be subtracted from U^{abs} as indicated in Eq. (5.12). It is one of main error sources in the work function based method, because the experimental estimate of Φ_{SHE} ranges from 4.28 to 4.85 V as mentioned in section 5.2. The choice of the value often depends on how well the computed results fit in with experiment, while the concern is that it may be just a cancellation of error from other sources. For example, another source of uncertainty in the work function based method may come from treatment of the water/air interface. As shown in Figure 5.4 the water/vacuum interface needs to be modeled for computing the surface potential of water (χ^S), which is a necessity in estimating the absolute electrode potential as indicated in Eq. (5.16). It has been shown that it requires quite a long time to reach the equilibrium for the water/vacuum interface [115, 116], while normal AIMD simulations are limited on the order of tens of ps. Therefore, due to the dynamic fluctuation of water/vacuum interface, the solid/liquid/vacuum model may be subjected to a large statistical uncertainty.

5.4.1.2 Vacuum reference in two steps

Although the vacuum is a good choice for potential reference, it would be beneficial if one could determine the electrode potential or other energy levels for models without a vacuum region, as shown in Figure 5.5(a). One could achieve this by alignment in two steps, which was initially developed to compute band alignment in solid heterojunctions [118]. This approach has been extended by Pham, Galli, and co-workers [117] for calculating

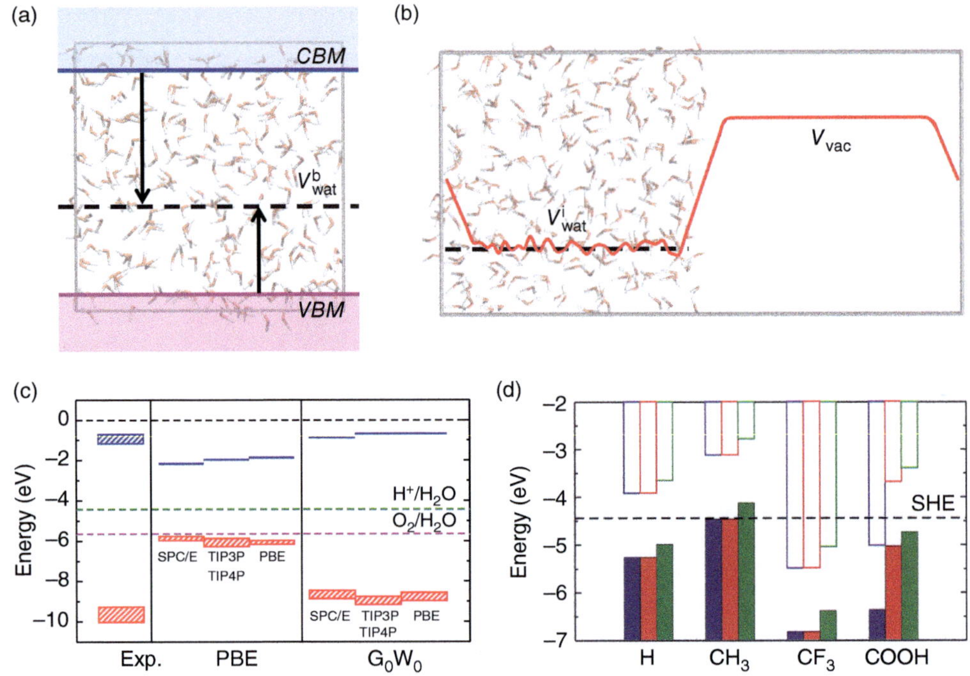

Figure 5.5 (a) Model of the bulk water. The blue and purple lines represent the CBM and VBM, respectively. The dashed line denotes the averaged electrostatic potential of bulk water V^{b}_{wat}. (b) Model of the water/air interface. The red line indicates the electrostatic potential profile, the dashed line is the average of potential in the bulk water region V^{i}_{wat}, and the potential at vacuum is expressed as V_{vac}. (c) Calculated band edges of bulk water with different models and electronic structure theories, the experimental values are used for reference [117]. (d) Calculated band edges of the functionalized Si surfaces in liquid water [48].

the positions of conduction band minimum (CBM) and valence band maximum (VBM) of liquid water. Liquid water widely exists in many physical and chemical processes, but its band structure has not reached consensus yet [67, 119–121]. Firstly, the unreferenced band edge of liquid water ϵ^{b}_{be} is obtained via the simulations of a bulk water model as shown in Figure 5.5(a). The first reference used in this method is the electrostatic potential of the bulk water V^{b}_{wat}, which is equal to zero as implemented in standard Ewald summation under periodic boundary conditions (PBC) [122]. Secondly, as the vacuum region is inevitable for determining the band edge in the absolute potential scale ϵ^{abs}_{be}, therefore a water/air interface is modeled as shown in Figure 5.5(b). Besides the vacuum, the interface also contains a water region, where the condition of the bulk water applies. Thus, the electrostatic potential of water phase can be used as internal reference. It is known that the electrostatic potential of the bulk water in the interface model V^{i}_{wat} is different from V^{b}_{wat} by an artificial term, the so-called Hartree potential shift (HPS) V_0, due to different compositions in the periodic cells. Moreover, the same HPS also affects other electronic energy levels, e.g. the band positions of bulk water. Therefore, we can obtain,

$$\epsilon^{b}_{be} - \epsilon^{i}_{be} = -e_0(V^{b}_{wat} - V^{i}_{wat}), \tag{5.19}$$

where $\epsilon_{\mathrm{be}}^{\mathrm{i}}$ is the band edge of the water/air interface. Combining Eq. (5.18) and (5.19), the band positions *vs* vacuum $\epsilon_{\mathrm{be}}^{\mathrm{abs}}$ can be calculated with

$$\epsilon_{\mathrm{be}}^{\mathrm{abs}} = \epsilon_{\mathrm{be}}^{\mathrm{b}} + e_0(V_{\mathrm{wat}}^{\mathrm{b}} - V_{\mathrm{wat}}^{\mathrm{i}} + V_{\mathrm{vac}}). \tag{5.20}$$

The computed results of $\epsilon_{\mathrm{be}}^{\mathrm{abs}}$ of liquid water with this method are shown in Figure 5.5(c). Clearly, the $\epsilon_{\mathrm{be}}^{\mathrm{abs}}$ computed with GGA-PBE functional has significant deviation from the experimental values, particularly for the VBM, which results from the notorious delocalization error in GGA functional [123, 124]. High-level electronic structure theories can be used to reduce this error, for instance, hybrid functionals [125–127], double hybrid functionals [128], and many body perturbation theory (MBPT) [117] like random phase approximation (RPA) [129]. In the work of Pham et al, the G_0W_0 approximation (MBPT) [117] is used to correct the error in the band edge caused by GGA functional,

$$\epsilon_{\mathrm{be}}^{\mathrm{G_0W_0}} = \epsilon_{\mathrm{be}}^{\mathrm{GGA}} + \Delta E_{\mathrm{corr}}, \tag{5.21}$$

where $\epsilon_{\mathrm{be}}^{\mathrm{G_0W_0}}$ and $\epsilon_{\mathrm{be}}^{\mathrm{GGA}}$ represent the band edges in the G_0W_0 approximation and GGA level, respectively. The ΔE_{corr} denotes the G_0W_0 correction. Figure 5.5(c) shows that the computed band edges in G_0W_0 level are close to the experimental values [130, 131].

This double reference scheme can be readily extended to determine the band edges of the solid water interfaces. The energy levels with reference to the electrostatic potential of water phase, as obtained in the periodic models of solid water interfaces in the absence of vacuum region, are then converted into absolute scale by shifting the electrostatic potential of water *vs* vacuum, as computed in the model of water vacuum interface. Pham et al. have used this scheme to compute the band edges of the functionalized Si surfaces in the water [48], as shown in Figure 5.5(d).

The double reference method is very general, and can also be applied to various scenarios to place energy levels on relevant scales. Taylor, Neurock and co-workers [107] have also taken the electrostatic potentials of water phase and vacuum as the double reference for determining the electrode potential of an electrified metal/water interface. For electrochemical interface models treating the electrode surface at the DFT level and electrolyte solution with a continuum model [22, 132], the electrostatic potential of continuum phase can be used as a reference to link the electrode potential of the charged electrode to the potential of the neutral electrode, with the latter corresponding to the PZC, which value can be taken from experiment. In a similar vein, the electrostatic potential of the solid at the electrochemical interface can also serve as reference for applying corrections of band edges with high-level electronic structure theories as calculated in bulk [32].

5.4.2 Reference electrode based methods

5.4.2.1 Computational standard hydrogen electrode

As mentioned above, work function based methods suffer from a notable uncertainty in the Φ_{SHE} and an additional, unnecessary complication of treating the water vacuum interface for comparing the computed electrode potentials to experiment. In contrast, the computational SHE method developed by Cheng and Sprik [31, 46, 67, 68, 133], places the computed electrode potential on the scale of SHE, directly comparable to electrochemical experiment and thus avoiding the uncertainties in the work function based methods.

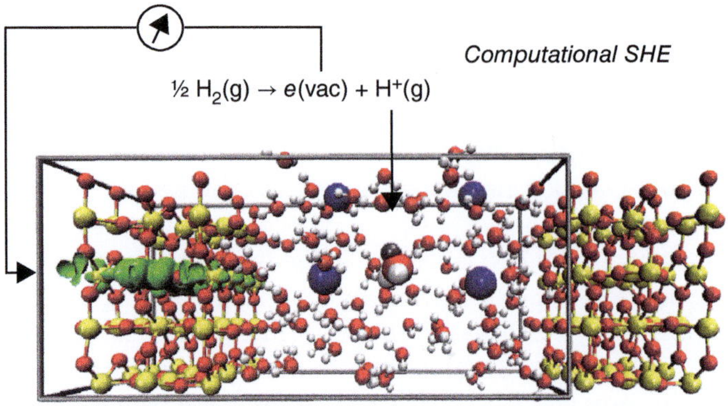

Figure 5.6 Representation of computational SHE for determining band positions (i.e. conduction band) of a TiO_2/H_2O interface. The Ti, O, H, and Na atoms are represented by yellow, red, white, and blue balls, respectively, and the grey ball indicates the inserted proton in the water phase. The green isosurface represents the excess electron in the conduction band of TiO_2 [30].

Based on Eq. (5.16), the key step for Φ_{SHE} is to calculate the W_{H^+}, which is approximated by the deprotonation free energy of a hydronium ion $\Delta_{dp}A_{H_3O^+}$ in the bulk water region of the interface model as shown in Figure 5.6 [133, 134].

The $\Delta_{dp}A_{H_3O^+}$ is calculated using the free energy perturbation (FEP) scheme, in which an auxiliary mapping Hamiltonian $H_\eta = (1-\eta)H_R + \eta H_P$ is constructed by linearly mixing the reactant Hamiltonian H_R and the product Hamiltonian H_P through coupling parameter η. As shown in Figure 5.7(a), the $\Delta_{dp}A^{(i)}_{H_3O^+}$ is obtained via the thermodynamic integration (TI),

$$\Delta A = A(\eta = 1) - A(\eta = 0) = \int_0^1 \langle \Delta E \rangle_\eta d\eta, \tag{5.22}$$

where $\langle \Delta E \rangle_\eta$, called as the vertical energy gap, is the ensemble average of total energy difference between the reactant and product state at fix configurations at a given η, as illustrated in Figure 5.7(b). Note that for proton/electron transfer, there is a change of net charge from reactant to product state, and thus the computed vertical energy gap is subjected to the uncertainty of potential reference under PBC, as mentioned above.

This uncertainty can be removed by coupling a half redox reaction X^-/X to the SHE, expressed as

$$X^-(aq) + H^+(aq) \rightarrow X(aq) + \tfrac{1}{2}H_2(g). \tag{5.23}$$

Thus, the redox potential of X^-/X on the scale of SHE ($U^{SHE}_{X^-/X}$) can be calculated from the free energy of Eq. (5.23),

$$e_0 U^{SHE}_{X^-/X} = \Delta_{ox}A_{X^-} + \Delta_{dp}A_{H_3O^+} - \Delta_f G^{g,o}_{H^+} - \Delta E_{ZP}. \tag{5.24}$$

The term $\Delta_{ox}A_{X^-}$ in Eq. (5.24) is the oxidation integral of X^- in water and calculated using the similar FEP scheme to $\Delta_{dp}A_{H_3O^+}$. Note that neither the $\Delta_{ox}A_{X^-}$ nor $\Delta_{dp}A_{H_3O^+}$ in Eq. (5.24) has physical meaning alone due to the artificial HPS, when the shifts in these two quantities cancel out when they are computed from the same simulation cell. The terms $\Delta_f G^{g,o}_{H^+}$

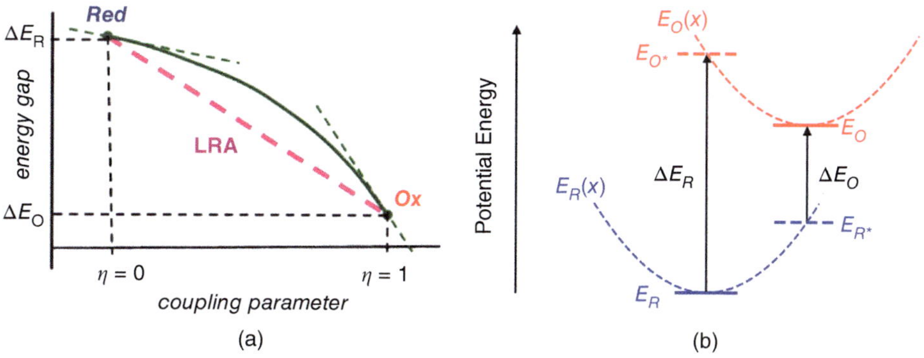

Figure 5.7 (a) Free energy profile of thermodynamic integration path along the coupling parameter connecting the reactant and product state, e.g. reduced and oxidized state in calculation of redox potentials. (b) Representation of the potential energy surfaces of a species (X) in the reduced and oxidized states. ΔE_R and ΔE_O denote the vertical energy gaps of the reduced and oxidized state, respectively. The notations can be readily converted into protonation states for proton insertion process, e.g. in pKa calculation.

and ΔE_{ZP} in Eq. (5.25) represent the formation free energy of a proton in the gas phase and the zero-point energy correction for the O–H bond in the hydronium ion, which are 15.81 eV and 0.35 eV, respectively [68].

It is straightforward to extend the computational SHE for calculating band alignment of a semiconductor/water interface with respect to SHE (U_{be}^{SHE}),

$$e_0 U_{be}^{SHE} = -\epsilon_{be} + \Delta_{dp} A_{H_3O^+} - \Delta_f G_{H^+}^{g,o} - \Delta E_{ZP}. \tag{5.25}$$

The term ϵ_{be} in Eq. (5.25) represents the pristine energy of the band edge of the semiconductor surface, as obtained from the interface model. In recent years, this computational SHE scheme has been applied for determining the CBM and VBM of various semiconductor/water interfaces [46, 47, 135].

5.4.2.2 Computational standard hydrogen electrode in two steps

One problem for computational SHE is the high computational cost, particularly for computation of the deprotonation free energy of the aqueous H_3O^+ ion in an interface model, which limits wide applications of this method to complex systems, such as metal/water interfaces. In the most recent development of the computational SHE [31], the interface model is supplemented by a pure water model as shown in Figure 5.8(a), to avoid expensive computation of $\Delta_{dp} A_{H_3O^+}$. To distinguish, the superscript (i) and (w) are used to denote the terms for the interface and pure water model, respectively. The key idea of the new approach is replacing the calculation of the term $\Delta_{dp} A_{H_3O^+}$ at interface with the pure water model. It should be clear that the $\Delta_{dp} A_{H_3O^+}$ calculated from different models are not same due to the different PBC. An internal reference, i.e. the electrostatic potential of bulk water $-e_0 \phi_{wat}$, can be used to realign the potential shift. As illustrated in Figure 5.8, we can have the following relation,

$$e_0 \Delta V = \Delta_{dp} A_{H_3O^+}^{(i)} - \Delta_{dp} A_{H_3O^+}^{(w)} = -e_0 \left(\phi_{wat}^{(i)} - \phi_{wat}^{(w)} \right). \tag{5.26}$$

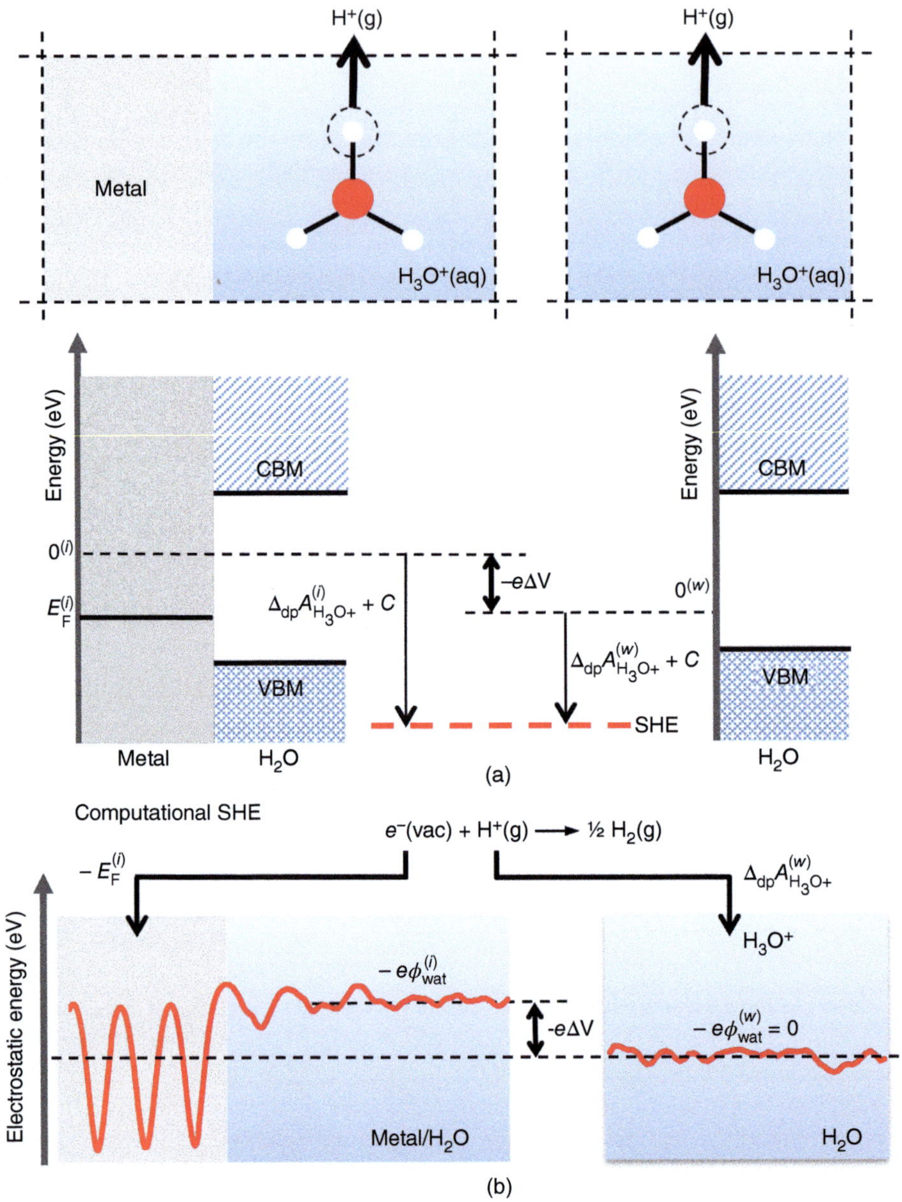

Figure 5.8 Representation of the computational SHE method [31]. (a) The deprotonation processes of a hydronium ion in the interface and pure water models, respectively. The deprotonation energies computed from these two models differ by $-e_0\Delta V$ due to the different PBC conditions. (b) The electrostatic potential profiles of interface and pure water models. The computed electrostatic potential of the bulk water region differ by the same $-e_0\Delta V$.

In the Ewald summation, the term $\phi_{\text{wat}}^{(\text{w})}$ in Eq. (5.26) is set to zero. Then, substituting Eq. (5.25) with Eq. (5.26), the electrode potential of the interface can be reformulated as

$$e_0 U^{\text{SHE}} = -\epsilon_F^{(\text{i})} + \Delta_{\text{dp}} A_{\text{H}_3\text{O}^+}^{(\text{w})} - e\phi_{\text{wat}}^{(\text{i})} - \Delta_{\text{f}} G_{\text{H}^+}^{\text{g,o}} - \Delta E_{ZP}. \quad (5.27)$$

The $\Delta_{\text{dp}} A_{\text{H}_3\text{O}^+}^{(\text{w})}$ in Eq. (5.27) denotes the deprotonation energy of a H_3O^+ calculated from a pure water model. Note that this value may vary with the DFT code used, particularly depending on the pseudopotentials. Our recommended value for the standard settings in the code CP2K is 15.35 eV, which has been verified against box size and various levels of electronic structure theories [67, 68, 136]. According to Eq. (5.27), two terms are required for computing the electrode potential; one is the Fermi level of the interface $\epsilon_F^{(\text{i})}$, and the other is the electrostatic potential of the bulk water region inside the interface model $-e\phi_{\text{wat}}^{(\text{i})}$. These two quantities can be easily reached with one single AIMD run. This scheme has been validated by computing the PZC of a series of transition metal/water interfaces [31].

5.4.2.3 Computational Ag/AgCl reference electrode

The SHE is often used for electrochemical cells in aqueous solution. However, in non-aqueous solutions (such as organic solvent and ionic liquids) its application often limited due to the unknown potential shift caused by the liquid-liquid junction. In experiments, one of the alternative reference electrodes for non-aqueous systems is the Ag/AgCl electrode [137–139], and thus the computational Ag/AgCl reference electrode has recently been developed [140] to directly compare the simulation results with the experiment. Similar to the computational SHE, the computational Ag/AgCl reference electrode should be coupled with another half reaction. For a redox couple X^-/X, its potential with respect to the Ag/AgCl reference electrode corresponds to the free energy of the full reaction,

$$\text{AgCl}(s) + \text{X}^-(aq) \rightarrow \text{Ag}(s) + \text{Cl}^-(aq) + \text{X}(aq). \quad (5.28)$$

As shown in Figure 5.9, the key step in the computational Ag/AgCl electrode is the calculation of the work function of Cl^- anion, W_{Cl^-}, which corresponds to the computed desolvation integral of aqueous Cl^- anion, ΔA_{Cl^-}, corrected for the total energy of the Cl^- anion in the gas phase, E_{Cl^-}. Thus, the equilibrium potential of X^-/X with respect to Ag/AgCl reference electrode ($a_{\text{Cl}^-} = 1$) $U_{\text{X}^-/X}^{\text{Ag/AgCl}}$ can be written as,

$$e_0 U_{\text{X}^-/X}^{\text{Ag/AgCl}} = \Delta_{\text{ox}} A_{\text{X}^-} - (\Delta A_{\text{Cl}^-} + E_{\text{Cl}^-}) + \Delta_{\text{f}} G_{\text{Cl}^-}^{\text{g,o}}. \quad (5.29)$$

The term $\Delta_{\text{f}} G_{\text{Cl}^-}^{\text{g,o}}$ is the Gibbs formation energy of Cl^- in the gas phase, as illustrated in Figure 5.9, which is −1.3 eV from the thermodynamic data [141]. This approach has been verified in aqueous solution with respect to the computational SHE, and the error is less than 0.2 eV.

5.5 Computation of potentials of zero charge

The PZC is an intrinsic property of the electrochemical interface, and is well defined as the potential at which there is no net charge on the electrode surface. To measure the PZC of

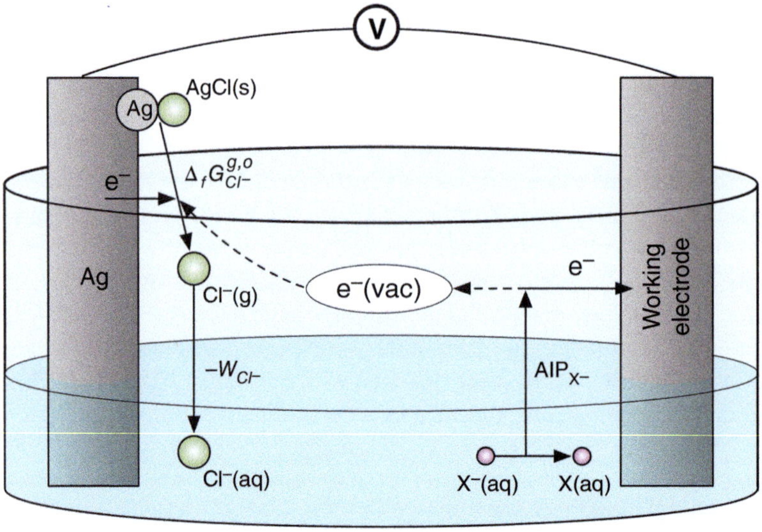

Figure 5.9 Thermodynamic cycle used to calculate the potential of redox couple X^-/X on the scale of the Ag/AgCl electrode. The electrostatic reference for electron and ions follows the convention in electrochemistry, which is the vacuum just outside the solution. $\Delta_f G^{g,o}_{Cl^-}$ and W_{Cl^-} are the Gibbs formation free energy of Cl^- in the gas phase and the work function of Cl^-, respectively. The AIP_{X^-} is the adiabatic ionization potential (AIP) of a solvated species, and can be considered as the electronic work function of an electron bound in solvated X^-. The standard state for the gas phase is $p° = 1$ bar, and the standard concentration of the solvated Cl^- is $c° = 1$ M.

metal/water interfaces, various methods and techniques have been developed in the past few decades, such as measurement of surface tension, impedance spectra, CO-charge displacement method and determination of maximum entropy [70, 142–144]. Using these methods, the PZC of *sp* metal and transition metal surfaces have been determined. For platinum-group metals, for instance Pt(111), the accurate PZC is still under some debate due to the surface contamination and etc., and most values measured fall in a range, 0.2~0.4 V vs SHE [145–147]. A very positive value of ~0.8 V vs SHE has even been proposed using an ex situ preparation/immersion technique [148], and consequently a recent work has suggested that there are two PZC for the Pt(111)/water interface [149]. Theoretical modeling can complement the experimental approaches to study the microscopic structures and potentials of the electrochemical interfaces. There are also certain advantages for computation, for example, surface contamination can be naturally avoided by modeling.

Some computed results of the PZC of Pt(111)/water interface have been summarized in Table 5.1, and it clearly shows the important effects of the choice of the density functional and the representation method of the solvent on the computed PZC. The computed PZC can vary by more than 2 V, depending on the setup. Comparing to the experimental value, we can find that the U^{SHE}_{PZC} computed in a full Pt(111)/water interface model using AIMD based on PBE-D3 functional shows the best agreement [31]. Considering that the functional error in computation of the metal work function W^M_e will be transferred to the U^{SHE}_{PZC}, a property unaffected by this error in W^M_e is introduced for comparison among the computational results, i.e. the outer potential difference $\psi^M - \psi^S$ coinciding with the difference between

Table 5.1 Computed PZC of Pt(111) surface in aqueous solution. Results obtained by different density functionals and various representations of solvent are summarized, and for comparison the experimental values of PZC of Pt(111) are also included. The absolute SHE potential is chosen as 4.44 V, as recommended by Trasatti [62]

DFT	Solvent	U_{PZC}^{SHE} (V)	W_e^M (eV)	$\psi^M - \psi^S$ (V)	Ref.
PBE-D3	bulk (AIMD)	0.2	5.8	−1.1	[31]
RPBE-D3	bulk (AIMD)	0.52	5.51	−0.55	[42, 65]
PBE	bilayer (H-down)	0.86	5.52	−0.22	[28]
PBE	bilayer (H-up)	−1.26	5.52	−2.34	[28]
PBE	continuum	0.31	5.52	−0.77	[22]
RPBE	continuum	1.13	5.96	−0.39	[150]
RPBE	0.5 monolayer + continuum	0.5	5.96	−1.02	[150]
Experiment		0.3	5.9	−1.2	[145–147]

work function and PZC at the absolute scale (see Eq. (5.14)). As can be seen in Table 5.1, the calculated value of $\psi^M - \psi^S$ from the same work [31] also well reproduces the experimental data, validating the simulation model and method used.

Theoretically, the term $\psi^M - \psi^S$ can be separated into two contributions,

$$\psi^M - \psi^S = \Delta\phi_{ori} + \Delta\phi_{el}. \tag{5.30}$$

The term $\Delta\phi_{ori}$ in Eq. (5.30) represents the potential change induced by the orientation dipole of water at interface, and $\Delta\phi_{el}$ is the potential shift at interface caused by electron redistribution due to interaction between water and the metal surface. Separation between these two factors is not directly accessible in experiment. AIMD simulation [31] shows that the contribution from water orientation is negligible on several metals including Pt(111), Pd(111), Ag(111), and Au(111), and the electron redistribution dominates the outer potential difference. Furthermore, it can be shown that the significant electron redistribution at the Pt(111)/water interface results from chemisorption of surface water as shown in Figure 5.10.

We note that U_{PZC}^{SHE} and $\psi^M - \psi^S$ computed with AIMD of similar Pt(111)/water interface models but by a different RPBE-D3 functional [42, 65] are different by ~0.3 and ~0.6 V from those by the PBE-D3 functional, respectively. Comparison between the structures of the interfacial water by the two functionals suggests that RPBE-D3 underestimates the effect of charge redistribution between water and Pt(111), leading to a smaller $\Delta\phi_{el}$, and thus less accurate U_{PZC}^{SHE} and $\psi^M - \psi^S$. Following the surface science approach, water bilayer models have been used sometime for representing the solvation effect on metals, e.g. calculating electrocatalytic reactions [109, 151, 152]. As listed in Table 5.1, water orientation has drastic effect on the U_{PZC}^{SHE}; neither H-down and H-up bilayer model gives correction U_{PZC}^{SHE} or $\psi^M - \psi^S$. Another common approach in literature is to represent water phase by implicit continuum solvation models. In the absence of explicit water, the continuum models would omit chemisorption of water on metal surfaces and thus miss the electronic effect on the interface potential. Moreover, the $\psi^M - \psi^S$ calculated with continuum solvation method is

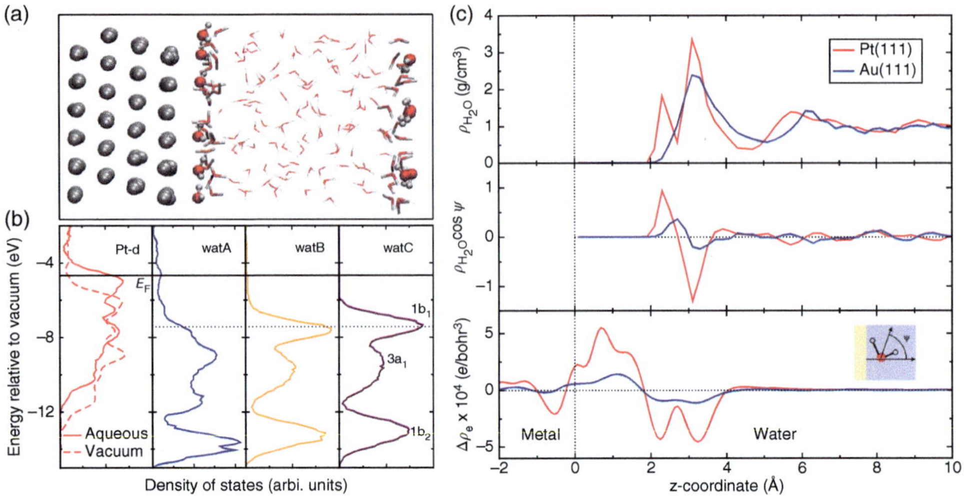

Figure 5.10 (a) Representative structure of the Pt(111)/water interface. Pt, O and H are colored with grey, red and white, respectively. The watA, watB and watC are represented by ball, stick and line models, respectively. (b) Projected density of states of the Pt(111)/water interface. (c) Profiles of water density, dipole orientation, electronic redistribution of the Pt(111)/water and Au(111)/water interface along the surface normal direction. The zero in *z* indicates the position of metal surface [31].

strongly dependent on the choice of the parameters used in the dielectric function [22, 150]. Inclusion of a fraction of explicit water may alleviate the neglection of water chemisorption, however, and the problem is then how to treat the orientation and hydrogen bond network of the explicit water in a consistent manner, particularly in the presence of electric field.

5.6 Summary

This chapter summarizes the theoretical background of electrochemistry, including concepts of reference electrode, electrode potential and absolute potential. We also reviewed the latest development in simulating electrochemical interfaces. Focusing on the *ab initio* computation of electrochemical interfaces, in which the electrodes are normally treated with electronic structure theory (e.g. DFT), and the electrolyte solution can be represented in various approaches including continuum models, JDFT method, classical and quantum explicit water molecules. It is still rather challenging on how to realistically model electrified interfaces and electric double layers (EDL). We stressed that the electrode potential is a key thermodynamic quantity for linking simulated electrochemical interfaces to experiment, and thus summarized a few computational methods for determining the electrode potentials of electrochemical interfaces. The significance of these methods lies on the fact that interfacial properties, such as spectroscopic information and differential capacitance, can be compared between theory and experiment at the same electrochemical conditions. At last, the PZC of the Pt(111)/water interface computed from various models and computational setup are compared among each other together with experiment. This also highlights the importance of the explicit water chemically interacted with the metal surface and statistical sampling of water configurations at the interface.

Acknowledgement

We thank the National Natural Science Foundation of China (Grant Nos. 21861132015 and 21621091) for funding support.

References

1 D. Strmcnik, K. Kodama, D. van der Vliet, J. Greeley, V. R. Stamenkovic, and N. M. Marković. The role of non-covalent interactions in electrocatalytic fuel-cell reactions on platinum. *Nat. Chem.*, 1(6):466–472, Sep 2009.

2 Yan-Yan Hu, Zigeng Liu, Kyung-Wan Nam, Olaf J Borkiewicz, Jun Cheng, Xiao Hua, Matthew T Dunstan, Xiqian Yu, Kamila M Wiaderek, Lin-Shu Du, Karena W Chapman, Peter J Chupas, Xiao-Qing Yang, and Clare P Grey. Origin of additional capacities in metal oxide lithium-ion battery electrodes. *Nat. Mater.*, 12(12):1130–6, Dec 2013.

3 Céline Merlet, Benjamin Rotenberg, Paul A. Madden, Pierre-Louis Taberna, Patrice Simon, Yury Gogotsi, and Mathieu Salanne. On the molecular origin of supercapacitance in nanoporous carbon electrodes. *Nat. Mater.*, 11(4):306–310, Apr 2012.

4 Kevin Leung and Craig M. Tenney. Toward first principles prediction of voltage dependences of electrolyte/electrolyte interfacial processes in lithium ion batteries. *J. Phys. Chem. C*, 117(46):24224–24235, Nov 2013.

5 Kevin Leung. Predicting the voltage dependence of interfacial electrochemical processes at lithium-intercalated graphite edge planes. *Phys. Chem. Chem. Phys.*, 17(3):1637–1643, Dec 2015.

6 Wolfgang Schmickler and Elizabeth Santos. *Interfacial Electrochemistry*. Springer Berlin Heidelberg, Berlin, Heidelberg, 2nd edition, 2010.

7 María Escudero-Escribano, Christopher Wildi, Jonathan A. Mwanda, and Angel Cuesta. Metallization of cyanide-modified Pt(111) electrodes with copper. *J. Solid State Electrochem.*, 20(4):1087, Apr 2016.

8 A. Cuesta and D.M. Kolb. The structure of bromide and chloride adlayers on Au(100) electrodes: an in situ STM study. *Surf. Sci.*, 465(3):310–316, Oct 2000.

9 Masatoshi Osawa, Minoru Tsushima, Hirokazu Mogami, Gabor Samjeské, and Akira Yamakata. Structure of water at the electrified platinum-water interface: a study by surface-enhanced infrared absorption spectroscopy. *J. Phys. Chem. C*, 112:4248, Mar 2008.

10 Ken-ichi Ataka, Takao Yotsuyanagi, and Masatoshi Osawa. Potential-Dependent Reorientation of Water Molecules at an Electrode/Electrolyte Interface Studied by Surface-Enhanced Infrared Absorption Spectroscopy. *J. Phys. Chem.*, 100(25):10664, Jan 1996.

11 Akira Yamakata and Masatoshi Osawa. Destruction of the hydration shell around tetraalkylammonium ions at the electrochemical interface. *J. Am. Chem. Soc.*, 131(20):6892, May 2009.

12 Jian-Feng Li,Yi-Fan Huang, Sai Duan, Ran Pang, De-Yin Wu, Bin Ren, Xin Xu, and Zhong-Qun Tian. SERS and DFT study of water on metal cathodes of silver, gold and platinum nanoparticles. *Phys. Chem. Chem. Phys.*, 12(10):2493–2502, Feb 2010.

13 Zhong-Qun Tian, Bin Ren, Yan-Xia Chen, Shou-Zhong Zou, and Bing-Wei Mao. Probing electrode/electrolyte interfacial structure in the potential region of hydrogen evolution by Raman spectroscopy. *J. Chem. Soc., Faraday Trans.*, 92(20):3829, 1996.

14 Chao-Yu Li, Jia-Bo Le, Yao-Hui Wang, Shu Chen, Zhi-Lin Yang, Jian-Feng Li, Jun Cheng, and Zhong-Qun Tian. In situ probing electrified interfacial water structures at atomically flat surfaces. *Nat. Mater.*, 18:697–701, Jul 2019.

15 Hidenori Noguchi, Tsubasa Okada, and Kohei Uosaki. Molecular structure at electrode/electrolyte solution interfaces related to electrocatalysis. *Faraday Discuss.*, 140(0):125, 2009.

16 Yujin Tong, François Lapointe, Martin Thamer, Martin Wolf, and R. Kramer Campen. Hydrophobic water probed experimentally at the gold electrode/aqueous interface. *Angew. Chem. Int. Ed.*, 56:4211, Apr 2017.

17 Zachary D. Schultz, Scott K. Shaw, and Andrew A. Gewirth. Potential dependent organization of water at the electrified metal-liquid interface. *J. Am. Chem. Soc.*, 127(45):15916–15922, Nov 2005.

18 Michael F. Toney, Jason N. Howard, Jocelyn Richer, Gary L. Borges, Joseph G. Gordon, Owen R. Melroy, David G. Wiesler, Dennis Yee, and Larry B. Sorensen. Voltage-dependent ordering of water molecules at an electrode-electrolyte interface. *Nature*, 368(6470):444, Mar 1994.

19 Christopher A. Lucas, Paul Thompson, Yvonne Gründer, and Nenad M. Markovic. The structure of the electrochemical double layer: Ag(111) in alkaline electrolyte. *Electrochem. Commun.*, 13(11):1205, Nov 2011.

20 Juan-Jesus Velasco-Velez, Cheng Hao Wu, Tod A Pascal, Liwen F Wan, Jinghua Guo, David Prendergast, and Miquel Salmeron. The structure of interfacial water on gold electrodes studied by x-ray absorption spectroscopy. *Science*, 346(6211):831, Nov 2014.

21 S Myneni, Y Luo, L Å Näslund, M Cavalleri, L Ojamäe, H Ogasawara, A Pelmenschikov, Ph Wernet, P Väterlein, C Heske, Z Hussain, L G M Pettersson, and A Nilsson. Spectroscopic probing of local hydrogen-bonding structures in liquid water. *J. Phys.: Condens. Matter*, 14(8):L213–L219, Mar 2002.

22 Ya-Hui Fang, Guang-Feng Wei, and Zhi-Pan Liu. Theoretical modeling of electrode/-electrolyte interface from first-principles periodic continuum solvation method. *Catal. Today*, 202:98, Mar 2013.

23 Ya Hui Fang and Zhi Pan Liu. Mechanism and tafel lines of electro-oxidation of water to oxygen on RuO2(110). *J. Am. Chem. Soc.*, 132(51):18214–18222, Dec 2010.

24 Oliviero Andreussi, Ismaila Dabo, and Nicola Marzari. Revised self-consistent continuum solvation in electronic-structure calculations. *J. Chem. Phys.*, 136(6):064102, Feb 2012.

25 G. Fisicaro, L. Genovese, O. Andreussi, N. Marzari, and S. Goedecker. A generalized Poisson and Poisson-Boltzmann solver for electrostatic environments. *J. Chem. Phys.*, 144:014103, Jan 2016.

26 Kiran Mathew, Ravishankar Sundararaman, Kendra Letchworth-Weaver, T. A. Arias, and Richard G. Hennig. Implicit solvation model for density-functional study of nanocrystal surfaces and reaction pathways. *J. Chem. Phys.*, 140(8):084106, Feb 2014.

27 A. Gross, F. Gossenberger, X. Lin, M. Naderian, S. Sakong, and T. Roman. Water structures at metal electrodes studied by ab initio molecular dynamics simulations. *J. Electrochem. Soc.*, 161(8):E3015, 2014.

28 Sebastian Schnur and Axel Groß. Properties of metal-water interfaces studied from first principles. *New J. Phys.*, 11:125003, 2009.

29 Tao Cheng, Alessandro Fortunelli, and William A Goddard. Reaction intermediates during operando electrocatalysis identified from full solvent quantum mechanics molecular dynamics. *Proc. Natl. Acad. Sci.*, 116:7718, Mar 2019.

30 J Cheng and M Sprik. The electric double layer at a rutile TiO(2) water interface modelled using density functional theory based molecular dynamics simulation. *J. Phys.: Condens. Matter*, 26(24):244108, 2014.

31 Jiabo Le, Marcella Iannuzzi, Angel Cuesta, and Jun Cheng. Determining potentials of zero charge of metal electrodes versus the standard hydrogen electrode from based on density-functional-theory-based molecular dynamics. *Phys. Rev. Lett.*, 119:016801, Jul 2017.

32 Zhendong Guo, Francesco Ambrosio, Wei Chen, Patrick Gono, and Alfredo Pasquarello. Alignment of redox levels at semiconductor-water interfaces. *Chem. Mater.*, 30(1):94–111, Jan 2018.

33 Assil Bouzid and Alfredo Pasquarello. Atomic-scale simulation of electrochemical processes at electrode/water interfaces under referenced bias potential. *J. Phys. Chem. Lett.*, 9:1880–1884, Apr 2018.

34 Yao Sha, Ted H. Yu, Yi Liu, Boris V. Merinov, and William A. Goddard. Theoretical study of solvent effects on the platinum-catalyzed oxygen reduction reaction. *J. Phys. Chem. Lett.*, 1(5):856–861, Mar 2010.

35 Xiaohang Lin, Ferdinand Evers, and Axel Groß. First-principles study of the structure of water layers on flat and stepped Pb electrodes. *Beilstein J. Nanotechnol.*, 7:533, Apr 2016.

36 Xiaohang Lin and Axel Groß. First-principles study of the water structure on flat and stepped gold surfaces. *Surf. Sci.*, 606(11-12):886, Jun 2012.

37 Jiabo Le, Angel Cuesta, and Jun Cheng. The structure of metal-water interface at the potential of zero charge from density functional theory-based molecular dynamics. *J. Electroanal. Chem.*, 819:87–94, Jun 2018.

38 Jiabo Le, Qiyuan Fan, Laura Perez-Martinez, Angel Cuesta, and Jun Cheng. Theoretical insight into the vibrational spectra of metal-water interfaces from density functional theory based molecular dynamics. *Phys. Chem. Chem. Phys.*, 20:11554–11558, May 2018.

39 Luana S Pedroza, Adrien Poissier, and M-V Fernández-Serra. Local order of liquid water at metallic electrode surfaces. *J. Chem. Phys.*, 142(3):034706, Jan 2015.

40 Jinggang Lan, Jürg Hutter, and Marcella Iannuzzi. First-Principles Simulations of an Aqueous CO/Pt(111) Interface. *J. Phys. Chem. C*, 122:24068, Oct 2018.

41 Luca Bellarosa, Rodrigo García-Muelas, Guillem Revilla-López, and Núria López. Diversity at the Water-Metal Interface: Metal, Water Thickness, and Confinement Effects. *ACS Cent. Sci.*, 2:109, Feb 2016.

42 Sung Sakong, Katrin Forster-Tonigold, and Axel Groß. The structure of water at a Pt(111) electrode and the potential of zero charge studied from first principles. *J. Chem. Phys.*, 144:194701, May 2016.

43 Jun Cheng and Michiel Sprik. Acidity of the aqueous rutile TiO_2 (110) surface from density functional theory based molecular dynamics. *J. Chem. Theory Comput.*, 6(3):880, Mar 2010.

44 Marialore Sulpizi, Marie-Pierre Gaigeot, and Michiel Sprik. The silica-water interface: how the silanols determine the surface acidity and modulate the water properties. *J. Chem. Theory Comput.*, 8(3):1037, 2012.

45 Oliver R. Gittus, Guido Falk Von Rudorff, Kevin M. Rosso, and Jochen Blumberger. Acidity constants of the hematite-liquid water interface from ab initio molecular dynamics. *J. Phys. Chem. Lett.*, 9(18):5574–5582, Sep 2018.

46 J Cheng and M Sprik. Alignment of electronic energy levels at electrochemical interfaces. *Phys. Chem. Chem. Phys.*, 14:11245, 2012.

47 Jun Cheng and Michiel Sprik. Aligning electronic energy levels at the TiO_2/H_2O interface. *Phys. Rev. B: Condens. Matter*, 82(8):1, 2010.

48 Tuan Anh Pham, Donghwa Lee, Eric Schwegler, and Giulia Galli. Interfacial effects on the band edges of functionalized si surfaces in liquid water. *J. Am. Chem. Soc.*, 136(49):17071–7, Dec 2014.

49 Tuan Anh Pham, Yuan Ping, and Giulia Galli. Modelling heterogeneous interfaces for solar water splitting. *Nat. Mater.*, 16:401–408, Apr 2017.

50 Jun Cheng, Xiandong Liu, John A Kattirtzi, Joost VandeVondele, and Michiel Sprik. Aligning electronic and protonic energy levels of proton-coupled electron transfer in water oxidation on aqueous TiO_2. *Angew. Chem. Int. Ed.*, 53(45):12046, Nov 2014.

51 Alexander Bagger, Logi Arnarson, Martin H. Hansen, Eckhard Spohr, and Jan Rossmeisl. Electrochemical CO reduction: a property of the electrochemical interface. *J. Am. Chem. Soc.*, 141(4):1506–1514, Jan 2019.

52 Olle Björneholm, Martin H. Hansen, Andrew Hodgson, Li-Min Liu, David T. Limmer, Angelos Michaelides, Philipp Pedevilla, Jan Rossmeisl, Huaze Shen, Gabriele Tocci, Eric Tyrode, Marie-Madeleine Walz, Josephina Werner, and Hendrik Bluhm. Water at interfaces. *Chem. Rev.*, 116:7698, 2016.

53 Robert Seidel, Stephan Thürmer, and Bernd Winter. Photoelectron spectroscopy meets aqueous solution: studies from a vacuum liquid microjet. *J. Phys. Chem. Lett.*, 2(6):633–641, Mar 2011.

54 J. E. B. Randles. The real hydration energies of ions. *Trans. Faraday Soc.*, 52(1):1573, Jan 1956.

55 W Ronald Fawcett. The ionic work function and its role in estimating absolute electrode potentials. *Langmuir*, 24:9868, Sep 2008.

56 John R. Farrell and Peter McTigue. Precise compensating potential difference measurements with a voltaic cell. The surface potential of water. *J. Electroanal. Chem.*, 139(1):37, Oct 1982.

57 Wilford N. Hansen and Galen J. Hansen. Absolute half-cell potential: A simple direct measurement. *Phys. Rev. A*, 36(3):1396, Aug 1987.

58 W.N. Hansen and D.M. Kolb. The work function of emersed electrodes. *J. Electroanal. Chem. Interfacial Electrochem.*, 100(1-2):493, Jun 1979.

59 Abdirisak A Isse and Armando Gennaro. Absolute potential of the standard hydrogen electrode and the problem of interconversion of potentials in different solvents. *J. Phys. Chem. B*, 114:7894, Jun 2010.

60 Michael D. Tissandier, Kenneth A. Cowen, Wan Yong Feng, Ellen Gundlach, Michael H. Cohen, Alan D. Earhart, James V. Coe, and Thomas R. Tuttle. The proton's absolute aqueous enthalpy and Gibbs free energy of solvation from cluster-ion solvation data. *J. Phys. Chem. A*, 102(40):7787, 1998.

61 J. Schneider, C. Franke, and D. M. Kolb. Image-potential-induced surface states at the Ag(111)-electrolyte interface. *Surf. Sci.*, 198(1-2):277, Jan 1988.

62 S. Trasatti. The absolute electrode potential: an explanatory note (Recommendations 1986). *Pure Appl. Chem.*, 58:955, Jan 1986.

63 Howard Reiss and Adam Heller. The absolute potential of the standard hydrogen electrode: a new estimate. *J. Phys. Chem.*, 89(20):4207, Sep 1985.

64 Liang Yu, Xiulian Pan, Xiaoming Cao, P. Hu, and Xinhe Bao. Oxygen reduction reaction mechanism on nitrogen-doped graphene: A density functional theory study. *J. Catal.*, 282(1):183–190, Aug 2011.

65 Sung Sakong and Axel Groß. The electric double layer at metal-water interfaces revisited based on a charge polarization scheme. *J. Chem. Phys.*, 149:084705, Aug 2018.

66 Axel Groß and Sung Sakong. Modelling the electric double layer at electrode-electrolyte interfaces. *Curr. Opin. Electrochem.*, 14:1–6, Apr 2018.

67 Jun Cheng and Joost VandeVondele. Calculation of electrochemical energy levels in water using the random phase approximation and a double hybrid functional. *Phys. Rev. Lett.*, 116(8):086402, Feb 2016.

68 Jun Cheng, Xiandong Liu, Joost VandeVondele, Marialore Sulpizi, and Michiel Sprik. Redox potentials and acidity constants from density functional theory based molecular dynamics. *Acc. Chem. Res.*, 47:3522–9, Dec 2014.

69 Wolfgang Schmickler. Electronic effects in the electric double layer. *Chem. Rev.*, 96:3177, Jan 1996.

70 Sergio Trasatti and Enn Lust. The potential of zero charge. In *Modern Aspects of Electrochemistry*. Kluwer Academic Publishers, Boston, 1999.

71 J. M. Heras and L. Viscido. The behavior of water on metal surfaces. *Catal. Rev.*, 30(2):281, Jun 1988.

72 Patricia A. Thiel and Theodore E. Madey. The interaction of water with solid surfaces: Fundamental aspects. *Surf. Sci. Rep.*,7(6-8):211, Oct 1987.

73 Kai-Ming Ho, B. N. Harmon, and S. H. Liu. Surface-state contribution to the electroreflectance of noble metals. *Phys. Rev. Lett.*, 44(23):1531–1534, Jun 1980.

74 E. Spohr. Some recent trends in computer simulations of aqueous double layers. *Electrochim. Acta*, 49(1):23, Dec 2003.

75 E. Spohr. Computer simulation of the water/platinum interface. *J. Chem. Phys.*, 93(16):6171, Aug 1989.

76 J Bocker, RR Nazmutdinov, Eckhard Spohr, and K Heinziger. Molecular dynamics simulation studies of the mercury-water interface. *Surf. Sci.*, 335:372, Jul 1995.

77 K. Raghavan, K. Foster, K. Motakabbir, and M. Berkowitz. Structure and dynamics of water at the Pt(111) interface: Molecular dynamics study. *J. Chem. Phys.*, 94(3):2110, Feb 1991.

78 K. Foster, K. Raghavan, and M. Berkowitz. A molecular dynamics study of the effect of temperature on the structure and dynamics of water between Pt walls. *Chem. Phys. Lett.*, 162(1-2):32, Oct 1989.

79 J. Ilja Siepmann and Michiel Sprik. Influence of surface topology and electrostatic potential on water/electrode systems. *J. Chem. Phys.*, 102(1):511, Jan 1995.

80 E. Spohr. Computer simulations of Electrochemical interfaces. In *Advances in Electrochemical Science and Engineering*. Wiley-VCH Verlag GmbH, 2008.

81 D. T. Limmer, A. P. Willard, P. Madden, and D. Chandler. Hydration of metal surfaces can be dynamically heterogeneous and hydrophobic. *Proc. Natl. Acad. Sci.*, 110(11):4200, Feb 2013.

82 Tianying Yan, Christian J. Burnham, Mario G. Del Pópolo, and Gregory A. Voth. Molecular dynamics simulation of ionic liquids: The effect of electronic polarizability. *J. Phys. Chem. B*, 108(32):11877–11881, 2004.

83 Zhen Cao, Revati Kumar, Yuxing Peng, and Gregory A. Voth. Hydrated proton structure and diffusion at platinum surfaces. *J. Phys. Chem. C*, 119(26):14675–14682, Jul 2015.

84 Dominik. Marx and Jurg Hutter. *Ab initio molecular dynamics: basic theory and advanced methods*. Cambridge University Press, 2009.

85 M. Born and R. Oppenheimer. Zur Quantentheorie der Molekeln. *Ann. Phys.*, 389(20):457, Jan 1927.

86 R. Car and M. Parrinello. Unified approach for molecular dynamics and density-functional theory. *Phys. Rev. Lett.*, 55(22):2471, Nov 1985.

87 Basile F. E. Curchod and Todd J. Martínez. Ab initio nonadiabatic quantum molecular dynamics. *Chem. Rev.*, 118(7):3305–3336, Apr 2018.

88 Thomas D. Kühne, Matthias Krack, Fawzi R. Mohamed, and Michele Parrinello. Efficient and accurate car-parrinello-like approach to born-oppenheimer molecular dynamics. *Phys. Rev. Lett.*, 98(6):066401, Feb 2007.

89 Wei-Na Zhao and Zhi-Pan Liu. Mechanism and active site of photocatalytic water splitting on titania in aqueous surroundings. *Chem. Sci.*, 5(6):2256, May 2014.

90 Ravishankar Sundararaman, William A. Goddard, and Tomas A. Arias. Grand canonical electronic density-functional theory: Algorithms and applications to electrochemistry. *J. Chem. Phys.*, 146(11):114104, Mar 2017.

91 Ismaila Dabo, Yanli Li, and Nicola Marzari. Ab initio electrochemical properties of electrode surfaces. In *Fuel Cell Science: Theory, Fundamentals, and Biocatalysis*, pages 415–431. John Wiley & Sons, Inc., 2010.

92 Sung Sakong and Axel Groß. The importance of the electrochemical environment in the electro-oxidation of methanol on Pt(111). *ACS Catal.*, 6(8):5575–5586, Aug 2016.

93 Nicolas G. Hörmann, Oliviero Andreussi, and Nicola Marzari. Grand canonical simulations of electrochemical interfaces in implicit solvation models. *J. Chem. Phys.*, 150(4):041730, Jan 2019.

94 Jun Haruyama, Tamio Ikeshoji, and Minoru Otani. Electrode potential from density functional theory calculations combined with implicit solvation theory. *Phys. Rev. Mater.*, 2(9):095801, Sep 2018.

95 Kendra Letchworth-Weaver and T. A. Arias. Joint density functional theory of the electrode-electrolyte interface: Application to fixed electrode potentials, interfacial capacitances, and potentials of zero charge. *Phys. Rev. B*, 86(7):075140, Aug 2012.

96 Lin Shen and Weitao Yang. Quantum Mechanics/molecular mechanics method combined with hybrid all-atom and coarse-grained model: theory and application on redox potential calculations. *J. Chem. Theory Comput.*, 12(4):2017–2027, Apr 2016.

97 Dorothea Golze, Marcella Iannuzzi, Manh Thuong Nguyen, Daniele Passerone, and J?rg Hutter. Simulation of adsorption processes at metallic interfaces: An image charge augmented QM/MM approach. *J. Chem. Theory Comput.*, 9(11):5086, Nov 2013.

98 Hyung Kyu Lim, Hankyul Lee, and Hyungjun Kim. A seamless grid-based interface for mean-field QM/MM coupled with efficient solvation free energy calculations. *J. Chem. Theory Comput.*, 12(10):5088–5099, Oct 2016.

99 Hyung-Kyu Lim, Youngkook Kwon, Han Seul Kim, Jiwon Jeon, Yong-Hoon Kim, Jung-Ae Lim, Beom-Sik Kim, Jina Choi, and Hyungjun Kim. Insight into the microenvironments of the metal-ionic liquid interface during electrochemical CO_2 reduction. *ACS Catal.*, 8(3):2420–2427, Mar 2018.

100 Hui-Fang Wang and Zhi-Pan Liu. Formic acid oxidation at Pt/H_2O interface from periodic DFT calculations integrated with a continuum solvation model. *J. Phys. Chem. C*, 113(40):17502, Oct 2009.

101 N. David Mermin. Thermal properties of the inhomogeneous electron gas. *Phys. Rev.*, 137(5A):A1441, Mar 1965.

102 A. Y. Lozovoi, A. Alavi, J. Kohanoff, and R. M. Lynden-Bell. Ab initio simulation of charged slabs at constant chemical potential. *J. Chem. Phys.*, 115(4):1661, Jul 2001.

103 Nicéphore Bonnet, Tetsuya Morishita, Osamu Sugino, and Minoru Otani. First-principles molecular dynamics at a constant electrode potential. *Phys. Rev. Lett.*, 109(26):266101, Dec 2012.

104 Assil Bouzid and Alfredo Pasquarello. Redox levels through constant fermi-level ab initio molecular dynamics. *J. Chem. Theory Comput.*, 13(4):1769–1777, Apr 2017.

105 Kazuya Shiratori and Katsuyuki Nobusada. Development of a finite-temperature density functional approach to electrochemical reactions. *J. Phys. Chem. A*, 112(42):10681, Oct 2008.

106 Wolfgang B Schneider and Alexander A Auer. Constant chemical potential approach for quantum chemical calculations in electrocatalysis. *Beilstein J. Nanotechnol.*, 5(1):668, Jan 2014.

107 Christopher D. Taylor, Sally A. Wasileski, Jean-Sebastien Filhol, and Matthew Neurock. First principles reaction modeling of the electrochemical interface: Consideration and calculation of a tunable surface potential from atomic and electronic structure. *Phys. Rev. B*, 73:165402, Apr 2006.

108 Jean-Sébastien Filhol and Matthew Neurock. Elucidation of the electrochemical activation of water over Pd by first principles. *Angew. Chem.*, 118(3):416, Jan 2006.

109 Jan Rossmeisl, Egill Skúlason, Mårten E Björketun, Vladimir Tripkovic, and Jens K Nørskov. Modeling the electrified solid-liquid interface. *Chem. Phys. Lett.*, 466(1-3):68, 2008.

110 Egill Skuulason, Vladimir Tripkovic, Marten E. Bjoorketun, Sigrídur Gudmundsdoottir, Gustav Karlberg, Jan Rossmeisl, Thomas Bligaard, Hannes Joonsson, and Jens K. Nørskov. Modeling the electrochemical hydrogen oxidation and evolution reactions on the basis of density functional theory calculations. *J. Phys. Chem. C*, 114(42):18182, Oct 2010.

111 Osamu Sugino, Ikutaro Hamada, Minoru Otani, Yoshitada Morikawa, Tamio Ikeshoji, and Yasuharu Okamoto. First-principles molecular dynamics simulation of biased electrode/solution interface. *Surf. Sci.*, 601(22):5237, 2007.

112 Tamio Ikeshoji, Minoru Otani, Ikutaro Hamada, and Yasuharu Okamoto. Reversible redox reaction and water configuration on a positively charged platinum surface: first principles molecular dynamics simulation. *Phys. Chem. Chem. Phys.*, 13(45):20223, Dec 2011.

113 Yumin Qian, Ikutaro Hamada, Minoru Otani, and Tamio Ikeshoji. Inhibition of water dissociation on a pitted Pt(111) surface: First principles study. *Catal. Today*, 202:163, Mar 2013.

114 Jörg Neugebauer and Matthias Scheffler. Adsorbate-substrate and adsorbate-adsorbate interactions of Na and K adlayers on Al(111). *Phys. Rev. B*, 46(24):16067–16080, Dec 1992.

115 Thomas D. Kühne, Tod A. Pascal, Efthimios Kaxiras, and Yousung Jung. New insights into the structure of the vapor/water interface from large-scale first-principles simulations. *J. Phys. Chem. Lett.*, 2(2):105–113, Jan 2011.

116 Francesco Ambrosio, Zhendong Guo, and Alfredo Pasquarello. Absolute energy levels of liquid water. *J. Phys. Chem. Lett.*, 9(12):3212–3216, Jun 2018.

117 T. Anh Pham, Cui Zhang, Eric Schwegler, and Giulia Galli. Probing the electronic structure of liquid water with many-body perturbation theory. *Phys. Rev. B*, 89(6):060202, Feb 2014.

118 Chris G. Van de Walle and J. Neugebauer. Universal alignment of hydrogen levels in semiconductors, insulators and solutions. *Nature*, 423(6940):626–628, Jun 2003.

119 Francesco Ambrosio, Giacomo Miceli, and Alfredo Pasquarello. Electronic levels of excess electrons in liquid water. *J. Phys. Chem. Lett.*, 8(9):2055–2059, May 2017.

120 Alex P. Gaiduk, Tuan Anh Pham, Marco Govoni, Francesco Paesani, and Giulia Galli. Electron affinity of liquid water. *Nat. Commun.*, 9(1):247, Dec 2018.

121 Wei Chen, Francesco Ambrosio, Giacomo Miceli, and Alfredo Pasquarello. Ab initio electronic structure of liquid water. *Phys. Rev. Lett.*, 117(18):186401, Oct 2016.

122 S. W. de Leeuw, J. W. Perram, and E. R. Smith. Simulation of electrostatic systems in periodic boundary conditions. I. Lattice sums and dielectric constants. *Proc. R. Soc. A*, 373(1752):27, Oct 1980.

123 Aron J Cohen, Paula Mori-Sánchez, and Weitao Yang. Challenges for density functional theory. *Chem. Rev.*, 112(1):289, Jan 2012.

124 Aron J Cohen, Paula Mori-Sánchez, and Weitao Yang. Insights into current limitations of density functional theory. *Science (New York, N.Y.)*, 321(5890):792, Aug 2008.

125 Carlo Adamo and Vincenzo Barone. Toward reliable density functional methods without adjustable parameters: The PBE0 model. *J. Chem. Phys.*, 110(13):6158, Mar 1999.

126 Jochen Heyd, Gustavo E. Scuseria, and Matthias Ernzerhof. Hybrid functionals based on a screened Coulomb potential. *J. Chem. Phys.*, 118(18):8207, Apr 2003.

127 Jochen Heyd, Gustavo E. Scuseria, and Matthias Ernzerhof. Erratum: Hybrid functionals based on a screened Coulomb potential (Journal of Chemical Physics (2003) 118 (8207)). *J. Chem. Phys.*, 124(21):219906, Jun 2006.

128 Igor Ying Zhang and Xin Xu. Doubly hybrid density functional for accurate description of thermochemistry, thermochemical kinetics and nonbonded interactions. *Int. Rev. Phys. Chem.*, 30(1):115, Jan 2011.

129 Andreas Hesselmann. Random-phase-approximation correlation method including exchange interactions. *Phys. Rev. A: At. Mol. Opt. Phys.*, 85(1):012517, Jan 2012.

130 B. Winter, R. Weber, W. Widdra, M. Dittmar, M. Faubel, and I. V. Hertel. Full valence band photoemission from liquid water using EUV synchrotron radiation. *J. Phys. Chem. A*, 108(14):2625–2632, 2004.

131 Annette Bernas, Christiane Ferradini, and Jean Paul Jay-Gerin. On the electronic structure of liquid water: Facts and reflections. *Chem. Phys.*, 222(2-3):151–160, Oct 1997.

132 Sung Sakong, Maryam Naderian, Kiran Mathew, Richard G Hennig, and Axel Groß. Density functional theory study of the electrochemical interface between a Pt electrode and an aqueous electrolyte using an implicit solvent method. *J. Chem. Phys.*, 142(23):234107, Jun 2015.

133 Jun Cheng, Marialore Sulpizi, and Michiel Sprik. Redox potentials and pKa for benzoquinone from density functional theory based molecular dynamics. *J. Chem. Phys.*, 131:154504, 2009.

134 Francesca Costanzo, Marialore Sulpizi, Raffaele Guido Della Valle, and Michiel Sprik. The oxidation of tyrosine and tryptophan studied by a molecular dynamics normal hydrogen electrode. *J. Chem. Phys.*, 134(24):244508, Jun 2011.

135 Xue-Ting Fan, Mei Jia, Ming-Hsien Lee, and Jun Cheng. Water effect on band alignment of GaP: A theoretical insight into pyridinium catalyzed CO2 reduction. *J. Energy Chem.*, 26(4):724–729, Jul 2017.

136 Martina Mangold, Leslie Rolland, Francesca Costanzo, Michiel Sprik, Marialore Sulpizi, and Jochen Blumberger. Absolute p K a values and solvation structure of amino acids from density functional based molecular dynamics simulation. *J. Chem. Theory Comput.*, 7(6):1951–1961, Jun 2011.

137 Rob Atkin, Natalia Borisenko, Marcel Drüschler, Sherif Zein El Abedin, Frank Endres, Robert Hayes, Benedikt Huber, and Bernhard Roling. An in situ STM/AFM and impedance spectroscopy study of the extremely pure 1-butyl-1-methylpyrrolidinium tris(pentafluoroethyl)trifluorophosphate/Au(111) interface: potential dependent solvation layers and the herringbone reconstruction. *Phys. Chem. Chem. Phys.*, 13(15):6849, Mar 2011.

138 Ove Oll, Tavo Romann, Piret Pikma, and Enn Lust. Spectroscopy study of ionic liquid restructuring at lead interface. *J. Electroanal. Chem.*, 778:41–48, Oct 2016.

139 Jaanus Kruusma, Arvo Tõnisoo, Rainer Pärna, Ergo Nõmmiste, Enn Lust, Jaanus Kruusma, Arvo Tõnisoo, Rainer Pärna, Ergo Nõmmiste, and Enn Lust. In situ x-ray photoelectron spectroscopic and electrochemical studies of the bromide anions dissolved in 1-ethyl-3-methyl imidazolium tetrafluoroborate. *Nanomaterials*, 9(2):304, Feb 2019.

140 Xiaohui Yang, Angel Cuesta, and Jun Cheng. A computational Ag/AgCl reference electrode from density functional theory-based molecular dynamics. *in revision by J. Phys. Chem. B*, 2019.

141 DR Lide. *Handbook of Chemistry and Physics*, 86th edn. CRC. CRC press, Boca Raton, FL., 2005.

142 D.M. Kolb and J. Schneider. Surface reconstruction in electrochemistry: Au(100-(5 × 20), Au(111)-(1 × 23) and Au(110)-(1 × 2). *Electrochim. Acta*, 31(8):929, Aug 1986.

143 A.M. El-Aziz, L.A. Kibler, and D.M. Kolb. The potentials of zero charge of Pd(111) and thin Pd overlayers on Au(111). *Electrochem. Commun.*, 4(7):535, Jul 2002.

144 Víctor Climent, Barry A. Coles, and Richard G. Compton. Coulostatic potential transients induced by laser heating of a Pt(111) single-crystal electrode in aqueous acid solutions. Rate of hydrogen adsorption and potential of maximum entropy. *J. Phys. Chem. B*, 106(23):5988–5996, 2002.

145 Angel Cuesta. Measurement of the surface charge density of CO-saturated Pt(111) electrodes as a function of potential: The potential of zero charge of Pt(111). *Surf. Sci.*, 572:11, 2004.

146 Michael J Weaver. Potentials of zero charge for platinum (111)-aqueous interfaces: a combined assessment from in-situ and ultrahigh-vacuum measurements. *Langmuir*, 14(14):3932–3936, 1998.

147 Rubén Rizo, Elton Sitta, Enrique Herrero, Víctor Climent, and Juan M. Feliu. Towards the understanding of the interfacial pH scale at Pt(111) electrodes. *Electrochim. Acta*, 162:138, Apr 2015.

148 U.W. Hamm, D. Kramer, R.S. Zhai, and D.M. Kolb. The pzc of Au(111) and Pt(111) in a perchloric acid solution: an ex situ approach to the immersion technique. *J. Electroanal. Chem.*, 414(1):85–89, Sep 1996.

149 Jun Huang, Tao Zhou, Jianbo Zhang, and Michael Eikerling. Double layer of platinum electrodes: Non-monotonic surface charging phenomena and negative double layer capacitance. *J. Chem. Phys.*, 148:044704, Jan 2018.

150 Ryosuke Jinnouchi and Alfred B. Anderson. Electronic structure calculations of liquid-solid interfaces: Combination of density functional theory and modified Poisson-Boltzmann theory. *Phys. Rev. B*, 77:245417, Jun 2008.

151 Vladimir Tripković, Egill Skúlason, Samira Siahrostami, Jens K Nørskov, and Jan Rossmeisl. The oxygen reduction reaction mechanism on Pt(111) from density functional theory calculations. *Electrochim. Acta*, 55:7975–7981, 2010.

152 Leanne D. Chen, Michal Bajdich, J. Mark P. Martirez, Caroline M. Krauter, Joseph A. Gauthier, Emily A. Carter, Alan C. Luntz, Karen Chan, and Jens K. Nørskov. Understanding the apparent fractional charge of protons in the aqueous electrochemical double layer. *Nat. Commun.*, 9(1):3202, Dec 2018.

6

Molecular Dynamics of the Electrochemical Interface and the Double Layer

Axel Groß[1,2]

[1] *Institute of Theoretical Chemistry, Ulm University, 89069 Ulm, Germany*
[2] *Helmholtz Institute Ulm (HIU), Electrochemical Energy Storage, 89069 Ulm, Germany*

6.1 Introduction

Electrochemistry is concerned with processes at the interface between an electron conductor – the electrode – and an ion conductor – the electrolyte [1]. This also means that in general the mean electrostatic potential across the interface between electrode and electrolyte changes. This leads to a redistribution of the electronic charges at the interface. Note furthermore that a liquid electrolyte typically contains a certain concentration of positive and negative ions, i.e., cations and anions, respectively, that in the bulk electrolyte are assumed to be uniformly distributed. However, at the electrode-electrolyte interface, due to electrostatic and chemical interactions the ion concentration might be vastly different from the one in the bulk electrolyte. In a simplified view, the reorganisation of electronic and ionic charge leads to the formation of two oppositely charged layers [2] forming the so-called electric double layer (EDL). The EDL "causes" the potential drop, but at the same time its formation is also a consequence of the difference in the potentials in the electrode and the electrolyte.

It is now well known that the simple picture of a capacitor-like double layer at the electrode-electrolyte interface does not capture the essence of the potential drop, in particular the dependence of the capacitance on the electrode potential. First it has been realized that the thermal distribution of the ions in the electrolyte leads to an additional diffuse layer [3–5]. Although based on an atomistic reasoning, the potential drop in these models is still expressed in a continuum approach. The correct atomistic description requires to perform proper statistical averages taking also polarization effects into account. Only very recently has it become possible to carry out the necessary demanding molecular dynamics simulations because of the development of more efficient codes and the increase in computer power. Hence the textbook understanding of the electric double layer is still based on continuum concepts that are more than one hundred years old. Here I will first review these concepts and then present some recent results of molecular dynamics simulations that are based on quantum chemical electronic structure calculations, so-called *ab initio* molecular dynamics (AIMD) simulations. As these simulations are still numerically rather demanding, a complete atomistic picture based on such simulations has not been

Atomic-Scale Modelling of Electrochemical Systems, First Edition.
Edited by Marko M. Melander, Tomi T. Laurila, and Kari Laasonen.

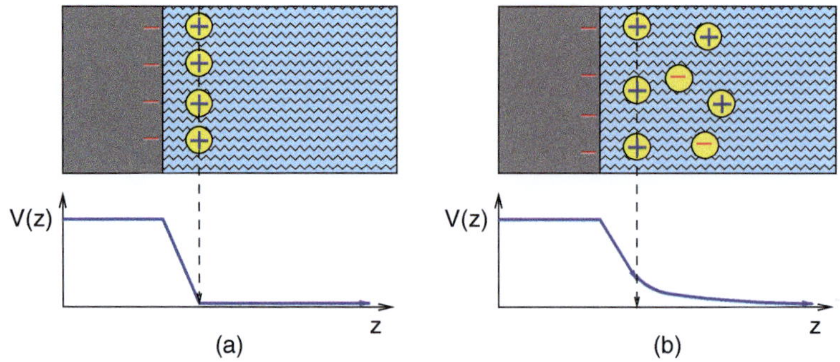

Figure 6.1 Illustration of possible scenarios for electric double layers. a) Charged electrode together with a layer of counter ions corresponding to the Helmholtz model. The dashed line indicates the Helmholtz layer. b) In addition to the layer of counter ions, an additional diffuse layer of anions and cations is present, corresponding to the combination of the Helmholtz and the Gouy-Chapman model.

established yet, but I will describe the new insights into the electronic and geometrical structure of the EDL that have been gained from these molecular dynamics simulations.

6.2 Continuum description of the electric double layer

At an electrified interface between an electrode and an electrolyte, the charged electrode repels co-ions of the same charge while attracting counterions of the opposite charge. Assuming that the counterions still have an intact solvation shell, they will form a layer at a certain distance from the electrode with the distance d given by the radius of the solvation shell. Such a scenario, which was first realized by Helmholtz [2], is illustrated in Fig. 6.1a. Thus the electrode surface and the ion layer form two layers of opposite charge, the electric double layer (EDL), resulting in a parallel-plate capacitor. The potential between the surface charge and the ions forming the so-called *Helmholtz layer* exhibits linear behavior as a function of separation from the electrode. The double-layer capacitance C of such an EDL is constant,

$$C = \frac{Q}{V} = \varepsilon\varepsilon_0 \frac{A}{d} , \tag{6.1}$$

where Q is the charge of the EDL, V the voltage drop accross the EDL, ε the permittivity, ε_0 the electric constant, A the area, and d is the distance between the two layers.

This means that the capacitance is independent of the electrode potential. However, this is at variance with experimental observations [1]. In order to allow for a varying capacitance, Gouy [3] and Chapman [4] refined the Helmholtz model early in the 20th century. They assumed that there is a diffuse thermal distribution of both anions and cations (note that any electrolyte has to be overall charge neutral, which means that the charge of one sort of ions has to be compensated by counterions). In fact, such a formulation is close to the macroscopic treatment of semiconductor interfaces between n-type and p-type semiconductors incorporating space charge layers. Assuming Boltzmann statistics for the distribution of both anions and cations of absolute charge $|ze|$ with a bulk density of n_0 leads to the

Poisson-Boltzmann equation [1] for the inner potential $\phi(x)$,

$$\frac{d^2\phi}{dx^2} = -\frac{zen_0}{\varepsilon\varepsilon_0}\left(\exp\left(-\frac{ze\phi(x)}{k_BT}\right) - \exp\left(\frac{ze\phi(x)}{k_BT}\right)\right). \tag{6.2}$$

For $ze\phi(x)/k_BT \ll 1$, which is valid for small charge densities on the electrode, the exponentials can be linearized. This yields an exponentially decreasing inner electric potential $\phi(x)$ and a charge density $\rho(x)$ of

$$\phi(x) = \frac{\sigma}{\varepsilon\varepsilon_0\kappa}\exp(-\kappa z)\ , \quad \rho(x) = -\sigma\kappa\exp(-\kappa z)\ , \tag{6.3}$$

where σ is the surface charge density of the electrode and

$$\kappa = \left(\frac{2(ze)^2n_0}{\varepsilon\varepsilon_0k_BT}\right)^{1/2} \tag{6.4}$$

is the inverse *Debye length*. Taking $U = \phi(x = 0)$ as the electrode potential [1], a double-layer capacitance per unit area of

$$C = \frac{Q}{V} = \varepsilon\varepsilon_0\kappa \tag{6.5}$$

results. So for small charge densities, the result for a parallel-plate capacitor is recovered with the distance of the two plates given by the Debye length, but still the capacitance is constant. However, the nonlinear Poisson-Boltzmann can in fact explicitly be solved [1]. Then a differential capacitance $C = \partial\sigma/\partial\phi$ results, which is given by

$$C = \varepsilon\varepsilon_0\kappa\cosh\left(\frac{z\varepsilon_0(U - U_{\text{pzc}})}{k_BT}\right)\ , \tag{6.6}$$

where U_{pzc} is the so-called potential of zero charge. This expression reproduces the experimental observation that the capacitance exhibits a minimum a the potential of zero charge. Still, at higher ion concentrations a much more complex dependence of the capacitance on the potential can be found. Furthermore, in the Gouy-Chapman approach all ions are assumed to be mobile. Thus it does not take into account that some ions adsorb specifically at the electrode surface and therefore it fails to describe highly charged double layers.

In order to address these limitations, Stern suggested a combination of the Helmholtz and Gouy-Chapman models [5], which is illustrated in Fig. 6.1b. Close to the electrode, it consists of an internal Stern layer corresponding to the Helmholtz layer in which the potential drops linearly. Further out, there is an outer diffuse layer corresponding to a Gouy-Chapman layer exhibiting an exponential potential drop. The overall capacitance C of these two layers can be evaluated according to

$$\frac{1}{C} = \frac{1}{C_{\text{GC}}} + \frac{1}{C_{\text{H}}} \tag{6.7}$$

where C_{GC} is the Gouy-Chapman capacitance and C_{H} is the Helmholtz capacitance, which is assumed to be independent of the electrolyte concentration.

The Stern model still has its limitations. The ions are described as point charges, the electrode is assumed to be a perfect conductor so that no electric fields can penetrate, and only electrostatic interactions are taken into account. In fact, there have been several attempts to develop a more realistic description [1, 6, 7], but the majority of these approaches have still been on a continuum level. Hence our basic understanding of the electric double layer at electrode-electrolyte interfaces is still based on concepts that have been established about a century ago. There is certainly a need to gain an atomistic understanding of the structure and properties of the electric double layer. Due to the liquid nature of the electrolyte, this requires appropriate statistical averages to be performed. Using empirical interaction potentials such as force fields, atomistic similations of electrode/electrolyte interfaces have already been possible some decades ago [8, 9]. However, it is important to realize that the formation of the electric double layer, especially when specifically adsorbed species are involved, is associated with a significant rearrangement of charges [10, 11] and strong polarization effects [12]. Force fields are not able to capture these effects, even interaction potentials derived from modern machine learning techniques [13] typically do not include such an electronic rearrangement. Thus a quantum chemical approach based on efficient electronic structure calculations is required to obtain a realistic model of the electric double layer. The still rather high computational cost of such calculations has prohibited progress in this respect.

However, due to the ever-increasing computer power and the development of more efficient algorithms it has now become possible to run such first-principles electronic structure calculations allowing *ab initio* molecular dynamics (AIMD) simulations based on periodic density functional theory (DFT) calculations [12, 14–19] that yield an appropriate statistical sampling. At the same time, it is important to realize that an electrolyte always acts as a reservoir for ionic species that can adsorb specifically or non-specifically at eletrode surfaces, as illustrated in Fig. 6.1. Thus it is also critical to take the proper adsorbate structure at the electrodes as a function of the crucial electrochemical parameters (electrode potential, ion concentration in the electrolyte, pH value, ...) into account. Interestingly enough, for a number of systems these adsorbate structures can be determined without taking the electrochemical environment explicitly into account. We will therefore first address the equilibrium coverages of metal electrodes and then discuss recent AIMD simulations of the electrochemical interface and the double layer.

6.3 Equilibrium coverage of metal electrodes

At electrochemical electrode-electrolyte interfaces, species from the electrolyte will adsorb at the electrode surface, if the energy gain upon adsorption from the electrolyte is sufficiently large. Thus, in order to evaluate the adsorption energy of an ion, the energy of the reference state in the electrolyte, the electrochemical potential $\tilde{\mu}$ of the solvated ion,

$$\tilde{\mu} = \mu + neU \,, \tag{6.8}$$

where n is the charge of the particle, needs to be known. This appears to be a demanding task, as it in principle requires to employ thermodynamic integration schemes, but in fact it can be done relatively easily based on the concept of the computational hydrogen electrode (CHE) [20, 21].

For pedagogical reasons, let us first consider adsorption at solid-gas interfaces. The impingement rate of gas-phase particles on the surface is typically rather small, even at ambient conditions. Thus the presence of the gas above the surface does not really need to be explicitly modelled in the calculations. Rather, it is sufficient to implicitly take the gas-phase molecules into account through their corresponding chemical potential.

Under equilibrium conditions, the Gibbs free energy of adsorption $\Delta\gamma$ for $N_{\rm ads}$ adsorbates bound to a surface area A_s at a given temperature T and pressure p can be expressed, using the so-called concept of *ab initio* thermodynamics [22], as

$$\begin{aligned}\Delta\gamma(T,p) &= \gamma(T,p,N_{\rm ads}) - \gamma_{\rm clean}(T,p,0)\\ &= \frac{1}{A_s}\Delta G^{ads}(T,p)\\ &\approx \frac{N_{\rm ads}}{A_s}\left(E_{\rm ads} - \Delta\mu_{\rm ads}(T,p)\right) . \end{aligned} \tag{6.9}$$

Here, the change in the free energy upon adsorption is denoted by $\Delta G^{ads}(T,p)$, whereas in the last line of Eq. (6.9) $\Delta\mu_{\rm ads}(T,p)$ corresponds to the temperature and pressure dependent part of the chemical potential of the adsorbate in the gas-phase. In principle, in this last line G_{ads} including the change in entropy and zero-point energies upon adsorption should appear. However, at solid-gas interfaces these contributions are typically small [22]. It is therefore typically replaced by the total adsorption energy per particle $E_{\rm ads}$ at the solid/vacuum interface. This number can relatively easily be evaluated using periodic DFT calculations. This means that the whole dependence of the free energy of adsorption $\Delta\gamma(T,p)$ on temperature and pressure is given by the corresponding dependence of the chemical potentials of the species in the gas phase which acts as a reservoir. Still, this approximation is often rather successful in reproducing surface phase diagrams, for example in oxidation catalysis [23].

Let us now consider the adsorption of protons on an electrode from an aqueous electrolyte. As mentioned above, in the interaction of ions with electrochemical interfaces, the reference state does not correspond to the atom or molecule in gas phase – whose energy is relatively easy to determine – but rather to the species in solution in the presence of an electrode potential U. However, the determination of solvation energies can be avoided by taking advantage of the fact that most ions can be related to a corresponding gas phase species through their standard electrode potential. This is the main idea underlying the concept of the computational hydrogen electrode [20].

It is based on the fact that at standard conditions (pH= 0, p = 1 bar, T = 298 K), the zero of the electrode potential with respect to the standard hydrogen electrode (SHE) $U_{\rm SHE} = 0$ is defined as the electrode potential at which there is an equilibrium between a proton and an electron in aqueous solution $H^+(aq) + e^-$ and molecular hydrogen in the gas phase, $\frac{1}{2}H_2(g)$. Now electrochemistry does not only occur at standard conditions, however, the variation of the electrochemical potential with the proton concentration and the electrode potential is well known [1]:

$$\tilde{\mu}(\mathrm{H^+(aq)}) + \mu(e^-) = \frac{1}{2}\mu(\mathrm{H_2(g)}) - eU_{\rm SHE} - k_{\rm B}T\ln(10)\mathrm{pH} , \tag{6.10}$$

This concept is not only valid for protons, it also works for any redox couple such as $\frac{1}{2}$ $A_2 + e^- \rightleftarrows A^-$ [24, 25] where A_2 is, e.g., a halogen molecule. Then the electrochemical potential of the solvated halide is given by

$$\tilde{\mu}(\mathrm{A^-(aq)}) - \mu(e^-) = \frac{1}{2}\mu(\mathrm{A_2(g)}) + e(U_{\rm SHE} - U^0) + k_{\rm B}T\ln a_{\rm A^-} , \tag{6.11}$$

where U^0 is the reduction potential of the corresponding halide vs. U_{SHE} and a_{A^-} its activity coefficient. Equivalently to Eq. (6.9), the surface free energy of adsorption can be expressed as

$$\Delta\gamma(T, a_i, U) = \frac{N_{ads}}{A_s}\left(G_{ads} - \Delta\tilde{\mu}_{ads}(T, a_i, U)\right) . \tag{6.12}$$

where $\Delta\tilde{\mu}_i(T, a_i, U)$ is the contribution to the electrochemical potential that depends on temperature, activity a_i, and electrode potential which can be derived using the expressions Eqs. (6.10) and (6.11).

Up to here everything is still exact. However, the determination of the free energy of adsorption G_{ads} in principle requries to evaluate the free energy of the adsorbate systems in the presence of the electrolyte at a given electrode potential. This is computationally rather demanding. Therefore in the majority of studies using the CHE, entropic effects and the presence of the electrochemical environment are neglected [20]. This means that, similar to most applications of the concept of *ab initio* thermodynamics [22], in Eq. (6.12) the free energy G_{ads} is replaced by the total energy of adsorption E_{ads} at the electrode-vacuum interface. Consequently, the electrochemical control parameters electrode potential and concentrations or activities do not enter the determination of the energy of the adsorbate system but only of the reference system of the solvated species in the electrolyte.

This appears to be a very rough approximation given the fact that adsorbed species at electrode-electrolyte interfaces are still exposed to the liquid environment and that the stability of adsorbate structures at electrochemical interfaces strongly depends on the electrode potential. However, this approach has been surprisingly successful as far as reproducing and explaining observed trends in structures and processes at electrochemical interfaces is concerned. Examples include electrocatalytic reactions such as oxygen reduction [20, 26] and hydrogen evolution [27] on metal electrodes. It can only be speculated why the neglect of the elecrochemical environment still often yields satisfactory results. This is due to the lack of computational studies that systematically increase the complexity of the description and thus allow the dominant factors that determine processes at electrochemical interfaces to be identified. It has been argued that the total energies of adsorbate systems should only weakly depend on the applied electrode potential because metallic electrodes exhibit very good screening properties that make binding energies hardly dependent on applied electric fields [20, 28–30]. As far as the influence of the presence of the aqueous electrolyte is concerned, it has been observed that water typically interacts rather weakly with metal electrodes [31] so that its presence also hardly modifies the binding properties.

We will illustrate this concept for the adsorption of hydrogen and hydroxyl on Pt(111). In passing we note that sometimes there is some confusion whether one deals with proton or hydrogen adsorption on metal electrodes. Recall that an electrode in principle corresponds to an semi-infinite reservoir of electrons with the chemical potential of the electrons basically given by the Fermi energy. Hence it does not matter whether a proton or a hydrogen atom becomes adsorbed on the metal electrode. The adsorbed species becomes embedded in the electronic system of the metal electrode, and as there is in principle an infinite number of electrons in the reservoir, the initial charge state of the adsorbate, i.e. whether it initially was a neutral atom or an ion, does not matter.

The surface free energies of adsorbed H and OH on Pt(111) calculated using Eq. (6.13) as a function of the electrode potential are plotted in Fig. 6.2, for a variety of coverages [32].

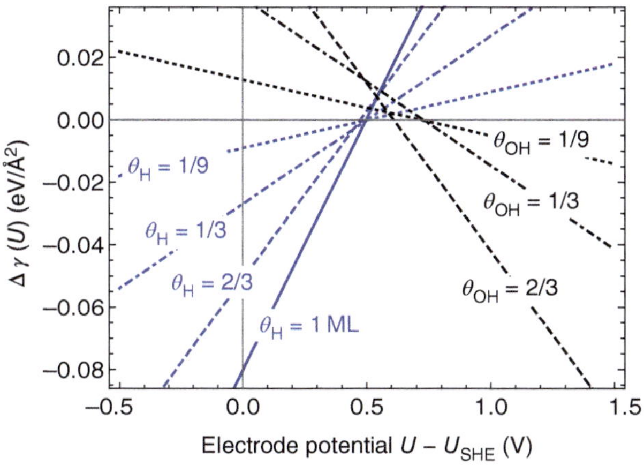

Figure 6.2 Surface free energies of adsorbed H and OH on Pt(111) as a function of the electrode potential U for different adsorbate coverages in the presence of implicit water (Courtesy of Dr. Sung Sakong, adapted from [32]).

For every considered surface structure, only one calculation of the total energy is necessary since the dependence of the total energy on the electrode potential is neglected. The slope of the curve is proportional to N_{ads}/A_S according to Eq. (6.12), i.e., to the coverage: the higher the coverage, the larger the slope. Within a certain range of electrode potentials, the structure with the lowest free energy of adsorption is thermodynamically stable.

In the particular calculations presented in Fig. 6.2, the electrochemical environment has actually not been entirely neglected, but the presence of water has been taken into account through a so-called implicit solvent model. This means that the water has been presented as a polarizable dielectric continuum [33]. In fact, adsorption energies have only been changed by at most 0.1 eV by considering the implicit solvent [32]. This suggests that for the adsorption of atoms and small molecules the presence of the aqueous electrolyte indeed does not substantially modify the energetics of adsorption.

In fact, the calculations presented in Fig. 6.2 almost quantitatively reproduce the well-known coverage of Pt(111) as a function of the electrode potential [34]: at low potentials below the potential of zero charge, i.e., up to potentials of about 0.5 V at which the surface is negatively charged, protons are attracted to the Pt(111) surface leading to a monolayer of so-called upd hydrogen at monolayer coverage, where upd stands for underpotential deposition. Between about 0.5 and 0.6 V, the so-called double layer region follows where the Pt(111) electrode remains uncovered. At higher potentials leading to a positively charged Pt(111) electrode surface, anions become attracted so that OH adsorption starts.

The adsorption energies entering Fig. 6.2 have been calculated just for one species that is present on the surface. However, this concept can also be easily generalized to more than one solvated species in the electrolyte [35]. The equilibrium adsorption structure is then given by those arrangements that yield the minimum free energy of adsorption

$$\Delta\gamma = \frac{1}{A_S}\left(G_{ads} - \sum_i n_i \Delta\tilde{\mu}_i(T, a_i, U)\right) \tag{6.13}$$

as a function of the electrochemical control parameters. This approach has been applied to study the coadsorption of protons and the halides chloride, bromide, and iodide on Pt(111) [35, 36]. The first step requires to calculate the energy of all possible co-adsorbate structures for various coverages. One has to be cautious not to overlook the possibly most stable structure. Using these energies, two-dimensonal phase diagrams of the thermodynamically stable surface adsorbate structure can be constructed as a function of the electrochemical potentials of the solvated species [35]. Note, however, that experimentally electrochemical potentials can not be directly specified. Rather, concentrations and the electrode potential are varied. Yet, the phase diagrams can be converted into something more useful, for example into Pourbaix diagrams, i.e., phase diagrams as a function of pH and the electrode potential.

Such a Pourbaix diagram for the co-adsorption of hydrogen and bromide [35, 36]) is shown in Fig. 6.3. Interestingly, the diagram mainly exhibits either pure hydrogen or pure bromide adsorption structures. There is only a small pocket of co-adsorption structures at pH values around 7. The absence of a broad range of co-adsorption phases is at first sight surprising because one might expect that there is some electrostatic attraction upon the adsorption of one cationic and one anionic species. However, as already mentioned above, upon chemisorption the adsorbates often loose their identity as either cations or anions. This is also reflected by the workfunction change upon adsorption. In general one expects that the adsorption of cationic species lowers the workfunction, whereas anionic adsorbates increase the workfunction because of the polarity of the resulting dipole layers.

Proton adsorption on Pt(111) indeed causes a small decrease in the work function [37], but in fact halogen adsorption at low to medium coverages also leads to a reduction of the

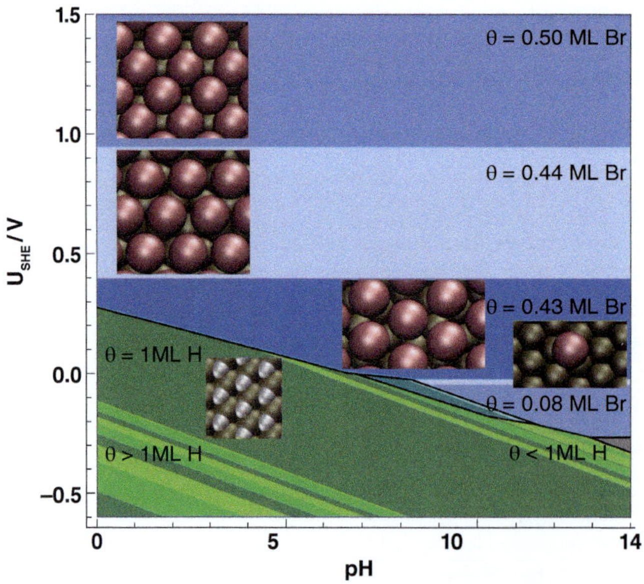

Figure 6.3 Calculated Pourbaix diagram showing the stable phases of co-adsorbed bromine and hydrogen on Pt(111) as a function of pH and electrode potential. The bromine concentration corresponds to an activity of 0.1. The insets illustrate the structures of the adlayers (adapted from [36]).

metal work function [38, 39]. This anomalous work function change is a consequence of the significant polarization of the adsorbed halogen atoms, as was shown by periodic DFT calculations [10, 11, 40], which is particularly strong for iodine, but also present for bromine and chlorine. Thus there is rather a dipole-dipole repulsion instead of attraction, which disfavors halogen-hydrogen co-adsorbate phases.

One consequence of the absence of co-adsorption phases is illustrated in Fig. 6.3. Upon lowering the electrode potentials, pure bromine adsorption phases become replaced by pure hydrogen adsorption phases for pH values below 7, i.e., in this regime there is a competetive adsorption of bromine and hydrogen. Such a competetive adsorption between halides and protons has in fact been observed experimentally on Pt(111) [41–44]. This agreement shows that the neglect of the electrochemical environment indeed seems to be justified in these cases, namely for the specific adsorption of atoms on metal electrodes.

There is another important aspect related to the equilibrium coverage of metal electrodes as a function of the electrochemical control parameters. As the insets of Fig. 6.3 illustrate, at low pH values Pt(111) becomes densely covered by either bromine atoms or by hydrogen atoms over the whole range of electrode potentials within the stability window of water. These adlayers modify the electronic, chemical and catalytic properties of the Pt(111) electrode significantly. Any simulation that would model the properties of an Pt(111) electrode in the presence of an aqueous electrolyte at low pH with bromide anions needs to take into account the particular equilibrium adsorbate structures because otherwise the simulations would not properly model the properties of the halide-covered Pt(111) electrode.

This is in fact true for any realistic simulation of electrochemical electrode/electrolyte interfaces. The electrolyte contains solvated salt ions in order to be conductive. Thus the electrolyte also acts as a three-dimensional reservoir for these solvated ions that are characterized by their electrochemical potentials. The two-dimensional electrode/electrolyte interface becomes covered by these ions depending on whether their specific or non-specific adsorption is associated with a gain in free energy.

Hence any theoretical study of processes at electrochemical electrode/electrolyte phases should start with an appropriate consideration of the stable adsorbate structure. The identification of the stable adsorbate structures under equilibrium can be quite cumbersome. For example, for the halide systems such as the one illustrated in Fig. 6.3, the adsorption energies of more than 100 different possible adsorbate structures had been evaluated [35]. As we will see in the next section, in any modeling of electrode/electrolyte interfaces in principle the statistical nature of liquid electrolytes has to been taken appropriately into account. Practically, this requires to run numerically demanding AIMD simulations. However, fortunately the stability of electrochemical adsorbate phases can often be estimated neglecting the explicit presence of the electrolyte in the determination of the total energies of the adsorbate structures [11, 35, 36], at least for many interfaces between metallic electrodes and aqueous electrolytes containing small ions such as halides.

6.4 First-principles simulations of electrochemical interfaces and electric double layers

In all the electronic structure calculations presented so far, the explicit presence of water has not been taken into account. The exact structure of water layers adsorbed on metal electrodes had in fact intensively been debated some time ago [45–47]. On close-packed hexagonal metal surfaces, it was assumed that water adsorbs in ice-like hexagonal water bilayers [45] whose structure is similar to that of the densest layer of ice [48]. In this structure, every second water molecule is adsorbed in a orientation parallel to the surface with the oxygen atom above a top site of the metal surface. This corresponds to the most favorable adsorption configuration for a water monomer on metal surfaces [45, 46]. However, in order to complete the ice-like hexagonal bilayer structure, every other water molecule can not be bound to the metal surface in the same fashion. Instead, there are two different possible configuration, the so-called H-down and H-up structures with one hydrogen atom either pointing towards or away from the surface. These structures are illustrated in the insets of Fig. 6.4.

In fact, the adsorption energies of both water structures are typically rather similar. On Pt(111), the H-down structure is more stable than the H-up structure by only 0.03 eV per water molecule [46]. However, there is in fact one property that differs significantly between the two structures, which is the dipole moment perpendicular to the surface. This also leads to significant changes of the work function change of Pt(111) upon the adsorption of one water layer. Whereas the H-up layer lowers the work function by $\Delta\Phi = -2.34$ eV, the H-down only lowers it by $\Delta\Phi = -0.22$ eV [46]. Still, this result is surprising. Because of the opposite orientation of the H-down and the H-up water molecules, one would have naively expected that the one configuration lowers the work function and the other increases it by roughly the same amount.

In order to understand why this is not the case, an analysis of the underlying electronic structure is rather instructive. In Fig. 6.4, the laterally averaged charge density difference $\Delta\rho$,

$$\Delta\rho = \rho(H_2O/Pt(111)) - (\rho(H_2O) + \rho(Pt(111))) , \tag{6.14}$$

is plotted as a function of the vertical height. $\Delta\rho$ corresponds to the charge rearrangement induced by the adsorption of the water bilayers on Pt(111).

Interestingly enough, for both the H-up and the H-down structure there is a strong polarization of the water molecules leading to a considerable electronic charge transfer from the water bilayer towards the region between the uppermost metal layer and the water molecules. Similar results have also been found for water layers on Ru(0001) [49]. The charge transfer is larger for the H-down water bilayer than for the H-up bilayer. In both cases, this transfer leads to the creation of an additional dipole layer with a orientation that reduces the work function. For the H-down bilayer, this additional dipole moment overcompensates the effect of the intrinsic dipole moment of the water bilayer causing a net work function reduction on Pt(111). In contrast, for the H-up bilayer, these two dipole moments add up.

According to experiments, however, a water layer on Pt(111) lowers the work function by about −1.0 eV [50, 51], a value, that is in between the work function changes caused by the H-up and the H-down bilayers. Such a work function change has in fact been reproduced by running *ab initio* molecular dynamics simulations of a water layer on Pt(111) at room temperature starting with both the H-up and the H-down configurations [46]. Along the trajectory, the water layers do not maintain their pure H-up or H-down configuration. The water layer becomes disordered, in particular as far as the orientation of the

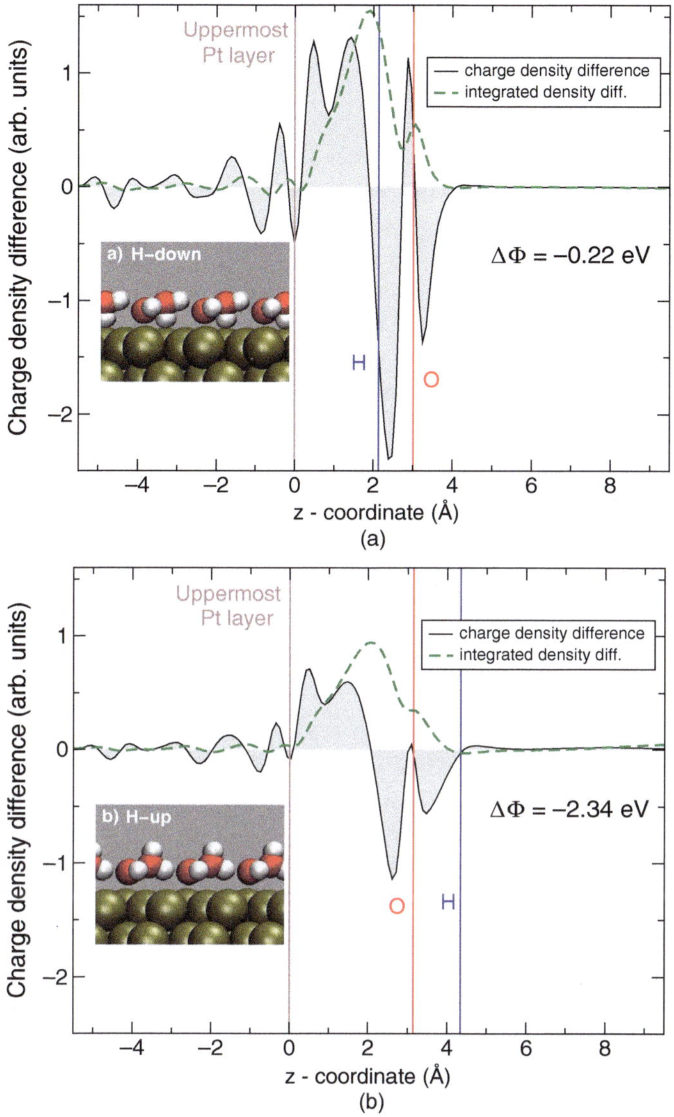

Figure 6.4 Laterally averaged electron density difference (Eq. (6.14)) and intergrated charge density difference as a function of the distance from the Pt surface for the H-down (a) and a H-up (b) water bilayer on Pt(111). The bilayer structures are illustrated in the insets. The vertical lines indicate the positions of the uppermost Pt layer and the hydrogen and oxygen atom of the oppositely oriented water molecules. Note that the same scale for the charge density difference in both figures has been chosen (adapted from [46]).

water molecules is concerned. This results in an averaged mean dipole moment that is independent of the initial configuration and leads to a work function change close to the experimental values [46, 47]. These findings provide a strong indication that water layers at close-packed metal surfaces are not ordered, but rather disordered.

There is also a consequence of these findings for the modeling of metal/water interfaces. Because of the still high computational demand of AIMD simulations, one might consider using numerically less expensive techniques such as force-fields as a basis for MD simulations. And indeed, parametrized classical interaction potentials are capable of reproducing liquid water properties quite well [52–54]. Such classical interaction potentials have also been used extensively to determine the structure of metal-water interfaces at finite temperatures [8, 9, 55, 56]. As these simulations allow to treat large systems, there are certainly merits of these simulations. However, the strong polarization of water layers at metal surfaces illustrated in Fig. 6.4 leads to additional electrostatic interactions that influence the structure of the water layers directly at the metal surface. Hence one has to take the results of these force-field simulations with caution as they do not capture these polarization effects.

On the other hand, this makes the determination of electrode/electrolyte interfaces with an explicit description of the liquid electrolyte computationally very demanding. However, due to the increase in computer power, it has become possible to run such demanding AIMD simulations for rather large systems and run times [12, 14, 16, 19, 57–59].

One particular example of such a simulation is illustrated in Fig. 6.5 which shows a snapshot of a 40 ps AIMD run for six layers of water on hydrogen-covered Pt(111) within a 6×6 geometry, resulting in total in 144 water molecules [12]. In addition, the average difference $\delta V(z)$ of the electrostatic potential of Pt(111) in vacuum and of hydrogen-covered Pt(111) with the water film is plotted so that the change of the electrostatic potential at the electrode/electrolyte interface due to interaction with the aqueous electrolyte becomes apparent. There are some interesting features of this electrostatic potential that are not captured in the schematic models shown in Fig. 6.1. First of all, these schematic models usually assume that the metal electrode is an perfect electron conductor which does not allow any electric fields to penetrate into the electrode. In contrast, Fig. 6.5 shows a significant change of the electrostatic potential upon introduction of the water film together with the hydrogen layer. This also indicates that there is a considerable charge rearrangement within the first three Pt layers caused by the presence of the aqueous electrolyte. This means that the picture of an ideal metallic electrode is not fully appropriate.

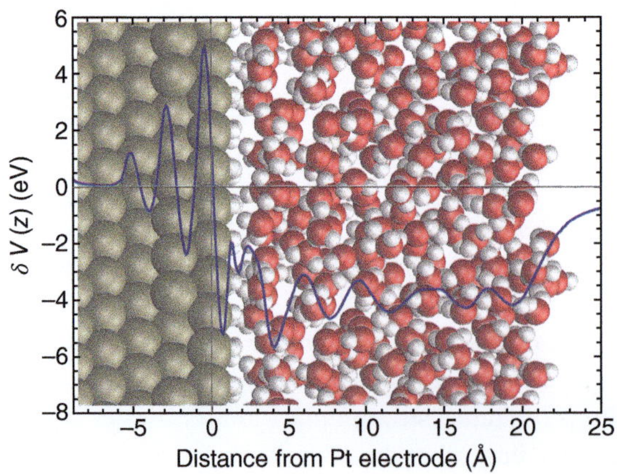

Figure 6.5 Difference $\delta V(z)$ of the electrostatic potential of Pt(111) in vacuum and of hydrogen-covered Pt(111) with a water film averaged over a 40 ps AIMD run [12] together with a snapshot of the AIMD trajectory with six layers of water on the hydrogen-covered Pt(111) electrode. Reprinted from [58] with permission from Elsevier.

Second, the electrostatic potential does not show a smooth decay, but rather exhibits some oscillatory structure due to the atomistic structure of the aqueous electrolyte, but even within the hydrogen layer. As a consequence, it is not trivial to assign a width to the electric double layer because of this oscillatory structure, only as a rough estimate a width of about 1 nm can be deduced. Note that the distribution of the hydrogen and oxygen atom of the water layers becomes already rather smooth beyond the first layer [12], or in other words, the water becomes almost bulk water-like already from the second layer on.

Hence such AIMD simulations provide valuable insights into the atomistic structure of the electric double layer. However, because of the high computational demand of these simulations, no complete picture of the structure of the EDL as a function of the electrode material and the presence of ions in the electrolyte has emerged. For example, in the simulations underlying Fig. 6.5, a full layer of hydrogen atoms adsorbed on Pt(111) has been adsorbed, so-called underpotential deposited (upd) hydrogen that is stable at low electrode potentials, as demonstrated in Fig. 6.3. The presence of such a upd layer lowers the interaction of water with the Pt electrode, it passivates the surface, so that the water layers are shifted 1 Å away from the Pt electrode [60]. And due to the weaker interaction with Pt the water-water interaction becomes somewhat stronger, leading to a stronger structuring of the first water layer [60].

Still, there have been hardly any AIMD studies of the electric double layer in the presence of other specifically adsorbed ions such as halides, sulfates, or nitrates which might influence the water structure much more significantly compared to hydrogen. The same is true for non-specifically adsorbed ions, typically cations, that adsorb with their solvation shell still intact. Their interaction with the electrode is weaker than the one of specifically adsorbed ions, but they might still considerably influence electrocatalytic processes at the electrodes. Hence there is certainly a strong demand for such AIMD simulations.

Recently the consideration of the electrolyte in so-called implicit solvent models, i.e., as a polarizable dielectricum, has become quite popular [18, 32, 61–64], as they do not demand to perform statistical averages. Still, their validity and reliability has not been fully assessed yet, again because of the lack of appropriate atomistic simulations that they can be compared with. Hence it is fair to say that in spite of all the progress made in recent years our knowledge about the atomistic structure of electric double layers is still in its infancy.

6.5 Electric double layers at battery electrodes

Finally we want to consider the effect of electric double layers at other electrochemical interfaces, for example as they occur in batteries. We will mainly rely on macroscopic arguments [58, 65, 66], but we will argue that EDLs are also essential in batteries. Let us first concentrate on the basic operation principle of a battery. In contrast to electrocatalytic systems the electrodes in batteries, anode and cathode, are also used to store ions. Hence ion mobility is a critical parameter for the performance of electrodes as the ions penetrating the electrode have to further propagate inside of the electrode. At the same time, the electrodes also have to be electronic conductors in order to allow the recombination of ions and electrons. Commonly we speak about charging and discharging of a battery. However, note that the electrodes still have to remain overall charge neutral as macroscopically charged

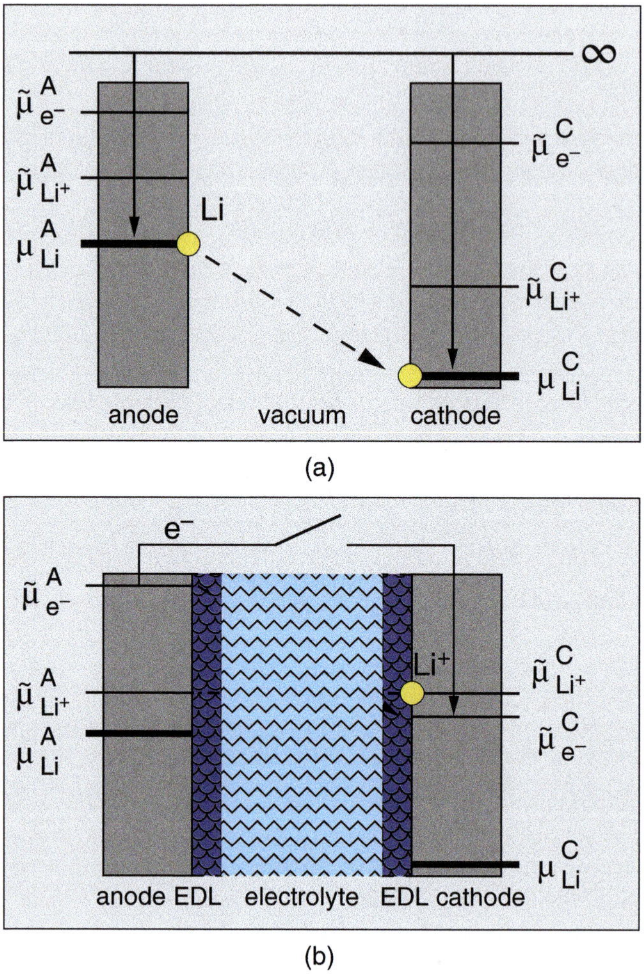

Figure 6.6 Illustration of alignment of the electrochemical potentials in the anode and cathode of a battery. a) Cathode and anode not coupled to each other, b) cathode and anode connected through an electrolyte (adapted from [58].)

materials are energetically very costly. Hence it is in principle more correct to denote these processes, e.g. in the case of Li-ion batteries, by lithiation and de-lithiation.

Figure 6.6 sketches a Li-ion battery rather schematically. It is crictical to realize that there must be a driving force for a battery to be operative. The driving force is the energy gain upon the transfer of an Li atom from the anode to the cathode, i.e., upon the de-lithiation of the anode and the lithiation of the cathode, which can be expressed as

$$x\text{Li (anode)} + \text{Li}_y\text{(cathode)} \rightarrow \text{Li}_{x+y}\text{(cathode)}\,. \tag{6.15}$$

This means that the binding energy of the lithium atom in the cathode has to be larger than the Li binding energy in the anode. Expressing these binding energies through the chemical potentials $\mu_{\mathrm{Li}}^{\mathrm{A}}$ and $\mu_{\mathrm{Li}}^{\mathrm{C}}$ of Li in the anode and the cathode, respectively [67], the gain in free energy upon the transfer can be expressed as

$$\Delta G = \mu_{\mathrm{Li}}^{\mathrm{C}} - \mu_{\mathrm{Li}}^{\mathrm{A}} \tag{6.16}$$

(see the illustration in Fig. 6.6a). However, in a battery anode and cathode are separated by an electrolyte which is an ion conductor but ideally not an electron conductor. This forces the electrons to propagate through an external circuit where they can provide useful power before Li-ions and electrons recombine at the cathode. Still, thermodynamically the specific propagation is irrelevant, only the initial and the final state are crucial. Therefore, the so-called open-circuit voltage (OCV) V_{OC} associated with this reaction is given by the energy gain $\Delta G < 0$ divided by the number x of charge carriers [68]

$$V_{\mathrm{OC}} = \frac{-\Delta G}{xF}\,, \tag{6.17}$$

where F is the Faraday constant. Note that V_{OC} corresponds to the maximum voltage that can be delivered by a battery. As it corresponds to a free-energy difference of two well-defined charge-neutral configurations, it can rather easily be calculated by periodic density functional theory calculations [68, 69]. Using the expression (6.16) for ΔG, the open-circuit voltage in a Li-ion battery can also be determined via [67]

$$V_{\mathrm{OC}} = (\mu_{\mathrm{Li}}^{\mathrm{A}} - \mu_{\mathrm{Li}}^{\mathrm{C}})/e\,, \tag{6.18}$$

where e is the elemental charge.

However, when anode and cathode are connected by an electrolyte, Li-ions can in principle move from the anode to the cathode as the electrolyte is an ion conductor. In such a situation, it is useful to decompose the chemical potential μ_{Li} of the Li atom into the electrochemical potentials of the Li^+ ion and the electron,

$$\mu_{\mathrm{Li}} = \tilde{\mu}_{\mathrm{Li}^+} + \tilde{\mu}_{\mathrm{e}^-}\,. \tag{6.19}$$

This decomposition has already been made in Fig. 6.6a where the anode and cathode are assumed to be uncoupled and where all levels are assumed to be given with respect to the vacuum level. Upon introduction of the electrolyte, Li ions will move from the anode to the cathode because of the gradient in their electrochemical potential that is initially present until the electrochemical potential of the Li ions in the anode and cathode become aligned. Still, under open-circuit conditions no electrons can move from the anode to the cathode so that the Li-ions can not penetrate the cathode in macroscopic amounts. Hence the alignment of the electrochemical potential of the Li ions can only be achieved by the formation of electric double layers at anode and cathode, which leads to potential changes at the electrodes associated with an electric field. However, this electric field also acts on the electrons, but because of the opposite charge of the electrons, it has the opposite effect on the electronic energy levels. The chemical potentials of the electrons $\tilde{\mu}_{\mathrm{e}^-}^{\mathrm{A}}$ and $\tilde{\mu}_{\mathrm{e}^-}^{\mathrm{C}}$ in the anode and cathode, respectively, corresponding to the Fermi levels of the anode and cathode, are shifted apart from each other by exactly the same amount the electrochemical potentials of the Li ions are shifted together. As a consequence, the difference between the Fermi energies in anode and cathode corresponds to the open-circuit voltage, i.e., the OCV can be expressed as

$$V_{\mathrm{OC}}^{\mathrm{el}} = (\tilde{\mu}_{\mathrm{e}^-}^{\mathrm{A}} - \tilde{\mu}_{\mathrm{e}^-}^{\mathrm{C}})/e\,, \tag{6.20}$$

where the superscript of V_{OC}^{el} denotes the fact that the electrodes are connected through an electrolyte. This also explains why the open-circuit voltage can be compared with the HOMO-LUMO gap of the electrolyte in order to assess the stability of the electrolyte during battery operation [67], but this is usually not properly justified in the literature.

In this derivation, we have entirely neglected the formation of the solid-electrolyte interphase (SEI) which is critical in Li-ion batteries and typically leads to an additional potential drop at the electrode. However, this is not crucial for the derivation given above. In fact, the importance of the presence of electric double layers for the formation of the solid-electrolyte interphase has been stressed before [70]. Yet, in the discussion of battery properties, electric double layers are often not taken into account. It is certainly fair to say that our atomistic knowledge about electrode/electrolyte interfaces in batteries is even more limited than the corresponding knowledge with respect to interfaces in electrocatalytic systems. Still, it might well be that electric double layers are also crucial for a full understanding of the electrochemical processes in batteries.

6.6 Conclusions

It is well established that electrochemical electrode-electrolyte interfaces are associated with the formation of electric double layers. Depending on the particular electrolyte, these electric double layers might have have a decisive influence on the chemical and/or electronic properties of the interface. However, our current understanding of the structure and the role of electric double layers is still to a large extent rooted in concepts that are more than 100 years old and that are based on point charge and continuum approaches. This is caused by the fact that an atomistic modeling of electric double layers including electronic effects can only be done with quantum chemical approaches and requires extensive statistical sampling. Up to recently such simulations had been prohibitively expensive, but now first studies of the structure of electric double layers at electrochemical interfaces between metal electrodes and aqueous electrolytes based on *ab initio* molecular dynamics simulations have been performed. These simulations demonstrated strong electronic polarization effects in the first metal and water layers. Still, the number of these studies is rather limited so that no complete picture of the properties of electric double layers has emerged yet. This is unfortunate as electric double layers do not only play an important role in electro-catalysis but also in devices for electrochemical energy storage such as batteries. We trust that soon further atomistic studies will be carried out so that a more complete picture of the geometric and electronic structure of electric double layers will be obtained.

Acknowledgement

This work contributes to the research performed at CELEST (Center for Electrochemical Energy Storage Ulm-Karlsruhe).

References

1 W. Schmickler, E. Santos, *Interfacial Electrochemistry*, 2nd edn. (Springer, Berlin, 2010).
2 H. Helmholtz, Ueber einige gesetze der vertheilung elektrischer ströme in körperlichen leitern mit anwendung auf die thierisch-elektrischen versuche, *Pogg. Ann.* 1853, 89, 211.
3 L.G. Gouy, Sur la constitution de la charge électrique à la surface d'un électrolyte, *J. Phys. Theor. Appl.* 1910, 9, 457.
4 D.L. Chapman, A contribution to the theory of electrocapillarity, *Philos. Mag. Series* 6 1913, 25(148),475.
5 O. Stern, Zur theorie der elektrolytischen doppelschicht, *Z. Elektrochem. Angew. P.* 1924, 30, 508.
6 S. Trasatti, Structure of the metal/electrolyte solution interface: new data for theory, *Electrochim. Acta* 1991, 36(11–12), 1659.
7 J. Huang, A. Malek, J. Zhang, M.H. Eikerling, Non-monotonic surface charging behavior of platinum: A paradigm change, *J. Phys. Chem. C* 2016, 120(25), 13587.
8 E. Spohr, Computer simulations of the water/platinum interface, *J. Phys. Chem.* 1989, 93, 6171.
9 E. Spohr, Effect of electrostatic boundary conditions and system size on the interfacial properties of water and aqueous solutions, *J. Chem. Phys.* 1997, 107, 6342.
10 T. Roman, A. Groß, Periodic density-functional calculations on work-function change induced by adsorption of halogens on Cu(111), *Phys. Rev. Lett.* 2013, 110, 156804.
11 F. Gossenberger, T. Roman, K. Forster-Tonigold, A. Groß, Change of the work function of platinum electrodes induced by halide adsorption, Beilstein J. *Nanotechnol.* 2014, 5, 152.
12 S. Sakong, A. Groß, The electric double layer at metal-water interfaces revisited, *J. Chem. Phys.* 2018, 149, 084705, doi: 10.1063/.
13 J. Behler, Representing potential energy surfaces by high-dimensional neural network potentials, *J. Phys.: Condens. Matter* 2014, 26, 183001.
14 S. Sakong, K. Forster-Tonigold, A. Groß, The structure of water at a Pt(111) electrode and the potential of zero charge studied from first principles, *J. Chem. Phys.* 2016, 144(19), 194701.
15 M.H. Hansen, A. Nilsson, J. Rossmeisl, Modelling ph and potential in dynamic structures of the water/Pt(111) interface on the atomic scale, *Phys. Chem. Chem. Phys.* 2017, 19, 23505.
16 J. Le, Q. Fan, L. Perez-Martinez, A. Cuesta, J. Cheng, Theoretical insight into the vibrational spectra of metal-water interfaces from density functional theory based molecular dynamics, *Phys. Chem. Chem. Phys.* 2018, 20, 11554.
17 A. Bouzid, A. Pasquarello, Atomic-scale simulation of electrochemical processes at electrode/water interfaces under referenced bias potential, *J. Phys. Chem. Lett.* 2018, 9, 1880.
18 A. Baskin, D. Prendergast, Improving continuum models to define practical limits for molecular models of electrified interfaces, *J. Electrochem. Soc.* 2017, 164(11), E3438.
19 H.H. Kristoffersen, T. Vegge, H.A. Hansen, OH formation and H_2 adsorption at the liquid water-pt(111) interface, *Chem. Sci.* 2018, 9, 6912.

20 J.K. Nørskov, J. Rossmeisl, A. Logadottir, L. Lindqvist, J.R. Kitchin, T. Bligaard, H. Jónsson, Origin of the overpotential for oxygen reduction at a fuel-cell cathode, *J. Phys. Chem. B* 2004, 108(46), 17886.

21 A.A. Peterson, F. Abild-Pedersen, F. Studt, J. Rossmeisl, J.K. Nørskov, How copper catalyzes the electroreduction of carbon dioxide into hydrocarbon fuels, *Energy Environ. Sci.* 2010, 3, 1311.

22 K. Reuter, M. Scheffler, Composition, structure, and stability of RuO_2(110) as a function of oxygen pressure, *Phys. Rev. B* 2001, 65, 035406.

23 J. Rogal, K. Reuter, M. Scheffler, CO oxidation at Pd(100): A first-principles constrained thermodynamics study, *Phys. Rev. B* 2007, 75, 205433.

24 H.A. Hansen, I.C. Man, F. Studt, F. Abild-Pedersen, T. Bligaard, J. Rossmeisl, Electrochemical chlorine evolution at rutile oxide (110) surfaces, *Phys. Chem. Chem. Phys.* 2010, 12, 283.

25 F. Gossenberger, T. Roman, A. Groß, Equilibrium coverage of halides on metal electrodes, *Surf. Sci.* 2015, 631, 17.

26 J.A. Keith, G. Jerkiewicz, T. Jacob, Theoretical investigations of the oxygen reduction reaction on Pt(111), *Chem. Phys. Chem.* 2010, 11, 2779.

27 J.K. Nørskov, T. Bligaard, A. Logadottir, J.R. Kitchin, J.G. Chen, S. Pandelov, U. Stimming, Trends in the exchange current for hydrogen evolution, *J. Electrochem. Soc.* 2005, 152, J23.

28 J. Rossmeisl, J.K. Nørskov, C.D. Taylor, M.J. Janik, M. Neurock, Calculated phase diagrams for the electrochemical oxidation and reduction of water over Pt(111), *J. Phys. Chem. B* 2006, 110, 21833.

29 H.F. Wang, Z.P. Liu, Formic acid oxidation at Pt/H_2O interface from periodic DFT calculations integrated with a continuum solvation model, *J. Phys. Chem. C* 2009, 113, 17502.

30 R. Jinnouchi, K. Kodama, Y. Morimoto, DFT calculations on H, OH and O adsorbate formations on Pt(111) and Pt(332) electrodes, *J. Electroanal. Chem.* 2014, 716, 31.

31 Y. Gohda, S. Schnur, A. Groß, Influence of water on elementary reaction steps in electrocatalysis, *Faraday Discuss.* 2008, 140, 233.

32 S. Sakong, M. Naderian, K. Mathew, R.G. Hennig, A. Groß, Density functional theory study of the electrochemical interface between a Pt electrode and an aqueous electrolyte using an implicit solvent method, *J. Chem. Phys.* 2015, 142(23), 234107.

33 K. Mathew, R. Sundararaman, K. Letchworth-Weaver, T.A. Arias, R.G. Hennig, Implicit solvation model for density-functional study of nanocrystal surfaces and reaction pathways, *J. Chem. Phys.* 2014, 140, 084106.

34 B. Genorio, D. Strmcnik, R. Subbaraman, D. Tripkovic, G. Karapetrov, V.R. Stamenkovic, S. Pejovnik, N.M. Marković, Selective catalysts for the hydrogen oxidation and oxygen reduction reactions by patterning of platinum with calix[4]arene molecules, *Nat. Mater.* 2010, 9, 998.

35 F. Gossenberger, T. Roman, A. Groß, Hydrogen and halide co-adsorption on Pt(111) in an electrochemical environment: a computational perspective, *Electrochim. Acta* 2016, 216, 152.

36 X. Lin, F. Gossenberger, A. Groß, Ionic adsorbate structures on metal electrodes calculated from first-principles, *Industrial & Engineering Chemistry Research* 2016, 55(42), 11107.

37 P. Ferrin, S. Kandoi, A.U. Nilekar, M. Mavrikakis, Hydrogen adsorption, absorption and diffusion on and in transition metal surfaces: A DFT study, *Surf. Sci.* 2012, 606, 679.

38 W. Erley, Chlorine adsorption on the (111) faces of Pd and Pt, *Surf. Sci.* 1980, 94, 281.

39 E. Bertel, K. Schwaha, F.P. Netzer, Adsorption of bromine on Pt(111) - observation of an irreversible order-disorder transition, *Surf. Sci.* 1979, 83, 439.

40 T. Roman, F. Gossenberger, K. Forster-Tonigold, A. Groß, Halide adsorption on close-packed metal electrodes, *Phys. Chem. Chem. Phys.* 2014, 16, 13630.

41 A. Bittner, J. Wintterlin, B. Beran, G. Ertl, Bromine adsorption on Pt(111), (100), and (110) — an STM study in air and in electrolyte, *Surf. Sci.* 1995, 335, 291, Proceedings of the IUVSTA Workshop on Surface Science and Electrochemistry.

42 N. Garcia-Araez, V. Climent, E. Herrero, J. Feliu, J. Lipkowski, Thermodynamic studies of chloride adsorption at the Pt(111) electrode surface from 0.1 m $HClO_4$ solution, *J. Electroanal. Chem.* 2005, 576, 33.

43 N. Garcia-Araez, V. Climent, E. Herrero, J. Feliu, J. Lipkowski, Thermodynamic studies of bromide adsorption at the Pt(111) electrode surface perchloric acid solutions: Comparison with other anions, *J. Electroanal. Chem.* 2006, 591, 149.

44 N. Garcia-Araez, J.J. Lukkien, M.T. Koper, J.M. Feliu, Competitive adsorption of hydrogen and bromide on Pt(100): Mean-field approximation vs. monte carlo simulations, *J. Electroanal. Chem.* 2006, 588, 1.

45 A. Michaelides, Density functional theory simulations of water-metal interfaces: Waltzing waters, a novel 2D ice phase, and more, *Appl. Phys. A* 2006, 85, 415.

46 S. Schnur, A. Groß, Properties of metal–water interfaces studied from first principles, *New J. Phys.* 2009, 11(12), 125003.

47 S. Schnur, A. Groß, Challenges in the first-principles description of reactions in electrocatalysis, *Catal. Today* 2011, 165(1),129.

48 P.A. Thiel, T.E. Madey, The interaction of water with solid surfaces: Fundamental aspects, *Surf. Sci.* Rep. 1987, 7, 211.

49 G. Materzanini, G.F. Tantardini, P.J.D. Lindan, P. Saalfrank, Water adsorption at metal surfaces: A first-principles study of the $p(\sqrt{3}\times\sqrt{3})r30^{\circ}$ H_2O bilayer on Ru(0001), *Phys. Rev. B* 2005, 71, 155414.

50 E. Langenbach, A. Spitzer, H. Lüth, The adsorption of water on Pt(111) studied by irreflection and UV-photoemission spectroscopy, *Surf. Sci.* 1984, 147, 179.

51 M. Kiskinova, G. Pirug, H. Bonzel, Adsorption and decomposition of H_2O on a K-covered Pt(111) surface, *Surf. Sci.* 1985, 150, 319.

52 J.I. Siepmann, M. Sprik, Influence of surface topology and electrostatic potential on water/electrode systems, *J. Chem. Phys.* 1995, 102, 511.

53 P.S. Crozier, R.L. Rowley, D. Henderson, Molecular dynamics calculations of the electrochemical properties of electrolyte systems between charged electrodes, *J. Chem. Phys.* 2000, 113, 9202.

54 R. Bukowski, K. Szalewicz, G.C. Groenenboom, A. van der Avoird, Predictions of the properties of water from first principles, *Science* 2007, 315, 1249.

55 K. Raghavan, K. Foster, K. Motakabbir, M. Berkowitz, Structure and dynamics of water at the Pt(111) interface: Molecular dynamics study, *J. Chem. Phys.* 1991, 94, 2110.

56 X. Xia, M.L. Berkowitz, Electric-field induced restructuring of water at a platinum-water interface: A molecular dynamics computer simulation, *Phys. Rev. Lett.* 1995, 74, 3193.

57 O. Björneholm, M.H. Hansen, A. Hodgson, L.M. Liu, D.T. Limmer, A. Michaelides, P. Pedevilla, J. Rossmeisl, H. Shen, G. Tocci, E. Tyrode, M.M. Walz, J. Werner, H. Bluhm, Water at interfaces, *Chem. Rev.* 2016, 116, 7698.

58 A. Groß, S. Sakong, Modelling the electric double layer at electrode/electrolyte interfaces, *Curr. Opin. Electrochem.* 2019, 14, 1.

59 J. Le, M. Iannuzzi, A. Cuesta, J. Cheng, Determining potentials of zero charge of metal electrodes versus the standard hydrogen electrode from density-functional-theory-based molecular dynamics, *Phys. Rev. Lett.* 2017, 119, 016801.

60 T. Roman, A. Groß, Structure of water layers on hydrogen-covered Pt electrodes, *Catal. Today* 2013, 202(0),183.

61 G. Fisicaro, L. Genovese, O. Andreussi, S. Mandal, N.N. Nair, N. Marzari, S. Goedecker, Soft-sphere continuum solvation in electronic-structure calculations, *J. Chem. Theory Comput.* 2017, 13(8), 3829.

62 R. Sundararaman, W.A. Goddard, T.A. Arias, Grand canonical electronic density-functional theory: Algorithms and applications to electrochemistry, *J. Chem. Phys.* 2017, 146(11), 114104.

63 J. Haruyama, T. Ikeshoji, M. Otani, Analysis of lithium insertion/desorption reaction at interfaces between graphite electrodes and electrolyte solution using density functional + implicit solvation theory, *J. Phys. Chem. C* 2018, 122, 9804.

64 N.G. Hörmann, O. Andreussi, N. Marzari, Grand canonical simulations of electrochemical interfaces in implicit solvation models, *J. Chem. Phys.* 2019, 150, 041730.

65 A. Latz, J. Zausch, Multiscale modeling of lithium ion batteries: thermal aspects, *Beilstein J. Nanotechnol.* 2015, 6, 987.

66 S. Hein, A. Latz, Influence of local lithium metal deposition in 3D microstructures on local and global behavior of lithium-ion batteries, *Electrochim. Acta* 2016, 201, 354.

67 J.B. Goodenough, The Li-ion rechargeable battery: A perspective, *J. Am. Chem. Soc.* 2013, 135, 1167.

68 M.S. Islam, C.A.J. Fisher, Lithium and sodium battery cathode materials: computational insights into voltage, diffusion and nanostructural properties, *Chem. Soc. Rev.* 2014, 43, 185.

69 F. Gschwind, G. Rodriguez-Garcia, D. Sandbeck, A. Gross, M. Weil, M. Fichtner, N. Hörmann, Fluoride ion batteries: Theoretical performance, safety, toxicity, and a combinatorial screening of new electrodes, *J. Fluorine Chem.* 2016, 182, 76.

70 K. Leung, A. Leenheer, How voltage drops are manifested by lithium ion configurations at interfaces and in thin films on battery electrodes, *J. Phys. Chem. C* 2015, 119, 10234.

7

Atomic-Scale Modelling of Electrochemical Interfaces through Constant Fermi Level Molecular Dynamics

Assil Bouzid and Alfredo Pasquarello*

Chaire de Simulation à l'Echelle Atomique (CSEA), Ecole Polytechnique Fédérale de Lausanne (EPFL), CH-1015 Lausanne, Switzerland

7.1 Introduction

Electrochemical reactions involving electron/proton transfer at metal/water interfaces occur in several technologically relevant environments, such as solar cells, energy conversion, and storage devices. In these systems, the electron/proton transfer is governed by the applied bias potential, which defines the amount of excess charge at the interface. Hence, modelling the metal/water electrochemical interface at constant electrode potential is of paramount importance towards controlling and understanding the electrochemical properties of the interface. In practice, this task is not trivial and several challenges are faced when density functional theory (DFT) based methods are used within periodically repeated cells. First, the control of the electrode potential is still at a pioneering stage and a universal grand canonical method is still not available. Second, the lack of an alignment scheme to simulate systems at well defined electrode potential represents a major issue toward the advancement of this field. In this context, an ideal modelling scheme should, on the one hand, allow one to reference the electrode potential to the standard hydrogen electrode (SHE), and, on the other hand, give access to the reaction mechanisms and the structural properties of the electrical double layer at the interface with the electrode.

Previous modelling schemes studied the metal/water interface through either an implicit [1–4] or an explicit but static [5, 6] solvent. In the latter case, only a few layers were considered. These models were used to access the hydrogen and oxygen evolution reactions. Recently, interface models from first-principles molecular dynamics (MD) with explicit water have given access to the potential of zero charge (pzc) at the Pt(111)/water interface [7, 8]. A further step consists in modelling the metal/water interface under variable bias potential. In a pioneering work, Lozovoi et al. [9] modelled electrified interfaces by allowing the slab to exchange electrons with a reference electrode under constant preset chemical potential. While this approach made possible the simulation of

*Present address: Institut de Recherche sur le Céramiques, Centre Européen de la Céramique, 12 Rue Atlantis, 87068 Limoges.

Atomic-Scale Modelling of Electrochemical Systems, First Edition.
Edited by Marko M. Melander, Tomi T. Laurila, and Kari Laasonen.

charged slabs like in electrochemical set-ups, it suffered from strong fluctuations of the electron density and electronic structure calculations were difficult to converge [9, 10]. Subsequently, Otani et al. [11] and Jinnouchi et al. [1] introduced alternative schemes to study the dynamics at the solvent/electrode interface at fixed excess charge [12, 13]. More recently, Bonnet et al. performed molecular dynamics simulations at constant electrode potential [10], in which the metal/electrolyte interface was simulated in a non-repeated cell. In this modeling scheme, an effective screening medium [11] was adopted to ensure the charge neutrality of the simulation cell. This technique revealed mechanistic aspects of electrochemical processes at metal/water interfaces [12, 14], but the energetics were affected and the band alignment was thus difficult to access.

As far as the band alignment is concerned, the standard hydrogen electrode (SHE) as set by Cheng and Sprik enables a direct comparison of the computed electrochemical energy levels to experiments either in bulk liquid water or at electrode/water interfaces [8, 15]. This technique is based on a Born-Haber cycle linking the aqueous H^+_{aq} to $H^+(g)$ in the gas phase and the associated energy level is evaluated through the Gibbs free energies of the reaction. Rossmeisl et al. introduced a generalized scheme, in which not only the potential but also the pH could be varied [16, 17]. This method constructs the grand canonical potential by performing several simulations at different pHs and electrode potentials and uses the experimental SHE to determine the calculated work function of each configuration on an absolute energy scale. While these methods advanced the field and improved our understanding of electrochemical interfaces, none of them provides at the same time a simulation at constant bias potential referenced to the SHE and a description of the mechanistic aspects of electrochemical reactions at the atomic scale.

In this contribution, we first review the methodology behind the constant Fermi level molecular dynamics. We then validate our theoretical framework in the determination of redox levels of aqueous species. Next, we apply the constant Fermi level technique to the study of metal-water interfaces. In particular, we focus on the study of the mechanistic aspects of the Volmer reaction. Furthermore, we extend the technique to reference the applied potential to the SHE in the case of metal-water interface. This procedure allows us to study, on the one hand, the macroscopic properties at the metal-water interface and, on the other hand, the atomic-scale processes at the metal-water interface.

7.2 Method

In this section, we review the dynamical equations for achieving constant Fermi-level simulations, as introduced by Bonnet et al. [10]. We consider a system of particles described by a set of atomic positions r_i and a total electronic charge N_e. To control the Fermi level, this system is connected to an external potentiostat at a fixed Fermi level $\bar{\epsilon}_F$ acting like an electron reservoir or a fictitious external electrode. The electronic charge N_e is considered as a dynamical variable with inertia M_e and the extended system is driven by the grand canonical potential [10]: $\Omega = E^{tot}(r_i, N_e) - N_e\bar{\epsilon}_F$, where $E^{tot}(r_i, N_e)$ is the total energy in the absence of the electron reservoir and $N_e\bar{\epsilon}_F$ is the energy of N_e electrons in the reservoir. According to this definition, the forces acting on the atoms and the electronic charges are given by [10]:

$$F_i = -\frac{\partial E^{\text{tot}}(r_i, N_e)}{\partial r_i} \quad \text{and} \quad F_e = -(\epsilon_F - \bar{\epsilon}_F), \tag{7.1}$$

where ϵ_F is the instantaneous Fermi energy. F_e drives electrons into the system when $\epsilon_F < \bar{\epsilon}_F$ and drives electrons into the reservoir when $\epsilon_F > \bar{\epsilon}_F$, so that the instantaneous Fermi energy ϵ_F averages out to the preset $\bar{\epsilon}_F$. The dynamical equations for the charge evolution then read:

$$\dot{N}_e = \frac{P_e}{M_e} \quad \text{and} \quad \dot{P}_e = F_e = -(\epsilon_F - \bar{\epsilon}_F), \tag{7.2}$$

where P_e is a fictitious momentum associated to the dynamical variable N_e.

In this technique, the electronic charge in the system is controlled in an analogous way as the temperature and the pressure in constant-temperature [18–21] and constant-pressure [22] molecular dynamics, respectively. To control the fluctuations of N_e and consequently those of ϵ_F, it is also possible to couple the charge dynamics to a separate thermostat set at the temperature T_e [10]. This does not create any significant thermal flow in the system, due to the weak coupling between the atoms and the charges [10].

For a good choice of M_e and T_e, the instantaneous Fermi level of the system is governed by the preset $\bar{\epsilon}_F$. In the case of redox reactions, performing MD at constant Fermi level corresponds to defining the net charge in the system. Hence, by controlling $\bar{\epsilon}_F$, one can drive the reaction from the reactants in a charge state q to the products in a charge state q', and get the structural transformations induced by the charge transfer.

7.3 CFL-MD in aqueous solution: Determination of redox levels

We focus on the determination of redox levels of aqueous ions. To this end, we resort to density functional theory (DFT) through the rVV10 functional [23, 24], which is based on a generalized gradient approximation for the exchange-correlation energy and augmented to include nonlocal van der Waals interactions. Within this DFT scheme, the short-range interactions are controlled via an empirical parameter b. We adopted the value of $b = 9.3$ [25]. The core-valence interactions were described by normconserving pseudopotentials according to the prescription of Troullier and Martins [26]. The wave functions of the valence electrons were expanded on a plane-wave basis set defined by a kinetic energy cutoff of 80 Ry. The Brillouin zone was sampled at the Γ point. We performed Born-Oppenheimer molecular dynamics using a time step of $\Delta t = 0.48$ fs to integrate the equations of motion. We used a velocity rescaling method to set the temperature at $T = 350$ K. The aqua ion was simulated with 31 water molecules in a cubic supercell with a side of 9.85 Å. We used the suite of codes provided in the quantum-ESPRESSO package [27].

The cell charge neutrality is ensured through a uniform neutralizing background. At each MD step, the electronic charge is first computed via Eq. 7.2. Then, the electronic structure is minimized at this fixed charge. Next, the atomic positions and the electronic charge are

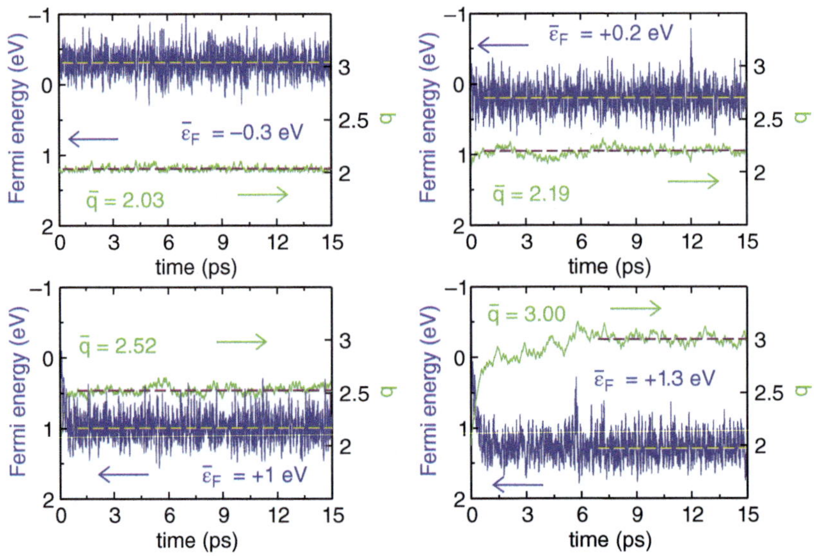

Figure 7.1 Time evolution of the instantaneous Fermi level ϵ_F (blue lines) and total electronic charge q (green lines) in constant Fermi-level molecular dynamics of the aqueous Fe^{+2}/Fe^{+3} redox couple. The panels correspond to four different preset values of the Fermi energy $\bar{\epsilon}_F$: −0.3, +0.2, +1.0, and +1.3 eV. The average values after equilibration are given for both ϵ_F and $\bar{q}$ (dashed lines). The energies are referred to the SHE.

updated. In this work, the energy levels were referred to the standard hydrogen electrode, for which we took the level calculated in [28] at the same level of theory. With this choice of alignment, the conduction band minimum and the valence band maximum of liquid water occur at −2.5 and 1.7 eV, respectively. A complete description of the computational setup is provided in [29].

We validate our method in the case of the reduction of the Fe^{+3} aqua ion:

$$Fe^{+3} + e^- \rightarrow Fe^{+2}. \tag{7.3}$$

After 3 ps of equilibration of Fe^{+3} at 350 K, the last configuration has been taken as the starting point of several runs at different Fermi energies $\bar{\epsilon}_F$, as shown in Fig. 7.1. The choice of the target Fermi energies is guided by the energy level of the lowest unoccupied single-particle state of Fe^{+3}. The occupation number of this state goes from 0 to 1 as the reduction proceeds. In practice, the instantaneous Fermi energy goes within ~1 ps to the target value $\bar{\epsilon}_F$ by taking on a fractional number of electrons from the potentiostat. After this period, the time averages of the instantaneous Fermi energy are taken over at least 10 ps and are found to agree with the target value within at most 0.02 eV. However, the charge dynamics are slower and require longer equilibration times (at most ~6 ps). The charge fluctuations depend directly on the choice of the fictitious mass parameter M_e and the fictitious electronic temperature T_e.

Statistical averages are taken over the equilibrated parts of the trajectories and correspond to periods of at least 10 ps. Figure 7.2a shows the charge transition as a function of the preset value of the Fermi energy $\bar{\epsilon}_F$. Through the appearance of fractional charges, the redox reaction is driven towards the formation of Fe^{+2}. We remark that the charge of the system is

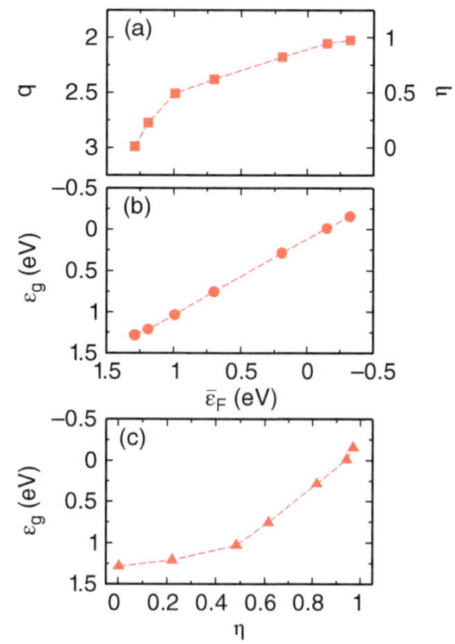

Figure 7.2 Constant Fermi-level molecular dynamics of the aqueous Fe^{+2}/Fe^{+3} redox couple: (a) evolution of the electronic charge (left axis) and of the occupation number (right axis) as a function of the preset Fermi energy $\bar{\epsilon}_F$; (b) evolution of the single-particle energy level ϵ_g as a function of $\bar{\epsilon}_F$; (c) ϵ_g as a function of occupation number η. The energies are referred to the SHE.

fully determined by the occupation η of the single-particle energy level in the gap, which is the only state undergoing an occupation change in our simulation. During this evolution, the single-particle energy level ϵ_g practically corresponds to $\bar{\epsilon}_F$ (cf. Figure 7.2b), the small differences resulting from the width of the employed Gaussian smearing. From Figs. 7.2a and 7.2b, it is possible to extract the dependence of ϵ_g on η (Figure 7.2c).

The commonly used method to compute redox levels is thermodynamic integration. Within this technique the redox level is given by:

$$\mu(q/q') = \frac{\Delta A}{q' - q} + \frac{E^q_{\text{corr}} - E^{q'}_{\text{corr}}}{q' - q} - \epsilon_{\text{v}}, \tag{7.4}$$

where ϵ_{v} is the valence band maximum of the bulk system, E^q_{corr} an electrostatic finite-size correction term, and ΔA the thermodynamic integral given by:

$$\Delta A = \int_0^1 \langle \Delta E \rangle_\eta d\eta, \tag{7.5}$$

where η is the Kirkwood coupling parameter and $\langle \Delta E \rangle_\eta$ is the total energy difference between charge states q and q' averaged over configurations achieved at fixed η.

We here resort to Janak's theorem [30], according to which the derivative of the DFT energy functional $E[N]$ with respect to the occupation f of the highest occupied Kohn–Sham (KS) state g gives the KS eigenvalue ϵ_g corresponding to that state:

$$\frac{\partial E[N]}{\partial f} = \epsilon_g. \tag{7.6}$$

In its integral form, Janak's theorem gives the change in energy of the system upon addition of one electron at fixed geometry,

$$E^{N+1} - E^N = \int_0^1 \epsilon_g(f) df. \tag{7.7}$$

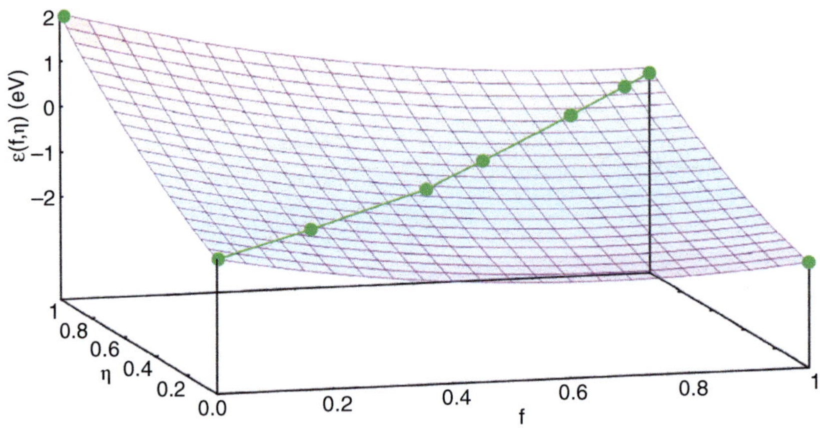

Figure 7.3 Schematic representation of the function $\epsilon(f,\eta)$ in the twodimensional (f,η) space. For capturing the quadratic dependence of $\epsilon(f,\eta)$, the dots shown in green are sampled in the present work.

The constant-Fermi-level technique consists in driving the electronic charge from state q to state q' through the realization of several intermediate systems at fixed Fermi level. Although it is the Fermi level that is controlled during the simulations, the parameters of the simulation can be set to allow only for minimal fluctuations of the electronic charge of the system. Then, taking the electronic charge of the system as Kirkwood coupling parameter allows one to draw a clear similarity with the TI method. We assume that during the reaction a single state g is involved and that its occupation goes from 0 to 1. The electronic charge then corresponds to the occupation η of state g. In this description, the TI method gives the redox level through the calculation of the thermodynamic integral ΔA, as given in Eq. 7.5. Using Janak's theorem in Eq. 7.7, the thermodynamic integral can be expressed as:

$$\Delta A = \int_0^1 \langle \Delta E \rangle_\eta d\eta = \int_0^1 \left\langle \int_0^1 \epsilon_g(f) df \right\rangle_\eta d\eta = \int_0^1 \int_0^1 \epsilon(f,\eta) df\, d\eta, \tag{7.8}$$

where we have used the following definition:

$$\epsilon(f,\eta) = \langle \epsilon_g(f) \rangle_\eta. \tag{7.9}$$

In the latter expression, the average is taken over structures corresponding to a state g with occupation η, while f indicates the occupation of g at fixed geometry.

As seen above, $\epsilon(f,\eta)$ is not necessarily a linear function. A more accurate approximation then consists in assuming a quadratic functional form in both η and f:

$$\epsilon(f,\eta) = \epsilon_S + \alpha\left(\eta - \frac{1}{2}\right) + \beta\left(f - \frac{1}{2}\right) + \alpha'\left(\eta - \frac{1}{2}\right)^2 + \beta'\left(f - \frac{1}{2}\right)^2 + \gamma\left(\eta - \frac{1}{2}\right)\left(f - \frac{1}{2}\right), \tag{7.10}$$

where ϵ_S, α, β, α', β' and γ are constants. Substituting eq 7.10 into Eq. 7.8 and carrying out the integrals leads to the following expression for the thermodynamic integral:

$$\Delta A = \int_0^1 \int_0^1 \epsilon(f,\eta) df\, d\eta = \epsilon_S + \frac{\alpha' + \beta'}{12}. \tag{7.11}$$

To determine ϵ_S, α' and β', we propose to calculate $\epsilon(f,\eta)$ in the following cases. First, it is natural to determine $\epsilon(f,\eta)$ for $f = \eta$ as this follows directly from the constant Fermi level simulations:

$$\begin{aligned}\epsilon(\eta,\eta) &= \epsilon_S + \underbrace{(\alpha+\beta)}_{A}(\eta - \frac{1}{2}) + \underbrace{(\alpha'+\beta'+\gamma)}_{B}(\eta-\frac{1}{2})^2 \\ &= \epsilon_S + A(\eta-\frac{1}{2}) + B(\eta-\frac{1}{2})^2,\end{aligned} \tag{7.12}$$

where the parameters ϵ_S, A, and B are easily obtained from a quadratic fit of $\epsilon(\eta,\eta)$. However, this fit is not sufficient to determine ΔA in Eq. 7.11. Therefore, we consider two additional points in the twodimensional (f,η) space:

$$\epsilon(0,1) = \epsilon_S - \tfrac{1}{2}(\alpha-\beta) + \tfrac{1}{4}(\alpha'+\beta'-\gamma), \tag{7.13}$$

$$\epsilon(1,0) = \epsilon_S + \tfrac{1}{2}(\alpha-\beta) + \tfrac{1}{4}(\alpha'+\beta'-\gamma). \tag{7.14}$$

Combining eqs 7.12, 7.13, and 7.14 with Eq. 7.11, we can express the thermodynamical integral as:

$$\Delta A = \int_0^1\int_0^1 \epsilon(f,\eta)df\,d\eta = \frac{5}{6}\epsilon_S + \frac{B}{24} + \frac{\epsilon(0,1)+\epsilon(1,0)}{12}. \tag{7.15}$$

If $\epsilon(f,\eta)$ is given by a quadratic function, the present expression for the thermodynamic integral leads to the exact value of the redox level defined in Eq. 7.4. Figure 7.3 schematically illustrates the function $\epsilon(f,\eta)$ in the two-dimensional (f,η) space and the points sampled in the present scheme.

We remark that the presently adopted second-order form (Eq. 7.10) goes beyond Marcus' linear regime [31, 32]. Indeed, on the basis of Janak's theorem, the integration of $\epsilon(f,\eta)$ over f produces the vertical energy gaps $\langle\Delta E\rangle_\eta$ required in the TI method (Eq. 7.5):

$$\langle\Delta E\rangle_\eta = \int_0^1 \Delta\epsilon(f,\eta)df = \epsilon_S + \alpha\left(\eta-\frac{1}{2}\right) + \alpha'\left(\eta-\frac{1}{2}\right)^2 + \frac{\beta'}{12}, \tag{7.16}$$

which by construction accounts for one higher order more than Marcus' linear regime. However, we note that having recourse to a quadratic form for $\epsilon(f,\eta)$ is not inherent to the constant Fermi level approach. Indeed, when strong nonlinearities occur, higher-order forms should be fitted to the data. Consequently, this entails a more detailed sampling of the function $\epsilon(f,\eta)$, in analogy to the more detailed sampling required for the vertical energy gaps in the TI method.

This procedure is applied to the determination of the redox potentials of aqueous ions. We carry out a detailed comparison between the redox potentials as obtained through the TI method and as extracted from the constant Fermi level simulations for three redox pairs: Fe^{3+}/Fe^{2+}, $O_2H^-/O_2H^\bullet$, and MnO_4^{2-}/MnO_4^-.

The simulation details described for the Fe^{3+}/Fe^{2+} were also adopted for the other redox pairs. However, to account for the larger solute sizes, we used 31 and 28 water molecules in the simulation cell for $O_2H^-/O_2H^\bullet$ and MnO_4^{2-}/MnO_4^-, respectively. For each redox pair, we performed constant Fermi-level simulations at fixed $\bar{\epsilon}_F$ corresponding to various values η of occupation of the state g. These simulations could be used directly for determining the redox potentials through the TI method. For the configurations obtained during the molecular

Table 7.1 Redox potentials vs. SHE (in eV) as obtained through constant Fermi level (CFL) simulations in the quadratic approximation, compared to those achieved with the thermodynamic integration (TI) method. Experimental values are shown for comparison.

System	CFL	TI	Expt.
$O_2H^\bullet/O_2H^-$	0.47	0.42	0.75
Fe^{+2}/Fe^{+3}	0.85	0.77	0.77
MnO_4^-/MnO_4^{-2}	−0.31	−0.28	0.56

dynamics at a given η, we calculated the average total energy difference $\langle \Delta E \rangle_\eta$ between occupations $f = 1$ and $f = 0$. Integration of $\langle \Delta E \rangle_\eta$ yielded the thermodynamic integral ΔA, from which the redox potential was derived through Eq. 7.4. This result is taken as reference value in the following. The obtained values are listed in Table 7.1.

Following the scheme outlined previously, we also determined the redox potentials via the averaged Kohn–Sham eigenvalues $\epsilon(f, \eta)$ of state g. The constant Fermi-level simulation provided us directly with $\epsilon(\eta, \eta)$. Within the quadratic approximation, we additionally calculated $\epsilon(0, 1)$ and $\epsilon(1, 0)$. and yielded through Eq. 7.15 an improved description of the thermodynamic integrals. The corresponding redox potentials are given in Table 7.1. Comparison with the reference TI values shows agreement within 0.08 eV, validating the proposed scheme for the calculation of redox potentials.

All redox couples give redox levels falling inside the reduced band gap achieved in the present DFT calculations. The calculated redox potentials can thus be compared with their experimental counterparts at 300 K (Table 7.1). Considering the reference results obtained with the TI method, we found an agreement with experiment within 0.3 eV for $O_2H^\bullet/O_2H^-$ and Fe^{+2}/Fe^{+3} and a larger deviation of 0.84 eV for MnO_4^-/MnO_4^{-2}. This level of agreement can be considered typical for DFT schemes at the present level of theory, as errors up to 0.9 eV have been found in redox potentials calculated with GGA functionals in previous studies [33, 34]. In particular, the large deviation observed for the permanganate redox potential could in part also arise from the choice of the volume that we allocated to the MnO_4^- ion, in the absence of any estimate of its partial molar volume.

7.4 CFL-MD at metal-water interface: The case of the Volmer reaction

In this section, we apply the constant Fermi energy molecular dynamic to the metal-water interface. In particular we focus on the reaction mechanism of the Volmer reaction. Our interface model consists of a 3×4 Pt slab of 3 layers, 31 water molecules, and one hydronium ion within a supercell of 9.79×8.48×19.50 Å^3. This size was chosen to ensure a water density of 1 g/cm^3 and a bulk-like $g_{OO}(r)$ partial correlation function. The adopted computation setup is similar to that used for the study of aqueous ions and is detailed in Ref. [35].

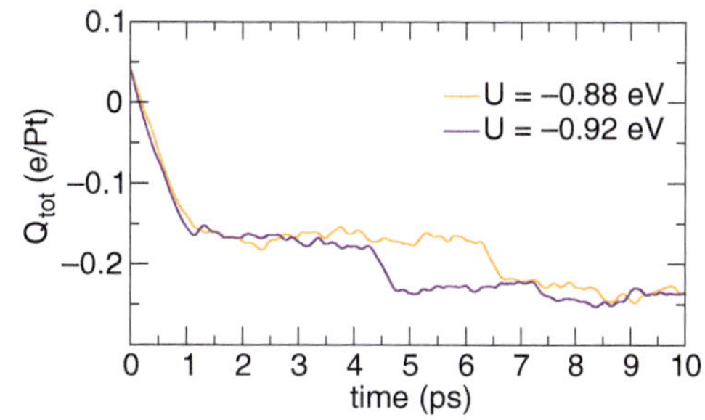

Figure 7.4 Time evolution of the total electronic charge Q_{tot} in constant Fermi-level molecular dynamics of the Pt(111)/water interface at an electrode potential of $U = -0.88$ eV and $U = -0.92$. Q_{tot} is defined as the total electronic charge and is here given per number of Pt surface atoms.

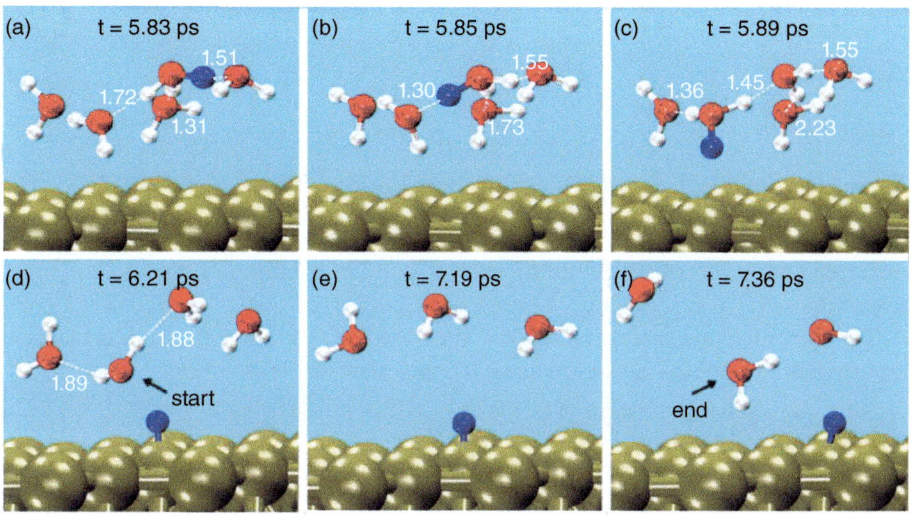

Figure 7.5 (a–d) Representative snapshots illustrating the atomistic mechanism of the Volmer reaction, in which the Eigen complex $H_9O_4^+$ transforms into a $H_7O_3^+$ complex before releasing the H atom to the Pt surface at $U = -0.88$ eV. (e-f) Snapshots illustrating the reorientation of the water molecule after H adsorption. The distances between H and O atoms (dashed white line) are given in Å.

We applied constant Fermi level MD to the Pt(111)/water interface by setting the Fermi level at $U = -0.88$ eV and $U = -0.92$ eV within the band gap of liquid water. Figure 7.4 represents the time evolution of the total electronic charge Q_{tot} of the system. We remark that Q_{tot} exhibits a jump at 6.2 and 4.3 ps for $U = -0.88$ and -0.92 eV, respectively. This change in the total electronic charge corresponds to the adsorption of the H_{aq}^+ ion at the Pt surface and signals electron transfer from the electron reservoir to the electrode according to the Volmer reaction: $H_{aq}^+ + e^- \rightarrow H_{ad}$. This reaction is expected to occur stochastically for electrode potentials lower than the SHE level provided that the simulation time is long enough.

By inspecting the atomistic structure we find in our simulations that the hydronium diffusion in bulk water is mediated by an Eigen complex $H_9O_4^+$. Figure 7.5 shows the reaction mechanism of the Volmer reaction, as obtained for $U = -0.88$ eV. This reaction occurs in three steps. First, the $H_9O_4^+$ complex diffuses into the vicinity of the Pt interface [Fig. 7.5(a–b)]. Second, the proton is transferred to a water molecule in a H_{down}

configuration and the Eigen complex transforms to a $H_7O_3^+$ complex [Fig. 7.5(c)]. Third, the proton is adsorbed at the metal surface [Fig. 7.5(e)] leaving the water molecule in a O_{down} configuration, which is energetically unfavorable. This leads to the reorientation of this water molecule resulting in a H_{down} configuration within a time span of ~ 1.15 ps. We note that the same reaction mechanism occurs at $U = -0.92$ eV. Otani et al. also investigated the Volmer reaction varying the electrode potential though the introduction of excess charge [13]. In their work, the hydronium ion was found to diffuse through a $H_5O_2^+$ complex. The involvement of a larger complex in our simulation might be due to the account of the van der Waals interactions, which strongly impact the hydrogen network. Nevertheless, the reported reorientation mechanism and its duration of ~ 0.8 ps are confirmed by our findings [13].

7.5 Referencing the bias potential to the SHE

One of the most difficult tasks in modelling electrified metal/water interface is to properly reference the electrode potential U to the SHE. We here define the electrode potential $\tilde{U}$ on a non-inverted scale and referenced to the valence band maximum of liquid water (ϵ_{vbm}) according to: $\tilde{U} = -U + \mu_{\text{SHE}} - \epsilon_{\text{vbm}}$.

We define the SHE as a sequence of two reactions (see fig. 7.6 in which the intermediate state corresponds to a neutral hydrogen atom H_{ad} adsorbed at the Pt surface:

$$H^+_{aq} + e^- \rightarrow H_{ad}, \tag{7.17}$$

$$H_{ad} \rightarrow \frac{1}{2}H_2(g) + \text{Pt}. \tag{7.18}$$

We obtain the Gibbs free energy difference ΔG of the full reaction as the sum of ΔG_1 and ΔG_2, which correspond to reactions (7.17) and (7.18), respectively.

To express ΔG_1, we use the grand canonical formulation for describing solutes and defects [28, 36]:

$$\Delta G_1 = G(H_{ad}) - G(H^+_{aq}) - (\epsilon_{\text{vbm}} + \mu_e), \tag{7.19}$$

where μ_e is the electron chemical potential. In our work, the difference $G(H_{ad}) - G(H^+_{aq})$ is evaluated within the constant Fermi level method, in which the energy is governed by the grand canonical potential $\Omega_{\tilde{U}} = E^{\text{tot}}(r_i, N_e) - N_e(\epsilon_{\text{vbm}} + \tilde{U})$, where E^{tot} is the DFT total energy and N_e the number of electrons in the system. Here, the Gibbs free energies

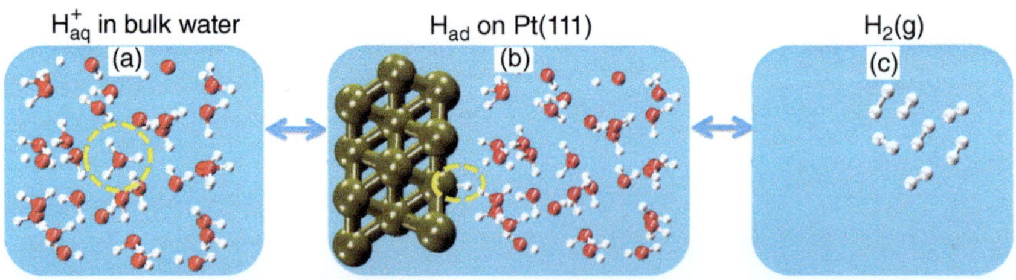

Figure 7.6 Schematic representation of the three steps of the standard hydrogen reaction considered in our work.

actually correspond to Helmholtz free energies, as volume variations are small and can be neglected [28]. We use the notation $F(\tilde{U})$ to indicate the free energy associated to the potential $\Omega_{\tilde{U}}$. Thus, we write:

$$\begin{aligned}\Delta G_1 &= F_{\tilde{U}}(\mathrm{H_{ad}}) - F_{\tilde{U}}(\mathrm{H^+_{aq}}) + \tilde{U} + \Delta_{\mathrm{zp}}^{\mathrm{H_{ad}}} - \Delta_{\mathrm{zp}}^{\mathrm{H^+_{aq}}} - \mu_e \\ &= \Delta F(\tilde{U}) + \tilde{U} + \Delta_{\mathrm{zp}}^{\mathrm{H_{ad}}} - \Delta_{\mathrm{zp}}^{\mathrm{H^+_{aq}}} - \mu_e,\end{aligned} \tag{7.20}$$

where we have isolated the zero point energies of $\mathrm{H_{ad}}$ and $\mathrm{H^+_{aq}}$, i.e. $\Delta_{\mathrm{zp}}^{\mathrm{H_{ad}}}$ and $\Delta_{\mathrm{zp}}^{\mathrm{H^+_{aq}}}$, as they are not explicitly accounted for in our DFT framework. $\Delta_{\mathrm{zp}}^{\mathrm{H^+_{aq}}}$ amounts to 0.36 eV [28], while $\Delta_{\mathrm{zp}}^{\mathrm{H_{ad}}}$ does not need to be evaluated as it drops in our final expressions for SHE. In Eq. (7.20), we introduced the free energy difference $\Delta F(\tilde{U}) = F(\mathrm{H_{ad}}) - F(\mathrm{H^+_{aq}})$ as described in our constant Fermi energy method. We determine $\Delta F(\tilde{U})$ through the Blue Moon scheme following recent literature [37, 38], in which redox reaction mechanisms at constant bias potential were studied. We use constrained molecular dynamics, in which the distance z between the proton and the Pt surface is taken as reaction coordinate [39, 40]:

$$\Delta F(\tilde{U}) = \int \langle f_c(z) \rangle dz, \tag{7.21}$$

where $f_c(z)$ is the constraint force computed within the constant Fermi energy scheme. Hence, the number of electrons in the system N_e depends on the spatial coordinate z.

The reaction coordinate in Eq. (7.21) goes from the position of the adsorbed $\mathrm{H_{ad}}$ to the bulk-like water region. In practice, we perform the Blue Moon simulation until the hydrogen atom reaches the first contact layer, where it is found in a singly positive charge state, and the remaining part of the integral is evaluated by considering the potential difference ΔW between the contact layer and the bulk-like water region. The latter is easily determined through the variation of O_{2s} levels and is given in Fig. 7.7(a). In Fig. 7.7(a), we show the obtained free energy profiles for four different values of $\tilde{U}$, equivalently expressed in terms of U in the figure. $\Delta F(\tilde{U}) + \tilde{U}$ should not depend on the value of $\tilde{U}$, and this is confirmed by the results of our simulations in Fig. 7.7(b), where its average value is shown to differ by less than 0.05 eV among the individual simulations. This result corroborates our scheme and supports its consistency.

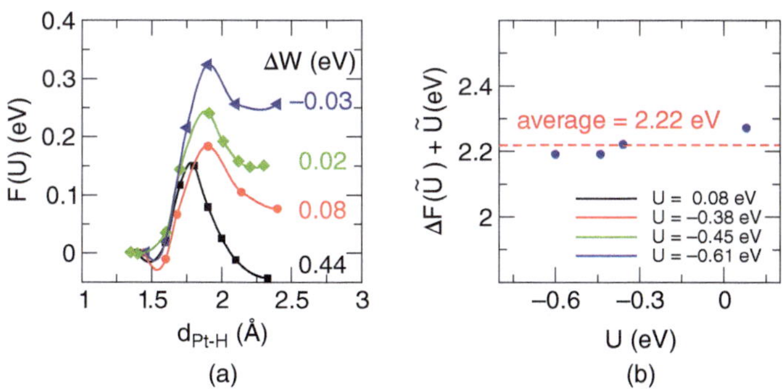

Figure 7.7 (a) Free energy profiles and (b) $\Delta F(\tilde{U}) + \tilde{U}$ as obtained through Blue Moon simulations for various electrode potentials expressed in terms of U. In (a), we also give the potential difference ΔW between the contact layer and the bulk-like region for each simulation.

For the second reaction (7.18), the free energy difference is given by:

$$\Delta G_2 = G\left[\frac{1}{2}H_2(g)\right] + G(Pt) - G(H_{ad}) = \mu_H + G(Pt) - G(H_{ad}), \tag{7.22}$$

where μ_H is the hydrogen chemical potential set to the free energy of gaseous H_2 in standard conditions and is evaluated as described in Ref. [28]. In our work, we express ΔG_2 as:

$$\Delta G_2 = \mu_H + \Delta A_{TI} - \Delta_{zp}^{H_{ad}}, \tag{7.23}$$

where ΔA_{TI} corresponds to the free energy cost of removing the H_{ad} from the Pt surface and placing it in the vacuum without considering the zero point energy of H_{ad}, which is expressed separately. We remark that, ΔG_2 does not depend on the electrode potential because there is no charge transfer involved in the reaction.

We evaluate ΔA_{TI} through the thermodynamic integration method, in which the Kirkwood parameter η evolves from $\eta = 0$ in the initial state with H_{ad} on the Pt surface to $\eta = 1$ in the final state in which the hydrogen atom has been removed. At $\eta = 0$, the vertical energy difference is obtained by removing the H atom from a MD run, in which H_{ad} is attached to the Pt surface in aqueous environment. At $\eta = 1$, the H atom is vertically inserted in configurations from a MD of the pristine Pt/water interface and allowed to relax while keeping all other atoms at fixed positions. We use two points at $\eta = 0$ and $\eta = 1$ to compute $\Delta A_{TI} + \mu_H$. We find that the vertical energy differences do not differ by more than 0.1 eV in the two cases, indicating that more intermediate values of η are not necessary for an accurate result. This behaviour should be ascribed to the fact that the reaction only involves neutral species and that the H occupies a position within the empty layer that separates the Pt surface from the water molecules. The integration of $\Delta A_{TI} + \mu_H$ over η yields 0.21 eV.

The redox level μ_e for which $\Delta G = \Delta G_1 + \Delta G_2 = 0$ can be isolated and gives:

$$\mu_e = \Delta F(\tilde{U}) + \tilde{U} + \Delta A_{TI} + \mu_H - \Delta_{zp}^{H^+_{aq}}. \tag{7.24}$$

We remark that this redox level has been calculated for the neutral interface, which corresponds to the pH at the point of zero charge (pzc). To retrieve the SHE level, a pH = 0 should be used. This leads to the consideration of a Nernstian shift from the pH_{pzc} to pH = 0:

$$\mu_e^{SHE} = \mu_e(pH = 0) = \mu_e - (0.059\ eV) \cdot pH_{pzc}, \tag{7.25}$$

where pH_{pzc} of the Pt(111)/water interface has experimentally been estimated at 3.4 [41, 42]. Combining Eqs. (7.24) and (7.25), we finally locate the SHE level at:

$$\mu_e^{SHE} = \Delta F(\tilde{U}) + \tilde{U} + \Delta A_{TI} + \mu_H - \Delta_{zp}^{H^+_{aq}} - (0.059\ eV) \cdot pH_{pzc}. \tag{7.26}$$

The SHE level calculated in this way falls at 1.87 eV with respect to the valence band of liquid water, as determined with the rVV10 exchange correlation functional.

7.6 Macroscopic properties at the metal-water interface

We first focus on the electric field established at the Pt interface. We consider the metal/water interface model of size $L_x \times L_y \times L_z$ shown in Fig. 7.8 with the water layer centered at $z = 0$. The boundary surfaces of the Pt electrode are at $z = \pm L_w/2$. In our simulation, we

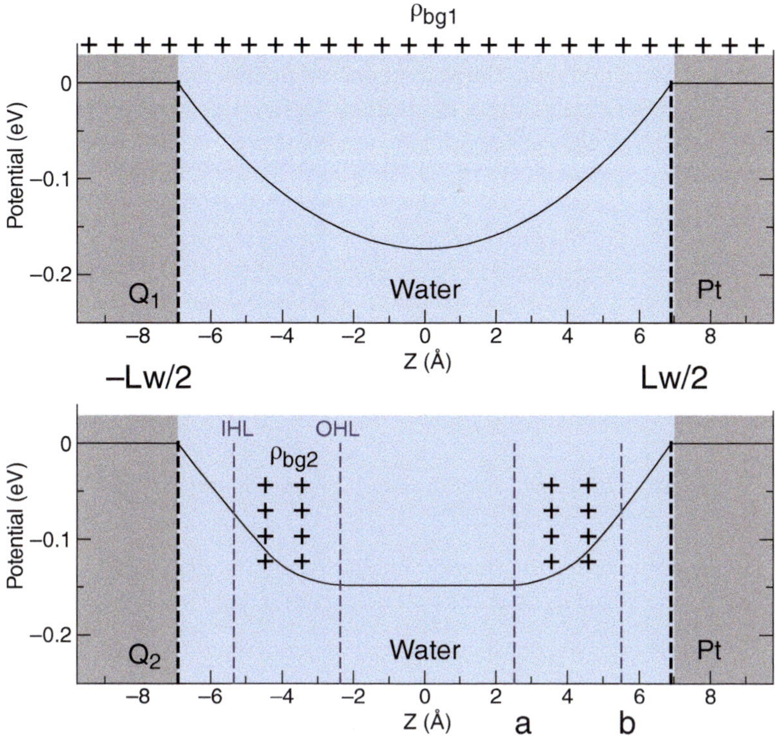

Figure 7.8 Schematic representation of the metal/water interface in two different classical models. In the top panel, the extra charge Q_1 added on the electrode is compensated by a uniform background charge ρ_{bg1} like in DFT supercell simulations. In the bottom panel, the extra charge on the electrode Q_2 is compensated by a background charge ρ_{bg2} confined between the inner (IHL) and outer (OHL) Helmholtz layers. Q_2 corresponds to the noncompensated charge on the electrode when a charge Q_1 is added to the model with a uniform background. The calculated potential corresponds to the case in which Q_1 equals four electrons. The cells of the classical models correspond to those used in the DFT simulations.

make use of a uniform neutralizing background charge (ρ_{bg1}), as schematically illustrated in the top panel of Fig. 7.8. However, in the physical situation that we would like to model the compensating charge (ρ_{bg2}) is found in the Helmholtz layer, as depicted in the bottom panel of Fig. 7.8. We account for these differences through classical electrostatics.

It should be realized that the uniform background entails effects on both the charge on the electrode and the potential difference ΔV between the electrode and bulk-like water. We first consider the effect on the charge. When the system is charged with a given nominal charge Q_1 on the electrode, a uniform background is established throughout the cell. Part of the added charge is directly compensated by the fraction of the uniform background charge that lies within the metal electrode. The actual charge Q_2 that contributes to the establishment of an electric field is therefore reduced, $Q_2 = \eta Q_1$, where $\eta = L_w/L_z$ is the fractional volume occupied by the water component expressed in terms of the width of the water layer L_w and the cell size L_z in the Z direction. In our DFT simulations $\eta = 0.71$.

Next, to correct the difference in potential, we compare two classical models showing the same cells as in our DFT simulations. In the first one, the added charge is compensated by a uniform background charge Q_1, while the second one represents the physical model with charge Q_2 (Fig. 7.8). In the case of the a uniform background charge, we perform a simulation at a total net charge Q_1. An opposite charge is induced in the neutralizing background and is uniformly distributed over the cell leading to a charge density ρ_{bg1}:

$$\rho_{\text{bg1}} = \frac{-Q_1}{L_x L_y L_z}. \tag{7.27}$$

The potential induced by this charge density is the solution of the Poisson equation:

$$\nabla^2 V = \frac{-4\pi\rho_{\text{bg1}}}{\epsilon}. \tag{7.28}$$

Considering the boundary conditions $V(-L_{\text{w}}/2) = V(L_{\text{w}}/2) = 0$, we find the following quadratic function for $V(z)$:

$$V(z) = \frac{-2\pi\rho_{\text{bg1}} z^2}{\epsilon} + \frac{2\pi\rho_{\text{bg1}}(L_{\text{w}}/2)^2}{\epsilon}. \tag{7.29}$$

In the case of the double layer model, the corresponding situation is obtained by considering an additional charge Q_2, which leads to the following surface charge density at each Pt surface:

$$\sigma_2 = \frac{Q_2}{2L_x L_y}, \tag{7.30}$$

and an opposite background charge confined in the double layer:

$$\rho_{\text{bg2}} = \frac{-Q_2}{2L_x L_y(b-a)} \tag{7.31}$$

Solving the Poisson equation in this system leads to the following potential:

$$V(z) = \frac{-2\pi\rho_{\text{bg2}}(a-b)^2}{\epsilon} + \frac{4\pi\sigma_2(a - L_{\text{w}}/2)}{\epsilon}; \quad \text{for } z < a \tag{7.32}$$

$$V(z) = \frac{-2\pi\rho_{\text{bg2}}(z-b)^2}{\epsilon} + \frac{4\pi\sigma_2(z - L_{\text{w}}/2)}{\epsilon}; \quad \text{for } a < z < b \tag{7.33}$$

$$V(z) = \frac{4\pi\sigma_2(z - L_{\text{w}}/2)}{\epsilon}; \quad \text{for } b < z < L_{\text{w}}/2 \tag{7.34}$$

We remark that the two models show very close electric fields in the region of the double layer, but the potentials bend to different values in the middle of the water region. Indeed, the model with a uniform background shows a parabolic behavior, whereas the physical one exhibits a flat potential after crossing the Helmholtz layer. In Fig. 7.8, these potentials have been calculated for a nominal charge Q_1 of four electrons. The screening of the water molecules is accounted for by a dielectric constant $\epsilon = 62.4$, corresponding to the experimental value at $T = 350$ K (Ref. [43]). In these conditions, the model with the uniform background charge yields $\Delta V_1 = 0.17$ eV, to be compared with $\Delta V_2 = 0.15$ eV in the physical model. The difference between these values should be used to correct the DFT potential.

For the particular simulation cell considered in our work, the numerical value of this difference is negligible. Hence, in our simulation, apart from the nominal correction of the extra charge on the electrode, the background charge does not produce any sizeable effect

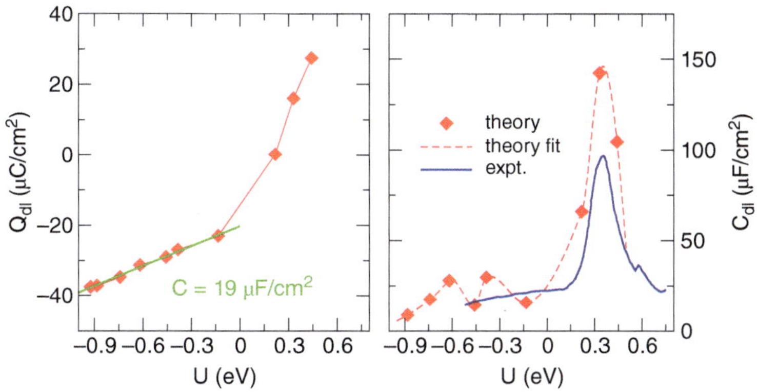

Figure 7.9 (left panel) Double layer charge Q_{dl} as a function of the applied electrode potential U. The red line represents a linear fit, from which the double layer capacitance C_{dl} is derived. (right panel) Double layer capacitance (blue dots) as a function of the applied electrode potential U. The simulation data are fitted through a polynomial function (blue dashed line) and are compared to experimental results (green line) from Ref. [44]. The energies are referred to the SHE. Source: Based on T. Pajkossy, D. Kolb, Double layer capacitance of Pt(111) single crystal electrodes Electrochim. Acta 46(20), 3063 (2001).

on the alignment. This implies that the uniform background effectively accounts for the counterions in an averaged way. However, we note that in general such corrections depend on the width of the water layer in the simulation cell.

We now focus on two electrochemical properties of the Pt(111)/water interface system, the double layer capacitance and the potential of zero charge. We applied constant Fermi level MD to the Pt(111)/water interface by setting the Fermi level at various positions within the band gap of liquid water. Within the alignment scheme detailed above, our applied potential covers a range from −0.92 to 0.44 eV versus the SHE. We define the excess charge at Pt surface Q_{dl} as $Q_{dl} = \frac{1}{2}(\eta Q_{tot} - 1)$, where η is the fractional volume corresponding to water in the supercell and accounts for the passivation of the background charge in the metal slab. In this definition, it is assumed that the aqueous hydronium ion H^+_{aq} always carries a positive charge and that the electronic charge is shared between the two Pt interfaces in the cell.

Figure 7.9 presents the evolution of the double layer charge as a function of the electrode potential. The pzc corresponds to the Fermi level of the neutral system with respect to the SHE level and is found at 0.22 eV in our simulation, in agreement with reported experimental values (0.28–0.37 eV vs. SHE [41, 42, 44]). Another accessible macroscopic electrochemical quantity is the double layer capacitance $C_{dl} = dQ_{dl}/dU$, and is compared to respective measured data in Fig. 7.9. The dependence of the calculated capacitance on U reproduces well the shape of the experimental data, and yields $C_{dl} \simeq 19\ \mu\text{Fcm}^{-2}$, in excellent agreement with the commonly accepted experimental value of $\sim 20\ \mu\text{Fcm}^{-2}$ (Refs. [45, 46] and [44]). We remark that the main peak in the C_{dl} at $U \simeq 0.33$ eV is associated with a rapid change in the double layer capacitance, which in turn is linked to the atomic-scale structural reorganization of the electrical double layer, as will be discussed in the analysis below. These results demonstrate the predictive power of our modelling scheme in accessing macroscopic electrochemical properties.

7.7 Atomic-scale processes at the metal-water interface

The atomic-scale structure of the electrical double layer is an important aspect in defining the metal/water interface and the electrochemical activity of the electrode [47]. Several experimental studies investigated the structure at interfacial water systems, either in neutral conditions or under bias potential. Water molecules have been observed to undergo reorientation depending on the applied potential in a variety of systems, including Pt(111) [48, 49], polycrystalline Pt [50], Au(111) [51], air/water and lipid/water charged interfaces. In particular, at metal surfaces, the interfacial water molecules are strongly hydrogen bonded with the O atoms pointing toward the metal surface for potentials above the pzc. At variance, for potentials below the pzc, the water molecules are weakly bonded and point their H atoms downwards. In this work, we take advantage of our theoretical scheme to study the structural reorganization of the water double layer at the Pt(111)/H_2O interface as a function of the bias electrode potential referenced to the SHE.

Figure 7.10 shows the distribution of O and H atoms, and the water dipole orientation as a function of a coordinate along the surface normal direction as the electrode potential U

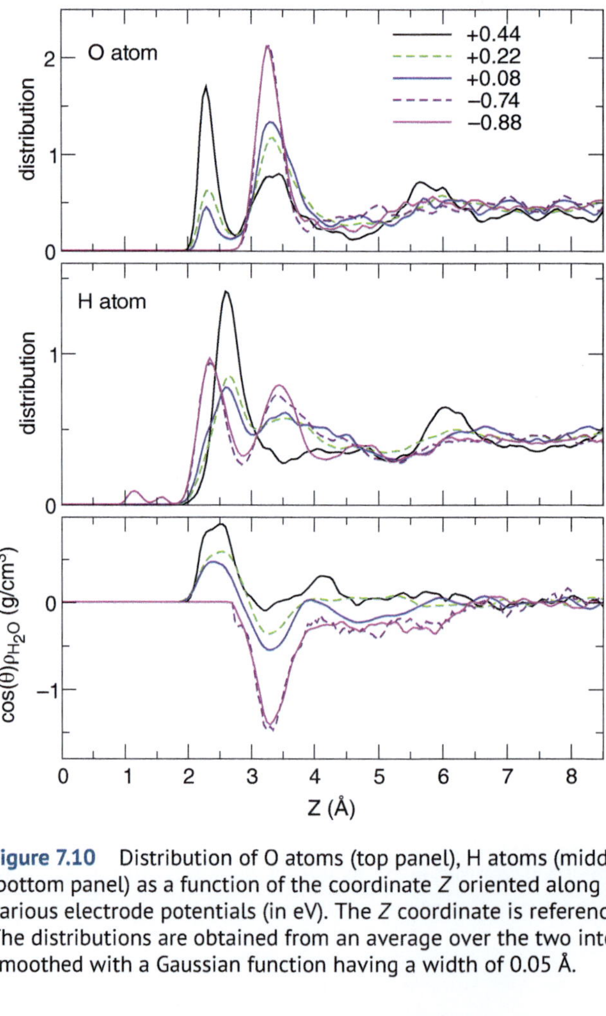

Figure 7.10 Distribution of O atoms (top panel), H atoms (middle panel), and dipole orientation (bottom panel) as a function of the coordinate Z oriented along the surface normal direction, at various electrode potentials (in eV). The Z coordinate is referenced with respect to the Pt surface. The distributions are obtained from an average over the two interfaces in our model and have been smoothed with a Gaussian function having a width of 0.05 Å.

is varied. For electrode potentials above the pzc ($U > 0.22$ eV), the O distribution features a first sharp peak at 2.3 Å and a second broad peak around 3.35 Å, while the H distribution exhibits a single principal peak centered at 2.6 Å. The orientation of the water molecules can be inferred from the dipole orientation, as presented in Figs. 7.10 and Fig. 7.11. The positive peak of the dipole distribution indicates that the water molecules in the first contact layer are in O_{down} configurations with the H atoms pointing towards the liquid. Specifically, we find two main dipole orientations at 55° and 86° for $U = 0.44$ eV (see Fig. 7.11) The dipole orientation of the water molecules in the second layer shows a very broad peak around 90° (see Fig. 7.11). As the potential decreases, the intensity of the first O peak diminishes, while that of the second peak increases (Fig. 7.10). This trend is accompanied by a reduction in the intensity of the first peak of the H distribution and the appearance of a second broad peak centered at 3.8 Å (Fig. 7.10). Focusing on the dipole orientation, we infer that some of the water molecules in the first contact layer reorient from O_{down} to H_{down} configurations, as the electrode potential drops. At the pzc (0.22 eV), we observe coexistence of the two configurations with a slight prevalence of the H_{down} configurations.

Under negative bias, all the water molecules are found in H_{down} configurations leading to the disappearance of the first O peak in Fig. 7.10. In addition, the first peak in the H distribution shifts toward 2.35 Å and a second H peak appears around 3.43 Å. The dipole orientation features a negative peak signifying that the water orientation is predominantly H_{down}. A further broad negative region appearing under strong negative bias indicates that also the molecules in the second water layer ($4 < Z < 6$ Å) orient their H atoms toward the interface. Accordingly, two main dipole orientations are found at 96° and 134° at $U = -0.92$ eV (see Fig. Fig. 7.11). However, deep in the water layer ($Z > 6.5$ Å) the dipole is found to be randomly oriented and the water structure can be taken as representative of the bulk phase.

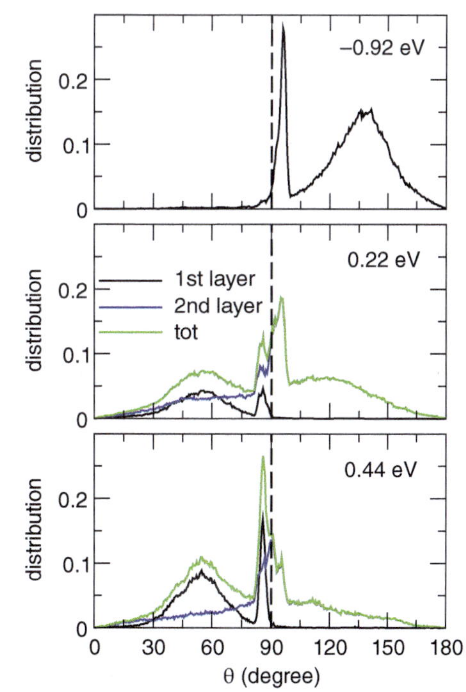

Figure 7.11 The dipole angular distribution of water molecules belonging to the electrical double layer at electrode potentials U= 0.44 eV (bottom panel), 0.22 eV (middle panel), and −0.92 eV (top panel). The dashed black line corresponds to $\theta = 90°$.

The atomistic description emerging from our work is fully consistent with the picture of water reorientation under bias potential inferred from experimental data [48–51].

7.8 Conclusion

In this contribution we first reviewed the constant Fermi energy molecular dynamics. Then, we applied this technique to the determination of redox levels. We showed that the redox reaction can be driven through the control of the Fermi level without any prior knowledge of the reaction products. For aqua ions, we demonstrated molecular dynamics with stable evaluations of the Fermi level and of the electronic charge, without any divergence or instability. Through Janak's theorem, a relation with the redox potential could be established. Introducing a quadratic function to model the dependence of the single-particle energy level on occupation, we found agreement within less than 0.1 eV with redox potentials calculated through the thermodynamic integration method.

Then we applied the constant Fermi-level molecular dynamics to study the Pt(111)/water interface under variable bias potential. First, we investigated the Volmer reaction mechanism involving the adsorption of a hydrogen atom at the Pt interface under negative bias. Second, we established an alignment scheme to determine the SHE reference level by considering the hydrogen adsorbed at the Pt surface as an intermediate step between the aqueous hydronium ion H^+_{aq} and the gas phase $H_2(g)$. Through the determination of the SHE level, our scheme allows one to reference the electrode potentials on an absolute scale. Third, we studied the double layer capacitance and the pzc, and found μ_{pzc} =0.22 eV and $C_{dl} \simeq 19\ \mu\text{Fcm}^{-2}$, in very good agreement with experiment. Finally, we investigated the atomic-scale structural reorganization of the electrical double layer as a function of the bias potential. Our results reveal that at potentials above the pzc the water structure is dominated by O_{down} configurations, in which the dipole of the water molecules is on average pointing towards the bulk water layer. At the pzc, H_{down} configurations coexist with O_{down} structures, but the latter disappear completely at negative bias potential, for which all the water dipoles are fully oriented toward the Pt surface. At the lowest potential considered, the water molecules orient both their H atoms toward the electrode.

Overall, our modeling scheme gives access simultaneously to macroscopic and microscopic properties of aqueous species and metal/water interfaces. This technique will considerably contribute to advancing the predictive power of the DFT modelling of electrode/electrolyte interfaces under bias potential.

Acknowledgements

This project has received funding from the European Union's Seventh Framework Programme for research, technological development and demonstration under grant agreement no. 291771 (call 2014). This work has been performed in the context of the National Center of Competence in Research (NCCR) "Materials' Revolution: Computational Design and Discovery of Novel Materials (MARVEL)" of the Swiss National Science Foundation. We used computational resources of CSCS, SCITAS, and CSEA-EPFL.

References

1 R. Jinnouchi, A.B. Anderson, *Phys. Rev. B* 77(24), 245417 (2008)
2 H.F. Wang, Z.P. Liu, *J. Phys. Chem. C* 113(40), 17502 (2009)
3 N. Bonnet, N. Marzari, *Phys. Rev. Lett.* 110(8), 086104 (2013)
4 S. Sakong, M. Naderian, K. Mathew, R.G. Hennig, A. Groß, *J. Chem. Phys.* 142(23), 234107 (2015)
5 V. Tripkovic, M.E. Björketun, E. Skúlason, J. Rossmeisl, *Phys. Rev. B* 84(11), 115452 (2011)
6 E. Skúlason, G.S. Karlberg, J. Rossmeisl, T. Bligaard, J. Greeley, H. Jónsson, J.K. Nørskov, *Phys. Chem. Chem. Phys.* 9(25), 3241 (2007)
7 S. Sakong, K. Forster-Tonigold, A. Groß, *J. Chem. Phys.* 144(19), 194701 (2016)
8 J. Le, M. Iannuzzi, A. Cuesta, J. Cheng, *Phys. Rev. Lett.* 119(1), 016801 (2017)
9 A.Y. Lozovoi, A. Alavi, J. Kohanoff, R.M. Lynden-Bell, *J. Chem. Phys.* 115(4), 1661 (2001)
10 N. Bonnet, T. Morishita, O. Sugino, M. Otani, *Phys. Rev. Lett.* 109(26), 266101 (2012)
11 M. Otani, O. Sugino, *Phys. Rev. B* 73(11), 115407 (2006)
12 M. Otani, I. Hamada, O. Sugino, Y. Morikawa, Y. Okamoto, T. Ikeshoji, *Phys. Chem. Chem. Phys.* 10(25), 3609 (2008)
13 M. Otani, I. Hamada, O. Sugino, Y. Morikawa, Y. Okamoto, T. Ikeshoji, *J. Phys. Soc. Jpn.* 77(2), 024802 (2008)
14 O. Sugino, I. Hamada, M. Otani, Y. Morikawa, T. Ikeshoji, Y. Okamoto, *Surf. Sci.* 601(22), 5237 (2007)
15 J. Cheng, M. Sprik, *Phys. Chem. Chem. Phys.* 14(32), 11245 (2012)
16 M.H. Hansen, C. Jin, K.S. Thygesen, J. Rossmeisl, *J. Phys. Chem. C* 120(25), 13485 (2016)
17 J. Rossmeisl, K. Chaŋ, R. Ahmed, V. Tripković, M.E. Björketun, *Phys. Chem. Chem. Phys.* 15(25), 10321 (2013)
18 S. Nosé, *Mol. Phys.* 52, 255 (1984)
19 S. Nosé, *J. Chem. Phys.* 81(1), 511 (1984)
20 W.G. Hoover, *Phys. Rev. A.* 31(3), 1695 (1985)
21 P.E. Blöchl, M. Parrinello, *Phys. Rev. B* 45, 9413 (1992). DOI 10.1103/PhysRevB.45.9413. URL http://link.aps.org/doi/10.1103/PhysRevB.45.9413
22 H.C. Andersen, *J. Chem. Phys.* 72(4), 2384 (1980)
23 O.A. Vydrov, T. Van Voorhis, *J. Chem. Phys.* 133(24), 244103 (2010)
24 R. Sabatini, T. Gorni, S. de Gironcoli, *Phys. Rev. B* 87(4), 041108 (2013)
25 G. Miceli, S. de Gironcoli, A. Pasquarello, *J. Chem. Phys.* 142(3), 034501 (2015)
26 N. Troullier, J.L. Martins, *Phys. Rev. B* 43(3), 1993 (1991). DOI 10.1103/PhysRevB.43.1993
27 P. Giannozzi, S. Baroni, N. Bonini, M. Calandra, R. Car, C. Cavazzoni, D. Ceresoli, G.L. Chiarotti, M. Cococcioni, I. Dabo, et al., *J. Phys. Condens. Matter* 21(39), 395502 (2009)
28 F. Ambrosio, G. Miceli, A. Pasquarello, *J. Chem. Phys.* 143(24), 244508 (2015)
29 A. Bouzid, A. Pasquarello, *J. Chem. Theory Comput.* 13(4), 1769 (2017)
30 J.F. Janak, *Phys. Rev. B* 18(12), 7165 (1978)
31 R.A. Marcus, *J. Chem. Phys.* 43(2), 679 (1965)

32 R.A. Marcus, *J. Chem. Phys.* 24(5), 966 (1956)
33 J. Cheng, X. Liu, J. VandeVondele, M. Sulpizi, M. Sprik, *Acc. Chem. Res.* 47(12), 3522 (2014)
34 J. Cheng, J. VandeVondele, *Phys. Rev. Lett.* 116(8), 086402 (2016)
35 A. Bouzid, A. Pasquarello, *J. Phys. Chem. Lett.* 9(8), 1880 (2018)
36 M. Todorova, J. Neugebauer, *Phys. Rev. Appl.* 1(1), 014001 (2014)
37 T. Ikeshoji, T. Uchida, M. Otani, M. Osawa, *J. Electroanal. Chem.* 800, 13 (2017)
38 T. Ikeshoji, M. Otani, *Phys. Chem. Chem. Phys.* 19(6), 4447 (2017)
39 M. Sprik, G. Ciccotti, *J. Chem. Phys.* 109(18), 7737 (1998)
40 G. Ciccotti, R. Kapral, E. Vanden-Eijnden, *Chem. Phys. Chem.* 6(9), 1809 (2005)
41 E. Gileadi, S. Argade, J.O. Bockris, *J. Phys. Chem.* 70(6), 2044 (1966)
42 R. Rizo, E. Sitta, E. Herrero, V. Climent, J.M. Feliu, *Electrochim. Acta* 162, 138 (2015)
43 C.G. Malmberg, A.A. Maryott, *J. Res. Nat. Bureau Stand.* 56, 1 (1956)
44 T. Pajkossy, D. Kolb, *Electrochim. Acta* 46(20), 3063 (2001)
45 T. Pajkossy, D. Kolb, *Electrochem. Commun.* 5(4), 283 (2003)
46 N. Garcia-Araez, V. Climent, E. Herrero, J.M. Feliu, J. Lipkowski, *Electrochim. Acta* 51(18), 3787 (2006)
47 D. Kolb, *Surf. Sci.* 500(1), 722 (2002)
48 T. Iwasita, X. Xia, *J. Electroanal. Chem.* 411(1-2), 95 (1996)
49 N. Garcia-Araez, V. Climent, J. Feliu, *J. Phys. Chem. C* 113(21), 9290 (2009)
50 M. Osawa, M. Tsushima, H. Mogami, G. Samjeske, A. Yamakata, *J. Phys. Chem. C* 112(11), 4248 (2008)
51 K.i. Ataka, T. Yotsuyanagi, M. Osawa, *J. Phys. Chem.* 100(25), 10664 (1996)

Part IV

8

From electrons to electrode kinetics: A tutorial review

Stephen Fletcher

The Fletcher Consultancy, Loughborough, Leicestershire LE11 3LU, United Kingdom

8.1 Global electro-neutrality

The concept of electric lines of force was introduced by Michael Faraday in 1837. Electric lines of force are smooth curves drawn in 3D space such that their tangents give the electric field direction and their line densities give the electric field magnitude. The field direction is then the direction that the electric force exerts on a positive test charge, while the field magnitude manifests as the electric force per unit charge. Thus, a line of force is the path followed by a positive test charge that is free to move in an electric field.

Electric lines of force always start from a positive charge and end on a negative charge. There are no "dangling lines of force". For this reason the Universe exhibits **global electro-neutrality**. There are many indications that this is true. Most prominently, since the electromagnetic force is much stronger than the gravitational force, violations of universal electro-neutrality would show up on cosmological scales of length, and they are not observed.

8.2 The electrochemical reference state

Electrochemical measurements require an **electrochemical reference state** for charged particles, consistent with global electro-neutrality. The most widely used reference state is the remote field-free vacuum. The word "remote" implies that the reference particle is at an infinite distance from the origin of the co-ordinate system, and the phrase "field-free" implies that the reference particle is effectively delocalized over all available space. The reference particle is also assumed to be at rest.

A convenient visualization of this model is a spherical capacitor consisting of two concentric plates (Fig. 8.1). The small inner plate of the capacitor represents the surface of an object of interest, while the large outer plate represents the remote field-free vacuum. By the principle of electro-neutrality, the charges on the outer plate exactly balance the charges on the inner plate. However, the particles on the outer plate are spread over an immeasurably greater area than those on the inner plate.

Atomic-Scale Modelling of Electrochemical Systems, First Edition.
Edited by Marko M. Melander, Tomi T. Laurila, and Kari Laasonen.

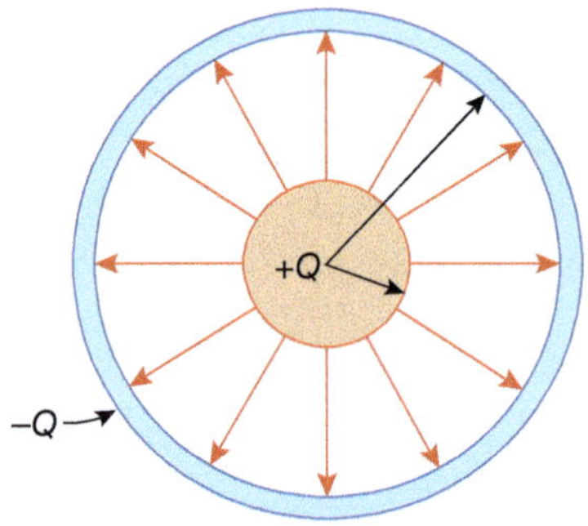

Figure 8.1 The Spherical Capacitor Model.

The work required to charge the surface of a particle is simply the work required to charge the capacitor as a whole. By a well-known formula of electric circuit theory, the mutual capacitance of the two plates of a spherical capacitor is

$$C = \frac{4\pi\varepsilon_0}{\left[\frac{1}{r_1} - \frac{1}{r_2}\right]} \tag{8.1}$$

and so the work w (joules; J) to charge the spherical plates by an amount of charge $\pm Q$ is:

$$w = \frac{Q^2}{2C}. \tag{8.2}$$

Here, ε_0 is the permittivity of free space (farads per meter; F m^{-1}), r_1 is the radius of the inner plate (meters; m), r_2 is the radius of the outer plate (meters; m), and Q is the magnitude of the charge on each plate of the capacitor (coulombs; C).

On open circuit the capacitor plates are noise-free. However, if the capacitor terminals are short-circuited by a conductor (i.e., connected to a reservoir of thermalized charge carriers) then both terminals experience random fluctuations of charge ("Johnson–Nyquist noise") of mean square magnitude

$$\langle Q^2 \rangle = k_B T C. \tag{8.3}$$

Here k_B is the Boltzmann constant (1.3806×10^{-23} J K^{-1}) and T is the absolute temperature (kelvin; K). In these circumstances w becomes a quadratic degree of freedom of the system, and, as such, is subject to the principle of equipartition. This states that all quadratic degrees of freedom store $k_B T/2$ of energy on time-average at thermodynamic equilibrium. This stored energy is the source of the charge fluctuations that drive electron transfer [1].

Returning to the open circuit case, if the vacuum is replaced by a dielectric medium having a relative permittivity (dielectric constant) of magnitude ε_r (dimensionless) then the simple mutual capacitance becomes

$$C = \frac{4\pi\varepsilon_0\varepsilon_r}{\left[\frac{1}{r_1} - \frac{1}{r_2}\right]} \tag{8.4}$$

and the work to charge the plates becomes:

$$w = \frac{Q^2}{8\pi\varepsilon_0\varepsilon_r}\left[\frac{1}{r_1} - \frac{1}{r_2}\right]. \tag{8.5}$$

Finally, letting $r_2 \to \infty$ we obtain:

$$w = \frac{Q^2}{8\pi\varepsilon_0\varepsilon_r r_1}. \tag{8.6}$$

This quantity is known as the **electrical self-energy** of an object inside a dielectric medium [2]. It corresponds to the work done on the total system to separate the charges of $+Q$ and $-Q$ at infinity and then to concentrate one of these charges on the surface of the object. The work of initially separating the charges of $+Q$ and $-Q$ at infinity is of course negligible if the charges are initially far apart, so with that proviso we may safely conclude that w consists only of the work required to concentrate the charge of $\pm Q$ on the surface of the object.

The question arises of possible energy losses during the act of concentrating charge on the central object. However, if the charging process takes place infinitesimally slowly, then any flow of electric current is negligibly small and any I^2R dissipation of heat can be neglected. In that case the energy losses are insignificant and the charging process may be considered to be thermodynamically reversible.

Now suppose the object of interest is an ion. The difference in electrical self-energy upon transferring it from a homogeneous medium of low dielectric constant ε_1 to one of high dielectric constant ε_2 is simply:

$$\Delta w = -\frac{Q^2}{8\pi\varepsilon_0 r}\left[\frac{1}{\varepsilon_1} - \frac{1}{\varepsilon_2}\right]. \tag{8.7}$$

This famous result, known as the **Born equation**, reveals the importance of electrical charge in the theory of ionic solutions. In particular, it tells us why ions dissolve preferentially in more polar solvents (i.e., in solvents having higher dielectric constants). It is because their self-energy is lower there. The dependence of Δw on $-1/r$ also indicates why smaller ions (such as Li^+) are more soluble than larger ions. This size-dependence makes lithium batteries possible.

The Born equation has the twin virtues of simplicity and elegance. But it also has some serious limitations. Most disconcertingly, by representing solvents as homogeneous dielectric media, it fails to capture the chemical and electrical interactions that occur on the molecular scale between ions and solvents. Among the missing chemical interactions are the loss of translational entropy of solvent molecules near to ions, and the work needed to form ion-sized cavities in the solvent. Among the missing electrical interactions are the short-range van der Waals forces between all the various components of the solution, and the increase in concentration of counter-ions that necessarily occurs around all charged species. In the latter case, the time-average charge of the ionic atmosphere exactly cancels the permanent charge of the central ion, and thus confers electro-neutrality on the solution as a whole. It is noteworthy that counter-ions are missing from both Born theory and Marcus theory [2–4]. This is a serious defect of both approaches, which renders them physically unrealistic.

8.3 The chemical potential

The chemical potential, μ, of a particle at a point (x, y, z) is the amount of **non-electrical work** (joules; J) that was expended in transferring the particle to that point from remote field-free vacuum, at constant temperature and pressure. (That is, all electrical work is excluded.) Thus, the chemical potential can be thought of as "how unaccepting the system is of a neutral particle". For an electrically charged particle the chemical potential simply measures the non-electrical part of the total work that must be done to force the particle into a system. The particle may be an atom, molecule, ion, electron, hole, or indeed any object that can be counted.

Constant temperature and pressure are specified in the definition of chemical potential because the point (x, y, z) may be located inside a medium whose phase composition and energy-level occupancies are determined by the temperature and pressure. Alternative thermodynamic formulations are of course possible using different constraints.

The chemical potential of a single particle is related to its **thermodynamic activity** a by the equation:

$$\mu(T, P) = \mu_0(T, P) + k_B T \ \ln(a). \tag{8.8}$$

Here T, P, and μ are all intensive functions of state. The constant $\mu_0(T, P)$ is called the standard chemical potential. While it is not a function of a, it may be a function of T and P. The choice of standard chemical potential is arbitrary, but is customarily chosen such that $a = 1$. The concept of "thermodynamic activity" was introduced by Gilbert Newton Lewis in 1907 [5]. At low concentrations, the activity of a solute is approximately equal to its concentration.

8.4 The electrostatic potential

The Electrostatic Potential, ϕ, of a particle at a point (x, y, z) is the amount of electrical work (in units of joules) that was expended in transferring the particle to that point from remote field-free vacuum, divided by its signed charge (in units of coulombs), at constant temperature and pressure. The unit of electrostatic potential ϕ is therefore the joule per coulomb, otherwise known as the **volt**. It is a scalar quantity. At a point (x_0, y_0, z_0) inside the bulk of a metal conductor, the electrostatic potential is sometimes referred to as the "Inner Electric Potential" or the "Galvani Potential". The electrostatic potential is always a single-valued function of location (x, y, z).

For ionic species the amount of electrical work (in units of joules) is therefore $ze\phi$, where the dimensionless integer z is the valence (charge number) of the ionic species, e is the unit positive charge ($+1.6021765 \times 10^{-19}$ coulomb), and ϕ is the local electrostatic potential (volts).

8.5 The electrochemical potential

The Electrochemical Potential, $\tilde{\mu}$, of a (possibly) multiply-charged particle at a point (x, y, z) is the *combined* work (chemical work plus electrical work, both measured in joules) that was expended in transferring the particle to that point from remote field-free vacuum, at constant temperature and pressure:

$$\boxed{\tilde{\mu} \overset{\text{def}}{=} \mu + ze\phi}. \tag{8.9}$$

Here μ is the chemical potential (joules), z is the valence (charge number) of the particle (dimensionless), e is the unit positive charge (coulombs), and ϕ is the local electrostatic potential (volts).

The electrochemical potential of a single particle is related to its thermodynamic activity a by the equation:

$$\tilde{\mu}(T,P) = \mu_0(T,P) + k_B T \ln(a) + ze\phi. \tag{8.10}$$

Here, T, P, and $\tilde{\mu}$ are all intensive functions of state. The constant $\mu_0(T, P)$ is called the standard chemical potential, while z is the valence (charge number) of the particle (dimensionless), e is the unit positive charge (coulombs), and ϕ is the local electrostatic potential (volts). Edward Armand Guggenheim popularized the concept of "electrochemical potential" in 1929 [6]. He was motivated to do so by the realization that it was not possible to remove the charge from an electron. There is no such thing as a "neutral electron". Hence, it is impossible to independently vary the charge and mass of a system of electrons. Accordingly, the electrochemical potential is the best available parameter for describing such a system.

Any particle will spontaneously move away from a point (x, y, z) if its electrochemical potential $\tilde{\mu}$ decreases as a result. The total rate of such particle movements can be quantified macroscopically as a flux, which is defined as the net flow of matter per unit time through a plane of unit area normal to the gradient of electrochemical potential. The totality of this flux is called **migration-diffusion**.

Migration-diffusion is the motion of particles driven by a gradient of electrochemical potential. The total migration-diffusion of material is described by the Nernst–Planck equation.

At constant temperature and pressure the electrochemical potential $\tilde{\mu}$ of a single particle is related to the derivative of the Gibbs energy of the system by the equation:

$$\tilde{\mu}(T,P) = \left(\frac{\partial G}{\partial n_i}\right)_{T,P,\phi,n_{j\neq i}}. \tag{8.11}$$

Here T is the absolute temperature, P is the pressure, G is the Gibbs energy, and n is the number of particles of type i. In plain English: the electrochemical potential is the Gibbs energy per particle. Thus, the increase of Gibbs energy $\mathrm{d}G$ caused by an increase of particle numbers $\mathrm{d}n_i$ is:

$$\mathrm{d}G = \sum_{1}^{n} \tilde{\mu}\,\mathrm{d}n_i. \tag{8.12}$$

From this equation it is clear that the Gibbs energy is an extensive variable whereas the electrochemical potential is an intensive variable.

8.5.1 The molar electrochemical potential

The molar electrochemical potential, $\bar{\tilde{\mu}}$, is the combined work [chemical work plus electrical work, both measured in joules per mole] that was expended in transferring one mole of (possibly) multiply-charged particles from remote field-free vacuum to the interior of a uniform phase at constant temperature and pressure:

$$\bar{\tilde{\mu}} = \bar{\mu} + zF\phi. \tag{8.13}$$

Here $\bar{\mu}$ is the molar chemical potential (joules per mole), z is the valence (charge number) of the particles (dimensionless), F is the Faraday constant (C mol^{-1}) and ϕ is the local electrostatic potential (volts).

Note. The mole, symbol mol, contains *precisely* 6.022 140 76 × 10^{23} particles [7]. This quantity is the Avogadro Constant, N_A, whose units are mol^{-1}.

8.5.2 The electrochemical potential of a single electron

The electrochemical potential of a single electron, $\tilde{\mu}_{e^-}$, is the combined work [chemical work plus electrical work, both measured in joules] needed to transfer a single electron to the point (x, y, z) from remote field-free vacuum, at constant temperature and pressure:

$$\tilde{\mu}_{e^-} = \mu_{e^-} - e\phi. \tag{8.14}$$

Here μ_{e^-} is the chemical potential of the electron (joules), e is the unit positive charge (coulombs), and ϕ is the local electrostatic potential (volts).

If the point (x, y, z) lies inside an uncharged piece of metal, and if the electron enters the lowest unoccupied molecular orbital (LUMO) of the metal at one atmosphere pressure and zero degrees kelvin, then the electrochemical potential of the added electron, $\tilde{\mu}_{e^-}$, defines the **Fermi Energy**, $\mathcal{E}_F$, of the metal:

$$\boxed{\mathcal{E}_F \overset{\text{def}}{=} \tilde{\mu}_{e^-} \quad \text{at 1 atm and 0 K}}. \tag{8.15}$$

The units of Fermi energy are joules, although electron volts are also widely used:

$$1\ \text{eV} = 1.602 \times 10^{-19}\ \text{joules per electron}. \tag{8.16}$$

8.5.3 The Nernst equation

As noted above, the electrochemical potential of a single particle is related to its thermodynamic activity a by the equation:

$$\tilde{\mu}(T, P) = \mu_0(T, P) + k_B T\ \ln(a) + ze\phi. \tag{8.17}$$

Now consider a redox reaction involving *two* particles, labelled Ox and Red. These could be cations, for example. Also let us assume that both particles are able to exchange n electrons with a metal electrode, in accordance with the reaction scheme:

$$\text{Ox} + ne^- \rightleftarrows \text{Red}. \tag{8.18}$$

Such a reaction may be thought of as one half of an electrolytic cell. At thermodynamic equilibrium, the electrochemical potentials of the reactants and products are necessarily equal:

$$\tilde{\mu}_{ox} + n\tilde{\mu}_{e^-} \overset{\text{def}}{=} \tilde{\mu}_{red} \quad \text{at equilibrium.} \tag{8.19}$$

However, although this equation defines the equilibrium state, it fails to explain the *mechanism* by which the particles and the electrons actually attain the equilibrium state. The central difficulty is that Ox and Red are both confined to the solution phase, whereas the electrons are confined to the metal phase, so classically they appear to be non-communicating. However, this problem is resolved magnificently by quantum mechanics, which allows the electrons to tunnel from the metal phase into individual particles in solution (and *vice versa*). When this happens, the electrons take with them the same electrochemical potential that they had in the metal phase. Indeed, this is the quantum equivalent of the conservation of energy, and it allows Ox and Red to restore their bulk concentrations (activities) after being subjected to small perturbations.

Expanding and rearranging the electrochemical potentials yields:

$$\begin{aligned}&\mu_{0(ox)} + k_B T \ln(a_{ox}) + z_{ox} e\phi_{ox}\\&\quad - \mu_{0(red)} - k_B T \ln(a_{red}) - z_{red} e\phi_{red}\\&= -n\tilde{\mu}_{e^-}.\end{aligned} \tag{8.20}$$

Now, since the electrostatic potential is single-valued:

$$\phi_{ox} = \phi_{red} = \phi \tag{8.21}$$

and since the electrochemical potential of each electron equals its Fermi energy $\mathcal{E}_F$ in the metal:

$$\tilde{\mu}_{e^-} = \mathcal{E}_F = -eE_F \tag{8.22}$$

we immediately deduce that:

$$(\mu_{0(ox)} - \mu_{0(red)}) + k_B T \ln\left(\frac{a_{ox}}{a_{red}}\right) + ne\phi = neE_F \tag{8.23}$$

where E_F is the potential (volts) of the exchangeable electrons in the electrode (i.e., the **Fermi Potential**). In arriving at this result we have used the simplified notation:

$$z_{ox} - z_{red} = n \tag{8.24}$$

and we have also made use of the elementary substitution:

$$z_e = -1 \tag{8.25}$$

Now defining a "*standard*" Fermi potential $E_{F(0)}$ (volts) corresponding to the standard activities $a_{ox} = a_{red} = 1$ we obtain:

$$E_{F(0)} \overset{\text{def}}{=} \frac{\mu_{0(ox)} - \mu_{0(red)}}{ne} + \phi. \tag{8.26}$$

Thus:

$$E_F = E_{F(0)} + \frac{k_B T}{ne} \ln\left(\frac{a_{ox}}{a_{red}}\right) \quad \text{at equilibrium.} \tag{8.27}$$

Or in molar quantities:

$$\boxed{E_{\rm F} = E_{\rm F(0)} + \frac{RT}{nF}\ln\left(\frac{a_{\rm ox}}{a_{\rm red}}\right) \quad \text{at equilibrium}}. \tag{8.28}$$

This is the celebrated **Nernst equation**. It relates the Fermi potential (on the vacuum scale, in volts) of the electrons in a metal electrode to the thermodynamic activities of the redox species in the adjoining solution. It has a long and colorful history, especially in the era before quantum mechanics was developed [8].

Unfortunately, values of $E_{\rm F}$ on the vacuum scale are difficult to measure. However, values of $E_{\rm F}$ with respect to other half-cells can be measured with ease. When two half-cells are combined into a full cell, the difference between their Fermi potentials is called the **electrode potential**. In the scientific literature, electrode potentials are often reported with respect to an internationally-agreed standard, such as the **standard hydrogen half-cell** (*q.v.*)

8.5.4 Fermi–Dirac distribution function

Particles fall into two classes, fermions and bosons. Electrons are fermions. This means that they obey the Pauli Exclusion Principle, and no two of them can occupy the same quantum state.

At temperatures above absolute zero, the distribution of electron energies inside a bulk piece of metal is thermally broadened. This broadening is described by the **Fermi–Dirac distribution function**. For fermions it takes the form

$$P(\mathcal{E}) = \frac{1}{\exp((\mathcal{E} - \tilde{\mu}_{\rm e^-})/k_{\rm B}T) + 1}, \tag{8.29}$$

where $\mathcal{E}$ is the energy of one particular quantum state and $P(\mathcal{E})$ is the probability that the state is occupied. Summing over all possible quantum states, the total number of electrons is

$$N = \int_0^\infty \frac{\rho(\mathcal{E})}{\exp((\mathcal{E} - \tilde{\mu}_{\rm e^-})/k_{\rm B}T) + 1}\, d\mathcal{E}, \tag{8.30}$$

where $\rho(\mathcal{E})$ is the **density of electronic states**. For a constant value of N it is clear from this equation that the electrochemical potential is a function of temperature. The Fermi energy is not.

Experimentally, it turns out that the dependence of the electrochemical potential $\tilde{\mu}_{\rm e}$ on temperature is weak at room temperature, which explains why many authors feel justified in replacing $\tilde{\mu}_{\rm e}$ in the Fermi–Dirac distribution function with its value at absolute zero, namely the Fermi energy, $\mathcal{E}_{\rm F}$. The error introduced by this approximation is often said to be slight (less than 0.1 eV at room temperature), but it is clearly not negligible.

Before the introduction of Fermi–Dirac statistics, it was a major puzzle why electrons in metals had different energies. The puzzle was solved (on the basis of Pauli Exclusion) by Enrico Fermi [9] and Paul Dirac [10] in 1926. Today, we know that only a small fraction of a metal's electrons are thermally excited at room temperature, due to the particular form of the Fermi–Dirac distribution.

The electrons that are thermally excited inside a piece of metal exhibit a surprisingly narrow range of energies around the Fermi energy $\mathcal{E}_{\rm F}$, of the order k_BT. The rest of the electrons below the Fermi energy are inert unless they are presented with empty states to tunnel into.

8.5.5 The molar electrochemical potential of an electron

The molar electrochemical potential of an electron, $\bar{\tilde{\mu}}_{e^-}$, is defined as the combined work (chemical work plus electrical work, both measured in joules per mole) needed to transfer one mole of electrons to the interior of a uniform phase from remote field-free vacuum, at constant temperature and pressure [11]:

$$\bar{\tilde{\mu}}_{e^-} = \bar{\mu}_{e^-} - F\phi. \tag{8.31}$$

Here $\bar{\mu}_{e^-}$ is the molar chemical potential of the electron (joules), F is the Faraday constant (C/mol), and ϕ is the local electrostatic potential (volts).

If the uniform phase is a metal, and the electrons enter the lowest unoccupied molecular orbitals at 0 K, then the molar electrochemical potential of the electrons, $\bar{\tilde{\mu}}_{e^-}$, defines the **molar Fermi energy,** $\overline{\varepsilon_F}$, of the metal:

$$\overline{\varepsilon_F} \stackrel{\text{def}}{=} \bar{\tilde{\mu}}_{e^-}. \tag{8.32}$$

8.5.6 Parsing the electrochemical potential. (I) Metal in a vacuum

In some special circumstances it is possible to separate the electrochemical potential into its constituent parts (namely the chemical potential and the electrostatic potential). This is possible, for example, near the surface of a piece of metal in a vacuum. To see how this works, consider a test electron that is gradually transported from the remote field-free vacuum to a point (x_0, y_0, z_0) deep inside the metal.

In the absence of physical matter, no chemical interactions can exist. Therefore, for as long as the test electron remains in empty space, its chemical potential is zero:

$$\mu_{e^-}^{\text{vac}} = 0. \tag{8.33}$$

By contrast, the electrostatic potential penetrates the whole of space. This means that every measurement of the total electrochemical potential $\tilde{\mu}_e$ of a stationary electron at a random point (x, y, z) inside a vacuum is *ipso facto* a measurement of the local electrostatic potential ϕ at the same point:

$$\tilde{\mu}_{e^-} = -e\phi \quad \text{(in vacuum)}. \tag{8.34}$$

As the test electron continues its progress towards the metal surface, however, it eventually reaches a point (x_1, y_1, z_1) where it makes first contact with the frontier orbitals of the metal surface. This typically occurs at a distance of ~3.00 nm. At this point of first surface contact the total electrochemical potential $\tilde{\mu}_{e^-}$ is defined as the **surface electrochemical potential** $\tilde{\mu}_{e^-}^{S}$. Thus:

$$\tilde{\mu}_{e^-}(x_1, y_1, z_1) \stackrel{\text{def}}{=} \tilde{\mu}_{e^-}^{S}(x_1, y_1, z_1) \quad \text{(at first surface contact)}. \tag{8.35}$$

Implying:

$$\tilde{\mu}_{e^-}^{S}(x_1, y_1, z_1) = -e\phi_1 \quad \text{(at first surface contact)}. \tag{8.36}$$

The electrostatic potential ϕ_1 at the point of first surface contact (x_1, y_1, z_1) is called the **outer potential** of the metal. Its units are volts.

As the test electron finally crosses the interface and reaches the point (x_0, y_0, z_0) inside the bulk of the metal it is stabilized by both electrical and chemical forces. As a result, its electrochemical potential drops precipitously from $-e\phi_1$ to the Fermi energy $\mathcal{E}_F$. The work required to escape this situation and return the electron back to the point of first surface contact (x_1, y_1, z_1) is called the **surface work function:**

$$w_{S1} = \tilde{\mu}_{e^-}(x_1, y_1, z_1) - \tilde{\mu}_{e^-}(x_0, y_0, z_0). \tag{8.37}$$

Or:

$$\boxed{w_{S1} = -e\phi_1 - \mathcal{E}_{F1}}. \tag{8.38}$$

Experimentally, the magnitude of the Fermi energy $\mathcal{E}_F$ is constant everywhere inside the bulk metal because no gradient of electrochemical potential can exist there at equilibrium. But the magnitude of the outer potential ϕ_1 is problematic. It depends sensitively on many hard-to-control factors, such as crystal orientation, alloy composition, surface films, etc. This makes accurate measurements of surface work functions a difficult task. Relative measurements are less demanding, however, and can often be made by Scanning Kelvin Probe Microscopy (*q.v.*)

8.5.7 The Volta potential difference

The **Volta potential difference** between two pieces of metal in a vacuum is defined as the difference in electrostatic potential between the point of first contact (x_1, y_1, z_1) at the surface of metal M1 and the point of first contact (x_2, y_2, z_2) at the surface of metal M2:

$$\Delta\phi_{12} \stackrel{\text{def}}{=} \phi_1 - \phi_2 \quad \text{(At first surface contact)}. \tag{8.39}$$

For obvious reasons, the Volta potential difference is commonly called the "contact potential difference".

In practical measurements of Volta potential difference, two dissimilar metals are brought together just beyond the electron tunneling distance (>6.00 nm) and allowed to equilibrate. During this process, the as-formed capacitor charges spontaneously until the electrochemical potentials ($\tilde{\mu}_{e^-}$) are the same on both metal plates. As we would expect, the net flow of electrons is away from the metal with the higher electrochemical potential and towards the metal with the lower electrochemical potential. The increment of charge (ΔQ_{12}) that builds up during the equilibration process is determined by the product of the mutual capacitance (C_{12}) of the metal surfaces and the Volta potential difference ($\Delta\phi_{12}$) between them. Thus:

$$\Delta Q_{12} = C_{12} \times \Delta\phi_{12}. \tag{8.40}$$

In practice the increment of charge involves a miniscule fraction of the total number of electrons inside both metals, so there is no significant change in the population of charge carriers. But the Volta potential difference can reach several hundred millivolts, which is a problem in precise measurements of voltage. The easiest way to eliminate this problem is to ensure that both terminals of any measuring device have the same alloy composition.

We should also mention that, despite their similar sounding names, voltmeters do not measure Volta potentials. In fact voltmeters measure the **electromotive force** (*emf*, volts) of the system to which they are attached:

$$emf \overset{\text{def}}{=} \frac{\Delta\tilde{\mu}_{e^-}}{-e}. \tag{8.41}$$

Here $\Delta\tilde{\mu}_{e^-}$ is the difference between the electrochemical potentials (joules) of the electrons in the two terminals of the measuring device, and e is the unit positive charge (coulombs). In electrical engineering, the term "motive force" is applied very loosely to any energy difference that causes particles to move.

8.5.8 Scanning Kelvin Probe Microscopy

Scanning Kelvin Probe Microscopy is a laboratory technique that is able to measure the difference in surface work function between two metals with high accuracy. One metal is formed into a sharp tip and the other is formed into a flat surface. The technique works by bringing the sharp tip very near to the flat surface, such that the points of first contact physically coincide.

The surface work function of the sharp metal tip (1) on the vacuum scale is:

$$w_{S1} = -e\phi_1 - \mathcal{E}_{F1}. \tag{8.42}$$

And the surface work function of the flat metal surface (2) on the vacuum scale is:

$$w_{S2} = -e\phi_2 - \mathcal{E}_{F2}. \tag{8.43}$$

The difference between them is therefore:

$$w_{S1} - w_{S2} = -e(\phi_1 - \phi_2) - (\mathcal{E}_{F1} - \mathcal{E}_{F2}). \tag{8.44}$$

Here, $(\phi_1 - \phi_2)$ is the Volta potential difference and $(\mathcal{E}_{F1} - \mathcal{E}_{F2})$ is the Fermi energy difference. As noted earlier, the Volta potential difference between two pieces of metal is just the difference in electrostatic potential between the point of first contact (x_1, y_1, z_1) at the surface of metal M1 and the point of first contact (x_2, y_2, z_2) at the surface of metal M2. If these points physically coincide, then

$$(x_1, y_1, z_1) = (x_2, y_2, z_2) \quad \text{at coincident contact.} \tag{8.45}$$

And therefore:

$$(\phi_1 - \phi_2) \overset{\text{def}}{=} 0 \quad \text{at coincident contact.} \tag{8.46}$$

Hence

$$w_{S1} - w_{S2} = -(\mathcal{E}_{F1} - \mathcal{E}_{F2}) \quad \text{at coincident contact.} \tag{8.47}$$

This is very elegant. At the point of coincident contact, the difference between the surface work functions of the metals (1) and (2) is just the negative of the difference between their Fermi energies (i.e., their externally measured *emf*).

In an alternative approach, the two pieces of metal may be connected by a short circuit and allowed to come to equilibrium. In that case

$$(\mathcal{E}_{F1} - \mathcal{E}_{F2}) \overset{\text{def}}{=} 0 \quad \text{at thermodynamic equilibrium.} \tag{8.48}$$

And therefore:

$$w_{S1} - w_{S2} = -e(\phi_1 - \phi_2) \quad \text{at thermodynamic equilibrium.} \tag{8.49}$$

In this case the difference between the surface work functions of the metals (1) and (2) manifests as a Volta potential difference.

8.5.9 The membrane potential

The **membrane potential** is synonymous with the **membrane *emf***. The membrane *emf* (volts) is defined as

$$emf \stackrel{\text{def}}{=} \frac{\Delta\tilde{\mu}_{e^-}}{-e}, \tag{8.50}$$

where $\Delta\tilde{\mu}_{e^-}$ is the electrochemical potential difference between a test electron on one side of the membrane and a test electron on the other side of the membrane, and e is the unit positive charge (coulombs). The *emf* is composed of two terms:

$$emf = \frac{\Delta\mu_{e^-}}{-e} + \Delta\phi. \tag{8.51}$$

Here $\Delta\mu_{e^-}$ is the difference in chemical potential (joules) between opposite sides of the membrane, and $\Delta\phi$ is the difference in electrostatic potential (volts) between opposite sides of the membrane. Electrochemical membrane potentials are readily measured by inserting matched reference half-cells on either side of the membrane and measuring the *emf* between them. However, electrostatic membrane potentials are much more elusive. They must generally be derived by extrapolation from molecular dynamics simulations or diffuse double layer theory. To summarize: the membrane potential is *not* the difference in *electric potential* between the interior and the exterior of a biological cell. The membrane potential is the difference in *electrochemical* potential (joules) divided by $-e$ (coulombs).

8.5.10 The electrochemical potential of a single proton

The electrochemical potential of a single proton, $\tilde{\mu}_{H^+}$, is the combined work [chemical work plus electrical work, both measured in joules] needed to transfer a single proton to the point (x, y, z) from remote field-free vacuum, at constant temperature and pressure:

$$\tilde{\mu}_{H^+} = \mu_{H^+} + e\phi. \tag{8.52}$$

Here μ_{H^+} is the chemical potential of the proton (joules), e is the unit positive charge (coulombs), and ϕ is the local electrostatic potential (volts). Alternatively, we can expand this equation into the form:

$$\tilde{\mu}_{H^+}(T,P) = \mu^0_{H^+}(T,P) + k_B T \ln(a) + e\phi. \tag{8.53}$$

Here, T, P, and $\tilde{\mu}_{H^+}$ are all intensive functions of state. The constant $\mu^0_{H^+}$ is the standard chemical potential, k_B is the Boltzmann constant, a is the thermodynamic activity of the proton (dimensionless), e is the unit positive charge (coulombs), and ϕ is the local electrostatic potential (volts).

8.5.11 The proton motive force

We can use the above results to derive the **proton motive force** (*pmf*) acting between two points in solution. As already shown, the electrochemical potential of a single proton at an arbitrary location (x_1, y_1, z_1) is:

$$\tilde{\mu}_{H^+(1)}(T, P) = \mu^0_{H^+}(T, P) + k_B T \ln(a_1) + e\phi_1. \tag{8.54}$$

Similarly, the electrochemical potential of a single proton at an arbitrary location (x_2, y_2, z_2) is:

$$\tilde{\mu}_{H^+(2)}(T, P) = \mu^0_{H^+}(T, P) + k_B T \ln(a_2) + e\phi_2. \tag{8.55}$$

Subtracting the former from the latter yields:

$$\Delta\tilde{\mu}_{H^+}(T, P) = k_B T \ln\left(\frac{a_2}{a_1}\right) + e(\phi_2 - \phi_1). \tag{8.56}$$

Now, since the proton motive force (volts) is defined as

$$pmf \stackrel{\text{def}}{=} \frac{\Delta\tilde{\mu}_{H^+}}{e} \tag{8.57}$$

we immediately obtain:

$$\boxed{pmf = \Delta\phi - Z\Delta\text{pH}}. \tag{8.58}$$

In this equation, the composite constant Z is defined as

$$Z \stackrel{\text{def}}{=} \frac{k_B T}{e}\ln(10) \tag{8.59}$$

and the pH of solution is given by the formula

$$\text{pH} = -\log_{10}(a) \tag{8.60}$$

where:

$$a = \gamma\left(\frac{c_{H^+}}{c_0}\right). \tag{8.61}$$

Here a is the proton activity (dimensionless), γ is the activity coefficient (also dimensionless, and close to unity in dilute solutions), c_{H^+} is the local concentration of protons (mol dm^{-3}) and c_0 is a universal constant known as the "*standard amount concentration*" of protons (1 mol dm^{-3}).

In the above derivation, $\Delta\phi$ is the electrostatic potential difference (measured in volts) between the two arbitrary locations, and ΔpH is the proton concentration difference (measured in units of pH) between the same two locations.

The above result for *pmf* with $\Delta\phi$ replaced by $\Delta\psi$ (the electrostatic component of the membrane potential), was Eq. (1) in Peter Mitchell's famous 1978 Nobel Prize lecture [12], in which he summarized his discovery that the proton motive force was the principal factor driving the mixed migration/diffusion of protons across certain biological membranes, such as the thylakoid membranes in chloroplasts.

8.5.12 The standard hydrogen half-cell

A hydrogen half-cell consists of an acid solution surrounding a platinized platinum electrode upon which two electrochemical reactions are proceeding in opposite directions. These reactions are the reduction of H^+ ions and the oxidation of H_2 molecules. At equilibrium, the platinized platinum adopts an electrode potential such that the rates of the oxidation and reduction reactions come into perfect balance. Thus, to the human eye, nothing visible seems to be happening. However, at the molecular level, the rates of both the forward reaction and the backward reaction are proceeding vigorously.

The stability of the half-cell potential is determined by the rates of the interfacial reactions (the faster the better) and by the surface area of the electrode (the larger the better). The need for stability explains why hydrogen half-cell electrodes are commonly "platinized" i.e., coated with high-surface-area platinum black. This treatment increases their surface area by a factor of ~25. (Much larger factors are actually possible but the resulting coatings lack adhesion.) The need for a high surface area also explains why it is difficult to miniaturize reference half-cells.

Surprisingly, there is no standard recipe for the deposition of platinum black. Many deposition solutions have been tried in the literature, with mixed success. In our laboratories we use 2% w/w chloroplatinic acid (H_2PtCl_6) in 2 M HCl (aq).

On another practical point, the internal resistance of a fully-assembled reference half-cell should be made as small as possible, to avoid near-field interference by the alternating mains supply. Commercial reference electrodes, which are designed for use in pH meters, often contain very high-resistance glass frits [porous membranes] which are prone to near-field interference. These need to be avoided in high quality measurements of electrode kinetics.

Strictly speaking, the **standard hydrogen half-cell** (commonly, but erroneously, referred to as the "the standard hydrogen electrode" or "S.H.E.") should contain unit activity of H^+ ions. However, since single ion activities are difficult to measure with accuracy, electrode kineticists generally "*cheat*" and use 1.0 M $H^+(aq)$ solutions instead. The error introduced by this procedure is thought to be small.

The equilibrium properties of the standard hydrogen half-cell are determined by the dynamic equilibrium:

$$2H^+(aq) + 2e^- \rightleftharpoons H_2(g) + 2H_2O \tag{8.62}$$

for which the corresponding Nernst equation is:

$$E = E_0 + \frac{RT}{2F} \ln \frac{(a_{H^+})^2}{a_{H_2} a_{H_2O}}. \tag{8.63}$$

Here E is the equilibrium potential of the half-cell, R is the universal gas constant, T is the absolute temperature, F is the Faraday constant, a_{H^+} is the activity of hydrogen ions, a_{H_2} is the activity of hydrogen gas, and a_{H_2O} is the activity of water. Under "standard" conditions, namely unit activity of all components at 298.15 K and 1 bar pressure, the standard equilibrium potential E_0 is defined as zero on the standard hydrogen scale, and this potential corresponds (roughly) to +4.44 V on the vacuum scale [13] (Fig. 8.2). Despite decades of effort, the value of the latter is still not known with accuracy.

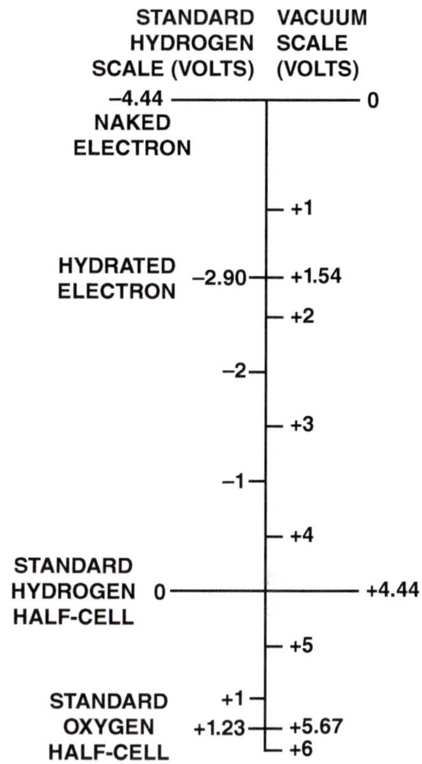

Figure 8.2 Two scales of potential (volts). Adapted from Trasatti [13].

8.5.13 The hydrated electron

The interaction of ionizing radiation with liquid water generates a number of highly reactive species, and the most important of these is the hydrated electron. Since its identification by Hart and Boag in 1962 [14, 15], the hydrated electron has fascinated chemists and physicists. When a free electron is added to water, it localizes in a cavity (electrostatic potential well) between two water molecules within less than 1 ps. There it forms two slightly asymmetric H-bonds with neighboring water molecules, which stabilize it for about 1ms. During its brief lifetime it exists in an *s*-like (near-spherical) ground state. Very rapidly, however, it decomposes by reaction with a proton or a water molecule to yield a hydrogen atom.

The standard potential of the reaction

$$e^-(\text{vac}) + H_2O(\text{aq}) \rightleftharpoons e^-(\text{aq}) \tag{8.64}$$

is −2.90 V vs. S.H.E. [16, 17]. Accordingly, hydrated electrons are not stable in the conventional range of electrode potentials (±2.50 V vs. S.H.E.). This means that they can be ruled out as reactants or products in conventional electrochemical reactions.

The Gibbs Energy of Hydration of the electron is −148.6 kJ/mol. This value was derived by means of electronic structure calculations [18]. It is identical to the value calculated from the standard potentials of the hydrated electron and the remote field-free vacuum. Based on a Born–Haber cycle and data from photo-emission threshold experiments, the enthalpy of hydration of the solvated electron in water has also been calculated, and found to be –166 kJ/mol [19].

Compiling the above data we have for the hydrated electron:

$$\Delta G_{\text{hyd}} = -148.6\,\text{kJ mol}^{-1} \tag{8.65}$$

$$\Delta H_{\text{hyd}} = -166.0\,\text{kJ mol}^{-1}. \tag{8.66}$$

Therefore, since

$$\Delta G = \Delta H - T\Delta S \tag{8.67}$$

we obtain:

$$\Delta S_{\text{hyd}} = -58.4\,\text{J K}^{-1}\text{mol}^{-1} \tag{8.68}$$

at 298.15 K. Based on the hypothesis [20] that $-28\,\text{J K}^{-1}\text{mol}^{-1}$ is the entropy loss per immobilized water molecule in the solvation shell of a hydrated ion at 298.15 K, we immediately deduce that the hydration number of the hydrated electron is $58.4/28 = 2.1$, in agreement with the quantum simulation studies of Schnitker and Rossky [21].

Further confirmation of the hydration number of 2 of the hydrated electron is provided by an OH stretch in the Raman spectrum, which shows clear evidence of hydrogen bonding between solvated electrons and nearby water molecules [22]. Also, the molecular simulation work of Zhan and Dixon has identified a hydration number of two as having the minimum energy configuration around the hydrated electron [23].

8.5.14 The hydrogen atom H*

The hydrogen atom (radical) is the conjugate acid of $e^-(aq)$ and is also a major decomposition product of $e^-(aq)$ in aqueous solution:

$$e^-(aq) + H^+(aq) \rightleftharpoons H^*(aq). \tag{8.69}$$

The standard potential of $H^*(aq)$ (–2.32 V *vs* S.H.E.) is considerably less negative than the standard potential of $e^-(aq)$ (–2.90 V *vs* S.H.E.), so the hydrogen atom is the more stable species [24]. Unfortunately, the hydrogen atom $H^*(aq)$ absorbs radiation only weakly (in the ultraviolet) and so it is almost invisible spectrophotometrically. However, it can be observed by Electron Paramagnetic Resonance (EPR) spectroscopy due to its electron spin of one-half. By contrast, $H^+(aq)$ and $H^-(aq)$ have spin-zero.

8.5.15 Parsing the electrochemical potential. (II) The co-sphere

The electrochemical potential, $\tilde{\mu}_{M^{\pm}}$, of a single ion (including its charge-balancing ionic atmosphere) in an electrolyte solution, consists of two separate work terms: (i) the *chemical* work needed to transfer the ion (complete with its ionic atmosphere) from remote field-free vacuum to the interior of the solution, and (ii) the *electrical* work needed to assemble charge fluctuations from remote field-free vacuum onto the ionic atmosphere. The parsing of the electrochemical potential in this way presupposes the existence of a single long-lived neutral species (the ion *plus* ionic atmosphere) and a number of short-lived charge fluctuations that perturb it. This combination of ion and ionic atmosphere has variously been called a "co-sphere" [25], a "Debye sphere" [26], or a "supermolecule" [27]. Here we adopt the name "co-sphere" as having historical priority. The identities of the species that are responsible for the charge fluctuations inside the co-sphere have not been identified with certainty, but doubtless consist of co-ions, counter ions, solvent dipoles, and various multipolar species that exist in electrolyte solutions. However, in concentrated electrolyte solutions it is most likely that the co-ions and counter ions are dominant [20].

A direct consequence of the above parsing is that the electrochemical potential of an ion and its co-sphere in an electrolyte solution is a parabolic function of its net charge fluctuations:

$$\boxed{\tilde{\mu}_{\mathrm{M^{n+}}} = \mu_{\mathrm{M^{n+}}} + \frac{(\pm Q)^2}{2C}}, \tag{8.70}$$

where $\tilde{\mu}_{\mathrm{M^{n+}}}$ is the electrochemical potential of the (ion + co-sphere) measured in joules, $\mu_{\mathrm{M^{\pm}}}$ is the chemical potential of the (ion + co-sphere) also measured in joules, C is the capacitance of the (ion + co-sphere) measured in farads, and $\pm Q$ is the net size of positive-or-negative charge fluctuations that perturb the co-sphere (coulombs). This equation is the basic building block of the Fletcher theory of electron transfer [27], in which Q plays the role of reaction co-ordinate.

As noted earlier, the Gibbs energy of a system of particles can be obtained simply by summing the electrochemical potentials of individual particles:

$$\mathrm{d}G = \sum_{i=1}^{i=n} \tilde{\mu}\,\mathrm{d}n_{\mathrm{i}}. \tag{8.71}$$

Thus, it is not difficult to construct Gibbs energy diagrams for ions and their co-spheres. In the case of electron transfer from an electrode into a solution species, the electrochemical potential of the reactants is:

$$\tilde{\mu}_{\mathrm{M^{n+}}} + \tilde{\mu}_{\mathrm{e^-}}. \tag{8.72}$$

While the electrochemical potential of the product is:

$$\tilde{\mu}_{\mathrm{M^{(n-1)+}}}. \tag{8.73}$$

8.5.16 Electron transfer (general introduction)

All electron transfer reactions are tunneling reactions. And all tunneling reactions obey the principle of conservation of energy. Thus, electrons may only travel horizontally on diagrams of Gibbs energy versus reaction co-ordinate. Accordingly, electron transfer is permitted only if the reactants and products have the same electrochemical potential (Gibbs energy).

A crude mental picture is that of "the two elevators". Imagine a multi-storey hotel having two elevators which we label A and B. The elevators move up and down at random between floors, but are on opposite sides of a foyer. A person in elevator A cannot exit the elevator, cross the foyer, and enter elevator B unless both elevators are on the same floor. So it is with electron transfer. Two nearby molecules must wait for random charge fluctuations to equalize their energies. Only then can electron transfer happen. The whole system behaves as a coincidence detector. Even though the elementary act of electron transfer is potentially very fast, it is necessary to wait for a comparatively long time for the random fluctuations of charge to bring A and B into energy alignment. This is why electron transfer is widely referred to as a "mixed" quantum-classical system. It is fundamentally a quantum tunneling effect, but one which is delayed by the necessary wait for classical charge fluctuations to equalize the energies of the reactants and products. Adding further complication, there is no guarantee that an electron will actually tunnel from A to B even when the energies are equalized. It takes time for an electron's wave function to build up on B, and the "elevator" may depart before the electron transfer is completed.

8.5.17 Johnson–Nyquist noise

Over the past century, many types of electrical noise have been identified in physical systems, including thermal noise, shot noise, flicker noise, and burst noise. However, only **thermal noise** is present at thermodynamic equilibrium.

In 1928, John Bertrand Johnson [28] and Harry Nyquist [29] showed that thermal noise in electrical circuits had a universal origin. It appeared whenever a complete circuit was formed between a reactance (a capacitance or an inductance) and a conductance. The principal manifestation of the noise was a spontaneous fluctuation of electric charge on the circuit nodes.

According to statistical thermodynamics, every reactance/conductance combination constitutes a degree of freedom of a system, and therefore (by the equipartition principle) stores $k_B T/2$ of energy on time-average. As a result, the mean square voltage across the terminals of a capacitor is $k_B T/C$ and the average stored energy is $\langle Q^2 \rangle/2C$.

In 2007, Stephen Fletcher [27] proposed that Johnson–Nyquist noise was the trigger for electron transfer, not Marcus–Pekar noise which had previously been widely accepted. The latter requires non-equilibrium fluctuations of the dielectric polarization of the medium [4, 30, 31], something that is very unlikely to occur in close-to-equilibrium systems.

8.5.18 The Molar Gibbs reorganization energy

In the theory of electron transfer, the molar Gibbs reorganization energy λ_m is the parameter that sets the energy scale for the problem. Stated formally, the molar Gibbs reorganization energy is the total amount of work required to convert the ionic atmosphere of the reactant species into the ionic atmosphere of the product species, while keeping the charge constant on the central ion. According to Marcus theory, this work equals the energy required to de-solvate the reactant. Thus, according to the Marcus theory, the molar Gibbs reorganization energy is just the negative of the molar Gibbs energy of solvation [32, 33].

Unfortunately, careful measurements of the molar Gibbs energies of solvation of univalent cations in water yield values in the range 250–475 kJ mol^{-1} [34], and these values far exceed the energies attainable by thermodynamic fluctuations, which are of the order $k_B T/2$ per particle, or 1.24 kJ mol^{-1} at 298 K. (The discrepancy is even worse for divalent and trivalent ions.) Thus, the Marcus theory presupposes the functioning of a degree of freedom which is actually "*frozen out*" at room temperature.

To overcome this difficulty, Fletcher has proposed an alternative model that requires less energy. He has suggested that the reorganization energy is actually the work required to assemble a unit positive charge on the co-sphere of an acceptor. Since the co-sphere is much larger than the ion at its center, less work is needed to charge it. Indeed, depending on the ionic strength of solution, Fletcher's reorganization energy is only a fraction of that required by the Marcus theory.

For non-interacting overlap of the molar Gibbs energy curves (the non-adiabatic case) the activation energy $\Delta\widehat{G}^*_{\mathrm{m}}$ for electron transfer is related to the molar Gibbs reorganization energy λ_{m} by the parabolic equation:

$$\Delta\widehat{G}^*_{\mathrm{m}} = \frac{(\lambda_{\mathrm{m}} + \Delta\widehat{G}^0_{\mathrm{m}})^2}{4\lambda_{\mathrm{m}}}. \tag{8.74}$$

where $\Delta\widehat{G}^*_{\mathrm{m}}$ is the molar Gibbs activation energy (kJ mol^{-1}), $\Delta\widehat{G}^0_{\mathrm{m}}$ is the difference in molar Gibbs energy between the reactant and product (kJ mol^{-1}), and λ_{m} is the molar Gibbs reorganization energy (kJ mol^{-1}).

8.5.19 The reaction co-ordinate

At thermodynamic equilibrium, the motions of particles (electrons, molecules) do not stop. Fluctuations persist in all the extensive parameters of a system's degrees of freedom. As a result of these fluctuations, the local Gibbs energy fluctuates too. As a way of simplifying this complex situation, a single **reaction co-ordinate** is commonly defined as the minimum energy pathway (path of shallowest ascent/steepest descent) through the manifold of all possible Gibbs energy curves that are generated by all possible fluctuations. The maximum energy along this pathway then defines the Gibbs energy of activation.

In general chemistry, reaction co-ordinates typically consist of bond angles, bond lengths, or partial electrical charges, or some linear combination thereof. In the case of outer-sphere electron transfer, however, the reaction co-ordinate consists solely of electrical charge.

For non-concerted reactions, beginning at thermodynamic equilibrium and ending at thermodynamic equilibrium, the reaction co-ordinate is the **extensive variable** of the single degree of freedom that takes the system to its transition state. Thus, in the case of outer-sphere electron transfer, the correct reaction co-ordinate is the net charge fluctuation Q on the co-sphere of the acceptor.

For concerted reactions, more complex co-ordinates are needed. These are widely referred to as "collective variables" or "order parameters" and they appear regularly in the literature on enzyme kinetics and phase transformations. They typically involve weighted sums of translational and rotational variables. However, complex co-ordinates of this type are not appropriate for outer-sphere electron transfer reactions, since bond-making and bond-breaking are absent.

8.5.20 The vertical energy gap

By parsing the electrochemical potential into chemical and electrical terms one finds that the reaction co-ordinate appropriate to outer-sphere electron transfer is the net charge fluctuation Q on the acceptor [20]. Accordingly, the Gibbs energy curves of ions in solution are quadratic functions of this charge.

The **vertical energy gap** is the name given to the difference between two Gibbs energy curves having the same configuration of ionic atmosphere but different unitary charges on their central ions. This parameter was introduced by Warshel in 1982 and has subsequently been adopted by many molecular simulators as a proxy for the reaction co-ordinate [35]. It was also endorsed by Marcus in 2000 as equivalent to his own reaction co-ordinate [33]. Let us therefore connect it with the present work.

We consider the Gibbs energy curves of reactants and products as a function of the reaction co-ordinate Q. Let us label these G_1 and G_2. Since the fluctuations of Q constitute quadratic degree of freedom of the system, we can write:

$$G_1 = \frac{(Q-a)^2}{2C_1} + b \tag{8.75}$$

$$G_2 = \frac{(Q-c)^2}{2C_2} + d, \tag{8.76}$$

where a, b, c and d are constants. For independent Gibbs energy curves (i.e., non-adiabatic kinetics) the energy gap is simply the geometric difference:

$$G_1 - G_2 = \left(\frac{1}{2C_1} - \frac{1}{2C_2}\right) Q^2 + \left(\frac{c}{C_2} - \frac{a}{C_1}\right) Q + \left(\frac{a^2}{2C_1} - \frac{c^2}{2C_2} + b - d\right). \tag{8.77}$$

It is immediately apparent that this function is *not* a linear transformation of the true reaction co-ordinate Q. There is a term in Q^2 which is always finite unless:

$$\left(\frac{1}{2C_1} - \frac{1}{2C_2}\right) = 0. \tag{8.78}$$

Accordingly, we deduce that the vertical energy gap $G_1 - G_2$ is not, in general, a reaction co-ordinate of the system. It only behaves as such if the Gibbs energy curves of reactants and products have the same curvature. Although this situation can be contrived in some special cases (such as self-exchange reactions in solution) it is unlikely to occur by chance in open systems. Thus, the use of the vertical energy gap as a reaction co-ordinate is unsafe. A better strategy is to estimate Q directly.

Reaching a consensus on the identification of the reaction co-ordinate for electron transfer continues to be difficult. Even today, the IUPAC "*Gold Book*" wrongly defines a reaction co-ordinate as a "geometric parameter" [36]. There is also a tendency among molecular simulators to label all reaction co-ordinates as q, leading to unnecessary confusion between Cartesian distance and electrical charge. And even leading experts have made some very dubious choices. Tibor Erdey-Grúz and Max Volmer believed that the reaction co-ordinate was the fractional distance across the double layer [37]. Noel Hush thought that the reaction co-ordinate was the extent of partial electron transfer between reactants and products [38], and Rudolph Marcus has consistently maintained that the reaction coordinate is derivable from a multidimensional set of Cartesian co-ordinates [33, 39]. In fact, none of these viewpoints is in full accordance with reality. As noted above, the correct reaction co-ordinate for outer-sphere electron transfer is the net charge Q on the ionic atmosphere of the acceptor, created by equilibrium noise.

8.5.21 Permittivity of solutions

The absolute permittivity of a physical substance is the total capacitance between the faces of a unit cube of that substance at constant temperature and pressure. Its units are farads per meter, F m^{-1}. The relative permittivity is the ratio of the absolute permittivity to that of empty space. Thus:

$$\varepsilon_r = \varepsilon / \varepsilon_0, \tag{8.79}$$

where ε_r is the relative permittivity (dielectric constant), ε is the absolute permittivity of the physical substance (F m^{-1}) and ε_0 is the permittivity of empty space (8.854×10^{-12} F m^{-1}). Solvents that have high relative permittivities ($\varepsilon_r \geq 15$) are called polar solvents.

Because ε_0 is the lowest possible permittivity we immediately deduce that:

$$\varepsilon_r \geq 1. \tag{8.80}$$

Experimentally, it is commonly observed that the relative permittivities of physical substances are frequency-dependent. This is because the atomic nuclei of molecules have such a large inertia that they cannot respond rapidly to high frequency perturbations.

8.6 Electrolytes and non-electrolytes

Solutions can be classified into two types: electrolyte solutions and non-electrolyte solutions. Electrolytes are soluble compounds that supply charge carriers for conduction. They do this by dissociating into mobile charged species (anions and cations). An example of an electrolyte solution is common salt dissolved in water:

$$\text{NaCl} \rightleftharpoons \text{Na}^+(\text{aq}) + \text{Cl}^-(\text{aq}). \tag{8.81}$$

By contrast, non-electrolytes are soluble compounds that do *not* supply charge carriers for conduction. They are unable to dissociate into mobile charged species.

Based only on this brief comparison, it might be thought that electrolyte solutions carry electrical current while non-electrolyte solutions do not. However, non-electrolyte solutions can also carry electric current if the conditions are right.

Suppose the non-electrolyte solution contains dipolar molecules. These may be from the non-electrolyte or from the solvent itself. In either case, the dipolar molecules have a net charge of zero and a small internal separation of polarity (i.e., they possess a finite dipole moment). When an external voltage is applied across a solution of these species, the dipolar molecules fail to sustain a **conduction current** due to their overall electro-neutrality. But they nevertheless tend to align with the direction of the applied electric field, a phenomenon which manifests externally as a **displacement current**. In effect, many microscopic displacements of charge add up to one macroscopic displacement of charge. By means of this mechanism, electric currents can transiently circulate through non-electrolyte solutions despite the absence of ions. A similar phenomenon is observed in solid-state dielectrics.

The theory of displacement currents was first developed by James Clerk Maxwell [40] in the mid-nineteenth century. Maxwell added displacement current to the conduction current of charge carriers in Ampère's Circuit Law. This allowed him to "save" the definition of electric current as a rate of flow of charge in a complete circuit. (For Maxwell, as for us, simply moving an electric charge from point A to point B does not constitute a current. A complete circuit is required.) The size of the displacement current passing between the plates of a capacitor is equal to the size of the conduction current in the wires leading up to the capacitor (Kirchhoff's Current Law).

The behavior of dipolar molecules inside non-electrolyte solutions was the subject of a classical study by Peter Debye, who identified the phenomenon now known as **Debye relaxation** [41]. Debye showed that the complex permittivity $\varepsilon(\omega)$ has the frequency-dependent form

$$\varepsilon(\omega) = \varepsilon_1 + \frac{\varepsilon_2}{1 + j\omega\tau}, \tag{8.82}$$

where ω is the angular frequency, ε_1 is the high-frequency permittivity (F m^{-1}), and $(\varepsilon_1 + \varepsilon_2)$ is the low-frequency permittivity (F m^{-1}). The parameter τ is the Debye relaxation time, here assumed to be the same for all dipoles. (In a more rigorous model, each dipole would have its own relaxation time, depending on its local environment.) Expressed as relative permittivities, it is found for water that

$$\varepsilon_1 \approx 5.4 \tag{8.83}$$

$$(\varepsilon_1 + \varepsilon_2) \approx 78.4 \tag{8.84}$$

and

$$\tau \approx 8.3 \text{ ps} \tag{8.85}$$

at 298.14 K. Although the relaxation process at 8.3 ps dominates the dielectric spectrum, there still remain some disagreements about its mechanism. However, molecular dynamics simulations generally ascribe this process to the re-orientation of water molecules embedded in a local cluster of hydrogen bonds [42].

Because the capacitance $C = \varepsilon A/d$, the Debye equation may also be written

$$C(\omega) = C_1 + \frac{C_2}{1 + j\omega R_2 C_2} \tag{8.86}$$

where:

$$R_2 C_2 = \tau_2. \tag{8.87}$$

This model corresponds exactly to the equivalent circuit shown in Fig. 8.3.

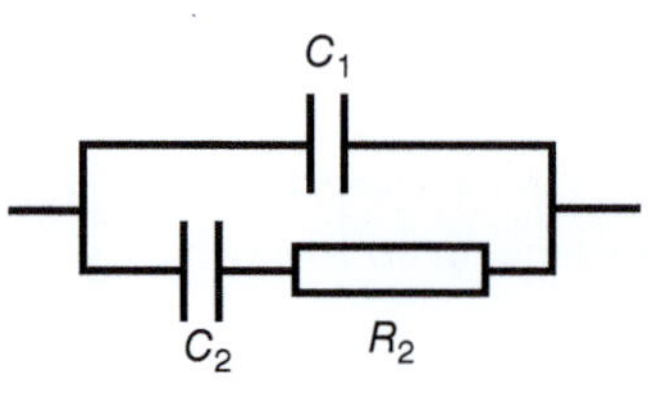

Figure 8.3 Equivalent circuit of a non-electrolyte solution.

8.6.1 Equivalent circuit of a non-electrolyte solution

Here the C_1 branch of the circuit is a model for the polarization of electron clouds (plus the polarization of empty space), while the R_2C_2 branch of the circuit is a model for the polarization of solvent dipoles. Since R_2C_2 is approximately 8.3 ps in water, it follows that the solvent dipoles can oscillate in synchrony with the applied electric field at all angular frequencies from zero hertz to 1 GHz.

In the theory of electron transfer, a key parameter is the amount of work (w_{NE}) needed to create charge fluctuations of magnitude $+Q$ or $-Q$ in the non-electrolyte (*NE*) solution, starting from a state of electro-neutrality. Unlike the permanent charges on ions, transient charges are not supplied externally from the remote field-free vacuum, but instead arise internally from the ionization of neutral species within the solution itself. In other words, the charge fluctuations are caused by **Johnson–Nyquist noise**. Figure 8.3 illustrates the fact that the noise current has only one route through the system, which is the closed circuit formed by R_2, C_2, and C_1. Thus, with regard to Johnson–Nyquist noise, the capacitors C_1 and C_2 act in series, and the total capacitance seen by the resistor R_2 is:

$$C_T = \left(\frac{1}{C_1} + \frac{1}{C_2}\right)^{-1}. \tag{8.88}$$

As always when capacitors are connected in series, the total capacitance is smaller than any individual capacitance. However, in the case that $C_2 \gg C_1$ (as in water) we have:

$$C_T \approx C_1. \tag{8.89}$$

In this case we find that the work w_{NE} needed to create a charge fluctuation of magnitude $+Q$ or $-Q$ in the non-electrolyte (NE) solution is:

$$w_{NE} \approx \frac{Q^2}{2C_1}. \tag{8.90}$$

8.6.2 Equivalent circuit of an electrolyte solution

The equivalent circuit of an electrolyte solution is shown in Fig. 8.4. It is the same as the equivalent circuit of a non-electrolyte solution, except that an additional branch of resistance R_3 (ohms) is included to account for electrolyte conduction. Equivalent circuits of this type are widely used in the field of dielectric spectroscopy [43].

Figure 8.4 Equivalent circuit of an electrolyte solution.

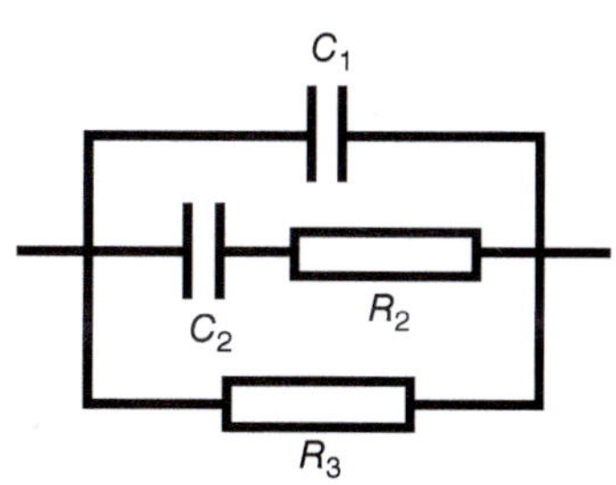

The complex permittivity $\varepsilon(\omega)$ as a function of angular frequency ω is

$$\varepsilon(\omega) = \varepsilon_1 + \frac{\varepsilon_2}{1 + j\omega\tau} + \frac{\sigma}{j\omega}, \tag{8.91}$$

where ε_1 is the high-frequency permittivity (F m^{-1}), and $(\varepsilon_1 + \varepsilon_2)$ is the low-frequency permittivity (F m^{-1}), the parameter τ is the relaxation time, and σ is the conductivity of the third branch in siemens per meter (S/m). Or, in terms of capacitance,

$$C(\omega) = C_1 + \frac{C_2}{1 + j\omega\tau} + \frac{1}{j\omega R_3}, \tag{8.92}$$

where R_3 is the resistance of the third branch in ohms (Ω).

As in the case of non-electrolyte (NE) solutions, a key parameter in electrolyte (E) solutions is the amount of work needed to create a charge fluctuation of magnitude $+Q$ or $-Q$, starting from an initial state of electro-neutrality. As in the non-electrolyte case, the charge fluctuations arise spontaneously from Johnson–Nyquist noise. However, this time, there are two resistances instead of one.

The capacitors C_1 and C_2 behave additively at all angular frequencies below 1 GHz. As a result, the noise current is constrained to flow through the resistance R_3 and into the **parallel** combination of capacitors C_1 and C_2. Thus, as regards Johnson–Nyquist noise, the total capacitance seen by the resistor R_3 is

$$C_T \approx (C_1 + C_2) \tag{8.93}$$

and hence the work (w_E) needed to create a charge fluctuation of magnitude $+Q$ or $-Q$ in the electrolyte (E) solution is:

$$w_E \approx \frac{Q^2}{2(C_1 + C_2)}. \tag{8.94}$$

Finally, comparison of w_E with w_{NE} shows that

$$w_E \approx w_{NE} \left(\frac{\varepsilon_1}{\varepsilon_1 + \varepsilon_2} \right). \tag{8.95}$$

Evidently, it requires less work to form a charge fluctuation in an electrolyte solution than it does in a non-electrolyte solution.

8.6.3 Probability of an electron jump

In order for an elementary act of electron transfer to occur with finite probability there must be some overlap between the frontier molecular orbitals of the two species concerned (i.e., between the donor species D and the acceptor species A). The presence of this overlap introduces a complication into electron transfer theory known as **electronic coupling**. During the lifetime of the transition state, the transferring electron necessarily experiences the electric fields of *two* species (the donor and the acceptor) not just the electric field of one species alone. As a result of this dual interaction, the potential energy of the transition state is less than it would otherwise be. The magnitude of this stabilizing effect is called **the electronic stabilization energy** (or "electronic coupling energy") and is denoted H_{DA}. It plays an important role in the theory. It also plays a decisive role in many electron transport chains in living cells.

The electronic stabilization energy is of interest to theorists because it is one of only two parameters that determine whether the elementary act of electron transfer is "quantum adiabatic" or not. (The other factor is the transition state lifetime.) A large electronic stabilization energy and a large transition state lifetime both favor quantum adiabaticity, i.e., a situation in which there is rapid and complete equilibration of the electronic wavefunction while the system traverses the reaction co-ordinate. Conversely, a small electronic stabilization energy and a small transition state lifetime both favor quantum non-adiabaticity, i.e., a situation in which there is non-equilibrium of the electronic wavefunction as the system traverses the reaction co-ordinate. In the latter case the rate of the elementary act of electron transfer must be described by an approximation technique, such as time-dependent perturbation theory [44] or Landau–Zener theory [45–48].

In real-world systems it is important to establish the existence (or non-existence) of quantum adiabaticity because it has a large effect on the transition probability of the elementary act of electron transfer. Let p be the transition probability that an electron will tunnel from the donor to the acceptor during the lifetime of the transition state. The only way that ($p = 1$) can occur is for the elementary act of electron transfer to be quantum adiabatic. Conversely, the only way that ($p < 1$) can occur is for the elementary act of electron transfer to be quantum non-adiabatic.

The physical meaning of quantum adiabaticity is now clear. If ($p = 1$) then an electron tunnels every time the transition state is reached. By contrast, if ($p < 1$) then an electron does not tunnel every time the transition state is reached. *In extremis*, the system may make millions of attempts to tunnel before an elementary act of electron transfer actually occurs. This is the case for spin-forbidden reactions, for example.

In solution-phase electrochemistry, electron transfer reactions are most commonly carried out in polar solvents. In such media the overlapping of frontier orbitals between donors and acceptors is generally weak because solvation shells form around all the solution components. As a result, the donor and acceptor species are physically remote from each other, and the elementary act of electron transfer is "quantum non-adiabatic". By contrast, quantum adiabatic electron transfer is rarer. In order for the latter to occur, the donors and acceptors must be organized into "contact ion pairs", or tethered by covalent bonds, in order to generate the necessary orbital overlap.

Similar reasoning applies to interfacial electron transfer. The electrical double layer at the electrode surface acts like a thick solvation shell, and prevents electrode orbitals from overlapping reactant orbitals. Thus, interfacial electron transfer is predominantly quantum non-adiabatic (i.e., long range). Only reactants that are "specifically adsorbed", or tethered to the electrode surface by covalent bonds, can achieve strong orbital overlap and thus achieve adiabatic status.

8.6.4 The Klopman–Salem equation

In general, the total energy change of the reactants as they enter the transition state is described by the **Klopman–Salem equation**, which derives from molecular orbital theory [49–51]. The Klopman–Salem equation has three principal terms. The first term describes the energy *increase* caused by the closed-shell repulsion between the occupied molecular orbitals of the donor and acceptor species. The second term describes the energy *decrease* caused by the overlap of occupied and unoccupied molecular orbitals (H_{DA}). The third term describes the energy *increase* associated with charge fluctuations in the ionic atmosphere of the reactants. The latter is the most important term and accounts for the majority of the activation energy of electron transfer reactions. The energy of the charge fluctuations is temporarily "loaned" from the surrounding heat bath.

$$\Delta E = \pm \sum_{ab} (n_a + n_b)\beta_{ab} S_{ab} + \left(\sum_{r}^{occ} \sum_{s}^{un} \pm \sum_{s}^{occ} \sum_{r}^{un} \frac{2\left(\sum_{ab} c_{ra} c_{sb} \beta_{ab}\right)^2}{E_r \pm E_s} \right) + \sum_{k} \frac{Q_D Q_A}{4\pi\varepsilon_0 \varepsilon_r r_{DA}}. \tag{8.96}$$

Here, n_a and n_b are the electron populations in orbitals a and b; β and S are resonance and overlap integrals; c_{ra} is the coefficient of atomic orbital a in molecular orbital r; E_r is the energy of molecular orbital r; Q_D and Q_A are the total charges on species D and A; $\varepsilon_0\varepsilon_r$ is the local permittivity; and r_{DA} is the distance between the species D and A.

Three important conclusions can be drawn from this equation: (i) the occupied orbitals of one species repel the occupied orbitals of the other, (ii) the highest occupied molecular orbitals (HOMOs) of one species attract the lowest unoccupied molecular orbitals (LUMOs) of the other, and (iii) the positive charge on one species attracts the negative charge on the other (and *vice versa*).

8.6.5 Electrode kinetics

Electrode kinetics typically involves two phases, a solid and a liquid, where the solid is a metal and the liquid is a solution of a conducting salt. Unlike heterogeneous kinetics, however, both the metal and solution form part of a complete electrical circuit. As current flows around the circuit, electron transfer occurs from the frontier orbitals of the electrode to the frontier orbitals of the reactant, to create a (reactant + electron) complex. This complex then rearranges into product. Electron transfer reactions occur so readily because electrons don't jump over energy barriers – they tunnel through them. They are able to do this because they have a very low mass.

All conventional electron transfer reactions are tunneling reactions. Moreover, all tunneling reactions (after relaxation of the perturbed wave function) obey the principle of conservation of energy. Thus any electron that tunnels into a reactant molecule takes with it the same energy that it had inside the electrode.

From this analysis we find that the total energy of the (reactant + electron) complex in solution is determined, in part, by the electrode potential of the electrode, because the free energy of the electron in the electrode is carried into the (reactant + electron) complex by the tunneling process. This is the remarkable feature of electrochemical reactions. It means that by changing the electrode potential of an electrode we can change the energy of all the reactants within electron tunneling distance of the electrode surface. As we shall soon see, this gives us control over the rates of electrochemical reactions, and therefore some control over the nature of the reaction products.

8.6.6 Homogeneous kinetics, first order

Consider two substances A and B, whose amount concentrations per unit volume of solution are c_A (mol m^{-3}) and c_B (mol m^{-3}), and which convert reciprocally via the reaction

$$\mathrm{A} \underset{k'_b}{\overset{k'_f}{\rightleftharpoons}} \mathrm{B}\,. \tag{8.97}$$

In this first-order reaction, the rate constants k'_f and k'_b have units of s^{-1}. (Indeed, the rate constants k'_f and k'_b are simply the reciprocals of the mean lifetimes of the species A and B, respectively.)

(*Note.* In a more general treatment, each rate constant would have units of (mol m^{-3})$^{1-n}$s^{-1} where n is the reaction order, but here we confine ourselves to the case $n = 1$.)

Neglecting any difference between concentration and activity, the rate of the forward reaction per unit volume of solution v'_f (mol m^{-3} s^{-1}) may be written

$$v'_f = k'_f c_A \tag{8.98}$$

and similarly the rate of the backward reaction v'_b (mol m^{-3} s^{-1}) may be written

$$v'_b = k'_b c_B. \tag{8.99}$$

The net conversion rate of the species A and B is therefore

$$\begin{aligned} v'_{net} &= v'_f - v'_b \\ &= k'_f c_A - k'_b c_B. \end{aligned} \tag{8.100}$$

At equilibrium the net conversion rate is zero, hence

$$\frac{c_B}{c_A} = \frac{k'_f}{k'_b} = K' \tag{8.101}$$

where the constant K' is called **the equilibrium constant**. Note that this is precisely the same result as that predicted by thermodynamics, namely a constant concentration ratio at equilibrium.

8.6.7 Homogeneous kinetics, second order

Consider two substances A and B, whose amount concentrations per unit volume of solution are c_A (mol m^{-3}) and c_B (mol m^{-3}), and which convert reciprocally via a second-order reaction

$$\mathrm{A} + \mathrm{B} \underset{k''_b}{\overset{k''_f}{\rightleftharpoons}} \mathrm{B} + \mathrm{A}\,. \tag{8.102}$$

This is commonly referred to as a *self-exchange reaction*. During a self-exchange reaction, an external observer sees no net change in the concentrations of A and B, and the rate constants k''_f and k''_b have units of (mol m^{-3})$^{-1}$s^{-1}. By symmetry,

$$k''_f = k''_b \tag{8.103}$$

and the Gibbs energy change for the reaction is

$$\Delta G_0 = G_{\text{products}} - G_{\text{reactants}} = 0. \tag{8.104}$$

Neglecting any difference between concentration and activity, the rate of the forward reaction per unit volume of solution v_f'' (mol m^{-3} s^{-1}) may be written

$$v_f'' = k_f'' c_A c_B \tag{8.105}$$

and similarly the rate of the backward reaction per unit volume of solution v_b'' (mol m^{-3} s^{-1}) may be written

$$v_b'' = k_b'' c_A c_B. \tag{8.106}$$

The net conversion rate of A to B, is therefore

$$\begin{aligned} v_{\text{net}}'' &= v_f'' - v_b'' \\ &= k_f'' c_A c_B - k_b'' c_A c_B \\ &= 0. \end{aligned} \tag{8.107}$$

Note that this system is always trapped in its initial state. Every time A reacts with B, it simply regenerates B and A, and so the equilibrium constant is always 1.

$$\frac{k_f''}{k_b''} = K'' = 1. \tag{8.108}$$

8.6.8 Homogeneous versus heterogeneous kinetics

In homogeneous kinetics, it is generally assumed that reactions occur with complete spatial uniformity. Given this assumption, the rate of reaction may be considered as the rate of change of the amount of substance A (mol) in any small element of volume dV (m^3), where dV may be located anywhere in the solution. In such circumstances, it becomes easy to evaluate the *total* rate of change of the amount of A by integrating (summing) the rate of the forward reaction per unit volume of solution over all the small elements of volume dV (m^3) in the reaction vessel. For example, for a homogeneous second-order reaction

$$\text{Total rate} = \int_0^V v_f'' \cdot dV \qquad (\text{mol s}^{-1})\,. \tag{8.109}$$

In heterogeneous kinetics, the analogous integration (summation) is more complicated. Reactions take place with spatial non-uniformity —they are confined to places where interfaces exist— and a different approach is called for. The approach favored by electrochemists is to focus on *the surface rate* of the forward reaction. The latter is defined as the total rate of change of the amount of A (mol s^{-1}) in the reaction vessel, but normalized by the total surface area of the electrode S (m^2). Henceforth we shall indicate this parameter by the symbol v_f (mol m^{-2} s^{-1}), and refer to it as *the rate of the forward reaction per unit area of interface.* The total rate of change of the amount of A is then given by the formula

$$\text{Total rate} = \int_0^S v_f.dS \qquad (\text{mol s}^{-1}). \tag{8.110}$$

Comparing the heterogeneous approach (un-primed symbols) with the homogeneous approach (primed symbols), we immediately observe that

$$\int_0^S v_f \cdot dS \quad \text{is the 2D analog of} \quad \int_0^V v_f''.dV. \tag{8.111}$$

This comparison emphasizes that the total rate of reaction in a reaction vessel is normalized in different ways in different approaches. In homogeneous kinetics, the total rate is normalized by the bulk volume. In heterogeneous kinetics, the total rate is normalized by the interfacial area.

8.6.9 Tunneling layer approximation

Although the relation between the *rates* of homogeneous and heterogeneous electron transfer reactions is straightforward, the relation between the *rate constants* is a little more complicated. This is particularly true for electrochemical reactions. To simplify the analysis, we here introduce a **tunneling layer approximation**, in which a finite layer of solution, of thickness x', is assumed to contain all the molecules that are within electron tunneling distance of the electrode surface.

Inside the tunneling layer, the concentrations of substances A and B are assumed to be uniform and equal to their bulk concentrations, i.e., neither positive adsorption nor negative adsorption is assumed to occur. The electron tunneling probability is also assumed to be spatially uniform inside the layer. Although this is clearly not true —the electron tunneling probability decays approximately exponentially with distance from the electrode surface— nevertheless the error introduced by this approximation will be negligible on the order-of-magnitude scale that we are considering.

Without loss of generality, we focus on the case of second-order homogeneous kinetics, i.e., the case of a "self-exchange" reaction, because it is a benchmark reaction scheme in the inorganic chemistry literature. (The simpler case of first-order homogeneous kinetics can readily be derived from the second-order case by replacing $k_f'' c_B$ with k_f'.) Evidently,

$$\begin{aligned}\text{Total rate inside tunneling layer} &= \int_0^{V_{TL}} v_f'' \cdot dV_{TL} \\ &= k_f'' c_A c_B V_{TL} \quad (\text{mol s}^{-1}).\end{aligned} \tag{8.112}$$

Here v_f'' (mol m^{-3} s^{-1}) is the rate of the forward reaction per unit volume of solution, and V_{TL} is the volume of the tunneling layer.

It follows that the rate of a second-order homogeneous reaction inside the tunneling layer, *per unit area of interface*, is

$$\frac{k_f'' c_A c_B V_{TL}}{A_{TL}} = k_f'' c_A c_B x' \quad (\text{mol m}^{-2}\,\text{s}^{-1}), \tag{8.113}$$

where A_{TL} is the area of the interface. This rate may now be compared directly with the rate of the heterogeneous reaction inside the tunneling layer, per unit area of interface, which is:

$$v_f = k_f c_A. \tag{8.114}$$

Comparing these two rates reveals the **equivalence relation**:

$$\boxed{k_f \Longleftrightarrow k''_f c_B x'}. \tag{8.115}$$

This relation indicates the value of the *heterogeneous* rate constant k_f (m s^{-1}) that is needed to generate the same total rate of reaction as that provided by a second-order *homogeneous* rate constant of value k''_f ((mol $m^{-3})^{-1}s^{-1}$).

It is instructive to note that, in the case of homogeneous kinetics, the amount concentrations c_A (mol m^{-3}) and c_B (mol m^{-3}) both refer to amount concentrations of *soluble* species per unit volume of solution; whereas, in the case of heterogeneous kinetics, only c_A (mol m^{-3}) does so. In the heterogeneous case c_B (mol m^{-3}) refers to an amount concentration of *insoluble* species (the surface sites where electron tunneling is occurring) calculated per unit volume of tunneling layer. It follows that, unlike the homogeneous rate constant for electrochemical reaction k''_f, the heterogeneous rate constant for electrochemical reaction k_f has the amount concentration of surface sites per unit volume of tunneling layer, and the mean tunneling layer thickness x', *already incorporated in its value*. This also explains why k_f and k''_f have such different units.

In an alternative (but equivalent) analysis, the product $c_B x'$ may be interpreted as the *surface concentration* of sites c_S at which electron transfer occurs:

$$c_B x' = c_S \text{ mol m}^{-2}. \tag{8.116}$$

In this case, k_f can be viewed as a *pseudo* first-order rate constant, in which the amount concentration of surface sites is assumed to remain un-depleted on the time scale of experiments.

8.6.10 The back of the envelope

In order to develop some physical insight into actual values of k_f let us make some back-of-the-envelope estimates of k''_f, c_B, and x'. The maximum observable value of k''_f is determined by the onset of diffusion control. In aqueous solutions $(k''_f)_{max}$ is typically of the order 10^6 (mol $m^{-3})^{-1}s^{-1}$. The amount concentration of surface sites c_B (mol m^{-3}) inside the electron tunneling distance of a transition metal electrode, is about 10^3 mol m^{-3}. The maximum thickness of the electron tunneling layer is about 1.4×10^{-9} m, here crudely rounded to 10^{-9} m.

[*Note.* Because the electron tunneling distance depends only on the Planck constant, the mass of the electron, and the height of the tunneling barrier (~1 eV in aqueous systems), the maximum thickness of ~1.4×10^{-9} m is quite robust, and unlikely to be more than an order of magnitude in error in any "normal" electrochemical system.]

Combining the above three estimates we get:

$$\begin{aligned}(k_f)_{max} &\approx (k''_f)_{max} \times c_B \times x' \\ &\approx 10^6 \times 10^3 \times 10^{-9} \\ &\approx 1 \text{ m s}^{-1}.\end{aligned} \tag{8.117}$$

This represents an order-of-magnitude estimate for the maximum observable rate constant $(k_f)_{max}$ for a heterogeneous electron transfer reaction.

The same result can be obtained by noting that

$$(k_\mathrm{f})_\mathrm{max} \approx \frac{D_\mathrm{A}}{(\delta)_\mathrm{min}}, \tag{8.118}$$

where D_A is the diffusion coefficient of substance A, and δ is the diffusion layer thickness. For small molecules in aqueous solution D_A is of the order 10^{-9} m^2 s^{-1}, and the minimum value of δ cannot be less than the mean thickness of the electron tunneling layer, namely 10^{-9} m. Hence:

$$(k_\mathrm{f})_\mathrm{max} \approx \frac{10^{-9}}{10^{-9}} \approx 1\ \mathrm{m\,s^{-1}}. \tag{8.119}$$

Finally, it may also be noted that the surface concentration of sites at which electron transfer occurs is

$$\begin{aligned} c_\mathrm{S} &\approx c_\mathrm{B} x' \\ &\approx 10^3 \times 10^{-9} \\ &\approx 10^{-6}\ \mathrm{mol\,m^{-2}} \\ &\approx 10^{-10}\ \mathrm{mol\,cm^{-2}}. \end{aligned} \tag{8.120}$$

If we now multiply this quantity by the Avogadro constant (~10^{24} sites mol^{-1}) then we obtain the following crude estimate of the *surface site density:*

$$N_\mathrm{S} \approx 10^{18}\ \mathrm{sites\ m^{-2}} \tag{8.121}$$

$$\approx 1\ \mathrm{site\ nm^{-2}}. \tag{8.122}$$

More precise estimates have been made by crystallography, ion adsorption, and scanning tunneling microscopy, and these typically yield results 1–15 times larger than this number. For comparison, the surface density of water molecules is about 10 molecules nm^{-2}. However, water is a small molecule when compared with most other molecules. Its diameter is only ~0.277 nm. So a surface site density of 1 site nm^{-2} is probably a realistic figure for large molecules.

From our earlier consideration of surface work function, we know that the magnitude of the outer potential ϕ_1 at a metal surface depends sensitively on many hard-to-control local factors, including crystal orientation, alloy composition, surface preparation, etc. Thus, on a solid electrode surface, there is likely to be a distribution of chemical potential, leading to the concept of active sites, i.e., some sites that have some special activity with respect to a particular reaction. However, the involvement of these sites requires a short-range chemical interaction (an adsorption process) to be effective.

8.6.11 The total rate constant of an electron transfer process

The total rate constant of an electron transfer process is frequently a composite of several elementary rate constants. To illustrate why this is so, let us analyze the situation where an elementary act of electron transfer is bounded by a pre-equilibrium and a post-equilibrium. In between, the rate constant k_2 of the elementary act is assumed to be pseudo-first order because it includes the constant concentration of surface sites at which electron transfer occurs. These sites are assumed to be identical.

$$\mathrm{X} \underset{k_{-1}}{\overset{k_1}{\rightleftharpoons}} \mathrm{X}^* \overset{k_2}{\rightarrow} \mathrm{Y}^* \underset{k_{-3}}{\overset{k_3}{\rightleftharpoons}} \mathrm{Y}. \tag{8.123}$$

The preceding and following equilibria might be, for example, fluctuations of charge in the ionic atmospheres of the reactant species [X] and product species [Y]. As pointed out by Fletcher [20], fluctuations of this type are needed to equalize the free energies of the intermediates [X^*] and [Y^*]. Also, we assume that the follow-up reaction k_3 is very fast, so that at all times

$$k_3 \gg (k_1, k_2). \tag{8.124}$$

The model now describes an electron transfer process in which a small but finite population of intermediates [X^*] can decay by two different mechanisms; by *backward* thermal decomposition (k_{-1}) and by *forward* electron transfer (k_2). This corresponds very closely to reality. The revised reaction scheme is

$$\mathrm{X} \underset{k_{-1}}{\overset{k_1}{\rightleftharpoons}} \mathrm{X}^* \overset{k_2}{\rightarrow} \mathrm{Y}. \tag{8.125}$$

For which the mean lifetime of the intermediate species [X^*] is simply

$$\langle \tau \rangle = \frac{1}{k_{-1} + k_2}. \tag{8.126}$$

An exact solution for the revised reaction scheme does exist. It can be derived using the eigenvalue method. However, for present purposes, it is convenient to remove solutions that are of no practical significance, in order that we remain focused on important special cases. One such case is the *pre-equilibrium approximation* [52]; another is the *steady state approximation* [53]. In the *pre-equilibrium approximation*, it is assumed that $(k_1, k_{-1}) \gg k_2$. One readily obtains

$$k_{\text{total}} \approx \frac{k_1 k_2}{k_{-1}} \approx K k_2, \tag{8.127}$$

where K is the equilibrium constant between [X] and [X^*]. This case is sometimes referred to as the "*non-adiabatic*" case. This terminology refers to the fact that the elementary electron transfer step fails to "*keep up*" with the arrival of the system at the transition state. Indeed, the system may arrive at the transition state many millions of times before electron transfer finally occurs.

By contrast, in the *steady state approximation*, it is assumed that $(k_1, k_{-1}) \ll k_2$. In this case, the intermediate [X^*] undergoes fast electron transfer, and equilibrium is never achieved. Instead one obtains

$$k_{\text{total}} \approx k_1 \tag{8.128}$$

and the intermediate disappears as rapidly as it is formed with rate constant k_1. This case is sometimes known as the technical literature as the "adiabatic" case. This terminology refers to the fact that the elementary electron transfer step occurs every time the transition state is reached. In other words, the electron transfer step "keeps up" with the arrival of the system at the transition state.

The fact that an electron is thousands of times lighter than an atom or molecule suggests that the adiabatic case should be the most widely observed in nature. However, the non-adiabatic case is important in cases of long-range tunneling and in spin-forbidden transitions.

In electrochemical terminology, the non-adiabatic case corresponds to a $C\hat{E}$ reaction (fast chemical equilibrium followed by slow electron transfer, while the adiabatic case corresponds to a $\hat{C}E$ reaction (slow chemical equilibrium followed by fast electron transfer). In the latter case, the activation of the electron transfer step is purely thermal (non-quantum).

8.7 Heterogeneous electron transfer

In the case of electron transfer across a phase boundary (e.g., electron transfer from a metal electrode to an acceptor species in solution), the law of conservation of energy dictates that the energy of the transferring electron must be added into that of the acceptor species, such that the sum equals the energy of the product species. At constant temperature and pressure the energy of the transferring electron is just its Gibbs energy.

Let us indicate by a **circumflex accent** those work terms which have the energy of the tunneling electron already added into them. Thus, for the Gibbs energy of the reactants we have:

$$\hat{G}_{\text{reactant}} = G_{\text{reactant}} - eE, \tag{8.129}$$

where e is the unit positive charge and E is the electrode potential (on the vacuum scale) of the tunneling electron. The term $-eE$ is the Gibbs energy of the tunneling electron.

For the conversion of reactant to product, the overall change in Gibbs energy is:

$$\begin{aligned}\Delta\hat{G}^0 &= \hat{G}_{\text{product}} - \hat{G}_{\text{reactant}} \\ &= \Delta G^0 + eE.\end{aligned} \tag{8.130}$$

In the "normal" region of electron transfer, for a metal electrode, Fermi–Dirac statistics compel the electron to tunnel from an energy level within $\pm k_B T$ of the Fermi energy, implying:

$$eE \approx eE_F. \tag{8.131}$$

Thus, for the elementary step of a heterogeneous electron transfer reaction from a metal electrode to an acceptor species in solution, we can use the Golden Rule directly:

$$k_{\text{et}} = \frac{2\pi}{\hbar}|H_{\text{DA}}|^2 \frac{1}{\sqrt{4\pi\lambda k_B T}} \exp\left(\frac{-(\lambda + \Delta G^0 + eE_F)^2}{4\lambda k_B T}\right). \tag{8.132}$$

Here λ is the reorganization energy and E_F is the Fermi potential of the electrons inside the metal electrode. This equation (with a slightly different notation) was first published by Revaz Dogonadze and Yurii Chizmadzhev in 1962 [54].

Converting to molar quantities:

$$\Delta\widehat{G}_m^0 = \Delta G_m^0 + FE \tag{8.133}$$

then yields:

$$\boxed{k_{\mathrm{et}} = \frac{2\pi}{\hbar}|H_{\mathrm{DA}}|^2 \frac{N_{\mathrm{A}}}{\sqrt{4\pi\lambda_{\mathrm{m}}RT}} \exp\left(\frac{-(\lambda_{\mathrm{m}} + \Delta G_{\mathrm{m}}^0 + FE_{\mathrm{F}})^2}{4\lambda_{\mathrm{m}}RT}\right)}. \tag{8.134}$$

Here k_{et} is the rate constant for the elementary act of electron transfer, $\hbar$ is the reduced Planck constant, H_{DA} is the electronic coupling matrix element between a single electron donor and a single electron acceptor, N_{A} is the Avogadro constant, λ_{m} is the molar Gibbs reorganization energy, ΔG_{m}^0 is the difference in molar Gibbs energy between the reactant and the product, and $-FE_{\mathrm{F}}$ is the molar Gibbs energy of the electron that tunnels from the Fermi level of the metal electrode into the acceptor. The parameter H_{DA} is complex, but may be considered (roughly) as a measure of orbital overlap.

The relative positions of the molar Gibbs energy curves are captured by the dimensionless parameter γ:

$$\gamma = \frac{1}{2}\left(1 - \frac{F\eta}{\lambda_{\mathrm{m}}}\right) . \tag{8.135}$$

This **gamma parameter** is simply the fractional distance along the reaction co-ordinate at which electron transfer occurs, at constant temperature and pressure. Thus, it indicates whether the transition state is "early" ($\gamma < 0.5$) or "late" ($\gamma > 0.5$) (Fig. 8.5). According to the **Hammond postulate** [55], early transition states are characteristic of rapid exergonic reactions, whereas late transition states are characteristic of slow endergonic reactions. We can see that this is certainly true for non-adiabatic electron transfer reactions.

It is clear by inspection that γ is *almost* a constant having the value 1/2, but it has a weak dependence on overpotential η determined by the size of the molar Gibbs reorganization energy λ_{m}. (A large value of λ_{m} means a weak dependence on η.) Most importantly,

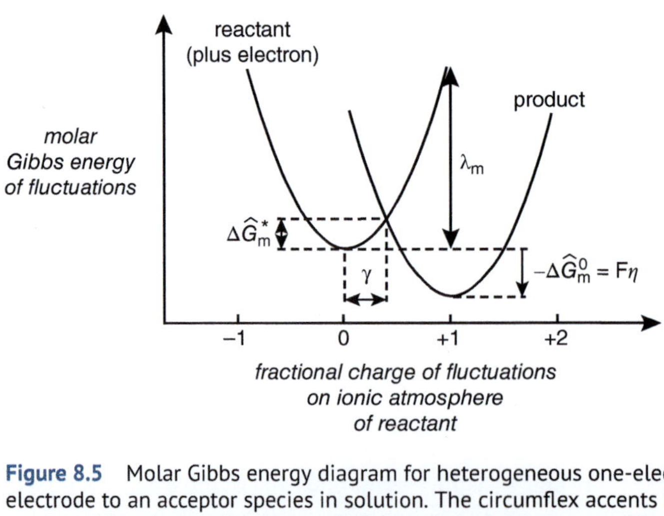

Figure 8.5 Molar Gibbs energy diagram for heterogeneous one-electron transfer from an electrode to an acceptor species in solution. The circumflex accents indicate that the molar Gibbs energy of the electron has been added into that of the reactant. The term η is the overpotential. The minima of the curves are locations of pure chemical potential (zero net charge).

γ supplies the link between the molar Gibbs energy of activation and the molar Gibbs reorganization energy:

$$\boxed{\Delta\widehat{G}^*_{\mathrm{m}} = \gamma^2\lambda_{\mathrm{m}}}. \tag{8.136}$$

Hence:

$$k_{\mathrm{et}} = \frac{2\pi}{\hbar}H_{\mathrm{DA}}^2\frac{N_{\mathrm{A}}}{\sqrt{4\pi\lambda_{\mathrm{m}}RT}}\exp\left(\frac{-\lambda_{\mathrm{m}}}{4RT}\right)\exp\left(\frac{(2\gamma+1)F\eta}{4RT}\right) \tag{8.137}$$

or:

$$k_{\mathrm{et}} = k_0\exp\left(\frac{(2\gamma+1)F\eta}{4RT}\right). \tag{8.138}$$

The same equation applies to the reverse reaction, except that γ is replaced by $(1-\gamma)$. Thus for the overall current-overpotential curve we can write:

$$\boxed{I = I_0\left\{\exp\left(\frac{(2\gamma+1)F\eta}{4RT}\right) - \exp\left(\frac{-(3-2\gamma)F\eta}{4RT}\right)\right\}}. \tag{8.139}$$

This equation is the macroscopic manifestation of the elementary act of electron transfer. It connects the quantum world with macroscopic observables.

The above result differs slightly from the "textbook" Butler–Volmer (BV) equation for a one-electron transfer reaction:

$$I = I_0\left\{\exp\left(\frac{\beta_{\mathrm{BV}}F\eta}{RT}\right) - \exp\left(\frac{-(1-\beta_{\mathrm{BV}})F\eta}{RT}\right)\right\}. \tag{8.140}$$

The reason for the difference is that the one-electron version of the Butler–Volmer equation is expressed in terms of a "symmetry factor" β_{BV} which connects the decrease in molar Gibbs energy of activation of the reaction to the increase in molar Gibbs energy of the electrons. The precise relationship between β_{BV} and γ is shown in Fig. 8.6 and is summarized by the equations:

$$\gamma + \frac{1}{2} = 2\beta_{\mathrm{BV}} \tag{8.141}$$

and:

$$\beta_{\mathrm{BV}} = \frac{1}{2}\left(1 - \frac{F\eta}{2\lambda_{\mathrm{m}}}\right). \tag{8.142}$$

Finally, for avoidance of ambiguity, we supply the following definitions:

- (Definition 1) The **symmetry factor** (dimensionless) for the elementary act of electron transfer is the ratio of the *decrease* in molar Gibbs energy of activation of the reactants to the *increase* in molar Gibbs energy of the electrons, at constant temperature and pressure.
- (Definition 2) The **gamma factor** (dimensionless) for the elementary act of electron transfer is the fractional distance along the reaction co-ordinate at which electron transfer occurs, at constant temperature and pressure.

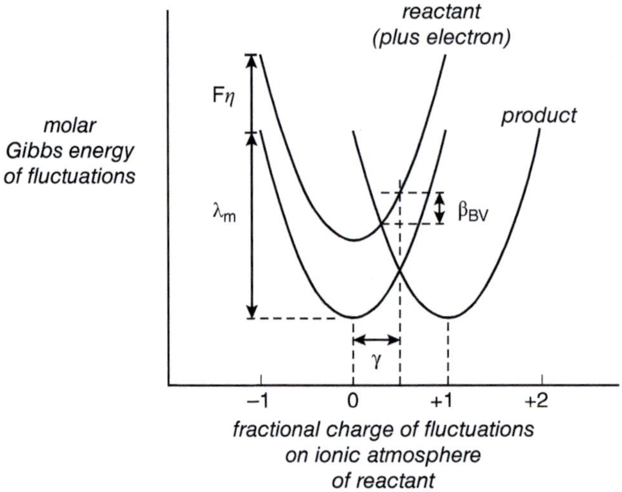

Figure 8.6 Molar Gibbs energy diagram for a one-electron transfer from an electrode to a reactant in solution, showing the relation between the symmetry factor β_{BV} and the gamma factor γ.

8.7.1 Tafel slopes for multi-step reactions

So far we have considered only a single-electron transfer process. Here we consider multi-electron transfer processes. We start by writing down the electron transfer steps individually:

$$\begin{aligned}
&\mathrm{A + e^- \rightleftharpoons B} \quad &&\text{[pre-step 1]}\\
&\mathrm{B + e^- \rightleftharpoons C} \quad &&\text{[pre-step 2]}\\
&\vdots \qquad \vdots &&\vdots\\
&\mathrm{Q + e^- \rightleftharpoons R} \quad &&\text{[pre-step } n_\mathrm{p}\text{]}\\
&\boxed{\mathrm{R} + n_\mathrm{q}\mathrm{e^-} \rightarrow \mathrm{S}} \quad &&\text{[rds]}\\
&\mathrm{S + e^- \rightleftharpoons T} \quad &&\text{[post-step 1]}\\
&\mathrm{T + e^- \rightleftharpoons U} \quad &&\text{[post-step 2]}\\
&\vdots \qquad \vdots &&\vdots\\
&\mathrm{Y + e^- \rightleftharpoons Z} \quad &&\text{[post-step } n_\mathrm{r}\text{]}
\end{aligned}$$

Next we adopt some simplifying notation. We define n_p as the number of electrons transferred *before* the rate-determining step. Then we define n_r to be the number of electrons transferred *after* the rate-determining step. In between, we define n_q to be the number of electrons transferred during one elementary act of the rate-determining step. (This is a stratagem to ensure that n_q can take only the values zero or one, depending on whether the rate-determining step (rds) is a chemical reaction or an electron transfer reaction.) Thus, the total number of electrons transferred is:

$$n = n_p + n_q + n_r. \tag{8.143}$$

Now we make some further assumptions. (i) The exchange current of the rate-determining step is at least one hundred times less than that of any other step, (ii) the rate-determining step of the forward reaction is also the rate-determining step of the backward reaction, (iii) no steps are concerted, (iv) there is no electrode blockage by adsorbed species, and (v) the reaction is in a steady state. Given these assumptions the total rate of reaction is:

$$I_{\text{total}} = I_0 \left\{ \exp\left(\frac{[n_p + n_q\beta_f]F\eta}{RT} \right) - \exp\left(\frac{-[n_r + n_q\beta_b]F\eta}{RT} \right) \right\}. \tag{8.144}$$

Or equivalently:

$$I_{\text{total}} = I_0 \left\{ \exp\left(\frac{\alpha_f F\eta}{RT} \right) - \exp\left(\frac{-\alpha_b F\eta}{RT} \right) \right\}, \tag{8.145}$$

where α_f and α_b are dimensionless "charge transfer coefficients". Although this equation strongly resembles the one-electron Butler–Volmer equation, it should be kept in mind that the charge transfer coefficients α_f and α_b do not behave like the symmetry factors β_{BV} and $(1 - \beta_{BV})$. Basically α_f and α_b are complex parameters describing the overall kinetics of a multi-electron process, whereas β_{BV} and $(1 - \beta_{BV})$ are fundamental parameters describing a single electron process.

In the case of multi-electron transfer processes, if the rate-determining step is purely chemical (i.e., *does not involve* electron transfer) then $n_q = 0$ and the symmetry factors disappear from the equations for α_f and α_b. Conversely, if the rate-determining step is an electrochemical step (i.e., *does involve* electron transfer) then $n_q = 1$ and the symmetry factors enter the equations for α_f and α_b.

The classical method of analyzing experimental data from steady-state current-overpotential curves is to plot graphs of $\log|I/I_0|$ versus η. The gradients of such graphs are called "Tafel slopes" and they provide very useful estimates of α_f or α_b. These data, in turn, afford insights into the interfacial kinetics and the rate-determining steps of complex reactions. Coupled with additional information from perturbative measurements (such as cyclic voltammetry) a detailed picture of the reactions at electrode surfaces can be constructed.

The use of logarithms to base ten means that Tafel slopes tell us how much we need to increase the overpotential in order to increase the reaction rate by a factor ten:

$$\alpha_f \text{ or } \alpha_b = \frac{2.303RT}{F} \left(\frac{\partial \log|I/I_0|}{\partial \eta} \right) \quad |I| > I_0. \tag{8.146}$$

A table of Tafel slopes is given below. The parameter Δ is shorthand for:

$$\Delta = \frac{F\eta}{2\lambda_m}. \tag{8.147}$$

In Table 8.1, E indicates an electrochemical step, C indicates a chemical step, D indicates a dimerization step, and a circumflex accent indicates a rate-determining step.

Table 8.1 Tafel slopes for selected reaction schemes at 25 °C. Reactions having chemical rate-determining steps have potential-independent Tafel slopes. Source: Adapted from Stephen Fletcher, "Tafel slopes from first principles", *Journal of Solid State Electrochemistry*, 13, 537–549 (2009).

Scheme	Tafel Slope mV decade^{-1}
$\hat{\text{C}}$E	∞
$\hat{\text{C}}$ED	∞
$\hat{\text{E}}$	118.4/(1 – Δ)
$\hat{\text{E}}$E	118.4/(1 – Δ)
$\hat{\text{E}}$EE	118.4/(1 – Δ)
$\hat{\text{E}}$C	118.4/(1 – Δ)
$\hat{\text{E}}$CE	118.4/(1 – Δ)
C$\hat{\text{E}}$	118.4/(1 – Δ)
C$\hat{\text{E}}$D	118.4/(1 – Δ)
E$\hat{\text{C}}$	59.2
E$\hat{\text{C}}$E	59.2
E$\hat{\text{E}}$	39.5/(1 – Δ/3)
E$\hat{\text{E}}$E	39.5/(1 – Δ/3)
EC$\hat{\text{E}}$	39.5/(1 – Δ/3)
EE$\hat{\text{C}}$	29.6
CE$\hat{\text{D}}$	29.6
EE$\hat{\text{E}}$	23.7/(1 – Δ/5)
EEE$\hat{\text{C}}$	19.7

8.8 The future: supercatalysis by superexchange

Carnarvon, "Can you see anything...?"
Carter, "Yes, wonderful things!" [56]

In electron transfer theory, the forward rate constant for an electron transfer reaction can be expressed in the following form:

$$k_{\text{ET}}^{+} = \kappa_{\text{E}}\kappa_{\text{N}}\nu \exp\left(\frac{-\Delta G^{*}}{RT}\right). \tag{8.148}$$

Here k_{ET}^{+} is the forward rate constant for the electron transfer reaction (s^{-1}), κ_{E} is the electronic transmission coefficient (dimensionless), κ_{N} is the nuclear transmission coefficient (dimensionless), ν is the pre-exponential factor (s^{-1}), ΔG^{*} is the Gibbs energy of activation (J mol^{-1}), R is the molar gas constant, (8.314 J K^{-1} mol^{-1}), and T is the absolute temperature (K).

On the orthodox view, catalysis is achieved by decreasing the Gibbs energy of activation ΔG^{*} below its catalyst-free value. However, it has recently been demonstrated that an alternative form of catalysis exists, in which the electron transmission coefficient κ_E is increased above its catalyst-free value [57]. All that is needed is to insert a suitable bridge molecule between the donor and the acceptor species. Under optimum conditions, the result can be a million-fold increase in the rate of electron transfer.

To illustrate how this new form of catalysis works, recall the Dogonadze–Chizmadzhev equation [54] for a heterogeneous electron transfer reaction (i.e., one involving electron transfer between the frontier orbitals of a metal electrode and a nearby species in solution):

$$k_{ET}^{+} = \frac{2\pi}{\hbar}|H_{DA}|^2 \frac{N_A}{\sqrt{4\pi\lambda_m RT}} \exp\left(\frac{-(\Delta G_m^0 + \lambda_m + FE_F)^2}{4\lambda_m RT}\right). \tag{8.149}$$

Here k_{ET}^{+} is the forward rate constant for electron transfer (s^{-1}), $\hbar$ is the reduced Planck constant (1.054×10^{-34} J s), H_{DA} is the electronic coupling matrix element between the donor and acceptor species (J), N_A is the Avogadro constant (6.022×10^{23} mol^{-1}), λ_m is the molar reorganization energy (J mol^{-1}), R is the molar gas constant (8.314 J K^{-1} mol^{-1}), T is the absolute temperature (K), ΔG_m^0 is the difference in molar Gibbs energy between the reactants and products, and E_F is the Fermi potential (on the vacuum scale) of the electrons in the frontier orbitals of the electrode surface.

The Dogonadze–Chizmadzhev equation tells us that the forward rate constant for heterogeneous electron transfer k_{ET}^{+} is controlled by five key parameters: the molar reorganization energy λ_m, the molar Gibbs energy difference between the reactants and products ΔG_m^0, the Fermi potential of the electrons E_F, the absolute temperature T, and the electronic coupling matrix element H_{DA}. The first four of these parameters determine the location, shape and population of the molar Gibbs energy surface on which the reaction takes place, while the fifth parameter quantifies the extent of orbital overlap between the donor species and the acceptor species. Adding a superexchange bridge greatly boosts the orbital overlap, and therefore greatly boosts the rate of electron transfer. We call this phenomenon "supercatalysis". The bridge molecular orbitals are readily able to participate in the electron transfer process provided the *Pauli Exclusion Principle* is not violated, i.e., provided any electron pairs that are formed do not have the same spin.

It has been known for a long time that superexchange can mediate electron transfer between remote orbitals [58–62]. In the case of a bridge molecule contributing a single linking orbital between a donor and an acceptor species, the electronic coupling matrix element H_{DA} takes the form:

$$H_{DA} = \frac{H_{DB}H_{BA}}{\Delta\varepsilon_{GAP}}. \tag{8.150}$$

Here H_{DB} and H_{BA} are the electronic coupling matrix elements for the donor-bridge and bridge-acceptor subsystems, and $\Delta\varepsilon_{GAP}$ is the energy gap between the bridge state and the paired donor-and-acceptor states, here assumed degenerate. It is clear that the greatest effect on the rate constant arises from the pair of orbitals that share the smallest value of $\Delta\varepsilon_{GAP}$. This pair of orbitals is likely to consist of the highest occupied molecular orbital of the donor and the lowest unoccupied molecular orbital of the bridge.

In recent work in our own laboratories, we have shown that the reduction of the pentasulfide ion on graphite electrodes can be speeded up by a factor of >10^5 by insertion of a cobalt phthalocyanine molecule between the ion and the electrode surface. The reason for this exceptional improvement is the presence of a half-filled d_{z^2} orbital orthogonal to the electrode surface, which provides a superexchange conduit for the electron transfer process.

Given the observation of superexchange in the catalyzed pentasulfide system, it is reasonable to suppose that the same phenomenon might be found in other systems for which cobalt phthalocyanine is a catalyst. Examples include oxygen reduction [63, 64], hydrogen peroxide activation [65] and carbon dioxide reduction [66].

Even more astonishing, we have discovered that a **single molecule transistor** regulates the flow of electrons through photosystem II [67]. Photosystem II is a large protein complex responsible for the first stage of oxygenic photosynthesis in plants, algae, and cyanobacteria. At the heart of the complex, electrons flow through an iron(II) *d*-orbital by a process of superexchange, at a rate which is gated by the ambient ligand field. The transistor operates in the negative feedback mode, and its existence suggests that single-molecule logic gates are technologically feasible. We believe this is the first recorded example of a single molecule electronic transistor in a living system.

Elsewhere, the role of spin-dependent mechanisms in electron transfer, and their application to oxygen electrochemistry, has been explored by Jose Gracia and colleagues, and guidelines for optimizing various exchange interactions have been suggested [68]. The future looks bright indeed for studies of superexchange and spin in electron transfer.

References

1 Stephen Fletcher, Tafel slopes from first principles, *Journal of Solid State Electrochemistry*, 13, 537–549 (2009) doi.org/10.1007/s10008-008-0670-8

2 Max Born, Volumen und Hydratationswärme der Ionen, *Zeitschrift für Physik*, 1, 45–48, (1920) doi.org/10.1007/BF01881023

3 Rudolph Arthur Marcus, Theory of oxidation-reduction reactions involving electron transfer.1, *J. Chem. Phys.*, 24, 966–978 (1956) doi.org/10.1063/1.1742723

4 Rudolph Arthur Marcus, Electrostatic free energy and other properties of states having nonequilibrium polarization I. *J. Chem. Phys.*, 24, 979–989 (1956) doi.org/10.1063/1.1742724

5 Gilbert Newton Lewis, Outlines of a new system of thermodynamic chemistry. *Proceedings of the American Academy of Arts and Sciences*, 43, 259–293 (1907) doi:10.2307/20022322

6 Edward Armand Guggenheim, The conceptions of electrical potential difference between two phases and the individual activities of ions. *J. Phys. Chem.*, 33, 842–849 (1929) doi:10.1021/j150300a003

7 Roberto Marquardt, Juris Meija, Zoltán Mester, Marcy Towns, Ron Weir, Richard Davis and Jürgen Stohner, Definition of the mole (IUPAC Recommendation 2017). *Pure Appl. Chem.*, 90, 175–180 (2018) doi.org/10.1515/pac-2017-0106

8 Fritz Scholz, Wilhelm Ostwald's role in the genesis and evolution of the Nernst equation. *Journal of Solid State Electrochemistry*, 21, 1847–1859 (2017) DOI: 10.1007/s10008-017-3619-y

9 Enrico Fermi, Sulla quantizzazione del gas perfetto monoatomico. *Rendiconti Lincei*, 3, 145–149 (1926)

10 Paul A. M. Dirac, On the theory of quantum mechanics. *Proceedings of the Royal Society A*, 112, 661–677 (1926) doi.org/10.1098/rspa.1926.0133

11 Sergio Trasatti and Roger Parsons, Interphases in systems of conducting phases (Recommendations 1985). *Pure and Applied Chemistry*, 58, 437–454 (1986) doi.org/10.1351/pac198658030437

12 Peter Mitchell, David Keilin's respiratory chain concept and its chemiosmotic consequences. December 8, *1978. Nobel Lectures, Chemistry 1971–1980, Editor-in-Charge Tore Frängsmyr*, Editor Sture Forsén, World Scientific Publishing Co., Singapore (1993)

13 Sergio Trasatti, The absolute electrode potential: an explanatory note. *Pure Appl. Chem.*, 58, 955–966 (1986) doi.org/10.1351/pac198658070955

14 Edwin James Hart and John Wilson Boag, Absorption spectrum of the hydrated electron in water and in aqueous solutions. *Journal of the American Chemical Society*, 84, 4090–4095 (1962) doi:10.1021/ja00880a025

15 John Wilson Boag and Edwin James Hart, Absorption spectra in irradiated water and some solutions: Absorption spectra of "hydrated" electron. *Nature*, 197, 45–47 (1963) doi:10.1038/197045a0

16 Albert John Swallow, Radiation chemistry. An introduction. Longman Group Ltd, London (1973) ISBN-10: 058246286X

17 Harold A. Schwarz, Free radicals generated by radiolysis of aqueous solutions. *J. Chem. Ed.*, 58, 101–105 (1981) doi:10.1021/ed058p101

18 Chang-Guo Zhan and David A. Dixon, The nature and absolute hydration free energy of the solvated electron in water. *J. Phys. Chem. B*, 107, 4403–4417 (2003) doi:10.1021/jp022326v

19 James V. Coe, Alan D. Earhart, Michael H. Cohen, Gerald J. Hoffman, Harry W. Sarkas and Kit H. Bowen, Using cluster studies to approach the electronic structure of bulk water: Reassessing the vacuum level, conduction band edge, and band gap of water. *J. Chem. Phys.*, 107, 6023–6031 (1997) doi.org/10.1063/1.474271

20 Stephen Fletcher, The theory of electron transfer. *Journal of Solid State Electrochemistry*, 14 (5), 705–739 (2010) doi:10.1007/s10008-009-0994-z

21 Jürgen Schnitker and Peter J. Rossky, Quantum simulation study of the hydrated electron. *The Journal of Chemical Physics*, 86, 3471–3485 (1987) doi.org/10.1063/1.452003

22 Michael J. Tauber and Richard A. Mathies, Structure of the aqueous solvated electron from resonance Raman spectroscopy: lessons from isotopic mixtures. *Journal of the American Chemical Society*, 125, 1394–1402 (2003) doi:10.1021/ja021134a

23 Chang-Guo Zhan and David A. Dixon, The nature and absolute hydration free energy of the solvated electron in water. *J. Phys. Chem. B*, 107, 4403–4417 (2003) doi:10.1021/jp022326v

24 David A. Armstrong, Robert E. Huie, Sergei Lymar, Willem H. Koppenol, Gabor Merényi, Pedatsur Neta, David M. Stanbury, Sten Steenken and Peter Wardman, Standard electrode potentials involving radicals in aqueous solution: inorganic radicals. *Bio-Inorganic Reaction Mechanisms*, 9, 59–61 (2013) doi.org/10.1515/irm-2013-0005

25 Ronald Wilfrid Gurney, *Ionic Processes in Solution*. McGraw- Hill Book Company, New York (1953) ISBN 10: 1296630676

26 Alexander Fridman and Lawrence A. Kennedy, *Plasma Physics and Engineering*. Taylor and Francis Books Inc., New York (2004) ISBN: 1-56032-848-7

27 Stephen Fletcher, A non-Marcus model for electrostatic fluctuations in long range electron transfer. *Journal of Solid State Electrochemistry*, 11, 965–969 (2007) doi.org/10.1007/s10008-007-0313-5

28 John Bertrand Johnson, Thermal agitation of electricity in conductors. *Phys. Rev.*, 32, 97–109 (1928) doi:10.1103/physrev.32.97

29 Harry Nyquist, Thermal agitation of electric charge in conductors. *Phys. Rev.*, 32, 110–113 (1928) doi:10.1103/physrev.32.110

30 Rudolph Arthur Marcus, Electrostatic free energy and other properties of states having nonequilibrium polarization. II. Electrode systems. In: *Special Topics in Electrochemistry* (Peter A. Rock, ed.) Elsevier, New York, NY, pp. 210–224 (1977) ISBN 978–0444416278

31 Rudolph Arthur Marcus, Free energy of nonequilibrium polarization systems. II. Homogeneous and electrode systems. *The Journal of Chemical Physics*, 38, 1858–1862 (1963) doi.org/10.1063/1.1733886

32 Rudolph Arthur Marcus, Electron transfer reactions in chemistry: theory and experiment. Nobel Lecture, December 8, 1992. In: *Nobel Lectures in Chemistry*, Vol 7 (1991–1995), Editor Bo G. Malmström, World Scientific Publishing Co., Singapore (1997) ISBN 10: 9810226802

33 Rudolph Arthur Marcus, Tutorial on rate constants and reorganization energies. *Journal of Electroanalytical Chemistry*, 483, 2–6, (2000) doi.org/10.1016/S0022-0728(00)00011-5

34 Yizhak Marcus, Thermodynamics of solvation of ions, Part 5. Gibbs free energy of hydration at 298.15 K. *J. Chem. Soc. Faraday Trans.*, 87, 2995–2999 (1991) doi:10.1039/FT9918702995

35 Arieh Warshel, Dynamics of reactions in polar solvents. Semi-classical trajectory studies of electron-transfer and proton-transfer reactions. *The Journal of Physical Chemistry*, 86, 2218–2224 (1982) doi:10.1021/j100209a016

36 IUPAC Compendium of Chemical Terminology, 2nd edition, (The "Gold Book"), compiled by A. D. McNaught and A. Wilkinson. Blackwell Scientific Publications, Oxford (1997) doi.org/10.1351/goldbook

37 Tibor Erdey-Grúz and Max Volmer, Zur theorie der wasserstoffüberspannung. *Zeitschrift für Physikalische Chemie*, A150, 203–213 (1930) doi:10.1515/zpch-1930-15020

38 Noel Sydney Hush, Adiabatic theory of outer sphere electron-transfer reactions in solution. *Transactions of the Faraday Society*, 57, 557–580 (1961) doi:10.1039/TF9615700557

39 Rudolph Arthur Marcus, Exchange reactions and electron transfer reactions including isotopic exchange. Theory of oxidation-reduction reactions involving electron transfer, Part 4. A statistical-mechanical basis for treating contributions from solvent, ligands, and inert salt. *Discussions of the Faraday Society*, 29, 21–31 (1960) doi:10.1039/DF9602900021

40 James Clerk Maxwell, On Physical Lines of Force, Part III. *The London, Edinburgh, and Dublin Philosophical Magazine and Journal of Science*, Series 4, 23, 12–24, (1862) doi.org/10.1080/14786446208643207

41 Peter Joseph Wilhelm Debye, Zur Theorie der anomalen Dispersion im Gebiete der langwelligen elektrischen Strahlung. *Berichte der Deutschen physikalischen Gesellschaft*, 15, 777–793 (1913)

42 Nguyen Q. Vinh, Mark S. Sherwin, S. James Allen, Deepu K. George, A. J. Rahman and Kevin W. Plaxco, High-precision gigahertz-to-terahertz spectroscopy of aqueous salt solutions as a probe of the femtosecond-to-picosecond dynamics of liquid water. *Journal of Chemical Physics*, 142, 164502 (2015) doi.org/10.1063/1.4918708

43 Vadim F. Lvovich, Impedance spectroscopy: *applications to electrochemical and dielectric phenomena*. John Wiley & Sons, Hoboken NJ (2012) ISBN: 978-0-470-62778-5

44 Paul Adrien Maurice Dirac. The quantum theory of the emission and absorption of radiation. *Proceedings of the Royal Society of London. Series A, Containing Papers of a Mathematical and Physical Character*, 114, 243–265 (1927) doi.org/10.1098/rspa.1927.0039

45 Lev Davidovich Landau, Zur theorie der energieübertragung. *II. Physikalische zeitschrift der Sowjetunion*, 2: 46–51 (1932)

46 Clarence Melvin Zener, Non-adiabatic crossing of energy levels. *Proceedings of the Royal Society of London A*, 137 (6): 696–702 (1932) doi:10.1098/rspa.1932.0165

47 Ernst Carl Gerlach Stückelberg, Theory of inelastic collisions between atoms. *Helvetica Physica Acta*, 5, 369–423 (1932) doi:10.5169/seals-110177.

48 Ettore Majorana, Atomi orientati in campo magnetico variabile. *Il Nuovo Cimento*, 9: 43–50 (1932) doi:10.1007/BF02960953.

49 Gilles Klopman, Chemical reactivity and the concept of charge- and frontier-controlled reactions. *Journal of the American Chemical Society*, 90, 223–234 (1968) doi:10.1021/ja01004a002

50 Lionel Salem, Intermolecular orbital theory of the interaction between conjugated systems. I. General theory. *Journal of the American Chemical Society*, 90, 543–552 (1968) doi:10.1021/ja01005a001

51 Ian Fleming, *Molecular Orbitals and Organic Chemical Reactions* (Reference Edition), John Wiley & Sons Ltd, Chichester, UK. (2010) ISBN 978-0-470-74658-5

52 Margaret Rae and Mário N. Berberan-Santos, A generalized pre-equilibrium approximation in chemical and photophysical kinetics. *Journal of Chemical Education*, 81 (3), 436–440 (2004) doi:10.1021/ed081p436

53 Chong-Wha Pyun, Steady-state and equilibrium approximations in chemical kinetics. *Journal of Chemical Education*, 48, 194–196 (1971) doi:10.1021/ed048p194

54 Revaz Romanovich Dogonadze and Yurii Aleksandrovich Chizmadzhev, Computation of the probability of an elementary act for certain heterogeneous redox reactions. *Doklady Akademii Nauk*, 144, 1077–1080 (1962) (In Russian)

55 George Simms Hammond, A correlation of reaction rates. *Journal of the American Chemical Society*, 77, 334–338 (1955) doi:10.1021/ja01607a027

56 Howard Carter and Arthur Cruttenden Mace, *The tomb of Tut-Ankh-Amen discovered by the late Earl of Carnarvon and Howard Carter*. Cassell & Co, London (1923–1933) ISBN 0486235009

57 Stephen Fletcher and Nicholas J. Van Dijk, Supercatalysis by superexchange. *J. Phys. Chem. C*, 120, 26225–26234 (2016) DOI: 10.1021/acs.jpcc.6b09099

58 P. George and John Stanley Griffith, Electron transfer and enzyme catalysis, Vol.1, Chapter 8 in *The Enzymes*, Eds. Paul D. Boyer, Henry Lardy, and Karl Myrbäck, Academic Press, NY and London, pp. *347–389* (1959)

59 Jack Halpern and Leslie Eleazer Orgel, The theory of electron transfer between metal ions in bridged systems. *Discuss. Faraday Soc.*, 29, 32–41 (1960) DOI: 10.1039/DF9602900032

60 Harden M. McConnell, Intramolecular charge transfer in aromatic free radicals. *J. Chem. Phys.*, 35, 508–515 (1961) doi.org/10.1063/1.1731961

61 Henry Taube, Electron transfer between metal complexes—a retrospective view (Nobel lecture). *Angew. Chem., Int. Ed. Engl.*, 23, 329–339 (1984) doi.org/10.1002/anie.198403293

62 Christopher C. Moser, Jonathan M. Keske, Kurt Warncke, Ramy S. Farid, and P. Leslie Dutton, Nature of biological electron transfer. *Nature*, 355, 796–802 (1992) doi.org/10.1038/355796a0

63 Raymond Jasinski, A new fuel cell cathode catalyst. *Nature*, 201, 1212–1213 (1964)

64 Raymond Jasinski, Fuel cell electrodes having a metal phthalocyanine catalyst. *Canadian Patent No.797924* (1965)

65 Nan Li, Wangyang Lu, Kemei Pei, Yuyuan Yao, and Wenxing Chen, "Ordered-mesoporous-carbon-bonded cobalt phthalocyanine: a bioinspired catalytic system for controllable hydrogen peroxide activation", *ACS Applied Materials & Interfaces*, 6, 5869–5876 (2014) DOI: 10.1021/am500766a

66 Charles M. Lieber and Nathan S. Lewis. Catalytic reduction of carbon dioxide at carbon electrodes modified with cobalt phthalocyanine. *Journal of the American Chemical Society*, 106, 5033–5034 (1984) DOI: 10.1021/ja00329a082

67 Stephen Fletcher, Discovery of a single molecule transistor in photosystem II. *Journal of Solid State Electrochemistry*, 19, 241–250 (2015) doi.org/10.1007/s10008-014-2567-z

68 Jose Gracia, Spin dependent interactions catalyse the oxygen electrochemistry. *Physical Chemistry Chemical Physics*, 19, 20451–20456 (2017) DOI: 10.1039/C7CP04289B

Manuscript Accepted 21 May 2019.

9

Constant potential rate theory – general formulation and electrocatalysis

Marko M. Melander

Nanoscience Center, Department of Chemistry, University of Jyväskylä, Finland

9.1 Kinetics at electrochemical interfaces

Using the electrode potential to control reaction thermodynamics and kinetics forms the backbone of all electrochemistry and electrocatalysis. Important reactions include the oxygen and hydrogen reduction/evolution reactions, nitrogen reduction to ammonia, and CO_2 reduction [1]. In all these electrochemical reactions small molecules are converted to products *via* successive proton-coupled electron transfer (PCET), electron transfer (ET), and proton transfer (PT) reactions as shown schematically in Figure 9.1. Therefore, controlling and understanding PCET, PT, and ET reactions is a central goal in the search for better electrocatalysts for a wide range of applications. PCET reactions can be concerted (CPET, electron and proton move at the same time) or sequential (ET-PT electron first, then proton or PT-ET). How to compute the CPET, ET, and PT reaction kinetics is the topic of this section.

Reactions at electrochemical interfaces are often divided to outer- and inner-sphere reactions [2]. Outer-sphere or electrochemical reactions are usually ET and PCET reactions taking place without specific adsorption on the electrode surface and the electrode acts only as an electron source [2]. Electrochemical kinetics for this class of reactions are often determined using non-adiabatic rate theory derived from a perturbative Fermi's Golden Rule formalism leading to theoretical treatments such as Dogonadze-Kutzentsov [3], non-adiabatic Schmickler [4], Gerischer [5], Soudackov-Hammes-Schiffer [6, 7] or non-adiabatic Marcus theories [8]. These theories are based on the reorganization of the reaction environment such that electron tunneling can take place. Influence of the electrode potential is usually added as a simple parameter in the Hamiltonian to change the free energy in a predefined way. Hence, these approaches usually utilize model Hamiltonians and it may be difficult to include realistic environments with complex electronic structure, solid-solvent interfaces, double-layer formation etc. in these models.

In inner-sphere reactions the reactants adsorb on the electrode surface and a covalent bond is formed [2]. Reaction kinetics are usually described in terms of a fully (i.e. electronically and vibronically) adiabatic (canonical) transition state theory with classical nuclei. The electrode potential can be included from the start using fixed-potential first principles

Atomic-Scale Modelling of Electrochemical Systems, First Edition.
Edited by Marko M. Melander, Tomi T. Laurila, and Kari Laasonen.

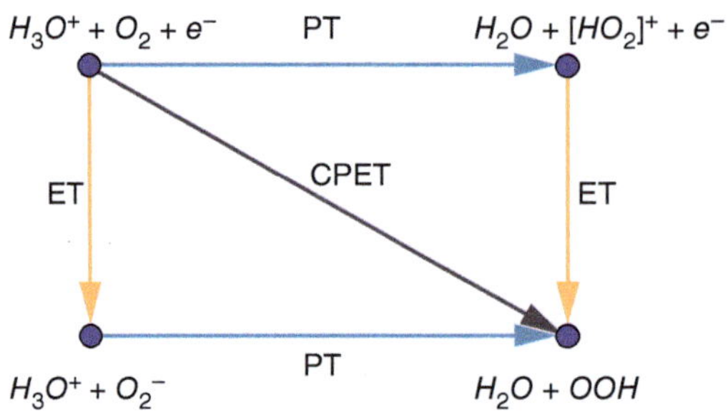

Figure 9.1 Depiction of a single PCET step showing both ET-PT, PT-ET, and CPET pathways for the initial step of acidic oxygen reduction reaction.

approaches such as grand canonical DFT. (see also the chapter by Peterson et al.) [9–12] Such approaches can naturally capture and mimic the complex reaction environments. However, until recently, only a small subset of reactions have been considered; only fully adiabatic CPET have been studied. All (vibronic and electronic) non-adiabaticity as well as nuclear tunneling in PT and PCET reactions have been largely neglected.

Recently, a unified rate theory of electrochemical (outer-sphere) and electrocatalytic (inner-sphere) reactions was put forward by the author of this chapter [13]. This constant potential rate theory is an extension of a very general canonical rate theory to the grand canonical ensemble (GCE). As will be shown below, the constant potential rate theory can naturally include electronic and vibronic non-adiabaticity as well as nuclear tunneling, for example. The general fixed-potential rate theory enables treating all potential-driven reactions within a single formalism instead of relying on separate theories for electrocatalysis and electrochemistry. This theory can (in most parts) be directly combined with modern, solid-state *ab initio* methods, such as DFT, without any modifications to the underlying codes. Hence, a realistic description including the electrode potential, solvation, and double-layer from first principles can be utilized. Combining well-defined, general rate theories with equally well-defined computational models paves way towards understanding electrocatalytic reactions from the atomic-level under realistic reaction conditions.

9.2 Rate theory in the grand canonical ensemble

As discussed extensively in this book and elsewhere [9–12, 14, 15], GCE provides the natural basis for modeling electrochemical interfaces from the atomic-level. Within GCE, the state of an electrochemical interface is defined in terms of fixed temperature (T), volume (V), and the (electro)chemical potentials ($\{\mu_i\}$) of all the species in the system. In particular, the electron chemical potential (μ_n) defines the Fermi level which again is directly related to the electrode potential [9]. Using this set of control parameters ($T, V, \{\mu_i\}$) ensures that all expectation values are computed for a chemical system at fixed temperature, volume/pressure and a set of chemical potentials. This also means that the particle number is allowed to fluctuate to keep the corresponding chemical potential constant.

Besides the presence of the electric potential, there is, in principle, nothing special about electrochemical or electrocatalytic kinetics as compared to reaction kinetics in other types of chemical systems. Gas-phase kinetics are usually determined in the microcanonical

EVT-ensemble while most condensed-phase kinetics are studied in the canonical Helmholtz *NVT*-ensemble or in the Gibbs *NPT*-ensemble. For atomic-scale modeling the most widely used is the canonical ensemble and in this ensemble the most general definition of a thermal rate constant is given by Miller [16–19]

$$k(T, V, N)Q_I = \int dEP(E)\exp[-\beta E] \tag{9.1}$$

where $P(E)$ is the transition probability at a given energy, Q_I is the canonical partition function of the initial state, and $\beta = (k_B T)^{-1}$. Despite its simple appearance, Eq. (9.1) is extremely powerful. Different choices for $P(E)$ provide the basis for computing reaction rates using common (canonical) transition state theory (TST), nuclear tunneling using semi-classical TST, instanton theory, or ring-polymer rate theory, non-adiabatic reactions etc. The Boltzmann distribution shows that the above equation gives a thermal rate constant.

The main idea behind the grand canonical rate theory is to extend Eq. (9.1) to an electrochemical GCE setting. This enables one to access a wide variety of different reactions and phenomena at electrochemical interfaces. Given that a thermodynamic (quasi)equilibrium is also reached at an electrochemical interface, extending the canonical rate theory to the GCE is straightforward and is given as [13]

$$k(T, V, \mu_n)\Xi_I = \sum_{N=0}^{\infty} \exp[\beta\mu_n N] \int dEP(E)\exp[-\beta E] \tag{9.2}$$

where $\Xi_I = \exp[\beta\mu_n N]Q_I$ is the initial state grand partition function at electron chemical potential μ_n and N is the number of electrons in the system. Equation (9.2) is still a thermal reaction rate but in an environment open to electrons at fixed chemical potential.

9.3 Adiabatic reactions

Electrochemical reactions can be roughly divided in two groups: adiabatic and non-adiabatic. In the former, reactions take place solely on the electronic/vibrational ground state. Non-adiabatic reactions on the other hand require accounting for electronically/vibrationally excited states as well. In this section electronically adiabatic reactions are treated while non-adiabatic reactions are treated in the next section.

9.3.1 Classical nuclei

The most commonly used rate theory is the transtition state theory (TST). In an electrochemical setting, an especially important TST describes classical nuclei and adiabatic reactions taking place at a fixed electrode potential potential energy surface (PES). The steps needed for obtaining a rate constant for this class of reactions involve invoking the TST approximation i.e. that i) there are no recrossings of the dividing surface separating reactants and products, ii) treating only the ground state and iii) using a classical Hamiltonian describing a fixed electrode potential potential energy surface (PES)

$$H_{cl}(\mu_n) = \sum_{i \in N_N} P_i^2 + V(\mathbf{q}_i, \mu_n) \tag{9.3}$$

where P_i is nuclear momentum and the grand free energy of the electrons is given by $\Omega_n(T, V, N_N, \mu_n; \mathbf{q}_i)$ for a system with N_N nuclei at positions $\mathbf{q}_i$. As the nuclei move on the effective PES provided by the electrons, one recognizes that $V(\mathbf{q}_i, \mu_n) = \Omega_n(T, V, N_N, \mu_n; \mathbf{q}_i)$ (see [9]). Using the above assumptions leads to the classical adiabatic TST rate. (for derivation details refer to [13])

$$k(T, V, \mu_n) = \langle\kappa\rangle_{\mu_n} \frac{k_B T}{h} \frac{\Xi^{\ddagger}}{\Xi_I} = \langle\kappa\rangle_{\mu_n} \frac{k_B T}{h} \exp\left[-\beta\Delta\Omega^{\ddagger}\right] \tag{9.4}$$

where $\Omega_i = -\ln(\Xi_i)/\beta$ is the grand free energy i.e. free energy at a fixed chemical potential. The $\langle\kappa\rangle_{\mu_n}$ prefactor term corrects for omissions in TST and the subscript reminds that this correction may depend on the electrode potential. In the field of heterogenous catalysis a harmonic TST rate constant is usually applied. An equivalent rate constant for electrocatalytic reactions is accomplished by Taylor expanding the grand free energies in terms of fixed potential vibrational frequencies. This yields the fixed potential harmonic TST

$$k_{HTST}(T, V, \mu_n) = \frac{\nu_N}{2\pi} \exp\left[-\beta\Delta\Omega_N^{\ddagger}\right] \tag{9.5}$$

where ν_N is the frequency along the reaction coordinate, and the total electronic grand free energy $\Omega_i^{\ddagger} = \Omega_n - TS_N$ is composed of the grand potential energy Ω_n and the harmonic vibrational entropy $S_i = -k_B \ln(\prod_i^N \nu_i)$ with N real vibrational frequencies.

9.3.2 Fixed potential empirical valence bond theory

The rate of adiabatic reactions is largely defined by the reaction grand energy barrier. This barrier can be computed using methods such as the nudged elastic band [20] (NEB) combined with grand canonical DFT (see e.g. the chapter by Peterson et al.). However, these calculations need to be performed for a range of potentials, which can become insurmountable. As a result, many have resorted to using interpolating or extrapolating reaction barriers between different potentials using (experimental, calculated, or guessed) Tafel coefficients or Brønstedt-Evans-Polanyi (BEP) free energy methods.

Another useful method called the grand canonical ensemble empirical valence bond method (GCE-EVB) was recently developed [13] to provide a link between full-ranged NEB-DFT calculations and Tafel-BEP interpolations. GCE-EVB is based on a Marcus-like picture [8] of presenting the initial and final states using diabatic states, effective wave functions, and grand free energies. The diabatic states are basically charge or spin-localized states. Unlike adiabatic states, diabatic states retain their localized nature at all nuclear geometries.

Formally, the GCE-EVB requires computing two diabatic grand energy curves defined along a suitable reaction coordinate (solvent or energy-gap coordinate) at a fixed electrode potential and enforcing the charge/spin localization at all geometries, see Figure 9.2. The complete rate constant development is lengthy and presented in [13], while a simple interpretation is provided here. In simple terms, the reaction barrier is found from the crossing of these two diabatic grand energy curves. Assuming classical nuclei, and equally curved parabolic grand energy along the reaction coordinate gives the adiabatic Marcus theory result but now at a fixed electrode potential

$$k_{Marcus}(T, V, \mu_n) = \frac{k_B T}{h} \exp\left[-\beta\frac{(\Delta\Omega_{FI} + \Lambda)^2}{4\Lambda}\right] = \frac{k_B T}{h} \exp\left[-\beta\Delta\Omega_{dia}^{\ddagger}\right] \tag{9.6}$$

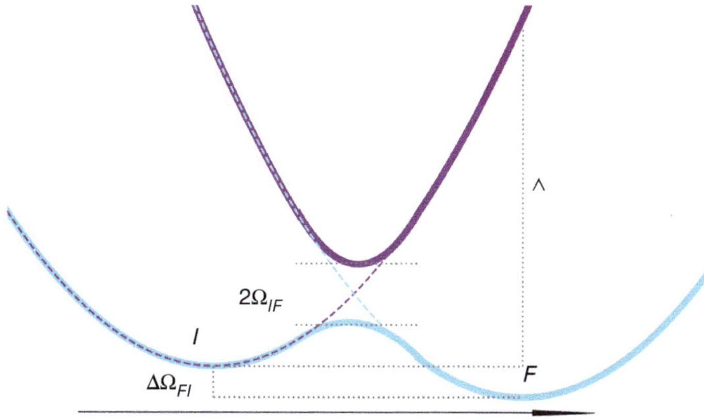

Figure 9.2 Depiction of the GCE-EVB curves. The solid blue (purple) line is the adiabatic ground (excited) state. The dashed blue (purple) line is the initial I (final F) diabatic state. Ω_{IF} is the coupling term between the diabatic states. $\Delta\Omega_{FI}$ and Λ are the fixed potential reaction and reorganization energies, respectively.

where $\Delta\Omega_{FI}$, Λ, and $\Omega^{\ddagger}_{dia}$ are the grand reaction, reorganization energy, and diabatic energy barrier, respectively (see also Figure 9.2). It can be shown [13, 21, 22] that the diabatic barrier can be corrected to give the adiabatic barrier using the diabatic coupling element Ω_{IF} to give

$$\Delta\Omega^{\ddagger}_{ad,EVB} \approx \Delta\Omega^{\ddagger}_{dia} - \Omega_{IF} + \frac{(\Omega_{IF})^2}{\Delta\Omega_{FI} + \Lambda} \tag{9.7}$$

This equation is most useful. Consider that the adiabatic barrier $\Delta\Omega^{\ddagger}_{ad,EVB}$ is computed at a single electrode potential using e.g. NEB-DFT. One can also compute the diabatic barrier $\Omega^{\ddagger}_{dia}$ at the same potential (see computational aspects). Then Ω_{IF} can be fitted to reproduce $\Delta\Omega^{\ddagger}_{ad,EVB}$ based on $\Omega^{\ddagger}_{dia}$. Now that all the parameters are known, the adiabatic barrier may be estimated using just the reaction grand free energy $\Delta\Omega$ at various electrode potentials. Eq. (9.7) provides a general way to compute and use non-linear Tafel-BEP relationships for evaluating the reaction grand energy as a function of the readily obtainable reaction grand energy. This evaluation can be done either be explicitly computing $\Delta\Omega$ as a function of the electrode potential using GCE-DFT or using the widely popular computational hydrogen electrode model [23].

9.3.3 Nuclear tunneling

As many electrocatalytic reactions involve a transfer of hydrogen, nuclear quantum effects may be important. Experimentally, it has been demonstrated that hydrogen tunneling takes place e.g. during alkaline ORR on Pt(111) [24] and HER on gold surfaces [25]. Therefore, it is important that nuclear quantum effects can be included in the general electrochemical/-electrocatalytic rate theory.

Modelling tunneling in the canonical ensemble can be roughly divided to two different different categories. The first one includes hydrogen motion as the reaction coordinate.

Theories in this direction include the instanton theory [17, 26], semiclassical TST [17, 27], and simple tunneling corrections such as the Bell correction [28]. Both deep (through the barrier) and shallow (corrections near the barrier top) can be treated using different approaches. Central to these treatments is that the proton coordinate is taken as the reaction coordinate and that proton motion is considered only for the proton ground state. The reaction barriers are given by the TST-like barrier for motion along the proton transfer coordinate. The prefactor introduces a tunneling correction along the proton coordinate.

Another branch of theories start from considering fluctuations in the reaction environment as the reaction coordinate and quantization of the proton movement [29–31] (similar to the non-adiabatic CPET below). In these curve crossing theories diabatic proton states are brought to resonance by fluctuations in the reaction environment. The proton then transfers in a non-adiabatic tunneling step. Both ground and excited proton vibrational states are included in the description. The reaction free energy barrier is provided by the environment reorganization leading to an activation barrier of the form Eq. (9.9). The prefactor is given by a Fermi's Golden Rule-like formula as in Eq. (9.9).

The curve crossing formula is expected to be valid for reactions where solvent reorganization dominates. Such reactions are likely to occur when significant charge reorganization takes place during the reaction. Also the solvent needs to interact strongly with both the reactants and products. On the other hand, the tunneling corrections to TST seem more appropriate for reactions where the role of the solvent is to stabilize the reaction intermediates but the solvent does not participate in the reaction dynamics. At the moment it is not clear which model is more appropriate for treating reactions at electrochemical interfaces.

Both approaches are easy to extend to reactions at the electrochemical interfaces. The key observation is that both formulations describe *electronically adiabatic* reactions. Then the nuclei, classical or quantized, move on an effective fixed electrode potential PES provided by Eq. (9.3). Using such a fixed electrode potential PES allows access to both electronically adiabatic tunneling models without any other modifications to the tunneling treatment.

9.4 Non-adiabatic reactions

9.4.1 Non-adiabatic reactions in electrochemistry

In this section an introduction to the rate theory of non-adiabatic reactions is provided. Non-adiabaticity is a hallmark for the breakdown of the Born-Oppenheimer approximation. In non-adiabatic reactions the electron (in the case of ET) or proton (in the case of PT/CPET) movement cannot be separated form nuclear motion.

If an electrochemical reaction is non-adiabatic, the electron/proton does not instantaneously respond to nuclear motion. Also the electronic/vibronic coupling (defined below) is smaller than thermal energy. Furthermore, the excited electronic/vibronic states cannot be neglected. PCET reactions can exhibit various flavors of non-adiabaticity: i) if the reaction is electronically non-adiabatic it is also vibronically non-adiabatic, ii) the electronically adiabatic can be vibrationally either adiabatic or non-adiabatic. For an excellent discussion on adiabaticity and non-adiabaticity see [32]

Both electronic, vibronic, and vibrational non-adiabaticity can be encountered in electrochemical systems. Electronically non-adiabatic reactions are expected to occur in outer-sphere ET and CPET in which the interaction between the electron donor and acceptor is weak or in which long-range electron transfer takes place. Reactions taking place on a surface, i.e. inner-sphere sphere reactions, are expected to be electronically adiabatic due to the strong electronic coupling between the electron donor and acceptor. Even at this electronically adiabatic limit, both vibrationally adiabatic [33] and non-adiabatic [25] reactions can take place in electrocatalytic systems.

9.4.2 Rate of ET and CPET reactions

The physical/chemical picture behind the rate constant depicted in Figure 9.3 can be explained as follows. First, the transferring proton is quantized. In ET (CPET), the initially localized diabatic electronic (vibronic) wave function transforms to a final localized diabatic electronic (vibrational) state. Reorganization of the environment leads to a point where the electronic (vibronic) states cross. At this crossing point a non-adiabatic transition takes place between the reactant and product degenerate electronic (vibronic) states – this corresponds to a tunneling of an electron (simultaneous tunneling of an electron and proton) from donors to acceptors. Finally, further reorganization of the environment stabilizes the reactant electronic (vibronic) state.

Kinetics of the above non-adiabatic transition can be described using a variation of Fermi's Golden Rule. This is also the direction taken in the derivation of fixed potential rate

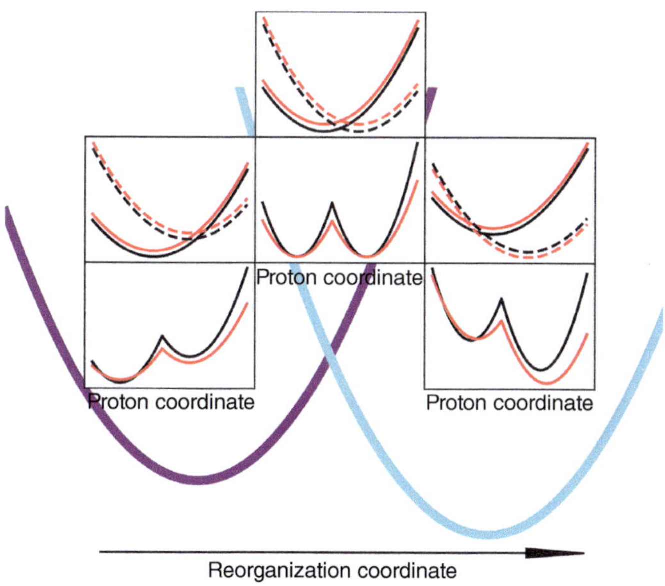

Figure 9.3 Schematic picture for electrochemical PCET. The blue (purple) line depicts an initial (final) vibronic diabatic state as a function of the environment reaction coordinate. The insets show two types of proton curves along the proton transfer coordinate at the initial, transition, and final solvent coordinates. The top (bottom) inset shows the electronically non-adiabatic (adiabatic) proton curves. The dashed solid (dashed) electronically non-adiabatic curve corresponds to electron localized at the initial (final) state. In each inset the black and red curves correspond to two different electrode potentials or states with different number of electrons.

constants of non-adiabatic reactions. The main difficulty in the fixed potential case comes from the computation of electronic transition matrix elements which are not defined for states with different number of electrons. Therefore, the transfer matrix element needs to be calculated for each canonical (fixed number) state. Afterwards, a weighted summation over particle number states and rates is performed.

The CPET kinetics are based on the CPET rate theory of Soudackov and Hammes-Schiffer [6, 7, 34, 35]. Within the canonical ensemble the relevant rate expressions were derived in [6, 7, 34, 35]. Recently this treatment was extended to the GCE, yielding CPET rate constants at fixed electrode potentials [13]. Non-adiabatic ET kinetics are treated in an analogous matter but proton transfer is neglected.

To derive the ET and CPET rates, electronic states are defined for initial (i) and final (f) states with a fixed number of particles (N). In the case of CPET, a geometric variable q denotes the position of the transferring proton. The fixed particle number electronic/vibronic energies at fixed particle number at nuclear geometry Q are then computed. This is written as

$$\langle iN|\hat{H}_N^{el}|iN\rangle = \varepsilon_{iN}(q,Q) \quad \text{and}$$
$$\langle fN|\hat{H}_N^{el}|fN\rangle = \varepsilon_{fN}(q,Q) \tag{9.8a}$$
$$[\hat{T}_q + \varepsilon_{iN}(q,Q)]\,|iuN\rangle = \epsilon_{uN}^{i}\,|iuN\rangle \quad \text{and}$$
$$[\hat{T}_q + \varepsilon_{fN}(q,Q)]\,|fvN\rangle = \epsilon_{vN}^{i}\,|fvN\rangle \tag{9.8b}$$
$$[\hat{T}_Q + \epsilon_{uN}^{i}]\,|mN\rangle = \mathcal{E}_{mN}\,|mN\rangle \quad \text{and}$$
$$[\hat{T}_Q + \epsilon_{vN}^{f}]\,|nN\rangle = \mathcal{E}_{nN}\,|nN\rangle \tag{9.8c}$$

where $\hat{H}_{el}$ is the electronic Hamiltonian, and $\hat{T}_q$ and $\hat{T}_Q$ are the kinetic energy operators for the proton and other nuclei, respectively. Again, the Eqs. (9.8b) are neglected in the case of ET. The initial (final) wave function is written within BOA as $|imN\rangle = |iN(Q)\rangle\,|mN\rangle$ and the energy is $E_{imN} = \varepsilon_{iN} + \epsilon_{mN}$ ($|fnN\rangle = |fN(Q)\rangle\,|nN\rangle$ and E_{fnN}). Correspondingly, the initial (final) vibronic state is $|iumN\rangle = |iN(q,Q)\rangle\,|uN(Q)\rangle\,|mN\rangle$ with an energy $E_{iumN} = \varepsilon_{iN} + \epsilon_{uN}^{i} + \mathcal{E}_{mN}$ ($|fvnM\rangle = |fN(q,Q)\rangle\,|vN(Q)\rangle\,|nN\rangle$ and $E_{fvnN} = \varepsilon_{fN} + \epsilon_{vN}^{i} + \mathcal{E}_{nN}$).

One notices that all the above wave functions depend explicitly on the number of electrons N. To arrive at a fixed potential rather than fixed particle number rate, the GCE Golden Rule rate is used [13]

$$k_{ET}^{NA} = \frac{2\pi}{\hbar\Xi_i}\sum_{N,m,n} e^{-\beta(E_{imN}-\mu N)}\left|\langle Nnf|\hat{V}_N|imN\rangle\right|^2 \delta(E_{imN} - E_{fnN}) \tag{9.9a}$$
$$k_{PCET}^{NA} = \frac{2\pi}{\hbar\Xi_i}\sum_{N,u,v,m,n} e^{-\beta(E_{iumN}-\mu N)}\left|\langle Nnvf|\hat{V}_N|iumN\rangle\right|^2 \delta(E_{iumN} - E_{fvnN}) \tag{9.9b}$$

where $\hat{V}$ couples the initial and final states. The quantity between the bars is the electronic/vibronic coupling term. It is also useful to define the weights $p_{iN} = e^{-\beta(E_{imN}-\mu N)}/\Xi_i$ and $p_{iuN} = e^{-\beta(E_{iumN}-\mu N)}/\Xi_i$. Starting from the general Golden Rule formula, various rate constants valid at different well-defined limits can be derived [6, 7, 13, 35, 36]. Though not shown above, it is important to stress that proton and CPET rates are very sensitive to the distance between the proton donor-acceptor distance, the R-coordinate. Various treatments of the R-coordinate are also available [30, 35]. Below, rate constants are shown only for when i) the short time approximation to the correlation is used, ii) high-temperature limit is taken, and iii) the R-coordinate is static. Applying these to the GCE Golden Rule rates above gives [13]

$$k_{ET}^{NA} = \sum_N p_{iN} \frac{\left|V_{N,if}\right|^2}{\hbar\sqrt{4\pi k_B T\lambda}} \exp\left[-\frac{(\Delta E_{fi}^N + \lambda)^2}{4k_B T\lambda}\right] \tag{9.10a}$$

$$k_{PCET}^{NA} = \sum_{N,u} p_{iu} \sum_v \frac{|V(R)_{uv}^N|^2}{\hbar\sqrt{4\pi k_B T\lambda_{uv}}} \exp\left[-\frac{\left(\Delta E_{uv}^N + \lambda_{uv}\right)^2}{4k_B T\lambda_{uv}}\right] \tag{9.10b}$$

where ΔE_{fi}^N (ΔE_{uv}^N) is the reaction energy between initial and final electronic (vibronic) states with N electrons. λ (λ_{uv}) is the reorganization energy required to reach the final electronic (vibronic) state. See Figures 9.2 and 9.3 for clarification. $V_{N,if}$ ($V(R)_{uv}^N$) is the electronic (vibronic) coupling term. For PCET reactions, the vibronic coupling term takes a different form for electronically adiabatic and non-adiabatic reactions [6, 7, 35, 37, 38]. A universally applicable semiclassical formula for the vibronic coupling can also be used [25, 37, 38].

Looking at Eq. (9.10), one recognizes that they bear remarkable resemblance with the canonical Marcus theory used for describing ET. They also share the same form with the PT rate constants within the curve crossing picture discussed in 9.3.3. The only real difference is that to obtain a rate at a given electrode potential, a summation over different canonical states is needed in the electronically non-adiabatic limit.

9.5 Computational aspects

The fixed potential rate theory does not make use of any empirical Hamiltonians. Instead, all the above rate equations can be parametrized and evaluated directly using first principles atomic simulations with general Hamiltonians. As there is no need to parametrize the model Hamiltonians, adoption and evaluation of the rates is straightforward (but potentially laborious).

There are a few special requirements for the software used for parametrizing the rate equations. First, simulation of charged systems is needed to sample the electrode potential. The overall cell neutrality can be accomplished using e.g. Poisson-Boltzmann or jellium like models detailed in other chapters of this book. Fixed potential calculations can be accomplished using either fixed potential SCF (see chapter by Sundararam and Arias), an iterative approach (see chapter by Peterson et al.), or using the potentiostat scheme in [39]. Second, the solvent effects should be included in the model. The solvent can be presented either explicitly or using an implicit solvent model; for details I refer to chapters by Arias, Peterson, and Andreussi in this the book. While adiabatic reactions can be in principle modeled without solvent contributions, the solvent is known to greatly affect the stability of reaction intermediates and should therefore be included for qualitatively and quantitatively accurate calculations. Computation of non-adiabatic reaction rates should always be performed in the presence of a solvent; the reaction barrier is directly related to the solvent/environment reorganization energy and neglecting the solvent contributions will most likely lead to incorrect results.

Given a software capable of handling charged systems and performing constant potential calculations, adiabatic rate constants can be readily evaluated. One only needs to compute the adiabatic constant potential reaction barrier using e.g. the NEB [20] method. One should also evaluate the free energy corrections using e.g. the harmonic vibrational frequencies which leads to a harmonic constant potential TST [13]. Tunneling contributions can be estimated using the same GCE potential energy surface using methods such as rate using Bell tunneling correction, one-dimensional WKB barrier penetration integral or harmonic semiclassical TST [40] or more advanced instanton approaches.

Evaluating non-adiabatic and GCE-EVB rate constants requires additional software capabilities. The main new requirement is the possibility to construct charge/spin localized diabatic states and to evaluate the electronic coupling between these states. Also the reorganization energy, which is an excited state quantity, needs to be computed. One widely implemented and available tool for evaluating the additional parameters is the constrained DFT methodology [41–44]. Evaluation of the vibronic/vibrational matrix elements is accomplished using e.g. a Fourier grid Hamiltonian [45] method which is easy to implement. GCE-EVB simulations should be accompanied with a constant potential simulation to compute fixed potential reaction and reorganization energies. Non-adiabatic rate constants need sampling at different charge states to evaluate the summation over the number of electrons. While this summation is in principle infinite, in practice only 5-10 charge states suffice because GCE weight is non-zero only for a few states [13].

A final computational aspect is the interpolation between the different rate equations. Such an interpolation is needed to bridge the adiabatic and non-adiabatic rate constants because the division between inner-sphere/electrocatalytic/adiabatic and outer-sphere/non-adiabatic/electrochemical is not always straigthforward. Also such a division depends on the reaction, reaction conditions as well as e.g. distance between the reactant and the electrode. Currently, a generally valid interpolation for fixed potentials has not yet been developed. In the canonical ensemble interpolation between electronically/vibrationally adiabatic and non-adiabatic reactions can be accomplished using the Landau-Zener formula [31, 46]. In CPET, a universally valid interpolation from a fully non-adiabatic to fully adiabatic reaction is accomplished using a semiclassical CPET prefactor [38].

9.6 Conclusions

In this chapter a general theoretical framework for computing reaction rates at fixed electrode potentials was presented. The basic approach is to extend (micro)canonical rate theories to the GCE to enable computation of rates at fixed potentials. Based on this strategy, it is shown that all rate theories developed within the canonical ensemble can be extended to GCE. Therefore, this theoretical construction enables a unified description of electrochemical and electrocatalytic reaction rates. Specifically, ways to address e.g. adiabatic, non-adiabatic, and tunnelling reactions can be formulated within GCE were addressed. Also, well-defined Tafel-BEP free energy relations were developed and analyzed.

The rate expressions can be readily combined with *ab initio* methods to study reaction kinetics at complex electrochemical interfaces as a function of the electrode potential. The rate equations have been constructed to enable straightforward parametrization using first principles calculations in realistic environments. Fully adiabatic reaction rates can be computed with any software providing simulation of charged electrodes. The GCE-EVB and non-adiabatic rate theories present an additional requirement of constructing diabatic states and computing the coupling matrix elements – both can be accomplished using e.g. the widely implemented constrained DFT.

The possibility to study reaction rates using general first principles methods gives direct access to atomistic electrochemistry. A wide variety of mechanistic, kinetic, and thermodynamic aspects of electrocatalytic reactions can be addressed on equal footing within the GCE and the complex interplay between the electrode potential, solvation, double-layer and electrocatalysis can be studied using atomistic simulations. Besides providing a rigorous and general theoretical framework for fixed potential kinetics, the work presented herein enables computational studies on pure ET, PT, and CPET with hydrogen tunnelling pathways to circumvent scaling relations often encountered in electrocatalysis.

References

1 Zhi Wei Seh, Jakob Kibsgaard, Colin F. Dickens, Ib Chorkendorff, Jens K. Nørskov, and Thomas F. Jaramillo. Combining theory and experiment in electrocatalysis: Insights into materials design. *Science*, 355(6321), 2017.

2 Elizabeth Santos Wolfgang Schmickler. *Interfacial Electrochemistry*. Springer-Verlag Berlin Heidelberg, 2010.

3 R.R. Dogonadze. 3. theory of molecular electrode kinetics. In N.S. Hush, editor, *Reactions of Molecules at Electrodes*, pages 135–228. Wiley-Intersciences, 1971.

4 Wolfgang Schmickler. Adiabatic and non-adiabatic electrochemical electron transfer in terms of green's function theory. *Russian Journal of Electrochemistry*, 53(10):1182–1188, Oct 2017.

5 Rudiger Memming. *Semiconductor Electrochemistry*. WILEY-VCH Verlag GmbH, 2001.

6 Alexander Soudackov and Sharon Hammes-Schiffer. Derivation of rate expressions for nonadiabatic proton-coupled electron transfer reactions in solution. *The Journal of Chemical Physics*, 113(6):2385–2396, 2000.

7 Charulatha Venkataraman, Alexander V. Soudackov, and Sharon Hammes-Schiffer. Theoretical formulation of nonadiabatic electrochemical proton-coupled electron transfer at metal-solution interfaces. *The Journal of Physical Chemistry C*, 112(32): 12386–12397, 2008.

8 R. A. Marcus. On the theory of electron transfer reactions. vi. unified treatment for homogeneous and electrode reactions. *The Journal of Chemical Physics*, 43(2):679–701, 1965.

9 Marko M. Melander, Mikael J. Kuisma, Thorbjørn Erik Køppen Christensen, and Karoliina Honkala. Grand-canonical approach to density functional theory of electrocatalytic systems: Thermodynamics of solid-liquid interfaces at constant ion and electrode potentials. *The Journal of Chemical Physics*, 150(4):041706, 2019.

10 N. David Mermin. Thermal properties of the inhomogeneous electron gas. *Phys. Rev.*, 137:A1441–A1443, Mar 1965.

11 Aurora Pribram-Jones, Stefano Pittalis, E. K. U. Gross, and Kieron Burke. Thermal density functional theory in context. In Frank Graziani, Michael P. Desjarlais, Ronald Redmer, and Samuel B. Trickey, editors, *Frontiers and Challenges in Warm Dense Matter*, pages 25–60. Springer International Publishing, 2014.

12 Georg Kastlunger, Per Lindgren, and Andrew A. Peterson. Controlled-potential simulation of elementary electrochemical reactions: Proton discharge on metal surfaces. *The Journal of Physical Chemistry C*, 122(24):12771–12781, 2018.

13 Marko Melander. Grand canonical rate theory for electrochemical and electrocatalytic systems I: General formulation and proton-coupled electron transfer reactions. *Journal of the Electrochemical Society*, 167(11):116518, 2020.

14 Kendra Letchworth-Weaver and T. A. Arias. Joint density functional theory of the electrode-electrolyte interface: Application to fixed electrode potentials, interfacial capacitances, and potentials of zero charge. *Phys. Rev. B*, 86:075140, Aug 2012.

15 Nicolas G. Hörmann, Oliviero Andreussi, and Nicola Marzari. Grand canonical simulations of electrochemical interfaces in implicit solvation models. *The Journal of Chemical Physics*, 150(4):041730, 2019.

16 William H. Miller. Quantum mechanical transition state theory and a new semiclassical model for reaction rate constants. *The Journal of Chemical Physics*, 61(5):1823–1834, 2018/08/20 1974.

17 William H. Miller. Semiclassical limit of quantum mechanical transition state theory for nonseparable systems. *The Journal of Chemical Physics*, 62(5):1899–1906, 2018/08/20 1975.

18 William H. Miller, Steven D. Schwartz, and John W. Tromp. Quantum mechanical rate constants for bimolecular reactions. *The Journal of Chemical Physics*, 79(10):4889–4898, 1983.

19 William H. Miller. Direct and correct calculation of canonical and microcanonical rate constants for chemical reactions. *The Journal of Physical Chemistry A*, 102(5):793–806, 1998.

20 Graeme Henkelman, Blas P. Uberuaga, and Hannes Jónsson. A climbing image nudged elastic band method for finding saddle points and minimum energy paths. *The Journal of Chemical Physics*, 113(22):9901–9904, 2000.

21 A. Warshel, J. K. Hwang, and J. Åqvist. Computer simulations of enzymatic reactions: examination of linear free-energy relationships and quantum-mechanical corrections in the initial proton-transfer step of carbonic anhydrase. *Faraday Discuss.*, 93:225–238, 1992.

22 Edina Rosta and Arieh Warshel. Origin of linear free energy relationships: Exploring the nature of the off-diagonal coupling elements in sn2 reactions. *Journal of Chemical Theory and Computation*, 8(10):3574–3585, 2012. PMID: 23329895.

23 J. K. Nørskov, J. Rossmeisl, A. Logadottir, L. Lindqvist, J. R. Kitchin, T. Bligaard, and H. Jónsson. Origin of the overpotential for oxygen reduction at a fuel-cell cathode. *The Journal of Physical Chemistry B*, 108(46):17886–17892, 2004.

24 Ken Sakaushi, Andrey Lyalin, Tetsuya Taketsugu, and Kohei Uosaki. Quantum-to-classical transition of proton transfer in potential-induced dioxygen reduction. *Phys. Rev. Lett.*, 121:236001, Dec 2018.

25 Zachary K. Goldsmith, Yan Choi Lam, Alexander V. Soudackov, and Sharon Hammes-Schiffer. Proton discharge on a gold electrode from triethylammonium in acetonitrile: Theoretical modeling of potential-dependent kinetic isotope effects. *Journal of the American Chemical Society*, 141(2):1084–1090, 2019.

26 Jeremy O. Richardson. Derivation of instanton rate theory from first principles. *The Journal of Chemical Physics*, 144(11):114106, 2016.

27 Rigoberto Hernandez and William H. Miller. Semiclassical transition state theory. a new perspective. *Chemical Physics Letters*, 214(2):129–136, 1993.

28 R. P. Bell. *The Tunnel Effect in Chemistry*. Springer, Boston, MA, 1980.

29 Daniel C. Borgis, Sangyoub Lee, and James T. Hynes. A dynamical theory of nonadiabatic proton and hydrogen atom transfer reaction rates in solution. *Chemical Physics Letters*, 162(1):19–26, 1989.

30 Daniel Borgis and James T. Hynes. Dynamical theory of proton tunneling transfer rates in solution: general formulation. *Chemical Physics*, 170(3):315–346, 1993.

31 Daniel Borgis and James T. Hynes. Curve crossing formulation for proton transfer reactions in solution. *The Journal of Physical Chemistry*, 100(4):1118–1128, 1996.

32 Sharon Hammes-Schiffer. Proton-coupled electron transfer: classification scheme and guide to theoretical methods. *Energy Environ. Sci.*, 5:7696–7703, 2012.

33 Yan-Choi Lam, Alexander V. Soudackov, Zachary K. Goldsmith, and Sharon Hammes-Schiffer. Theory of proton discharge on metal electrodes: Electronically adiabatic model. *The Journal of Physical Chemistry C*, 123(19):12335–12345, 2019.

34 Alexander Soudackov and Sharon Hammes-Schiffer. Multistate continuum theory for multiple charge transfer reactions in solution. *The Journal of Chemical Physics*, 111(10):4672–4687, 1999.

35 Alexander Soudackov, Elizabeth Hatcher, and Sharon Hammes-Schiffer. Quantum and dynamical effects of proton donor-acceptor vibrational motion in nonadiabatic proton-coupled electron transfer reactions. *The Journal of Chemical Physics*, 122(1):014505, 2005.

36 Sharon Hammes-Schiffer, Elizabeth Hatcher, Hiroshi Ishikita, Jonathan H. Skone, and Alexander V. Soudackov. Theoretical studies of proton-coupled electron transfer: Models and concepts relevant to bioenergetics. *Coordination Chemistry Reviews*, 252(3):384–394, 2008. The Role of Manganese in Photosystem II.

37 Jonathan H. Skone, Alexander V. Soudackov, and Sharon Hammes-Schiffer. Calculation of vibronic couplings for phenoxyl/phenol and benzyl/toluene self-exchange reactions: implications for proton-coupled electron transfer mechanisms. *Journal of the American Chemical Society*, 128(51):16655–16663, 2006. PMID: 17177415.

38 Yuri Georgievskii and Alexei A. Stuchebrukhov. Concerted electron and proton transfer: Transition from nonadiabatic to adiabatic proton tunneling. *The Journal of Chemical Physics*, 113(23):10438–10450, 2000.

39 Nicéphore Bonnet, Tetsuya Morishita, Osamu Sugino, and Minoru Otani. First-principles molecular dynamics at a constant electrode potential. *Phys. Rev. Lett.*, 109:266101, Dec 2012.

40 Justin T. Fermann and Scott Auerbach. Modeling proton mobility in acidic zeolite clusters: Ii. room temperature tunneling effects from semiclassical rate theory. *The Journal of Chemical Physics*, 112(15):6787–6794, 2000.

41 Qin Wu and Troy Van Voorhis. Direct optimization method to study constrained systems within density-functional theory. *Phys. Rev. A*, 72:024502, Aug 2005.

42 Qin Wu and Troy Van Voorhis. Extracting electron transfer coupling elements from constrained density functional theory. *J. Chem. Phys.*, 125(16):164105, 2006.

43 Benjamin Kaduk, Tim Kowalczyk, and Troy Van Voorhis. Constrained density functional theory. *Chem. Rev.*, 112(1):321–370, 2012.

44 Marko Melander, Elvar Ö. Jónsson, Jens J. Mortensen, Tejs Vegge, and Juan Maria García Lastra. Implementation of constrained dft for computing charge transfer rates within the projector augmented wave method. *Journal of Chemical Theory and Computation*, 12(11):5367–5378, 2016. PMID: 27749068.

45 D. J. Tannor. *Introduction to Quantum Mechanics: A Time-Dependent Perspective.* University Science Books, 2007.

46 Jochen Blumberger. Recent advances in the theory and molecular simulation of biological electron transfer reactions. *Chemical Reviews*, 115(20):11191–11238, 2015. PMID: 26485093.

Part V

10

Thermodynamically consistent free energy diagrams with the solvated jellium method

Georg Kastlunger, Per Lindgren, and Andrew A. Peterson

School of Engineering, Brown University, Providence, Rhode Island, 02912, USA

10.1 Computational studies of electrochemical systems – Recent advances and modern challenges

Density functional theory has proven very useful for studying solid-state systems in vacuum. However, conventional DFT is not directly applicable to electrochemical systems. The most prominent issue is that conventional DFT works within the canonical ensemble, where the number of particles is constant. Electrochemical reactions, however, demand an ensemble where the chemical potential of the active electrons – and not the number of electrons – is constant [1]. To see why electrochemical reactions in canonical DFT are problematic, one can consider the Volmer step of HER on Pt(111) in Fig. 10.1. Structural relaxation of the initial state, consisting of a Pt(111) surface and a protonated water bilayer, induces a charge separation where a fractional number of electrons from the additional hydrogen atom in the water bilayer transfers to the metal surface. The metal and water bilayer are thus negatively and positively charged, respectively. However, the final state contains a neutral water bilayer and an adsorbed hydrogen atom on the metal surface, where the transferred proton has received the excess charge from the surface to form hydrogen. The final state therefore consists of two neutral subsystems. This drastic change in the surface dipole along the reaction trajectory leads to a corresponding change in the electrode Fermi level, causing the work function to change according to Eq. 10.1,

$$\phi = \phi_{\text{ref}} - \frac{\mu}{e}, \tag{10.1}$$

where μ and ϕ_{ref} are the Fermi level of the metal and the reference potential, respectively, and e is the positive elementary charge.

This computational setup introduces an artificial potential bias where the initial and final states (and any states along an electrochemical reaction trajectory) have different applied potentials. This is evident in Fig. 10.1, where the work function changes by over 2 V in a nudged elastic band (NEB) [2, 3] calculation performed under a boundary condition defined by a constant number of electrons in the unit cell. Any realistic simulation of an electrochemical system should thus correspond to the (electronically) grand canonical ensemble.

Atomic-Scale Modelling of Electrochemical Systems, First Edition.
Edited by Marko M. Melander, Tomi T. Laurila, and Kari Laasonen.

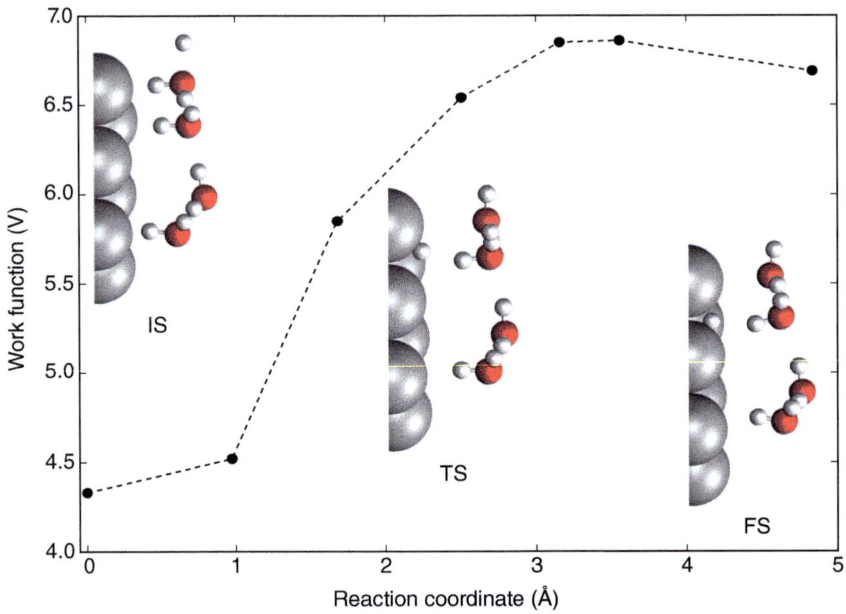

Figure 10.1 Work function in a canonical (constant-charge) nudged elastic band calculation of the Volmer step of HER on Pt(111). The fifth state (fourth interior) represents the transition state. None of the states along the reaction trajectory has the same work function; the work function changes by over 2 V in the course of the reaction.

A multitude of methods to study electrochemical systems within the DFT framework are available in the literature. The most commonly used method is the computational hydrogen electrode (CHE) [4], which allows for potential-dependent studies of reaction thermodynamics. The fundamental basis of CHE is the equilibrium between a proton-electron pair and gaseous hydrogen at 0 $V_{SHE/RHE}$. Any deviation from the equilibrium potential can be dealt with in post-processing steps according to Eq. 10.2 [5],

$$\mu(H^+) + \mu(e^-) = \frac{1}{2}\mu(H_2(g)) - eU, \tag{10.2}$$

where $\mu(H^+)$ and $\mu(e^-)$ are the chemical potentials of the proton and electron, respectively, e is the positive elementary charge and U is the applied potential. This method predicts that for a change in potential by 1 V, the free energy difference between two consecutive elementary states in an electrochemical reaction will change by 1 eV. The CHE model has proven incredibly useful for mechanistic studies of electrochemical systems [5], but is limited to the analysis of thermodynamically stable intermediates. Other methods have attempted to deal with the modeling of not only stable intermediates, but also transition states. These range from modeling electrochemical reactions as non-electrochemical counterparts [6] to treating the electrochemical interface explicitly. Among the latter, one method deals with increasingly large unit cells to minimize the artificial potential bias introduced by the metal charge depletion described above [7, 8]. This so-called "unit cell extrapolation" method allows for calculations of electrochemical reactions at constant potential, but is hampered by the computational expense associated with increasingly large unit cells, since DFT scales as $O(N_e^2)$-$O(N_e^3)$ [9]. Chan and Nørskov [10, 11] used a capacitor argument to model electrochemical systems based on canonical simulations and extrapolate the charge *a posteriori* to keep the metal work function constant. All these methods have the advantage that they avoid explicitly charging the electrode by adding or subtracting electrons. However, none of these methods provides a straightforward and rigorous way to study an electrochemical reaction at any arbitrary potential. Furthermore, some intermediate states may only be stable within a certain potential range, and thus require methods with appropriate potential control.

Other methods explicitly deal with applied charges [12–20]. This introduces an additional degree of freedom; electrons (or fractions thereof) must be added to or subtracted from the unit cell in order to reach a target potential. The variability in the number of electrons introduces yet another challenge in the study of metals. Metals are extended systems, and periodic boundary conditions must be used to study the semi-infinite nature of metal slabs. However, periodically repeating any unit cell with a net charge will cause the energy to diverge. One must therefore introduce a counter-charge in the unit cell to enforce charge neutrality. Several approaches have been developed to deal with this issue. The simplest approach is to use a homogeneous background charge. As the name suggests, this approach introduces a uniformly distributed background charge in the unit cell with the same magnitude as the explicit charge but opposite sign. This method, while computationally efficient, does not capture the physics of an electrochemical system. Since the background charge is smeared out in the entire unit cell, it interacts with all the atoms. Furthermore, the uniformly distributed background charge prevents any field-free region from being defined. The lack of a field-free region in the unit cell prohibits the definition of the constant reference potential in Eq. 10.1.

Another commonly used charge neutralization scheme is a localized counter charge. This counter charge can take any shape, but has been modeled as a Gaussian in previous works [13, 14]. Unlike the homogeneous background approach, this method does not suffer from a lack of field-free region in the unit cell. However, in the absence of a screening medium, this approach represents two isolated capacitor plates in vacuum. The absence of a screening medium creates an artificially large electric field between the working electrode and the counter charge, contradicting the physical picture of an electrochemical interface where the electric field decays according to the Gouy–Chapman–Stern model [21]. It is important to emphasize that any method that explicitly deals with charge should have a physically reasonable charge neutralization scheme that reproduces the physical picture of an electrochemical interface.

10.2 Thermodynamic consistency with a decoupled computational electrode model

The computational hydrogen electrode (CHE) model [4, 5] has been the most widely used methodology in the computational electrocatalysis community for the past 15 years.

The fundamental relation of CHE is the equilibrium of the standard hydrogen electrode (SHE) at any temperature,

$$\mu^0_{\mathrm{H^+}} + \mu^{0\mathrm{V}}_{\mathrm{e^-}} = \frac{1}{2}\mu^0_{\mathrm{H_2}}, \tag{10.3}$$

where $\mu^0_{\mathrm{H^+}}$, $\mu^{0\mathrm{V}}_{\mathrm{e^-}}$ and $\frac{1}{2}\mu^0_{\mathrm{H_2}}$ correspond to the chemical potential of the proton in solution, the electron in the circuit and H_2 in gas phase, respectively, in their respective standard states (pH 0, 0 V_{SHE}, p = 101325 Pa). It is challenging to accurately calculate $\mu^0_{\mathrm{H^+}}$ and $\mu^0_{\mathrm{e^-}}$, but $\mu^0_{\mathrm{H_2}}$ is directly accessible within electronic structure methods as the vibrational free energy of H_2. Therefore, substitution of the free energy of the proton-electron pair by H_2 is an elegant way to describe the thermodynamics of electrochemical reactions.

The electrode potential with respect to 0 V_{SHE} can be accounted for in a post-processing step by adding $-e\phi$ to $G_{\mathrm{H_2}}$. This accounts for the deviation of $\mu_{\mathrm{e^-}}$ from its standard state. Hence, the final equilibrium at any temperature and any given potential on a SHE scale (ϕ) changes into

$$\mu^0_{\mathrm{H^+}} + \mu^{\phi}_{\mathrm{e^-}} = \frac{1}{2}\mu^0_{\mathrm{H_2}} - e\phi_{\mathrm{SHE}}, \tag{10.4}$$

where

$$\mu^{\phi}_{\mathrm{e^-}} = \mu^{0\mathrm{V}}_{\mathrm{e^-}} - e\phi_{\mathrm{SHE}}. \tag{10.5}$$

The free energy change of an electrochemical reaction, such as $H^+ + e^- + * \rightarrow H*$ studied at pH 0, can now elegantly be written as

$$\Delta G_{\mathrm{rxn}} = \mu_{\mathrm{H}*} - \mu_* - \left(\frac{1}{2}\mu_{\mathrm{H_2(g)}} - e\phi\right), \tag{10.6}$$

where the chemical potentials of the initial (μ_*) and final states ($\mu_{\mathrm{H}*}$) are generally calculated as the vibrational free energies at charge neutral conditions. Applications of the CHE model have proven useful in numerous studies. However, the scheme is only applicable for calculating reaction thermodynamics. The reason for this limitation is that CHE does not account for changes in the interface dipole perpendicular to the electrode surface, $p_\perp$, in the course of a reaction. The electrode potential is a function of $p_\perp$, which (besides other system parameters, such as the dielectric properties of the solvent) determines the shape of the electrostatic potential profile between the bulk of the electrode and the electrolyte. In a complete electrochemical proton-coupled electron transfer reaction ($H^+ + e^- + * \rightarrow H*$), $p_\perp$ does not change significantly if one only compares the stable endstates of the reaction. Hence, constant-potential conditions are present in ΔG_{rxn} by default. For intermediate states of along a reaction trajectory, e.g. a saddle point, $p_\perp$ often differs from its value in the initial and final states. This causes the constant-potential condition to be broken. There are two reasons for this: (i) rotation of implicit dipoles at the interface, i.e., rotation of water molecules, and (ii) (partial) charge transfer that creates an induced dipole in the unit cell.

Here, we describe how this limitation of CHE can be lifted by changing from a constant-particle description to a constant-potential treatment. We note that this demands a scheme for treating charged electrochemical interfaces, such as the solvated jellium method [22] or Poisson–Boltzmann schemes [20, 23–27]. In such schemes, the free energy at constant applied potential is $G^{\phi} = \mu^{\phi} - N_e\mu^{\phi}_{\mathrm{e}}$ (often referred to as the (semi-)grand canonical free energy Ω).

We start the derivations by rewriting Eq. (10.6) into

$$\begin{aligned}\Delta G^{\phi}_{\mathrm{rxn}} &= \mu_{\mathrm{H}*} - \mu_{*} - (\mu^{0}_{\mathrm{H}^+} + \mu^{0\mathrm{V}}_{\mathrm{e}^-} - e\phi) \\ &= \mu_{\mathrm{H}*} - \mu_{*} - \mu^{0}_{\mathrm{H}^+} - \mu^{\phi}_{\mathrm{e}},\end{aligned} \tag{10.7}$$

where μ_{*} and $\mu_{\mathrm{H}*}$ represent the chemical potentials of the states where the proton is residing in the bulk of the solution (not included in the unit cell) and adsorbed on the surface, respectively, at charge-neutral conditions. $\mu^{0\mathrm{V}}_{\mathrm{e}^-}$ is the chemical potential of the electron at 0 V on the SHE scale and e is the (positive) elementary charge. μ^{ϕ}_{e} is the chemical potential of the electron in the electrode at ϕ, defined as the Fermi level of the electrode (referenced to the field free potential in the solvent near surface).

From Eq. 10.7, the reason for CHE's limitation to thermodynamic endstates is apparent. First, μ_{*} and $\mu_{\mathrm{H}*}$ are calculated at charge-neutral conditions, and the difference between them is only meaningful if their interface dipoles, $p_{\perp}$, and Fermi levels are equivalent. Additionally, an equivalent surface capacity is assumed, which makes it possible to vary ϕ linearly by changing μ_{e}.

Moreover, integer electron transfer is assumed, which does not allow for comparisons to reaction intermediates, whose numbers of electrons generally vary gradually due to the adiabatic nature of the reactions.

Constant-potential methods are able to lift both these limitations. By monitoring the Fermi level of the electrode and adapting the number of electrons in the unit cell, the chemical potential of the compared states can be calculated at any applied potential, ϕ. Additionally, the amount of transferred charge is easily accessible from the difference in the number of electrons at ϕ, ΔN_{e}. With these two changes, Eq. 10.7 can now be rewritten,

$$\Delta G^{\phi}_{\mathrm{rxn}} = \mu^{\phi}_{\mathrm{H}*} - \mu^{\phi}_{*} - \mu^{0}_{\mathrm{H}^+} - \Delta N_{\mathrm{e}}\mu^{\phi}_{\mathrm{e}}. \tag{10.8}$$

Equation 10.8 can be applied to any arbitrary intermediate of the reaction and gives the possibility of calculating activation energies for electrochemical reactions referenced to the initial state of a bulk solvated proton. The leap to constant-potential free energies is also apparent. Substituting $G^{\phi} = \mu^{\phi} - N_{e}\mu^{\phi}_{\mathrm{e}}$ into Eq. (10.8) now leads to

$$\Delta G^{\phi}_{\mathrm{rxn}} = G^{\phi}_{\mathrm{H}*} - G^{\phi}_{*} - \mu^{0}_{\mathrm{H}^+}, \tag{10.9}$$

which is a direct consequence of the electronically open systems that are compared; the difference in the number of electrons between the two states is directly accounted for in the constant-potential free energies. However, the systems are not open for protons, and the energetic cost of introducing protons into the system has to be added explicitly. $\mu^{0}_{\mathrm{H}^+}$, is not directly accessible within standard DFT methods, since it would demand sampling of the proton-water complex in the solution via e.g. molecular dynamics. One can, however, make use of the SHE equilibrium along with the absolute value of the chemical potential of the electron at 0 V (≈ -4.44 eV [28]). With this scheme, the definition of the free energy change along the reaction coordinate – including proton shuttling from bulk to interface – can be calculated as

$$\Delta G^{\phi}_{\mathrm{rxn}} = \Delta G^{\phi}_{\mathrm{H}*} - \Delta G^{\phi}_{*} - \frac{1}{2}G_{\mathrm{H}_2(\mathrm{g})} + \mu^{0}_{\mathrm{e}^-}, \tag{10.10}$$

This makes it possible to calculate the free energy change of a proton shuttling step directly from constant-potential free energies and the definition of $\mu^{0}_{\mathrm{H}^+} = \frac{1}{2}G_{\mathrm{H}_2} - \mu^{0}_{\mathrm{e}^-}$. Expressing

Eq. 10.10 in terms of the chemical potentials of the involved species leads to the elegant final solution,

$$\Delta G^{\phi}_{\rm rxn} = \mu^{\phi}_{\rm H*} - \mu^{\phi}_{*} - \frac{1}{2}\mu_{\rm H_2(g)} + (1 - \Delta N_{\rm e})\mu^{\phi}_{e}. \tag{10.11}$$

This equation directly reverts into the CHE equation if a reaction involves integer electron transfer and has a negligible change in the intrinsic dipole of the electrode-solution interface. Thus, it can be interpreted as a generalization of CHE that is applicable to any free energy change along a reaction trajectory involving a proton-electron pair.

Interestingly, the methodology is not only applicable to electrochemical reactions, but can be applied in a straightforward manner for any reaction at constant potential. For example, (de-)protonation of an acidic molecule adsorbed on an electrode does not involve electron transfer [42]. Equations 10.10 and 10.11 automatically capture the nature of the reaction; if a reaction does not involve electron transfer, $\Delta N_{\rm e}$ will be close to zero. Similarly, a reaction that does not involve proton transfer can be treated by not including $\mu^{0}_{\rm H+}$ in Eq. 10.10. This changes the description into a pure difference in constant-potential free energies.

10.3 Solvated jellium method (SJM)

10.3.1 Introduction

To address the issues associated with computational studies of electrochemical systems, we have developed the solvated jellium method (SJM). This method – which is implemented in the GPAW [29, 30] electronic structure code – allows for first-principles calculations of electrochemical systems at constant potential. The solvated jellium method can be partitioned into two sections: charging and screening. The system under study is explicitly charged using a Jellium slab [31]. Jellium is a charge-neutral entity that includes explicit charge and a background charge of equal magnitude but opposite sign. While the distribution of explicit charge is optimized within the self-consistent field (SCF) cycle of DFT, the background charge is spatially constrained in the SCF cycle. By itself, this charging approach would closely resemble that of the isolator capacitor approach described above [13, 14]. In order to create a more realistic behavior, we screen the electric field with explicit and implicit solvents. The resulting electrostatic problem is solved in the generalized Poisson equation

$$\nabla(\epsilon(\mathrm{r})\nabla\Phi(\mathrm{r})) = -4\pi(\rho_{\rm explicit}(\mathrm{r}) + \rho_{\rm jellium}(\mathrm{r})) \tag{10.12}$$

where the charge density $\rho(\mathrm{r})$ includes the neutral atomic system electron density and compensation charge ($\rho_{\rm explicit}(\mathrm{r})$), as well as the charge density from the jellium electron/hole density and its jelly background charge ($\rho_{\rm jellium}(\mathrm{r})$).

The electron density in the unit cell, $\rho_{\rm jellium}(\mathrm{r})$, is initially localized in the region of the homogeneous jellium counter charge. However, the electron/hole density transfers into the resulting Kohn–Sham states in the course of the SCF cycle, provided that the Fermi level of the atomic system lies well below the electrostatic potential of the solvent. This is generally the case, since the Fermi level is ≈4.4 V lower than the mean potential in the solvent at 0 V on the standard hydrogen electrode (SHE) scale. This large gap cannot be closed by the diffuse layer, described by the jellium counter charge, at reasonable applied charges or potentials. This localization only fails at extremely low potentials, deviating from 0 V_{SHE} by several volts, where the Fermi level is artificially high. This will be discussed in a later section of this chapter.

A description of the electrostatic problem on the basis of Eq. 10.12 is certainly approximative in comparison to more realistic Poisson–Boltzmann (PB) descriptions of the electrochemical interface [25–27, 32–34]. However, the appeal of SJM lies in its efficiency and stability for treating the electrostatic problem; the improved efficiency compared to PB methods stems from the fact that the generalized Poisson equation only needs to be solved once per SCF step of DFT. (Non-linearized) Poisson–Boltzmann schemes, on the other hand, demand a self-consistent treatment of the electrostatic problem. Here, the generalized Poisson equation has to be solved in each iteration while the ionic charge density is optimized. [26].

Regarding stability, it has been shown that PB schemes can lead to spikes in the electrostatic potential due to artificial accumulation of ions in the interface [35]. (This behavior can be avoided by applying the size-modified Poisson–Boltzmann equation [25–27].) In SJM, the homogeneous charge background is evenly distributed in the implicit solvent region, and the potential smoothly approaches the value in the solvent without spikes in the electrostatic potential. This smooth decay of the electrostatic potential leads to rapid and reliable convergence of the finite difference Poisson solver.

10.3.2 Electrostatic potential profiles and charge localization

The charging-screening scheme described in the last section addresses the issues associated with homogeneous background charges and localized background charges in vacuum. The solvated background charge allows for the definition of a reference potential in a field-free region of the unit cell, and the inclusion of explicit and implicit solvent between the working electrode and counter charge scales down the electric field considerably with respect to the vacuum case.

Figure 10.2 shows the electrostatic potential profile perpendicular to the surface plane for Au(111) covered by an H-down water bilayer. Here, we describe five distinct regions in the unit cell:

- **Electrode backside (I)**: This region, which extends from the far left of the unit cell until the back of the electrode is reached, is a vacuum region. This is a field-free region (constant electrostatic potential) that is independent of the applied charge/potential; the dipole-correction perpendicular to the lateral dimensions of the electrode ensures that the vacuum work function is constant. This behavior is not only valid for static structures with varying applied charges/potentials, but also for structures with changing geometries. An example of the latter is saddle point searches, where the geometry changes along the reaction coordinate. This behavior is unique to the solvated jellium method.
- **Electrode bulk (II)**: This region contains the metallic electrode. The electrostatic potential varies significantly due to the large number of electrons and the applied compensation charges close to the atomic cores. Additionally, the conducting nature of metallic systems screens the electric field on the reactive side of the electrode. This leads to a constant *xy*-averaged electrostatic potential inside the electrode.

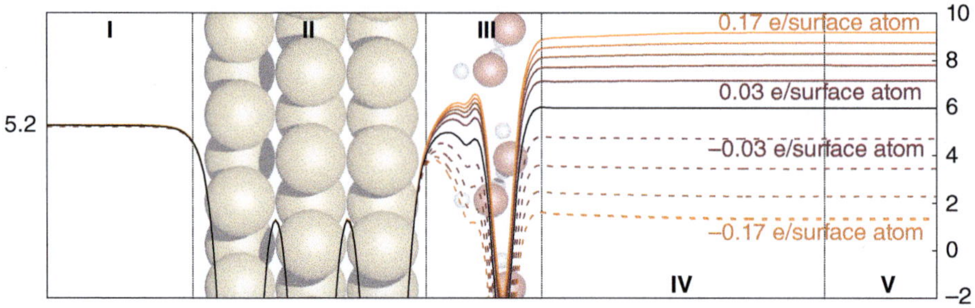

Figure 10.2 Plane-averaged electrostatic potential profile perpendicular to the electrode surface for a Au(111) slab covered with a hexagonal H-down water bilayer. The different lines represent varying applied charges in the SJM scheme, where the charge increment is 0.03 electrons per surface atoms. Source: Adapted from Georg Kastlunger, Per Lindgren, and Andrew A. Peterson. Controlled-potential simulation of elementary electrochemical reactions: proton discharge on metal surfaces *J. Phys. Chem. C* 2018, 122, 24, 12771–12781.

- **Stern layer (III)**: This region, located between the metallic electrode and the explicit water layer, is crucial for simulations of electrochemical reactions, since proton and electron transfer occurs here. As seen in Fig. 10.2, the applied potential has the greatest effect in this region. As shown later in this section, this behavior leads to a shift of the frontier orbitals of the explicit water molecules with respect to the metallic Fermi level.
- **Diffuse layer (IV)**: Behind the explicit water layer, the implicit solvent – with jelly background charge embedded in it – is present. Due to the dielectric constant of the chosen solvent the electric field is scaled down in this region, leading to comparatively minor changes in the electrostatic potential. When moving away from the surface, the net monopole in the system is gradually neutralized. Consequently, the electric field reduces exponentially, leading to a flat electrostatic potential profile at the end of the region. This homogeneous diffuse charged region, which is screened by the solvent, approximates the diffuse layer within which the electrode potential attracts ionic species from the bulk electrolyte.
- **Field free solvent (V)**: This region corresponds to the (mean) electrostatic potential in the solvent bulk. The nature of the charging scheme of SJM ensures that no monopole moment is present. Moreover, the dipole-correction ensures that no dipole moment is present in this region of the unit cell, leading to a flat electrostatic potential profile. This constant electrostatic potential value acts as the reference for the Fermi level of the slab, which determines the (absolute) electrode potential [28].

Figure 10.3 shows the electrostatic potential of charged systems relative to the neutral case within the SJM framework. The innermost region of the electrochemical interface has a high, constant electric field, while the electric field decays exponentially farther away from the electrode surface. These two regions correspond to the Stern layer and diffuse plane, respectively, in the Gouy–Chapman–Stern model [21] of an electrochemical interface. This relatively simple charge-neutralization scheme thus produces interfacial electrostatic potential profiles that are in good qualitative agreement with electrochemical theory and experiments.

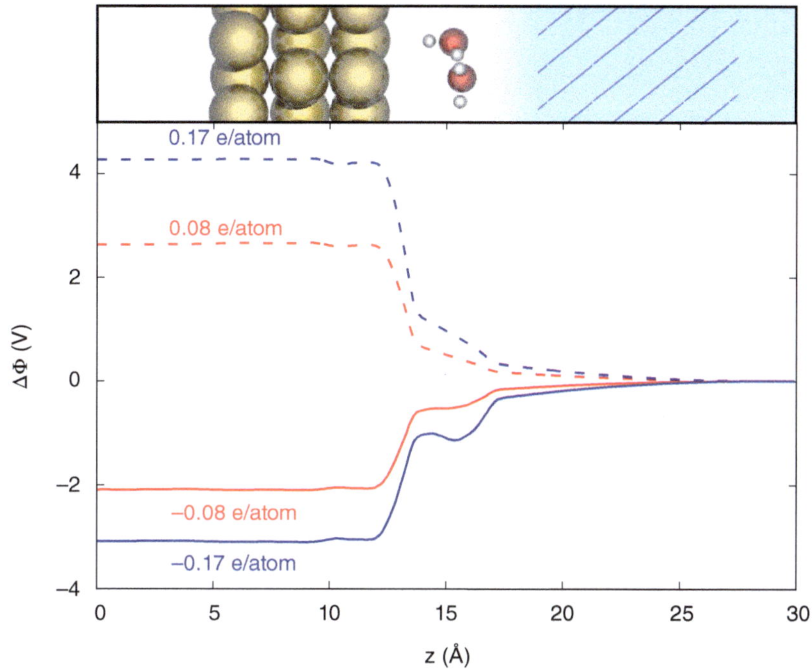

Figure 10.3 Electrostatic potential difference between charged systems and a neutral system with the same geometry. Copyright 2018 American Chemical Society. Source: Adapted from Georg Kastlunger, Per Lindgren, and Andrew A. Peterson. Controlled-Potential Simulation of Elementary Electrochemical Reactions: Proton Discharge on Metal Surfaces. *J. Phys. Chem. C* 2018, 122, 24, 12771–12781.

When explicitly charging a unit cell, one must ensure that the charge localizes in the region of interest; that is, the metal surface at the interface. The charge localization depends on the energetic alignment between metal surface and explicit water, and should be carefully monitored. SJM allows for selective charging of the electrochemical interface. Figure 10.4 shows the *xy*-averaged integration of electron density in an H-up water geometry on Pt(111) of charged systems relative to a neutral system of the same geometry.

In this representation, a slope indicates a change in the number of electrons relative to the neutral system. The back of the electrode is not solvated, and any applied charge does not affect the occupation of states. This is an important result, since the back of the electrode is fixed at the bulk lattice constant and is meant to represent the bulk of the electrode infinitely far away from the interface. However, the number of electrons changes significantly in the interface region, with the majority of the electrons either added to or subtracted from the metal surface. This charging scheme works selectively for charges of low and intermediate magnitude. However, Fig. 10.4 shows a deviation from charge-selectivity for charges of higher magnitude. For these high-magnitude charges, electrons are not only added to or subtracted from the metal at the interface, but also to or from the water bilayer. This is a result of the energetic alignment between metal and the frontier orbitals of water, and is in agreement with the findings of Björketun et al. [36], who found that the Fermi level of the metal will only change relative to the solvent if it is located between the highest

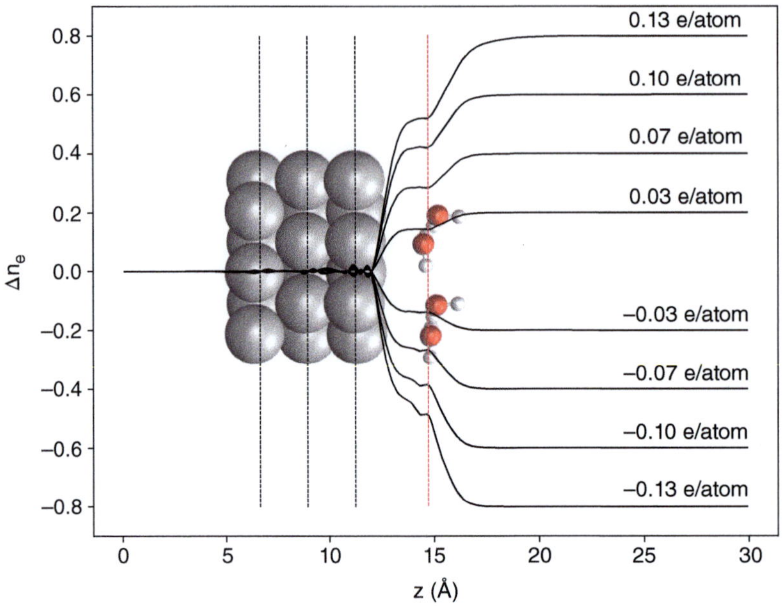

Figure 10.4 *xy*-averaged integration of electron density in an H-up water geometry on Pt(111) of charged systems relative to a neutral system of the same geometry. The vertical black lines represent the *z*-location of each Pt layer, and the vertical red line represents the average *z*-location of oxygen atoms in the water bilayer. The labels denote the surplus/deficiency of electrons normalized to the number of surface atoms.

occupied molecular orbital (HOMO) and lowest unoccupied molecular orbital (LUMO) of the solvent. The metal can therefore be selectively charged if its Fermi level is located in the HOMO-LUMO gap of the solvent. However, selectively adding electrons to the metal causes the Fermi level to increase relative to the frontier orbitals of the solvent. Conversely, subtraction of electrons from the metal decreases the Fermi level relative to the molecular eigenstates of the solvent. For high, positive charges (subtraction of electrons) the Fermi level of the metal decreases to a point where it is in resonance with the HOMO of the solvent. Addition of electrons on the other hand increases the Fermi level relative to the (solvent) reference potential until the electron density starts to spill out it into the implicit solvent region above the interface. When the two subsystems are in resonance, they are degenerate and further addition/subtraction of electrons will be added to or subtracted from both the metal and solvent. This is shown in Fig. 10.5, which shows the projected density of states (PDOS) of the p_z states of an H-down water geometry on Au(111) with varying charges. For positive charges of higher magnitude, the Fermi level of the metal approaches the HOMO of the water until the two subsystems are in resonance when ≈ 0.07 e/atom are removed from the unit cell.

While the charging for low and intermediate magnitudes is selective, it is not perfectly selective. The reason for this could be that there is always some degree of hybridization between the metal and the water bilayer. This degree of hybridization is highly dependent on the degree of solvation between the two subsystems. If one were to completely solvate the interface between metal and water bilayer, the two subsystems would be electrostatically decoupled. This would result in highly selective charging, but this is not a physically

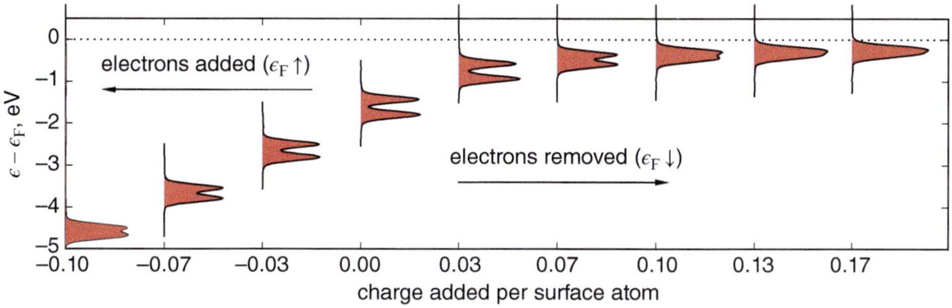

Figure 10.5 Density of states projected onto the p_z states of the water bilayer in an H-down geometry on Au(111). Source: Adapted from Georg Kastlunger, Per Lindgren, and Andrew A. Peterson. Controlled-potential simulation of elementary electrochemical reactions: proton discharge on metal surfaces *J. Phys. Chem. C* 2018, 122, 24, 12771–12781.

reasonable model since the electrostatic potential at the interface would deviate from that shown in Fig. 10.3. Rather than completely solvating the interface, the implicit solvent module has been modified to partially solvate the interface. Here, the innermost part of the interface – corresponding to the Stern layer of the Gouy-Chapman-Stern model – is not solvated to allow for a high, constant electric field. The remaining regions of the unit cell are solvated based on the location of the explicit atoms.

For all applied charges, we find that the majority of the applied charge localizes on the metal surface at the interface. The next logical step is to study the potential response as a function of applied charge, since potential is one of the variables of interest in electrochemistry. As one might expect for a systematic charging scheme, the potential changes linearly over a wide window of applied charges. The linear response, which is shown in Fig. 10.6, extends over a range of $\approx$ 6 V on Pt(111). This makes constant-potential calculations relatively inexpensive, since the charge optimization algorithm needs a maximum of three static calculations to reach a target potential. As noted in subsequent sections, the charge equilibrion scheme in SJM rarely needs more than two consecutive SCF cycles to reach a target potential. However, the linear response does not hold for high-magnitude charges. For high, positive charges (subtraction of electrons) the Fermi level is in resonance with the HOMO of the water bilayer, causing the charge selectivity to decrease as noted above. The potential response deviates from linearity for high, negative charges. However, these potentials are not of interest in HER (or other electrocatalytic reactions of interest), since the absolute standard hydrogen electrode potential is ~4.44 V [28].

10.3.3 Workflow of potential equilibration

One major advantage of the use of an implicit electrolyte is the reduction in computational cost, while providing a reasonable physical description of the solvent. In fact, explicitly accounting for the solution molecules and ions in the electrolyte would not only increase the computational demand tremendously, but also creates the need for thermodynamic sampling via e.g. molecular dynamics simulations. Hence, applying the averaged values for the solvent dielectric constant and neglecting the quantum and dipolar nature of the solvent is desirable, if not necessary. This is particularly important for simulations of electrode-electrolyte interfaces at constant potential. Since the total number of electrons

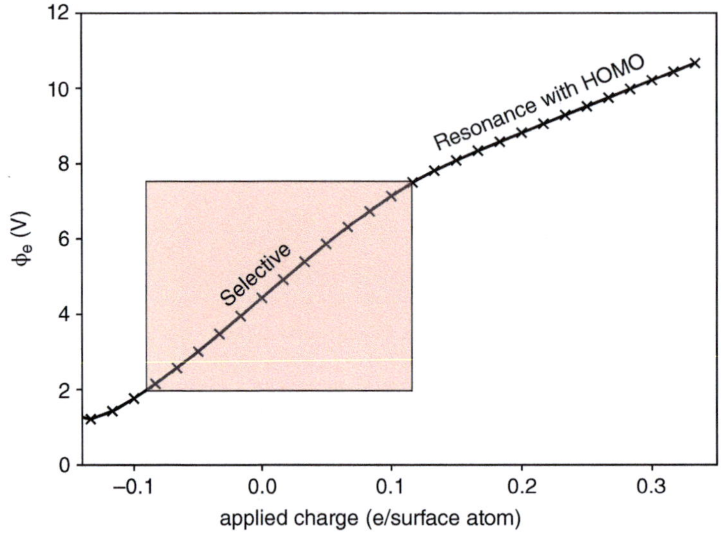

Figure 10.6 Electrode potential as a function of applied charge for an H-up water bilayer on Pt(111). Charging is selective in the range ≈ 2–8 V. The metal is in resonance with the HOMO of the water bilayer for high-magnitude positive charges.

in the system is allowed to vary, a strategy for reaching the desired potential by means of standard DFT calculation has to be developed. Three strategies for constant-potential structure optimizations are commonly pursued:

- **Constant charge relaxation followed by adaptation of the charge:** Here, the structure optimization is performed at charge neutral conditions. Once the forces in the system are minimized the charge is varied iteratively until the desired potential is reached [37, 38].
- **Alternatively optimizing the geometry and electrode potential:** In this methodology, two loops are used in series. The electrode potential is calculated after each ionic step of the structure relaxation. If the potential does not correspond to the desired value, it is iteratively equilibrated. Since the relationship between the electrons in the unit cell and the electrode potential is linear at constant geometry, a maximum of three iterations is generally needed in order to reach the target potential [22, 35, 39].
- **Constant-potential self consistent cycle:** In this scheme, the variational principle in the SCF cycle is adapted in order to satisfy the constant-potential condition. This is achieved by altering the Lagrange multiplier term from $-\mu_e(\int n(\mathbf{r})d\mathbf{r} - N_e)$, which satisfies the constant charge condition, to $-\mu_e \int n(\mathbf{r})d\mathbf{r}$, which automatically leads to the constant-potential free energy. While this scheme is very elegant and efficient, it can lead to numerical instability, expressed as a spatial oscillation of the electron density between SCF steps. Modifications to the SCF cycle have been proposed to correct this behavior [25].

Two algorithms are implemented in the solvated jellium method. The first follows the iterative scheme of the second strategy described above in an automated fashion. The standard workflow of this method is shown in the left panel of Fig. 10.7 for a structure optimization.

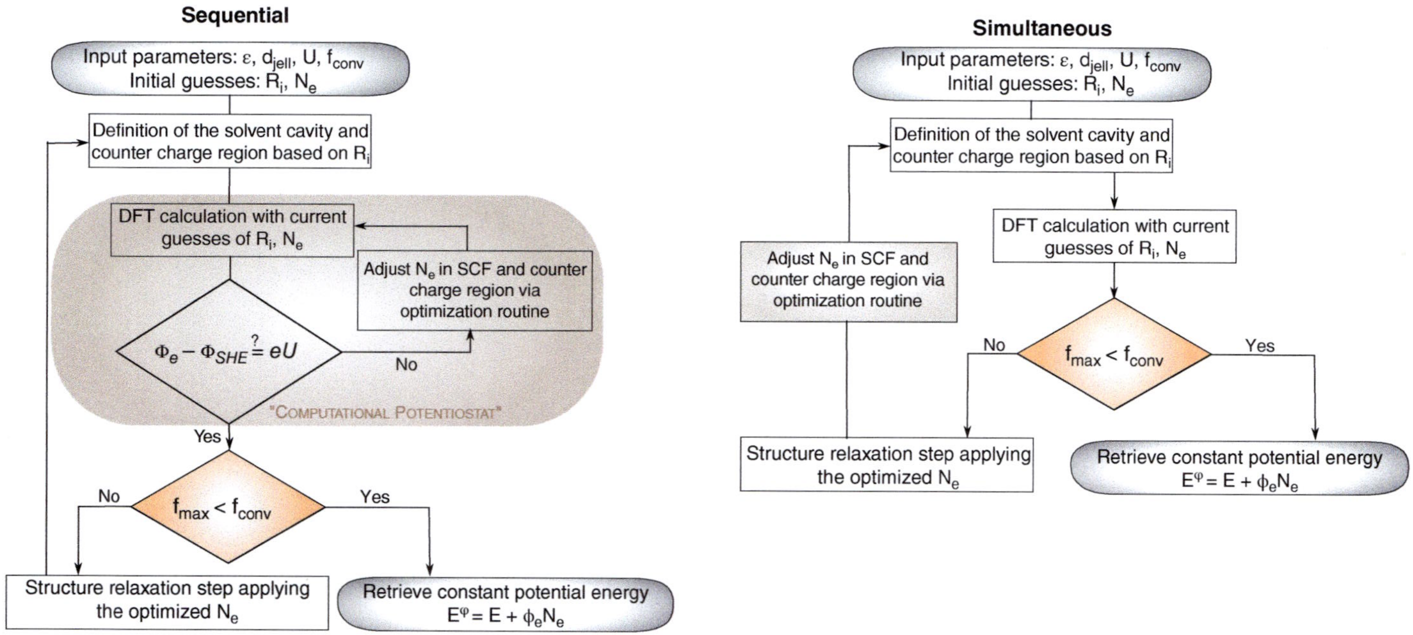

Figure 10.7 Workflow in the sequential (left) and simultaneous charge and geometry optimization applied in SJM.

It can, however, be applied to e.g. transition state searches in a straightforward way. The crucial component of the algorithm is highlighted in Fig. 10.7, namely the "computational potentiostat". This loop sets the potential by varying the number of electrons in the system. A detailed description of the workflow can be found elsewhere [22, 39]. However, a few aspects are worth highlighting for the scheme implemented in SJM:

- Using the converged wavefunctions from the previous calculation in a potential equilibration significantly reduces the number of iterations needed to reach convergence.
- The change in electrode potential with the number of electrons for a given geometry is nearly constant. Hence, reaching the target potential in three steps is generally possible. However, by remembering the slope from a previous charge equilibration, the algorithm can reach the target potential within a single charge equilibration step.
- The design of the algorithm suggests that the number of SCF cycles increases manifold compared to the number of force calls. In practice, the number of SCF cycles rarely reaches a factor of two compared to the number of force calls. This is due to the fact that ionic steps are generally small and only lead to small perturbations of the electrode potential. Hence, the number of electrons in the system do not need to be adapted after each step.

These aspects considerably decrease the computational cost of the method; the computational expense is generally well below 1.5 times that of a constant-charge optimization. In the upcoming section, we discuss the computational effort in detail.

In addition to the sequential charge and geometry optimization, SJM also includes a simultaneous scheme. In this algorithm, outlined in the right panel of Fig. 10.7, the potential is not equilibrated in a separate loop of SCF cycles, but is adapted directly when an ionic step happens. Here, the method uses the slope from the last step and redefines it at the same time. As shown in more detail in the upcoming section, this scheme does not lead to any penalty compared to a constant-charge structure optimization.

Comparison of the implemented potential equilibration schemes

In order to make the workflow shown in Fig. 10.7 more apparent and to estimate the numerical performance of the two described schemes, in the following an application to a structure optimization is shown. The system of choice for this example is an Au(111) slab covered by a water layer. In the initial geometry the H_2O molecules, included as a hexagonal ice-like layer, are rotated for the OH-bonds to be aligned with the electrode surface plane, except for one H_2O molecule which has been slightly tilted, in order to avoid artifacts arising from the high symmetry. In the course of the structure optimization the water layer will rotate and form the well-known H_2O bilayer structure in its H-down form. This particular setup was chosen because the change in the surface dipole of the electrode-interface is very pronounced in the course of the relaxation, due to the rotation of the polar OH-bonds. In addition to comparing the two charge equilibration schemes, a comparison to a constant charge relaxation, where the charge was set to the determined final charge from the constant potential schemes, will be given as well. Thus, the additional computational effort to constant charge DFT geometry optimizations is accessible. For both the sequential and simultaneous potential equilibration schemes the initial guess for the charge was set to 0 and automatically relaxed before the geometry optimization started.

Table 10.1 Final results of described geometry optimization. The shown results are final constant potential energy (E^{Φ}) referenced to the sequential charge equilibration scheme, Number of added electrons with respect to the uncharged system, deviation of the final electrode potential from the target potential (Φ^0=4.40V), number of force calls (#FC), number of SCF cycles (#SCF), total CPU time (Walltime x Number of cores), CPU time for 100 force calls.

	ΔE^{Φ} [eV]	$(N_e - N_e^0)$ [e^-]	$\Phi - \Phi^0$ [V]	#FC	#SCF	CPU [h]	CPU (100 FC) [h]
Constant charge	0.024	0.340	0.039	106	106	578	546
Sequential	0	0.347	0.002	103	178	810	786
Simultaneous	−0.012	0.331	−0.006	117	120	663	580

Figure 10.8 and Table 10.1 show the results of the comparison of the structure optimization schemes described above. Both potential equilibration schemes reached the same final geometry and energy save for some noise arising from the chosen convergence criterion (0.05eV/Å).

The target potential has been reached by both charge equilibration schemes, where the deviation is in the order of single digit mV. As can be seen in the upper right panel of Fig. 10.8 the sequential scheme never leads to a significant deviation from the target potential. The electrode potential in the simultaneous scheme on the other hand oscillates substantially and can sometimes even lead to spikes of strong deviation (as it is the case at the 44th SCF step). This is a consequence of the fact the change in electrode potential due to a change in the system's geometry is not estimated by the scheme. However, towards the end of the geometry optimization, where the change in geometry is small and the system's inherent dipoles do not vary their orientation significantly, the oscillation is strongly reduced and the target potential can be reached reliably. The constant charge scheme differs from the target potential by a slightly higher amount (39mV), which is a consequence of the fact that the charge has been chosen at the beginning of the optimization and has not been optimized in the course of it.

In this specific example, the number of ionic steps – or number of force calls – needed for reaching the final convergence was about 10% higher for the simultaneous charge and potential equilibration than for the sequential scheme. This, however, is not a general trend. In practical applications the relative number of force calls can vary and convergence can often be reached in the same of less number of steps when applying the simultaneous scheme. Counting the number of SCF cycles needed for convergence shows a different picture. While the number of cycles for simultaneous scheme is the same as the number of force calls, save for 3 initial SCF cycles in the beginning used to pre-equilibrate the potential. In the sequential scheme, on the other hand, the number of SCF cycles was 68% higher than in the constant charge scheme. The reason for this behavior is the rather tight potential equilibration criterion. Each time the potential varies for the target potential by more than 10mV after an ionic step, the charge is equilibrated without changing the molecular geometry. This behavior is shown in the lower right panel of the Fig. 10.8. Open circles represent SCF steps resulting in a potential outside of the set tolerance. Although the number of SCF cycles was 68% higher than in the compared constant charge scheme, the total CPU time only increased by 40%. This is a consequence of the implementation, which applied

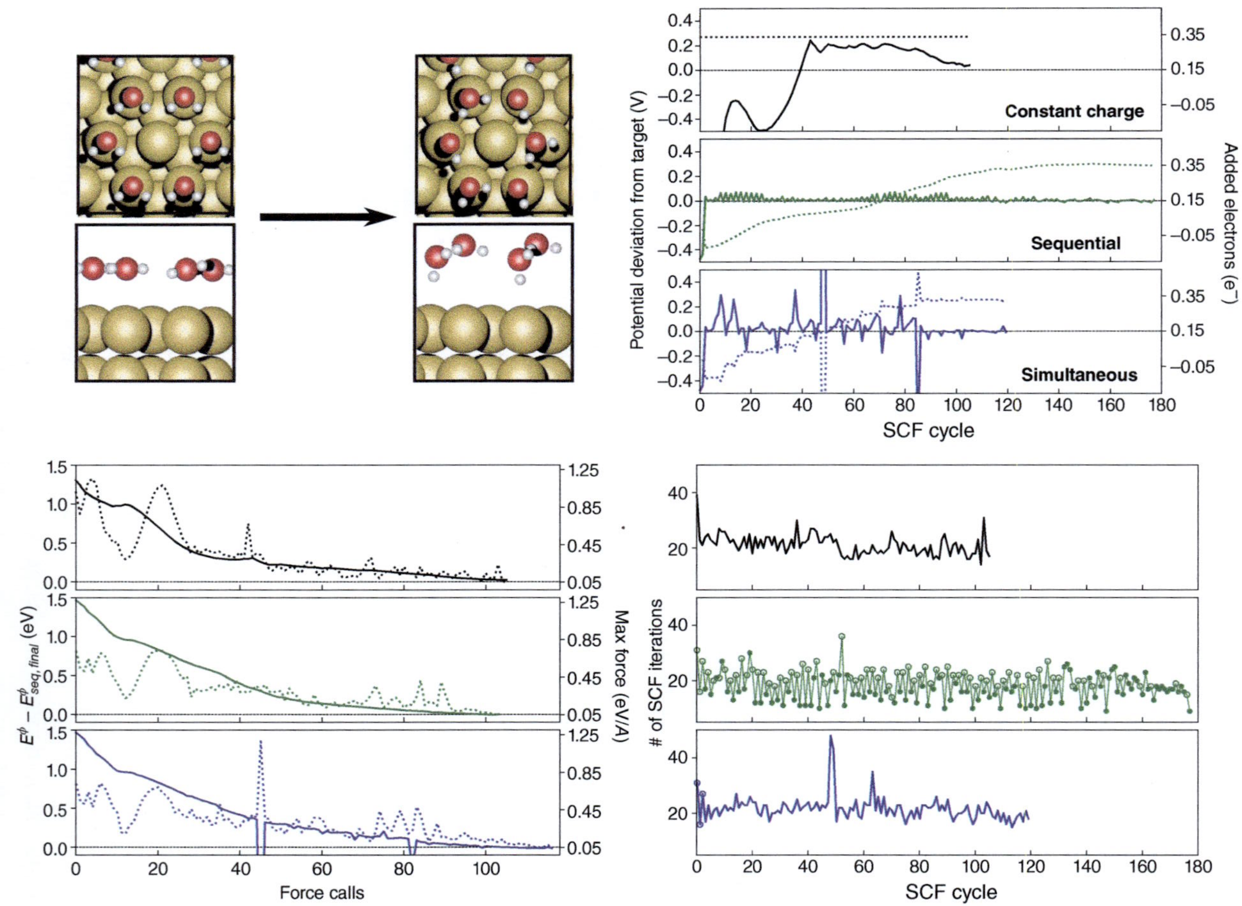

Figure 10.8 Comparison of the two constant potential geometry optimization schemes outlined in Fig. 10.7 and a constant charge relaxation, where the final surface charge determined via the constant potential schemes at an absolute potential of 4.4V. Left top panel: Initial and final geometry, Left bottom panel: E^ϕ referenced to the final E^ϕ of the sequential scheme optimization (left axis, solid line) and maximum force (right axis, dotted line), Right top panel: Electrode potential deviation from the target potential (left axis, solid line) and added charge (right axis, dotted line) in the SCF cycles of the optimization, Right bottom panel: Number of iterations in the SCF cycles, where, for the sequential scheme, SCF cycles resulting inside the potential tolerance of 10mV are shown as filled points, while open circles represent calculation triggering a potential equilibration.

the previously converged wavefunctions for the initial guess of the electron density, when the charge is equilibrated. As shown in the lower right panel of Fig. 10.8 the additional SCF cycles generally converge well within 20 iterations.

In summary, both charge equilibration schemes are viable for constant potential geometry optimizations. While the sequential scheme, applying the previous wavefunctions, leads to an increase in computational cost of about 40%, the method is robust and does not lead to any oscillation of the potential in the course of the relaxation. The simultaneous scheme is more efficient, leading to almost no added cost compared to a constant charge relaxation. However, the potential during the structure optimization can oscillate.

10.3.4 Shape of the jellium background charge

The description of the diffuse layer as a homogeneously charged region is clearly a simplification of a real electrochemical interface. However, this choice of description is very versatile and also leads to numerical stability and speed. The choice of the counter charge region is not necessarily limited to a large block-like counter charge region. Various parameters can be tuned in order to adapt the model to the simulated system.

For approximately planar interfaces, the block like counter charge region, which is completely uniform in the directions parallel to the electrode surface, is a reasonable approximation. However, the thickness of the electrochemical double layer is a strong function of the ionic strength of the electrolyte. The jellium background charge can be adapted to approximate this behavior. In principle, this corresponds to a change of the Debye screening length of the interface.

For non-planar interfaces, one can make use of the smooth permittivity function of the implicit solvent. In this scheme, the counter charge is scaled in dependence of the value of the dielectric shape function. This description of the jellium counter charge allows for the simulation of e.g. self-assembled monolayer structures, which are not close-packed and therefore allow the electrolyte to penetrate the adsorbate layer.

10.4 Example: Mechanistic studies of the hydrogen evolution reaction (HER)

The hydrogen evolution reaction (HER) – the simplest of electrocatalytic reactions – is one of the key reactions for a scalable hydrogen economy. In this half-cell reaction, protons from the electrolyte and electrons from the external circuit combine to form hydrogen gas at the cathode. In this section, we describe the elementary steps of hydrogen evolution and calculate potential-dependent reaction barriers. Moreover, we use SJM and the decoupled computational electrode described in Section 10.2 to account for the *complete* reaction trajectory of all elementary electrochemical reactions. This allows us to calculate current densities and coverage-dependent Tafel slopes.

HER is a two-step mechanism with three possible elementary reactions. The Volmer step is a proton discharge step from the electrolyte coupled with electron transfer to form adsorbed hydrogen. The second elementary reaction is either Heyrovsky or Tafel. The former describes hydrogen gas formation from an adsorbed hydrogen atom and a proton-electron pair from the electrolyte/circuit, whereas the latter is a purely chemical step between two adsorbed hydrogen atoms. Since the Volmer and Heyrovsky steps are electrochemical reactions, they require an electronically grand canonical treatment for the reasons outlined in the introduction.

The Tafel reaction, though non-electrochemical in nature, can also be a complicated elementary step. HER on platinum group metals is accompanied by underpotential deposition of hydrogen. This capacitive event occurs on the positive side of the equilibrium potential, and introduces a monolayer (or close to a monolayer) of hydrogen in the most stable adsorption sites. On the (111) facet of Pt, the fcc sites are occupied by these strongly bound hydrogen atoms before the surface begins to evolve H_2. The next available surface sites are the top sites, onto which the overpotential deposited hydrogen atoms adsorb. The nature of these weakly bound hydrogen atoms is vastly different from the strongly bound ones. Using constant-potential reaction barrier calculations and the decoupled computational electrode outlined above, we calculate thermodynamically consistent free energy diagrams for two possible HER mechanisms.

10.4.1 Potential dependence of the elementary steps of HER

We have calculated potential-dependent reaction barriers for all elementary steps of HER on Pt(111) with SJM and the nudged elastic band (NEB) method [2, 3], shown in Fig. 10.9. The fcc sites on the platinum lattice are saturated with a monolayer of hydrogen to simulate underpotential deposited hydrogen. The first two panels represent the Volmer and Heyrovsky steps, respectively. The electrochemical nature of these elementary steps manifests itself in the splitting of the potential-dependent minimum energy pathways; since electrons are transferred between the initial and final states, both the reaction energy and barrier change with applied potential. We find that the reaction barrier and reaction energy change quadratically and linearly, respectively, with applied potential. The latter validates the thermodynamic consistency of the model, if one takes into account the fractional charge transfer between a hybridized initial state and the final state [40]. The fractional charge transfer will be addressed in subsequent sections. The right panel of Fig. 10.9 shows the minimum energy pathways of the Tafel reaction. Here, we consider two configurations of reactive intermediates: (i) one neutrally adsorbed hydrogen in an fcc site and a weakly bound hydrogen in an ontop site, and (ii) two weakly adsorbed hydrogen atoms in adjacent ontop sites. Regardless of the nature of the reactive intermediate, we find that the potential dependence of the reaction barrier and energy is negligible, consistent with the non-electrochemical nature of this elementary step. Note that we do not makes assumptions regarding the nature of the reaction *a priori*; all elementary steps are treated identically, and the potential dependence is captured by the model.

The Volmer reaction is essentially barrierless at the equilibrium potential; the Butler–Volmer symmetry factor is close to unity. Moreover, the reaction barrier is low relative to those of the hydrogen-liberating steps. We note that the barrier height of the Heyrovsky step (electrochemical H_2 liberation) is significantly higher than for either Tafel mechanism at most potentials of interest.

The functional forms of the reaction energy and barrier can be explained by a quadratic relationship, functionally similar to Marcus theory [41] or empirical valence bond

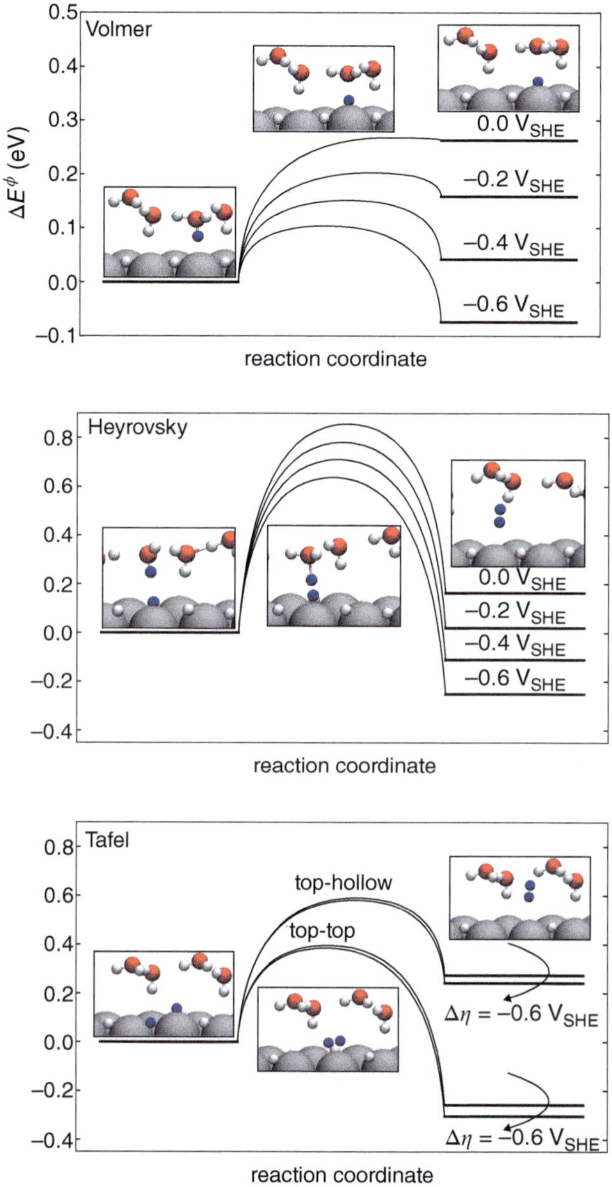

Figure 10.9 Potential-dependent potential energy diagrams for the elementary steps of HER on Pt(111). The Tafel insets represent optimized geometries for the top–hollow mechanism.

theory. This is shown in Eq. 10.13, where the reaction barrier changes quadratically for intermediate thermodynamic driving forces. This model requires one parameter, b, which we take to be the reaction barrier at thermoneutral conditions: the intrinsic reaction barrier. For high thermodynamic driving forces ($U \ll 0$ V_{SHE}), the reaction approaches an activationless state and the Butler–Volmer symmetry factor approaches zero. Conversely, the reaction approaches a barrierless region for low thermodynamic driving forces. Here, the Butler–Volmer symmetry factor is unity, and the reaction barrier and energy have identical responses to the applied potential.

$$E^{\ddagger} = \begin{cases} 0, & \Delta E < -4b \\ \dfrac{(\Delta E + 4b)^2}{16b}, & -4b \leq \Delta E \leq 4b \\ \Delta E, & \Delta E > 4b \end{cases} \tag{10.13}$$

The data from our constant-potential calculations on Au(111) and Pt(111) and the functional form in Eq. 10.13 are plotted in Fig. 10.10. As noted in subsequent sections, the Volmer reaction on Au(111) proceeds via ontop adsorption followed by exothermic diffusion into an fcc site. The latter part of the reaction trajectory is not accompanied by electron transfer, and is independent of the applied potential. Thus, the thermodynamic driving force in Fig. 10.10 corresponds to hydrogen adsorbed in an ontop site. This functional form captures the constant-potential data well, and properly describes the limiting cases for high-magnitude driving forces. Moreover, we find that the intrinsic reaction barrier, b, is higher on Au(111) than on Pt(111).

10.4.2 Charge transfer along reaction trajectories

SJM is an electronically grand canonical method, and electrons are added (subtracted) to (from) the electrode in order to keep the potential constant. Thus, the charge transfer along the reaction trajectory can be monitored for each elementary step. This analysis is helpful for partitioning the reaction trajectory into electrochemical and non-electrochemical trajectories. Here, we show the minimum energy pathways and charge transfer profiles for the Volmer and Heyrovsky steps on Au(111). First, we note that these are adiabatic

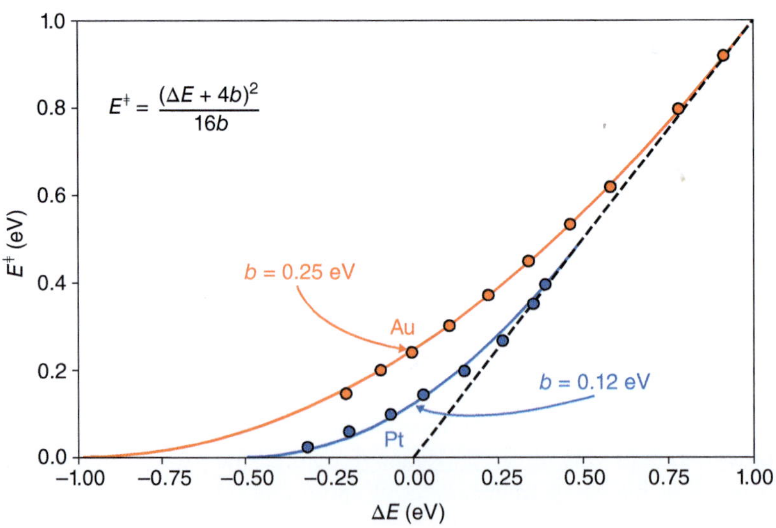

Figure 10.10 Constant-potential data for the Volmer reaction on Au(111) and Pt(111) fitted to the functional form in Eq. 10.13. The intrinsic reaction barrier, b, is lower on Pt(111).

calculations, and that the charge transfer is not a step-wise function; the electron transfer process is distributed between the intermediate states. However, the most dramatic change in electron transfer, i.e., the inflection point, coincides with the saddle point of the electrochemical reaction. The left panel of Fig. 10.11 shows the Volmer step at the equilibrium potential. Unlike platinum group metals, gold electrodes are not accompanied by underpotential deposition of hydrogen. Moreover, the weak hydrogen binding energy of Au results in low hydrogen coverage at moderate overpotentials. Thus, the most stable sites, i.e., the fcc sites, are available. Interestingly, we find that the reaction proceeds via adsorption into an ontop site, followed by exothermic diffusion into a neighboring fcc site; that is, the

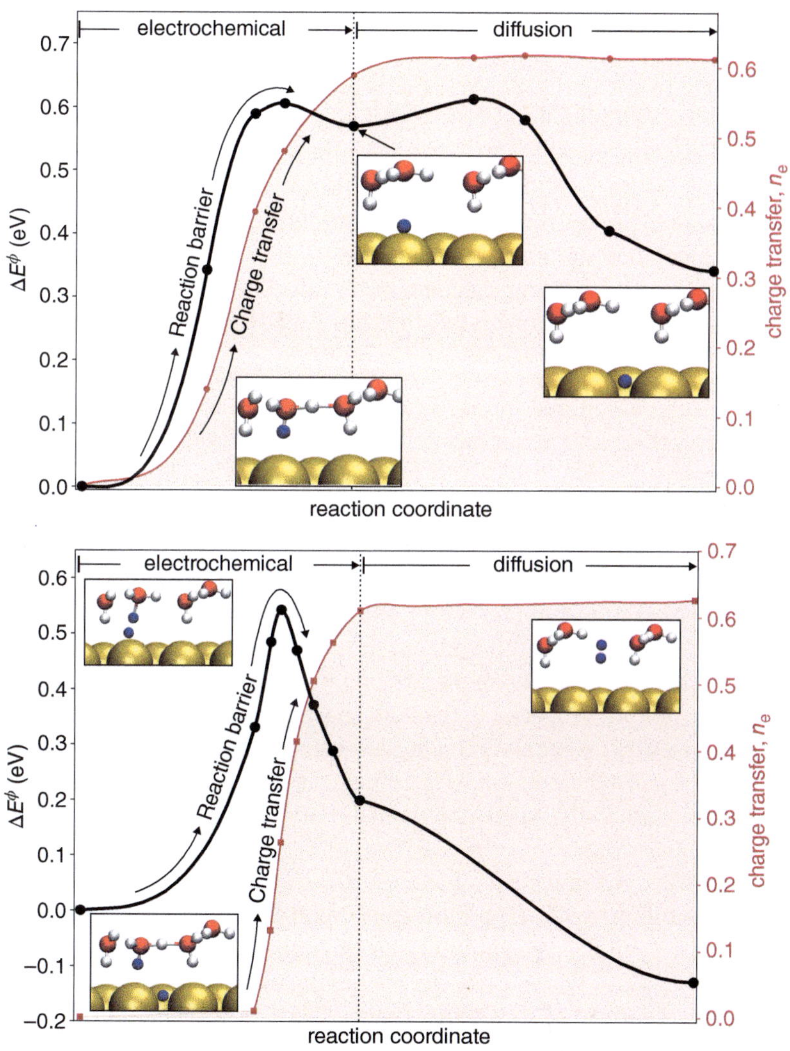

Figure 10.11 Minimum energy pathways (circles) and charge transfer profiles (squares) of the (left) Volmer and (right) Heyrovsky reactions on Au(111). The inflection point of the charge transfer curve coincides with the saddle point in both elementary steps.

electron transfer process ends when the hydrogen adsorbs in the ontop site. One could therefore create shorter pathways for these types of reactions, where a saddle point search is conducted between the initial state and the top-adsorbed hydrogen. The subsequent thermodynamic shift into an fcc site could be accounted for with a separate calculation.

The right panel of Fig. 10.11 shows the minimum energy pathway of the Heyrovsky step at the equilibrium potential. The saddle point is well-defined, and the charge transfer curve reflects this; the electron transfer process only occurs at or around the saddle point. The two diffusion processes – surface diffusion of hydrogen from an fcc site to an ontop site and diffusion of H_2 gas – exhibit negligible electron transfer and are independent of the applied potential.

One common characteristic of the two reactions is the fractional charge transfer; none of the two reactions includes the transfer of a full electron. A recent study [40] has investigated this phenomenon with self-interaction free methodologies, and concluded that the fractional charge transfer is a consequence of hybridization between solvated protons and the electrode in the electrochemical double layer. We account for this with our decoupled computational electrode model (section 10.2), where all states are referenced to the relevant initial state, i.e., a bulk solvated proton and an electron in the electrode. Thus, although the explicit reaction barrier calculations only account for part of the trajectory, we can combine them with thermodynamic proton shuttling calculations to calculate the energetics of the complete electrochemical reaction. We explain this in greater detail in section 10.4.3.

10.4.3 Thermodynamically consistent free energy diagrams from first principles

As noted in section 10.4.2, a recent study by Chen et al. has concluded that the fractional charge transfer observed in interfacial kinetic studies stems from hybridization of solvated protons and the electrode in the electrochemical double layer [40]. Thus, a complete electrochemical reaction trajectory has to account for proton shuttling from bulk solution into the electrochemical interface. We account for this with our decoupled computational electrode model outlined in Section 10.2; all states along the reaction trajectory are referenced to the true initial state, where the proton is solvated in the bulk and the electrode–water interface is calculated at constant potential. Since the pseudo-initial state is hybridized, the proton shuttling step also includes partial charge transfer. In net, the charge transfer from bulk solvation to adsorption is close to unity, where any small deviation from unity is accounted for in the model.

Thus, each elementary electrochemical step includes two reactions, both of which are accompanied by partial charge transfer: (i) proton shuttling from bulk to interface, and (ii) proton transfer from the hybridized pseudo-initial state to the electrode. This analysis is evident in free energy diagrams, where each electrochemical step includes a shoulder before the reaction barrier. Figure 10.12 shows free energy diagrams for the Volmer–Tafel and Volmer–Heyrovsky mechanisms, respectively, on Au(111). The Volmer and Heyrovsky steps are electrochemical, and thus require an additional proton shuttling step to ensure thermodynamic consistency. The Tafel step, however, is non-electrochemical, and does not require explicit consideration of proton shuttling from bulk into the interface.

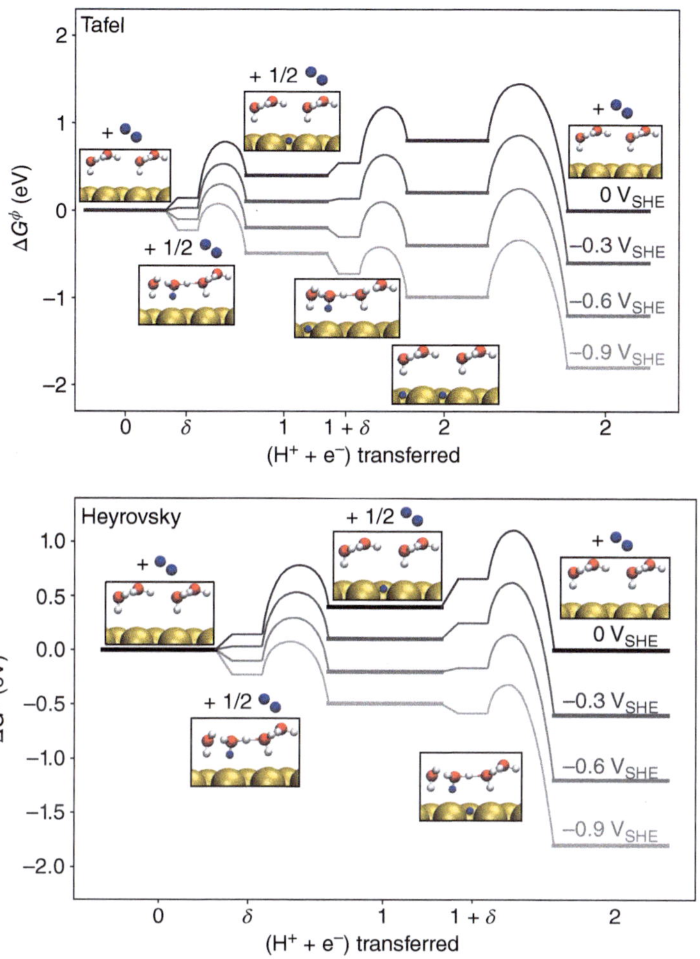

Figure 10.12 Potential-dependent free energy diagrams on Au(111). Note that each elementary electrochemical reaction includes a proton shuttling step from bulk solution into the interface.

Figure 10.12 shows two examples of thermodynamically consistent free energy diagrams. Each state along the reaction trajectory is properly referenced to the initial state with a bulk solvated proton. This analysis therefore accounts for the hybridization issue associated with interfacial studies, and ensures that the *complete* reaction trajectory is described. The results from these calculations can be used in potential-dependent microkinetic models to calculate current densities and coverage-dependent Tafel slopes.

References

1 Axel Gross. Theory of solid/liquid interfaces. *Psi-k Newsletter*, (125):1–20, 2014.

2 Graeme Henkelman and Hannes Jónsson. *Improved tangent estimate in the nudged elastic band method for finding minimum energy paths and saddle points*. 113(22):9978–9985, 2000.

3 Graeme Henkelman, Blas P. Uberuaga, and Hannes Jónsson. A climbing image nudged elastic band method for finding saddle points and minimum energy paths. *The Journal of Chemical Physics*, 113(22):9901–9904, 2000.

4 J.K. Nørskov, J. Rossmeisl, A. Logadottir, L. Lindqvist, J.R. Kitchin, T. Bligaard, and H. Jónsson. *Origin of the overpotential for oxygen reduction at a fuel-cell cathode.* 108(46):17886–17892, 2004.

5 Andrew A. Peterson, Frank Abild-Pedersen, Felix Studt, Jan Rossmeisl, and Jens K. Nørskov. How copper catalyzes the electroreduction of carbon dioxide into hydrocarbon fuels. *Energy & Environmental Science*, 3(9):1311, 2010.

6 Sneha A. Akhade, Nicole J. Bernstein, Monica R. Esopi, Michael J. Regula, and Michael J. Janik. A simple method to approximate electrode potential-dependent activation energies using density functional theory. *Catalysis Today*, 288(Supplement C):63–73, 2017.

7 Jan Rossmeisl, Egill Skúlason, Mårten E. Björketun, Vladimir Tripkovic, and Jens K. Nørskov. Modeling the electrified solid–liquid interface. *Chemical Physics Letters*, 466(1–3):68–71, 2008.

8 Egill Skúlason, Vladimir Tripkovic, Mårten E. Björketun, Sigrídur Gudmundsdóttir, Gustav Karlberg, Jan Rossmeisl, Thomas Bligaard, Hannes Jónsson, and Jens K. Nørskov. Modeling the electrochemical hydrogen oxidation and evolution reactions on the basis of density functional theory calculations. *J. Phys. Chem. C*, 114(42):18182–18197, 2010.

9 Stefan Goedecker. Linear scaling electronic structure methods. *Rev. Mod. Phys.*, 71(4):1085–1123, jul 1999.

10 Karen Chan and Jens K. Nørskov. *Electrochemical barriers made simple.* 6(14): 2663–2668, 2015.

11 Karen Chan and Jens K. Nørskov. Potential dependence of electrochemical barriers from ab initio calculations. *The Journal of Physical Chemistry Letters*, 7(9):1686–1690, 2016.

12 Christopher D. Taylor, Sally A. Wasileski, Jean-Sebastien Filhol, and Matthew Neurock. First principles reaction modeling of the electrochemical interface: Consideration and calculation of a tunable surface potential from atomic and electronic structure. *Phys. Rev. B*, 73(16):165402, apr 2006.

13 A.Y. Lozovoi, A. Alavi, Jorge Kohanoff, and R.M. Lynden-Bell. Ab initio simulation of charged slabs at constant chemical potential. *The Journal of Chemical Physics*, 115(4):1661–1669, jul 2001.

14 Alexander Y. Lozovoi and Ali Alavi. Reconstruction of charged surfaces: General trends and a case study of Pt(110) and Au(110). *Phys. Rev. B*, 68(24):245416, dec 2003.

15 M. Otani and O. Sugino. First-principles calculations of charged surfaces and interfaces: A plane-wave nonrepeated slab approach. *Phys. Rev. B*, 73(11):115407, mar 2006.

16 Kendra Letchworth-Weaver and T.A. Arias. Joint density functional theory of the electrode-electrolyte interface: Application to fixed electrode potentials, interfacial capacitances, and potentials of zero charge. *Physical Review B - Condensed Matter and Materials Physics*,86(7):1–16, 2012.

17 Kiran Mathew, Ravishankar Sundararaman, Kendra Letchworth-Weaver, T.A. Arias, and Richard G. Hennig. Implicit solvation model for density-functional study of nanocrystal surfaces and reaction pathways. *The Journal of Chemical Physics*, 140(8):84106, 2014.

18 Sung Sakong, Maryam Naderian, Kiran Mathew, Richard G. Hennig, and Axel Groß. Density functional theory study of the electrochemical interface between a Pt electrode and an aqueous electrolyte using an implicit solvent method. *The Journal of Chemical Physics*, 142(23):234107, 2015.

19 Ravishankar Sundararaman and Kathleen Schwarz. Evaluating continuum solvation models for the electrode-electrolyte interface: Challenges and strategies for improvement. *The Journal of Chemical Physics*, 146(8):84111, 2017.

20 Jason D. Goodpaster, Alexis T. Bell, and Martin Head-Gordon. Identification of possible pathways for C-C bond formation during electrochemical reduction of CO_2: New theoretical insights from an improved electrochemical model. *Journal of Physical Chemistry Letters*, 7(8):1471–1477, 2016.

21 Hainan Wang and Laurent Pilon. Accurate simulations of electric double layer capacitance of ultramicroelectrodes. *Journal of Physical Chemistry C*, 115(33):16711–16719, 2011.

22 Georg Kastlunger, Per Lindgren, and Andrew A. Peterson. *Controlled-potential simulation of elementary electrochemical reactions : proton discharge on metal surfaces.* 122:12771–12781, 2018.

23 Ryosuke Jinnouchi and Alfred B. Anderson. Electronic structure calculations of liquid-solid interfaces: Combination of density functional theory and modified Poisson-Boltzmann theory. *Physical Review B*, 77(24):245417, jun 2008.

24 Kendra Letchworth-Weaver and T.A. Arias. Joint density functional theory of the electrode-electrolyte interface: Application to fixed electrode potentials, interfacial capacitances, and potentials of zero charge. *Phys. Rev. B*, 86(7):75140, aug 2012.

25 Ravishankar Sundararaman, William A. GoddardIII, and Tomas A. Arias. Grand canonical electronic density-functional theory: Algorithms and applications to electrochemistry. *The Journal of Chemical Physics*, 146(11):114104, 2017.

26 G. Fisicaro, L. Genovese, O. Andreussi, N. Marzari, and S. Goedecker. A generalized Poisson and Poisson-Boltzmann solver for electrostatic environments. *The Journal of Chemical Physics*, 144(1):014103, jan 2016.

27 Francesco Nattino, Matthew Truscott, Nicola Marzari, and Oliviero Andreussi. Continuum models of the electrochemical diffuse layer in electronic-structure calculations. *Journal of Chemical Physics*, 150(4), 2019.

28 Sergio Trasatti. International union of pure and applied chemistry commission on electrochemistry * the absolute electrode potential: An explanatory note. *Pure and Applied Chemistry*, 58(7):955–966, 1986.

29 J.J. Mortensen, L.B. Hansen, and K.W. Jacobsen. *Real-space grid implementation of the projector augmented wave method.* 71(3):35109, jan 2005.

30 J. Enkovaara, C. Rostgaard, J.J. Mortensen, J. Chen, M. Dułak, L. Ferrighi, J. Gavnholt, C. Glinsvad, V. Haikola, H.A. Hansen, H.H. Kristoffersen, M. Kuisma, A.H. Larsen, L. Lehtovaara, M. Ljungberg, O. Lopez-Acevedo, P.G. Moses, J. Ojanen, T. Olsen, V. Petzold, N.A. Romero, J. Stausholm-Møller, M. Strange, G. A Tritsaris, M. Vanin, M. Walter, B. Hammer, H. Häkkinen, G.K.H. Madsen, R.M. Nieminen, J.K. Nørskov, M. Puska, T.T. Rantala, J. Schiøtz, K.S. Thygesen, and K.W. Jacobsen. *Electronic structure calculations with GPAW: a real-space implementation of the projector augmented-wave method.* 22(25):253202, 2010.

31 N.D. Lang and W. Kohn. Theory of metal surfaces: charge density and surface energy. *Physical Review B*, 1(12):4555–4568, 1970.

32 Kiran Mathew, Ravishankar Sundararaman, Kendra Letchworth-Weaver, T.A. Arias, and Richard G. Hennig. Implicit solvation model for density-functional study of nanocrystal surfaces and reaction pathways. *The Journal of Chemical Physics*, 140(8):84106, 2014.

33 Marko M. Melander, Mikael J. Kuisma, Thorbjørn Erik Køppen Christensen, and Karoliina Honkala. Grand-canonical approach to density functional theory of electrocatalytic systems: Thermodynamics of solid-liquid interfaces at constant ion and electrode potentials. *The Journal of Chemical Physics*, 150(4):041706, jan 2019.

34 Nicolas G. Hörmann, Oliviero Andreussi, and Nicola Marzari. Grand canonical simulations of electrochemical interfaces in implicit solvation models. *Journal of Chemical Physics*, 150(4), 2019.

35 Joseph A. Gauthier, Stefan Ringe, Colin F. Dickens, Alejandro J. Garza, Alexis T. Bell, Martin Head-Gordon, Jens K. Nørskov, and Karen Chan. Challenges in modeling electrochemical reaction energetics with polarizable continuum models. *ACS Catalysis*, 9(2):920–931, jan 2019.

36 Mårten E. Björketun, ZhenHua Zeng, Rizwan Ahmed, Vladimir Tripkovic, Kristian S. Thygesen, and Jan Rossmeisl. *Avoiding pitfalls in the modeling of electrochemical interfaces*. 555:145–148, 2013.

37 Ya-Hui Fang, Guang-Feng Wei, and Zhi-Pan Liu. Catalytic role of minority species and minority sites for electrochemical hydrogen evolution on metals: surface charging, coverage, and tafel kinetics. *The Journal of Physical Chemistry C*, 117(15):7669–7680, apr 2013.

38 Maxime Van Den Bossche, Egill Sk, Christoph Rose-Petruck, and J. Hannes. Assessment of constant-potential implicit solvation calculations of electrochemical energy barriers.

39 Jason D. Goodpaster, Alexis T. Bell, and Martin Head-Gordon. Identification of possible pathways for c–c bond formation during electrochemical reduction of CO_2: New theoretical insights from an improved electrochemical model. *The Journal of Physical Chemistry Letters*, 7(8):1471–1477, 2016.

40 L.D. Chen, M. Bajdich, J.M.P. Martirez, C.M. Krauter, J.A. Gauthier, E.A. Carter, A.C. Luntz, K. Chan, and Jens K. Nørskov. Understanding the apparent fractional charge of protons in the aqueous electrochemical double layer. *Nature Communications*, (2018):To be published, 2018.

41 R. A. Marcus. On the theory of electron–transfer reactions. VI. Unified treatment for homogeneous and electrode reactions. *The Journal of Chemical Physics*, 43(2):679–701, jul 1965.

42 Aimin Ge, Georg Kastlunger, Jinhui Meng, Per Lindgren, Jia Song, Qiliang Liu, Alexander Zaslavsky, Tianquan Lian, and Andrew A. Peterson. On the coupling of electron transfer to proton transfer at electrified interfaces. *Journal of the American Chemical Society*, 142(27):11829–11834.

11

Generation of Computational Data Sets for Machine Learning Applied to Battery Materials

Arghya Bhowmik, Felix Tim Bölle, Ivano E. Castelli, Jin Hyun Chang, Juan Maria García Lastra, Nicolai Rask Mathiesen, Alexander Sougaard Tygesen, and Tejs Vegge

Department of Energy Conversion and Storage, Technical University of Denmark, Anker Engelunds Vej, 301, Kgs. Lyngby, Denmark

11.1 Introduction

Understanding and improving the performance of a battery requires deep knowledge of a plethora of phenomena that occur on a wide range of length and time scales. For predictive modelling of batteries, the length scales span from the atomistic level, where quantities such as the open circuit voltage (OCV) can be determined, to the device level, which is employed in battery management systems. The modelling time scales start at the pico to nanosecond scale, to capture phenomena such as electron transfer and ionic diffusion, and reaches the year-scale in order to describe battery aging. Similarly, the length scales span from angstrom to meters. Obviously, there is no a single modelling framework that can cover all these scales simultaneously, but specific simulation tools have been developed to treat each of them independently. The biggest challenge to build a reliable multiscale approach to model batteries is to feed the larger scale levels, both length and time, with appropriate parameters. Ideally, these parameters for mesoscopic and macroscopic simulations would be derive from the output of microscopic scale calculations. In this regard, Machine Learning algorithms (MLAs) and related data-driven approaches are showing great promise to accelerate the coupling between the different simulation scales.

Machine learning is a subfield of AI, where an algorithm learns from examples to establish a functional mapping from input to output, and improves the mapping upon training. MLAs therefore rely on a 'training set' of systems from which the algorithm learns. Each computation in the training set is described by a 'vector descriptor' which contains a unique and meaningful way to describe the computed material. The training set also contains a number of target properties for each material. If the size of the training set is large enough, the MLA can figure out how the vector descriptors and the targeted properties are correlated. In order to obtain reliable outputs from the MLA, the data in the training set must be reliable. On the other hand, producing data to train the MLAs can be time-consuming. Thus, the method to produce the data in the training set must also be affordable. The delicate balance between reliability and affordability depends on the targeted property. In some cases, it is better to have a vast amount of data with moderate fidelity, while in other

Atomic-Scale Modelling of Electrochemical Systems, First Edition.
Edited by Marko M. Melander, Tomi T. Laurila, and Kari Laasonen.

situations it is more convenient to use a limited amount of high-fidelity data (here fidelity is understood as the degree to which a simulation reproduces the state and evolution of a set of given properties of a physically real entity). In this chapter, we present examples of the two situations in the context of microscopic modelling of batteries. First, we show how to produce a large set of data with moderate fidelity by means of a computational workflow. That workflow, based on Density Functional Theory (DFT) simulations, is able to predict open circuit voltages (OCV) and diffusivities of electrode materials that can be later used as input parameters in macroscopic models based on Finite Difference Elements (FDE). Secondly, we present how to produce high-fidelity data on the formation of solid electrolyte interphase (SEIs) by means of *ab initio* molecular dynamics. We conclude by illustrating how these data sets are employed to train the MLAs.

This chapter is arranged as follows. Section 11.2 presents the global structure of the workflow which creates moderate-fidelity data and describes the workflow to produce a large set of moderate-fidelity data on OCVs, mechanical stability, and cation diffusivity in intercalation electrodes. This workflow relies on several novel computational techniques which contribute to accelerate the data production and enhance its reliability. In subsection 11.2.1 we show how diffusivity is calculated within the workflow explaining how reflective symmetry can be exploited to boost Nudged Elastic Band (NEB) calculations (11.2.1.1) and discussing the importance of the choice of the right exchange-correlation functionals (11.2.1.2). Subsection 11.2.2 deals with the modelling of disorder in battery electrode materials, which is also part of the workflow. Section 11.3 shows one example of computational production of high-fidelity data, namely the use of *ab initio* molecular dynamics to understand the reduction reactions that bring to the first stages in the formation of SEIs. We conclude with a section 11.4 MLAs explaining how they can help to predict synthesizability and structure of battery materials and the evolution of interfaces based on high- and moderate-fidelity computational data.

11.2 Computational workflows for production of moderate-fidelity data sets

Given the tools that are able to accelerate calculations combined with the ever increasing computational resources available, a great interest lies in how to efficiently combine separate computational tasks into organized workflows that would make it easier to perform calculations and store data. Some of the goals of such an efficient workflow scheme include:

(a) consistent electronic structure calculations
(b) reproducible and transparent data
(c) reusability of the code scripts and ideas developed throughout the whole creation of the workflow.

The need for such workflows as well as the requirement of reproducibility of the calculations arose due to the lack of consistent data warehousing in the materials science community [1]. The reliability of the databases is also strongly related to the underlying data creation workflow. Within the materials community the Automated Interactive Infrastructure and Database for Computational Science (AiiDA) [2] as well as the workflow

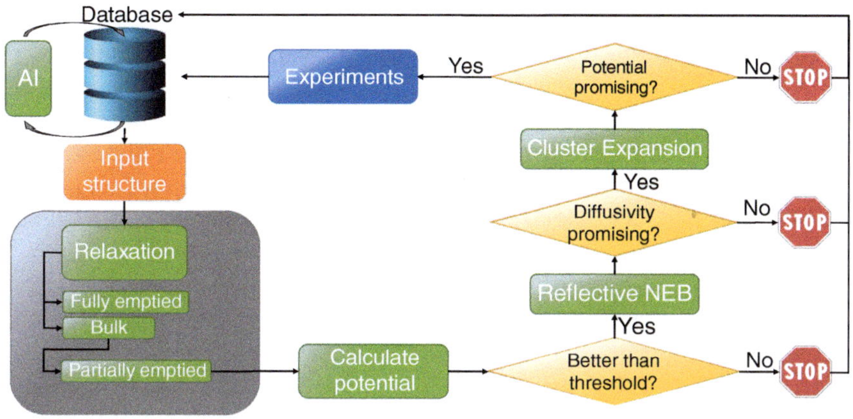

Figure 11.1 Generic workflow for finding ion-insertion cathode battery materials. Detailed explanation of the different steps are found in the text.

management system "Fireworks" [3] are good examples for how to handle thousands of electronic structure calculations distributed over different computational resources as well as users.

We will now show how to create a generic workflow for ion-insertion cathode battery materials while focusing on unraveling different parts of a scientific workflow. The main idea of the workflow is to keep a modularity that is easy to understand for users not familiar with computational workflow management concepts.

The physical properties of interest in the workflow are the mechanical stability, the OCV, and cation diffusivity (see Figure 11.1). These properties are obtained by combinig Density Functional Theory (DFT) calculations combined with other techniques, such as the Nudged Elastic Band or the Cluster Expansion methods. The goal is to find potential materials which could outperform the ones currently used in the state-of-the-art battery electrodes and test them experimentally. Key parameters that determine the performance of a material as battery electrode are the OCV at different charge states and the corresponding cation diffusivity. Possible ions could, for example, include lithium, sodium, or magnesium, as the workflow will not change its underlying structure.

The workflow (Figure 11.1) starts with reading in a starting structure taken from the Inorganic Crystal Structure Database (ICSD), ensuring synthesizability of the material. Two subsequent calculations relax the unit cell as well as a unit cell where all ions have been removed (labeled as "empty") to obtain the changes in volume change upon charging/discharging which can give insight into the mechanical stability. After the unit cell has been relaxed, the symmetry inequivalent ion sites are extracted using Atomic Simulation Environment (ASE) [4]. The information about inequivalent sites is subsequently used for creating starting structures for the OCV as well as diffusion barrier (NEB) calculations. Subsequent DFT calculations aim at calculating the OCV at very high/low ion concentrations. The prediction of the OCV at very high/low charge states comprises the smallest possible phase space including symmetry considerations. For instance, the phase space of investigating structures where only one ion has been removed from the supercell is often very small due to symmetry considerations. Calculating the

OCV at very high/low concentrations (i.e. removing a single ion from the structure or removing all but one ion from the structure) give a first indication on the performance of the cathode material. Depending on the calculated OCV, a subsequent step calculates the cation diffusivity. Here, computational time is further reduced to a minimum by using reflective NEB calculations (as discussed in subsection 11.2.1) disregarding the computation of symmetry equivalent structures. Cathode materials that show promising cation diffusivities as well as an adequate OCV will be part of a more in depth study of the OCV. The Cluster Expansion method (as discussed in subsection 11.2.2) helps to find the the minimum energy configuration at different charge states coping with the large possible phase space that arises due to the many possible ion arrangements within the supercell. In case of promising OCVs over different charge states the material will be part of further experimental investigations. All the data produced during the run is stored in a database. Especially materials that do not meet the specified requirements in terms of the OCV and the cation diffusivities should not be disregarded since machine learning (ML) approaches using data from the database can help to learn whether a material is promising even before running a single DFT calculation (as discussed in section 11.4).

The core of any computational workflow is its workflow management systems (WMS) responsible for submitting, observing and post processing the calculations. Due to the high likelihood of calculation errors, the WMS has to be capable of visualizing the status of the submitted calculations to the researcher enabling to fix failed calculations efficiently. While Fireworks is a good example of a WMS developed specifically for the materials science community, here we use an in-house tool with similar features called MyQueue (https://myqueue.readthedocs.io/en/latest/). The different calculation parts are further divided into sub-tasks: retrieving the starting structure, manipulating the structure, submitting the DFT calculation to the cluster, and finally post processing and storing the data in a database. Every DFT calculation is carried out in a separate folder. The folder structure does not only ensure uniqueness of the calculation but also carries information on the calculation input. The WMS will give information about every single step in the workflow and its status. In case of a failed workflow step, the user can readily check the error message in the calculation folder. This also makes it possible for the user to create automatic rules on how to restart a failed calculation, which can be specific depending on the ion used.

After discussing the structure of the workflow, we now explain how the goals of the computational work being (a) consistent, (b) reproducible and transparent, as well as (c) reusable have been met. Consistency is given by having all the code scripts collected in the version control tool GitLab while ensuring that the input and electronic structure calculation schemes are consistent within the whole workflow chain. While storing all calculation scripts in GitLab makes the workflow also transparent, a detailed documentation of how the results were obtained in the created database enables full reproducibility given by the used DFT input as well as the version of the used programs. Finally, the reusability will be of different advantage to the researchers developing the workflow compared to researchers trying to achieve similar results. The researchers in charge can easily modify the workflows and make use of them in other projects as the workflow forces the development of generic modules. In the workflow at hand, for example, running the same workflow to investigate possible lithium or magnesium cathode materials requires solely the change of a single input variable and reading in the appropriate testing materials. Since material scientists may not be familiar with all electronic structure codes (e.g. VASP [5]) or packages that help manipulating crystal structures (e.g. ASE), a great value lies in how the workflow was build up, as well as which general tools have been used. For instance, the ion-insertion cathode battery workflow could inspire researchers to include R-NEB calculations in future projects simply by looking at the workflow scheme (Figure 11.1).

Having created a consistent database that allows full reproducibility also opens the door towards using MLAs that rely heavily on good training data. The presented workflow gives the researcher confidence in the created dataset and prevents involved machine learning studies on DFT created datasets that lack the needed fidelity as this may have not been the major objective when the dataset has been created.

11.2.1 Ionic diffusion: NEB calculations

The NEB method is used to search for minimum energy paths (MEPs) between two fixed points on a potential energy surface (PES) [6–9]. This is a frequent problem in battery research, e.g., in intercalation materials where the diffusivity of the intercalated species is determined by the energy cost associated with jumps between local energy minima positions in the host material. In the NEB method the path of such a *jump* is represented by a set of points where each point refers to the full structure as it is at some point during the transition. Each point is also referred to as an image. The traditional NEB calculation is usually done following these steps:

1. Define initial and final images of the path.
2. Optimize the initial and final images.
3. Make an initial guess of the path.
4. Optimize the path.

In the case of intercalated lithium, defining the initial and final images usually just involves putting the lithium at the initial and final position of the jump that is being studied. Following this, each of the images should also be optimized down to the nearest local minimum. The simplest approach for making an initial guess of the path is to make a geometrical linear interpolation between the initial and final images. Besides being simple and relatively robust it is also often a good guess for bulk systems. If an unconstrained optimization of the path is carried out the intermediate images will all move to the nearest local minimum of the PES (which corresponds to the initial or final image), thus providing no information about the transition. To avoid this scenario, a constrained optimization is performed. This makes sure the images stay well separated and span the interesting part of the PES. The three main constraints needed to do an NEB calculation are: (a) the initial and final images are fixed; (b) spring forces parallel to the direction of the path are introduced between neighbouring images; (c) from the true force, only the part of the force perpendicular to the path at each image is used. In this way it is ensured that the images are covering the relevant part of the PES while they are free to converge on the MEP.

11.2.1.1 Symmetric NEB

Many of the fundamental jumps of defects relevant to batteries such as polarons, vacancies, intercalated species, etc. are reflection-symmetric. When reflection symmetry is present

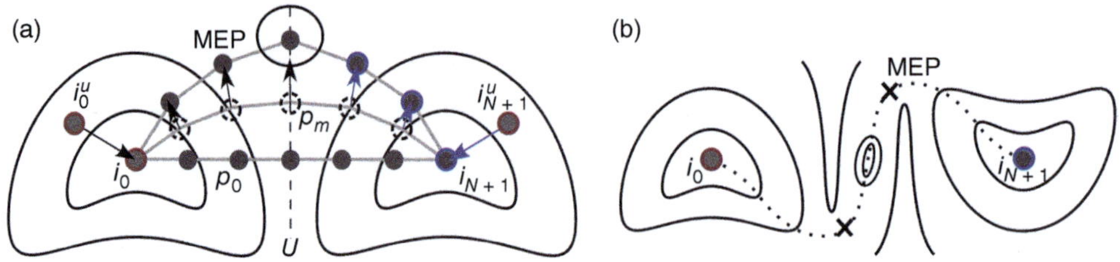

Figure 11.2 An illustration of the R-NEB method. (a) The solid black lines are contour lines of the potential energy surface. The circles represent images, i.e. atomic configurations. They are connected by gray lines to indicate a path. The bottom path, p_0, is the initial guess, the middle path, p_m, is an intermediate, mth, step of the optimization, and the top path is the MEP. The dashed vertical line shows the mirror *plane* of the reflection U. Circles highlighted with a red border are images given as input while circles with blue borders are images constructed through symmetry operations. Black arrows represent calculated forces and blue arrows represent forces obtained by symmetry. i_0 and i_{N+1} are the optimized initial and final images, respectively, and i_0^u and i_{N+1}^u are the un-optimized initial and final images, respectively. (b) An example of reflection with respect to a point. The inversion center is at the center of the figure and the MEP is shown with the dotted line. The central contour lines mark a local minimum on the PES and the crosses indicate the maximum energy points on the MEP. Adapted with permission from [10], copyright 2019 American Chemical Society.

only half of the barrier is unique and as a result at most half the MEP needs to be found. This opens up the possibility of exploiting the reflection symmetry to speed up NEB calculations. For this purpose we developed the Reflection-NEB (R-NEB) method which automatically identifies reflection symmetric NEB paths and avoids explicit force evaluations on half the images [10]. The missing forces are obtained by applying the appropriate symmetry operation to the forces calculated for the symmetry equivalent images (see Figure 11.2). The revised procedure for doing an NEB calculation following the R-NEB method is:

1. Define initial and final images of the path.
2. Optimize the initial image and obtain the optimized final image by symmetry.
3. Make an initial guess of the path.
4. Optimize the path with the usual NEB algorithm but carry out force evaluations for only one half of the path and get energy and atomic forces of the other half by applying the appropriate symmetry operation.

Additionally, the presence of a reflection symmetry makes sure that images placed half way between the initial and final images are positioned directly in a mirror plane. Images placed in a mirror plane are constrained to move only within this plane since any force taking the image out of the mirror plane will be exactly countered at the other side of the mirror. This means that, if it is known *a priori* that the maximum of the barrier is at the halfway point, the barrier can be determined using just a single image, thus gaining several factors of speed-up. We call this approach the Reflection-Middle-Image-NEB (RMI-NEB). It is important to stress that both R-NEB and RMI-NEB calculations agree perfectly with the traditional NEB method for any reflection symmetric barrier. However, when using the RMI-NEB method there is a risk that the maximum of the barrier is not halfway, in which case the RMI-NEB underestimates the barrier.

The RMI-NEB method can typically be used for sufficiently *simple* paths and for studies involving many similar paths with small variations, e.g. when studying the diffusion of different intercalated ions in materials with the same symmetry. Another use of the RMI-NEB method is for screening studies where the workflow only needs rough estimations of the barriers to identify the low energy barriers to be selected for further studies. In this case the only risk of using the RMI-NEB method is that it underestimates the true barrier. Thus, no important barriers will be discarded and if there is any doubt about the shape of the barrier a R-NEB calculation can subsequently be used to verify the result. As regards producing training data for MLAs, the RMI-NEB method presents the possibility to choose between a relatively computationally expensive dataset with high-fidelity values (using R-NEB or regular NEB) or a cheaper dataset with some risk of small errors. Recently, the NEB method has also been combined with neural network and Gaussian process regression algorithms to significantly reduce the computational cost [11, 12]. Interestingly, the speed up gained from both the MLAs and the R-NEB method could be harvested at the same time if they were to be integrated. The R-NEB and RMI-NEB methods are implemented in the open-source Python library ASE [4].

11.2.1.2 Choice of functionals for NEB

When modelling diffusivities with the NEB method, the choice of DFT functional will naturally play a large role. The diffusivity depends exponentially on the obtained diffusion barriers, $D \sim \exp\left(-\frac{E_b}{kT}\right)$, where E_b is the barrier of diffusion. [13] Therefore, small errors in the obtained DFT barriers will be exponentionally amplified on the diffusivities.

In layered 2D battery materials, the van der Waals bonding plays a critical role, as the vacuum between the layers provide a natural cavity for Li-ions, and thus also for the diffusion processes. These long-range dispersion forces must be explicitly accounted for in DFT, either by inclusion in the design of the xc-functional, or as an ad hoc correction. These methods, however, do not identically determine the long-range interactions between the layers, and are thus a source of error in the model.

An emerging class of materials called MXenes ($M_{n+1}X_nT_x$ for $n = 1, 2, 3$), where M is an early transition metal (e.g. V, Ti, etc.), $X = \{C, N\}$, and T is a functional termination group, (e.g. O, OH or F), are an example of such potential 2D battery materials. The large compositional space of MXenes allow for a high degree of tunability in electronic properties, which in turn allows for optimization of relevant energy storage parameters, such as energy and power densities as well as mechanical stability.

Using Ti_2CO_2 as a prototype MXene, we studied the effects of different long-range dispersion functionals on the corresponding NEB barriers of the intercalated Li ions. [14] Three functionals were chosen, based on some of the popular choices for studying MXenes with DFT: The vdW-DF2 functional, the optB88-vdW functional; and the PBE functional combined with a D3 correction (named PBE+D3). The diffusion barrier of lithium in bulk MXenes for the chosen functionals are seen in Figure 11.3a. This diffusion process is a good example the R-NEB, or even the RMI-NEB, could have saved a considerable amount of computational resources, however, these methods were still under development at the time of this study, so the conventional NEB algorithm was employed instead. Here we find a quite significant quantitative difference in the barrier height, depending on the functional choice. As the diffusion process takes place in the vacuum between MXene layers, simple

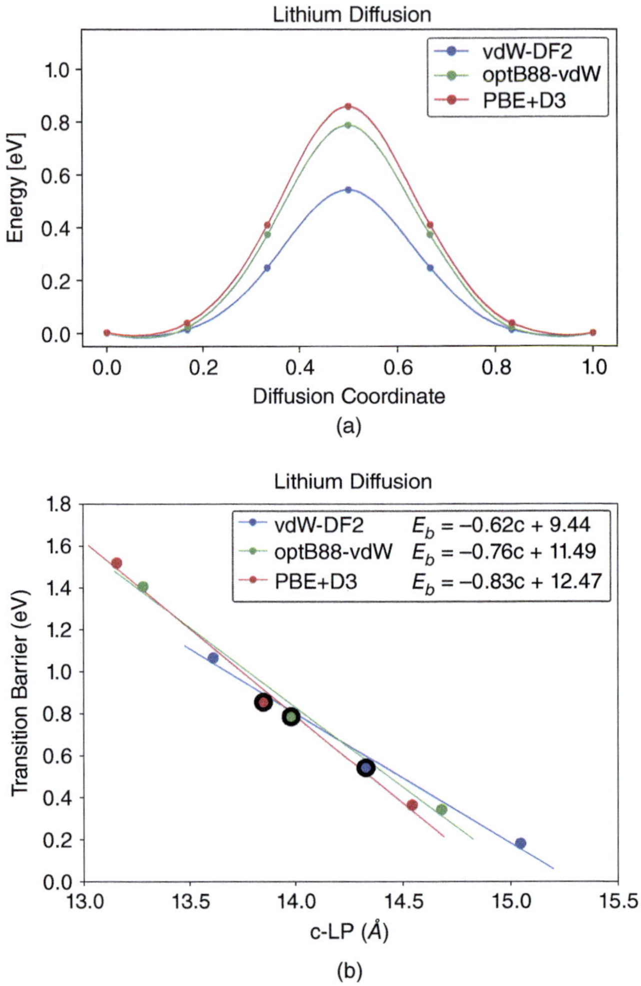

Figure 11.3 a) Diffusion in bulk MXene, of a single Li-ion. b) Diffusion barrier dependency on the out-of-plane lattice parameters. Optimized parameters highlighted by black marker for each functional. Adapted with permission from [14], copyright 2019 American Chemical Society.

electrostatics would suggest that a crucial parameter for the diffusion barrier is the interlayer distance. And indeed, if we tune the interlayer distance between the stacked layers, we find a linear relation between the out-of-plane lattice parameter and the energy barrier, as seen in Figure 11.3b. This reveals that the primary source of error in the energy barriers stems from an initial disagreement on that interlayer spacing of the MXenes between the functionals.

The consequence of this, is that a reliable estimate of the lattice parameters of the material plays an integral role. However, for technical reasons, getting experimental lattice parameter data for the MXenes has proved difficult, and thus we cannot compare our calculated values with experimental ones. Instead, we benchmarked materials of similar character with experimentally known lattice parameters, in order to estimate

Table 11.1 Mean absolute errors from benchmarking lattice parameters on similar materials, and applying those error estimates to the linear relation model from Figure 11.3b. Δr is the ratio of diffusion rates with ΔE_b as the difference in barrier at $T = 300$ K. Adapted with permission from [14], copyright 2019 American Chemical Society.

	E_b MAE (eV)	Δr
vdW-DF2	0.37	6.0×10^{-7}
optB88-vdW	0.12	8.7×10^{-3}
PBE + D3	0.15	2.5×10^{-3}

the reliability of the lattice parameter of each of the functionals. Assuming a general transferability between such benchmarking results of the lattice parameters, and using the linear relationship between the out-of-plane lattice parameter and energy barrier, as shown in Fig 11.3b, we estimated the errors in the diffusion barriers, and the corresponding errors in the diffusion rates, as seen in Table 11.1. The ratio of diffusion rates, $\Delta r = r_2/r_1 = \exp(-\Delta E_b/kT)$, where $\Delta E_b = E_{b,2} - E_{b,1}$, gives us an estimate of the order of magnitude the errors in the diffusion barriers induces on the diffusivity, meaning that we lose 7 orders of magnitude on the accuracy of the diffusion coefficient, with just an average error of 0.37 eV.

Intrinsic improvement upon the the functions is would naturally be the most desirable approach to this issue, e.g. by applying machine learning techniques to correct for the intrinsic errors in the functionals, as demonstrated by Mezei and Lilienfeld [15].

11.2.2 Disordered materials: Cluster Expansion

Cluster Expansion (CE) [16–21] is an elegant method that maps the physical property calculated from *ab initio* simulations onto a simpler Hamiltonian. CE formalism is often viewed as a generalized Ising or lattice-gas model [16, 18]. Although CE can be used to map physical properties such as band gap and density of states, the current discussion is limited to energy as it is the most widely used in the context of CE. CE has been used successfully in the past few decades to study crystalline materials exhibiting configurational disorder. The configurational dependence of physical properties is parameterized in the CE scheme, and the resulting lattice model allows rapid evaluations of the parameterized properties of configurations consisting of tens of thousands of atoms. The evaluation process is sufficiently fast to be used in Monte Carlo simulations, which enables one to assess the parameterized properties of the system at finite temperatures.

Early examples of employing CE for investigating materials for battery applications date back to 1998 where Van der Ven et al. [22, 23] and Wolverton and Zunger [24, 25] studied Li_xCoO_2. CE played a vital role in analyzing the ordering of cation and lithium vacancies in Li_xCoO_2. The relevance of CE in investigating battery materials is on the rise. In particular, Li-rich materials with disordered rocksalt structures are attracting much attention in the past few years [26–28]. Several software packages that allow users to construct CE,

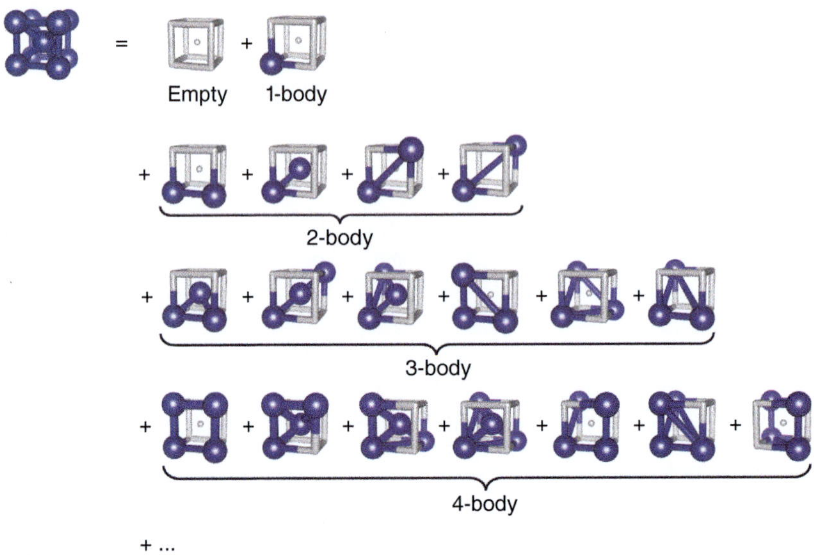

Figure 11.4 A simplified illustration of the decomposition of a body-centered cubic lattice into clusters. Reproduced from [32], copyright 2019 Institute of Physics. Creative Commons Attribution 3.0 licence. DOI: 10.1088/1361-648x/ab1bbc.

such as Alloy Theoretic Automated Toolkit (ATAT) [29, 30], UNiversal CLuster Expansion (UNCLE) [31], Cluster Approach to Statistical Mechanics (CASM), and CLuster Expansion in Atomic Simulation Environment (CLEASE) [32].

The core concept of the CE is to express the energy of a crystalline material to its configuration. A general CE formalism for the multicomponent alloy is discussed in [32], and a binary case (i.e., materials consisting of two types of elements) is used for illustration herein for simplicity. Each atomic site of the lattice is occupied by one of the two types of atoms, and the site variable s_i with a value of either +1 or −1 is assigned to each site i of the lattice. Any configuration of material with N atomic sites is then expressed with a vector $\boldsymbol{\sigma} = \{s_1, s_2, \ldots, s_N\}$. The total energy of a configuration $\boldsymbol{\sigma}$ is expanded in terms of the energy contribution of clusters, α. The clusters are a collection of atoms in the lattice as illustrated in Figure 11.4.

The energy contribution of each cluster is expressed via a correlation function, ϕ_α, which is the product of s_i for all the sites that the clusters contain (i.e., $\phi_\alpha = s_i \cdot s_j \cdot \ldots \cdot s_k$, where i, j and k are the sites within the corresponding cluster). The cluster expansion of the configurational energy, E, is written as

$$E = J_0 + \sum_\alpha J_\alpha \phi_\alpha, \tag{11.1}$$

where J are the coefficients describing the configurational energy dependence of clusters called effective cluster interactions (ECIs). Constructing the CE model ultimately amounts to finding the numerical values of the ECIs based on the energies obtained from *ab initio* calculations. An arbitrarily large number of clusters can be generated for an infinite crystal (as it is most often the case due to periodic boundary conditions used in many DFT codes). However, the accuracy of the CE model typically converges rapidly with respect to the size of clusters and the number of sites the clusters include. A relatively low truncation threshold, e.g. up to four-body clusters with the maximum diameter of 5 Å, is sufficient in practice for many materials systems.

It can be seen that Eq. 11.1 takes the same form as linear regression in machine learning. Consequently, CE is also regarded as a form of machine learning formalism [33], and several machine learning techniques such as compressive sensing [34] and Bayesian compressive sensing [35] have been introduced to CE. In the context of machine learning, CE is the method that maps the physical properties onto the occupation of sites in the lattice. Such formalism is naturally suited for investigating crystalline materials with a substitutional disorder to study phenomena such as thermal properties, order-disorder phase transition and OCV at finite temperatures.

For battery modelling, OCV at finite temperatures is one of the most desirable properties to simulate. We describe how one can simulate OCV based on the CE formalism with an example battery system consisting of Li_xCrO_2F ($0 \leq x \leq 2$) cathode and Li anode. The OCV of the system is defined as

$$\mathrm{OCV} = -\frac{\mu_{\mathrm{Li}}^{\mathrm{cathode}} - \mu_{\mathrm{Li}}^{\mathrm{anode}}}{e}, \tag{11.2}$$

where μ_{Li} is the chemical potential in eV per Li atom and e is an electron charge. As Li metal is used as an anode, $\mu_{\mathrm{Li}}^{\mathrm{anode}}$ is constant. Therefore, evaluating OCV is essentially calculating the chemical potential of Li_xCrO_2F relative to the reference Li metal. A direct method of computing the chemical potential is to use grand canonical Monte Carlo simulations based on the CE model where the chemical potential as a function of x at a given temperature is computed. The OCV plot generated from the grand canonical Monte Carlo simulation results is shown in Figure 11.5. Alternatively, one can utilize the fact that the chemical potential of cathode is the derivative of its free energy with respect to its Li content, i.e.,

$$\mu_{\mathrm{Li}}^{\mathrm{cathode}} = \mu_{\mathrm{Li}}^{\mathrm{Li_xCrO_2F}} = \frac{\mathrm{d}G_{\mathrm{Li_xCrO_2F}}}{\mathrm{d}x}, \tag{11.3}$$

where $G_{\mathrm{Li_xCrO_2F}}$ is the free energy of Li_xCrO_2F in eV per formula unit. The OCV can alternatively be written as

$$\mathrm{OCV} = -\frac{\frac{\mathrm{d}G_{\mathrm{Li_xCrO_2F}}}{\mathrm{d}x} - \mu_{\mathrm{Li}}^{\mathrm{anode}}}{e}. \tag{11.4}$$

One can then calculate the free energy of Li_xCrO_2F at different Li concentration using canonical Monte Carlo simulations and calculate its derivative with respect to x as shown in Figure 11.5.

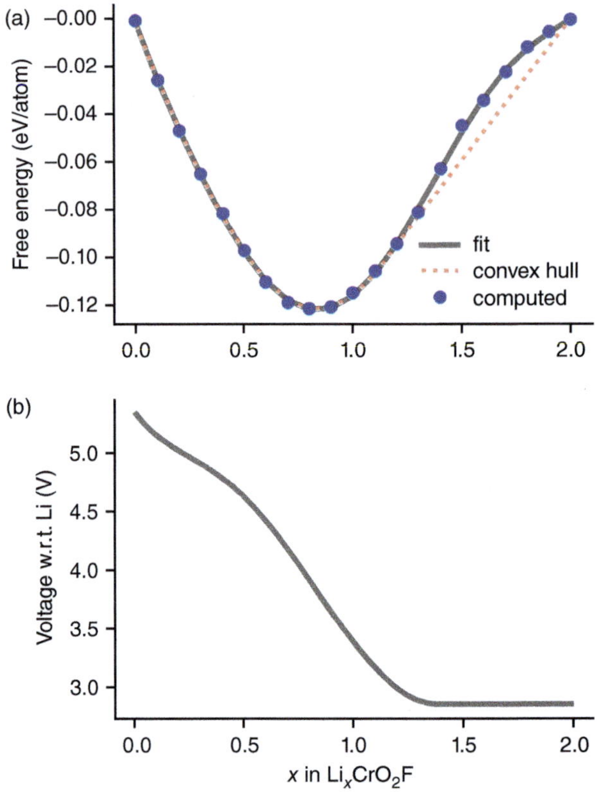

Figure 11.5 (a) Free energy of Li_xCrO_2F and (b) simulated OCV of a battery system consisting Li_xCrO_2F cathode and Li anode at T = 293 K. Reproduced from [32], copyright 2019 Institute of Physics. Creative Commons Attribution 3.0 licence. DOI: 10.1088/1361-648x/ab1bbc.

11.3 High-Fidelity data sets: *Ab initio* molecular dynamics simulations

The solid/electrolyte intefaces play a key role to determine the performance and the safety of a battery. In a Li-ion battery, for example, the high voltage obtainable from these devices, which is one of their advantages compared to other rechargeable batteries, is also the reason why the organic electrolyte is thermodynamically unstable at the working potential and decomposes into a mixture of organic and inorganic species. These layers, known as the solid electrolyte interphase (SEI) on the anode and cathode-electrolyte interphase (CEI) on the cathode, are formed during the initial cycles of charge/discharge and is essential for a battery to work.

In more detail, the SEI starts forming at the atomic scale level, with its nucleation through chemical reactions between the electrode surface and molecules from the electrolyte. These reactions involve fast transfer of electrons at time-scale down to femtoseconds. During growth, different nanometric inorganic (LiF, Li_2O, ...) and organic (Li_2CO_3, polymers, ...) compounds are formed and their interactions are crucial to allow ionic migration and to block electron conductivity. The SEI layer, as a passivating film formed on the anode, protects the electrolyte from further decomposition. However, the SEI film cannot be too thick or compact since it allows the transport of Li-ions from the electrolyte to the anode and vice versa, and should have good ionic conductivity. At the larger scale, the mechanical stability of the SEI is responsible the aging of the battery. Due to the complexity of this electrochemical system as well as the different time and length scales involved, the relation between reactions and interfaces is not completely clear yet and there is a lack of understanding of the nature of the SEI [36, 37].

Atomic- and multi-scale models are used to address different phenomena occurring at various length- and time-scale at the interface and explain experimental results [38, 39]. These models have different levels of fidelity and computational cost. At the atomic-scale level, for example, high-fidelity models are molecular dynamic (MD) simulations, in which the electrolyte is explicitly included. When each step of the MD simulation is calculated in the framework of DFT, we are dealing with *ab initio* molecular dynamics (AIMD). This approach can be efficiently used to investigate the evolution of the interfaces as a function of the potential of the battery and, for example, to understand the electrochemical reduction reactions that happen during the first cycles at the battery and that lead to the formation of SEI compounds. This methodology allows to model solid/electrolyte interfaces in LIB [40] and beyond batteries, such as the Pt(111)/water interface [41]. At a lower-fidelity, we can find simulations of interfaces, where the electrolyte is represented by a continuous medium (implicit solvation model) or by classical potentials, sometimes generated by machine learning algorithm trained from AIMD calculations. Although these models allow to treat larger systems [42, 43], the accuracy is reduced compared to AIMD. At the bottom of the fidelity ladder, we find simulations of the interfaces in vacuum, i.e. without the electrolyte. These simulations allow complex problems to be tackled in a simple and time-efficient way, which can then be verified using a higher-fidelity model.

In the following, we report an example of high-fidelity calculations to study the interface between metal single crystals and EC:EMC 3:7w, an organic electrolyte commonly used in LIB. In particular, we investigate the reduction of HF impurities that leads to the evolution of H_2 and formation of LiF. The use of single crystals have the advantage of simplify the interface, which allows the multiple reactions happening simultaneously at the interface to be decoupled. The realistic (i.e. the electrolyte si explicitly included), but idealized (i.e. model surfaces and lack of defects/impurities) calculations with high fidelity experiments is the key to understand this complex interface. The MD simulations at 300K produce an ensamble of structures that are used to calculate physical and chemical averages. Each computed structure is composed of different regions (as shown in Figure 11.6a): (a) fixed metal slab, (b) electrolyte, and (c) fixed top layer (ice structure of EC[44]), which has a total dipole along the direction normal to the surface equal zero. The top layer has the function of confining the electrolyte at the correct experimental density and of reproducing the bulk structure of the electrolyte. In addition, it needs a zero dipole to avoid effects on the orientation of the molecules in the electrolyte. Different concentrations of lithium, salts (like hexafluorophosphate, PF_6), and additives (like VC and PC) can be added to the electrolyte molecules. The electrolyte in the simulation is composed of eight EMC and four EC molecules. An equivalent definition of the generalized computational hydrogen electrode has been developed for Li to calculate the phase diagram of Li in the electrolyte. According to this description, the Gibbs free energy, G^{int}, is obtained from the equation: $G^{\text{int}} = \Delta E - n_{Li}(\Phi - U_e)$, where ΔE is the enthalpy of the system, n_{Li} the number of Li atoms in the configuration, U_e the calculated work function with respect to the Li/Li^+ scale,

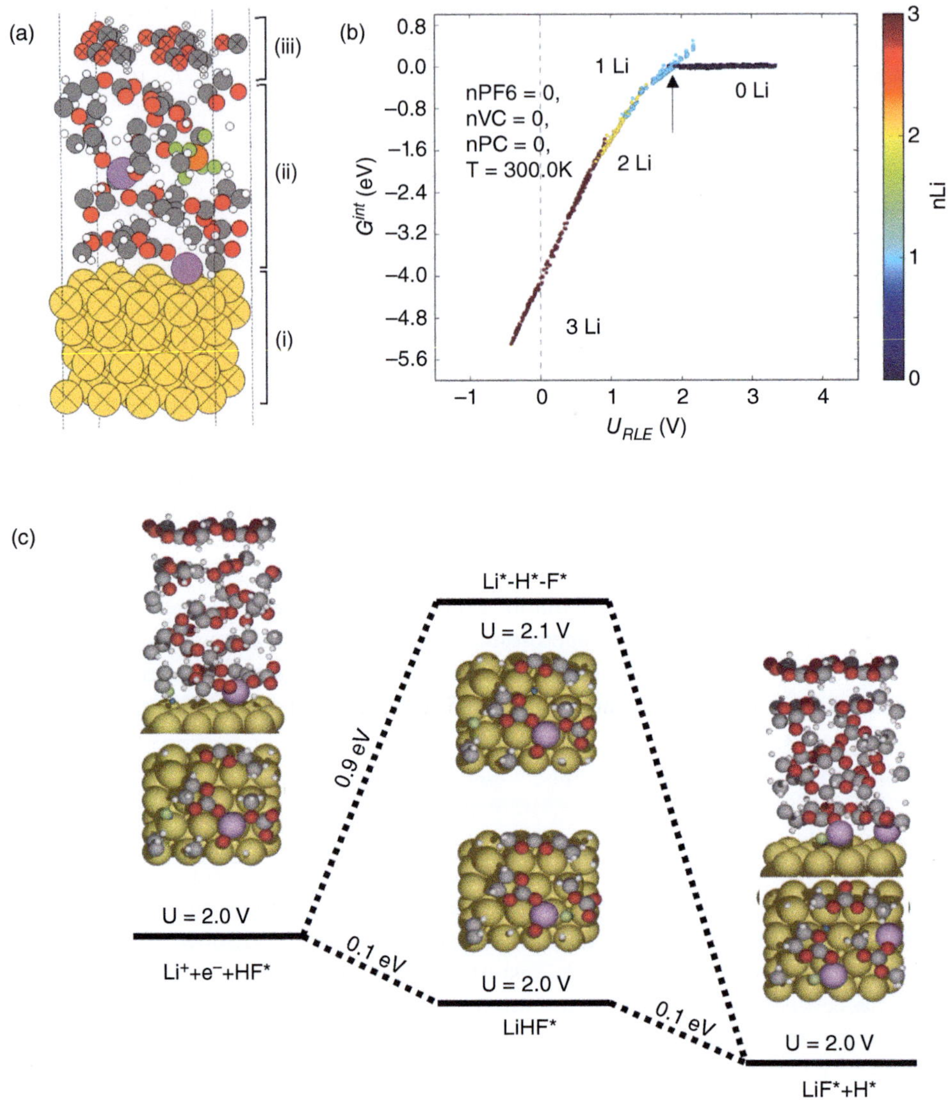

Figure 11.6 a) structure of the MD simulations (H atoms are represented in white, Li in purple, C in gray, O in red, F in green, P in orange, and Au in gold). A frozen position is indicated by an x. b) phase diagram of Li in EC:EMC. The Li coverage increasing with decreasing of the electrostatic potential. c) reaction path for the dissociation of HF and formation of LiF grains. Figure adapted from [40].

Φ (U_{RLE}). A grand canonical Metropolis Monte Carlo simulation at 300K is performed to ensure that each state could be visited with the same probability. More details on the methodology can be found in the literature [40, 41, 45, 46]. The phase diagram of Li in EC:EMC on Au(111) is shown in Figure 11.6b. At high potential, no Li atoms are present at the interface, and the coverage of Li increases with a decrease of the electrostatic potential. It has been shown that the Li adsorption potential, indicated with a black arrow in the figure, correlates with the electrochemical response in cyclicvoltammetry experiments. The structure of the interface at that potential can be then used to understand the reaction paths. The rigorous approach to do so would be to run AIMD simulations for the different steps of the chemical reaction. This would, however, be extremely expensive. The more feasible solution is to take the structure of the interface, freeze the atoms of the electrolyte far away from the reaction fragments, and run standard DFT calculations. With this approach, it has been shown that Li plays a key role in the formation of LiF from HF impurities, working as a catalyst for the dissociation of the HF molecule, as shown in Figure 11.6c. In the inital configuration, HF and Li are present at the interface (indicated as $Li^{+}+ e^{-}+HF^{*}$) at a potential corresponding to the adsorption potential of Li (2 V vs Li/Li^{+}). From there, we could follow two path: the highest energy path proceed by splitting HF in H^{*} and F^{*} (which has an energy 0.9 eV above the initial state) and the lowest path in which HF adsorbs on top of Li ($LiHF^{*}$, 0.1 eV below the initial state). The final state is a grain of LiF^{*} and H^{*}, both adsorbed on the surface, which can be obtained by combining Li^{*} and F^{*} (highest energy path) or removing an H from LiHF (lowest energy path). The dissociation of HF and formation of LiF is 0.2 eV downhill in energy when the reaction uses Li as a catalyst. The formation of rock-salt LiF from LiF grains is also favorable in energy and H_2 evolves from the H atoms.

The AIMD approach, in principle, could be also used to understand the formation of the CEI layer. Although because of the complexity of the cathode materials (oxide interface, lithiation rate, etc.), it is currently not feasible to perform AIMD simulations. More conventional approaches based on "static" DFT and investigation of reaction paths have been followed [47].

11.4 Machine Learning

Machine learning (ML) can be defined as the branch of artificial intelligence where functional mapping is done between input and output variable and the mapping improves by training with examples [48, 49]. As an illustration, a model can be developed to take the behaviour of a secondary battery cell during the first few cycles as input and predict the long-term capacity degradation of the cell [50]. The power of ML comes from the improving accuracy when larger data sets are available for training. Specifically in the area of battery material modelling, data for ML training can come from both experimental and simulation-based sources. Accuracy of ML models is dependent on the quality and quantity of data that is used for training [51]. Often ML models are trained assuming absolute accuracy of the training data. In the materials discovery area, both experiments and simulations attempt to gather information on real systems with precision that is associated with the method used. Thus it is important to consider the fidelity of data and cost of acquisition. Ideally data procurement must be optimized such that the incremental benefit of more data comes at the lowest cost. ML methods (especially Bayesian process regression type, which are very popular in the materials modelling community [52]) can work with uncertainties in input data [53, 54]. The uncertainties in data from physics-based simulations can thus be taken into account while estimating the reliability of ML-based models. Uncertainty in ML-based materials modelling additionally comes from that of the model itself. It can be especially significant if the model is used in a subspace not sampled during training [55]. This is called systemic uncertainty and can be alleviated by tracking the uncertainty and generating a new training data point through active learning to lower it [56].

Application of ML methods have accelerated materials discovery [57–62]. The bulk of ML application in battery research has been in prediction of state of charge (SoC) [63–66]. Battery material discovery has benefited from this trend as well [67–70] during the last few years, as ML has been utilized for predicting cathode materials' intercalation potential; electrolyte ionic conductivity, redox potential, and dendrite suppression. A broad area of ML application in battery material modelling is development of cheap surrogate models for atomic-scale simulations in lieu of electronic structure methods like DFT. For example, machine learned Gaussian approximation potential (GAP) [71] can be used for larger time and length scale simulation of anode materials and lithium intercalation process [72, 73]. These type of inter-atomic potential are derived from energies and forces generated by quantum mechanical calculation of smaller systems. The flexibility of the functional form of ML potentials like GAP- or neural-network-based ones [74, 75] allow modeling complex potential energy landscapes and systematic improvement with more data. Using *ab initio* methods for MD- or NEB-based dynamical property estimation is computationally very expensive, while classical potentials' lack of electronic description severely impede accuracy as battery material properties are electrochemical and electronic. ML potentials constructed from DFT can potentially match DFT accuracy.

Application of ML in battery material modelling goes beyond expediting molecular simulations. ML models can provide cheap surrogate models replacement for finite element method-based system-level simulations. Neural networks trained with finite element simulation results from a thermo-electrochemical model can verify design feasibility and predict specific energy and specific power from battery structural details like electrode thickness, particle radius, electrolyte concentration etc. [76]. Similarly, results from a large number of pseudo two-dimensional (P2D) model simulations (favored in battery design community due to its balance between accuracy and cost when solved using finite difference techniques) can be used to construct ML models based on random forest (RF) or gradient-boosted machine (GBM) or decision tree (DT) to predict battery behaviours like voltage decay curve much faster [77].

ML models can directly predict key electrode/electrolyte properties from structure/chemistry or other easily accessible material parameters or even the crystal structure itself after learning from data sets validated by experiments or simulations [68, 69, 78–80]. This class of machine learning is known as supervised machine learning. Kernel ridge regression (KRR), Gaussian process regression (GPR), support vector machine (SVM), and artificial neural network (ANN) models are popular methods for this purpose. ANN-based ML models [81]), which have become popular during last few year, employ many layers of interconnected neurons. The structure of the model provides immense flexibility in the functional space and is thus well suited for approximating complex chemical/physical processes in battery materials. The use of probabilistic weights in deep neural network is gaining popularity to understand the fidelity of the model itself [82]. Novel approaches like noise-contrastive priors can also be used to obtain an uncertainty estimate [83].

High-fidelity data sets for supervised ML are essential for model accuracy and can be obtained by automated high throughput simulations and experiments [84, 85]. Open availability of data sets is steadily growing and facilitating ML-based material design efforts [86–89]. While data agglomeration improves data quantity, challenges originate in fidelity of data from multiple sources even for the same method. Furthermore, reported data often is incomplete in terms of the details of the simulation parameters or exact structural details. The lack of standardization and completeness of the information can be addressed through the establishment of guidelines [2, 90]. For DFT-based training data, the exchange-correlation functional used have large implications on data fidelity. The development of the Bayesian error estimation functional with van der Waals correlation (BEEF-vdW) [91] can do uncertainty estimation in atomic-scale simulations. It has been used to predict high-fidelity lithium-graphite phase diagrams [92] important for lithium ion battery systems. It has also been used to identify intrinsic DFT-errors associated with the formation and breaking of specific chemical bonds in metal air batteries [93].

Especially for supervised ML, the input data need to be represented in the form of fingerprints. For example, fingerprinting for prediction of total energy from a molecular or crystal structure can be done with fourier series of atomic radial distribution functions, smooth overlap of atomic positions, atomic symmetry functions, or simply the atomic structure graph network [94–100]. The optimal choice of representation depends on the property and the specific MLA. Improved ML methods can learn pattern-like symmetry operations from data. Methods like KRR [101], needs symmetry to be built into the structural fingerprints, whereas convolutional neural networks can automatically detect symmetries [100]. Battery being an electrochemical system, the variety of molecules, materials and phases present is enormous and often key properties rise at the interface of different material phases present. The few existing application of ML in battery material modelling has been focused on bulk/molecular properties, but going forward, future applications aiming at complex interfacial properties need to move towards automated fingerprinting process. One of the most powerful methods for representation into latent variables from raw data is variational auto encoder (VAE) [102]. VAE would allow creation of unique representations that are appropriate for the specific property of the battery materials. VAE is a powerful unsupervised learning approach to model an unlabeled distribution of the data.

In unsupervised learning, we only use the data set for feature vectors or fingerprints and model the data distribution based on patterns between data points. For battery materials modelling, semi-supervised learning [103] might be most useful, where data distribution in feature space is learned unsupervised but data with known target variable is used for identifying sub-spaces in feature spaces. As for some of the materials, all structural and property details are available but more often target property data is incomplete. But the material data points without property data are still useful for learning the distribution of all materials features of interest.

Ideally we would like to tailor battery materials for a specific set of properties simultaneously. But the lack of a one-to-one mapping between chemical/structural details and the observable properties makes the functional relationship between them 'non-invertible', necessitating a generative inverse design approach [104]. Generative modelling is of practical use for inverse design, where the goal is to predict hitherto unknown material with the desired property. Unsupervised ML has the inherent advantage of enabling generative modelling. Generative modelling enables the prediction of new materials in the form of new data points in the feature space. The new data points can be used for qualitative model evaluation by actual physical simulation after evaluating whether the generated material structure looks realistic. For inverse materials design, when one-to-one mapping does exist and the design space is immensely large, the traditional screening strategies do not work and generative models have to be used [105, 106]. Successes of inverse computational design of materials are very rare as it is needed to identify and suggest material compositions and structures, which can, in fact, be synthesized. Recent work by Aykol et al. uses a network representation to identify synthesizable materials [107] from the OQMD database [87].

Directly fitting the material details to the target property can result in property predictions very fast but it loses on the physical understanding as well as fidelity, as changes in the underlying physical mechanism between the training and the test data would lead to low fidelity. An ML model trained on charge transport data from materials with polaronic transport [108] will be erroneous in predicting that for materials with electron tunneling [38] phenomenon. A hybrid of experiments, physical modelling, and a machine learning approach might be especially useful for battery material development. It would allow optimal utilization of all available resources based on the data fidelity and cost. For example, while designing intercalation cathode material with greater stability during cycling, cheap cycling experiments, synchrotron-based XRD investigation, expensive atomic scale characterization, a handful of high-accuracy DFT simulations, many low-accuracy ML-based force field simulations, or a ML model directly fitted to existing data can be used. Each of these methods has an inherent level of uncertainty. Intelligent selection based on the estimated fidelity should be used for accelerating the innovation process. A long-term goal can be complete integration of all synthesis, characterization, simulation, and data modelling methods towards a unified platform for closed-loop materials innovation.

Acknowledgements

The authors acknowledge support from the Villum Foundation's Young Investigator Program (4th round, project: In silico design of efficient materials for next generation batteries. Grant number: 10096) and Villum experiment project DeepDFT (grant number 0023105).

References

1 Joanne Hill, Gregory Mulholland, Kristin Persson, Ram Seshadri, Chris Wolverton, and Bryce Meredig. Materials science with large-scale data and informatics: unlocking new opportunities. *Mrs Bulletin*, 41(5):399–409, 2016.

2 Giovanni Pizzi, Andrea Cepellotti, Riccardo Sabatini, Nicola Marzari, and Boris Kozinsky. Aiida: automated interactive infrastructure and database for computational science. *Computational Materials Science*, 111:218–230, 2016.

3 Anubhav Jain, Shyue Ping Ong, Wei Chen, Bharat Medasani, Xiaohui Qu, Michael Kocher, Miriam Brafman, Guido Petretto, Gian-Marco Rignanese, Geoffroy Hautier, et al. Fireworks: a dynamic workflow system designed for high-throughput applications. *Concurrency and Computation: Practice and Experience*, 27(17):5037–5059, 2015.

4 Ask Hjorth Larsen, Jens Jørgen Mortensen, Jakob Blomqvist, Ivano E Castelli, Rune Christensen, Marcin Dułak, Jesper Friis, Michael N Groves, Bjørk Hammer, Cory Hargus, Eric D Hermes, Paul C Jennings, Peter Bjerre Jensen, James Kermode, John R Kitchin, Esben Leonhard Kolsbjerg, Joseph Kubal, Kristen Kaasbjerg, Steen Lysgaard, Jón Bergmann Maronsson, Tristan Maxson, Thomas Olsen, Lars Pastewka, Andrew Peterson, Carsten Rostgaard, Jakob Schiøtz, Ole Schütt, Mikkel Strange, Kristian S Thygesen, Tejs Vegge, Lasse Vilhelmsen, Michael Walter, Zhenhua Zeng, and Karsten W Jacobsen. The atomic simulation environment—a python library for working with atoms. *Journal of Physics: Condensed Matter*, 29(27):273002, 2017.

5 Georg Kresse and Jürgen Furthmüller. Efficient iterative schemes for ab initio total-energy calculations using a plane-wave basis set. *Physical Review B*, 54(16):11169, 1996.

6 Greg Mills and Hannes Jónsson. Quantum and thermal effects in H2 dissociative adsorption: Evaluation of free energy barriers in multidimensional quantum systems. *Physical Review Letters*, 72(7):1124–1127, 1994.

7 Hannes Jónsson, Greg Mills, and Karsten W Jacobsen. *Nudged elastic band method for finding minimum energy paths of transitions*, pages 385–404.

8 Graeme Henkelman and Hannes Jónsson. Improved tangent estimate in the nudged elastic band method for finding minimum energy paths and saddle points. *The Journal of Chemical Physics*, 113(22):9978–9985, 2000.

9 Graeme Henkelman, Blas P. Uberuaga, and Hannes Jónsson. Climbing image nudged elastic band method for finding saddle points and minimum energy paths. *The Journal of Chemical Physics*, 113(22):9901–9904, 2000.

10 Nicolai Rask Mathiesen, Hannes Jónsson, Tejs Vegge, and Juan Maria Garcia Lastra. R-neb: Accelerated nudged elastic band calculations by use of reflection symmetry. *Journal of Chemical Theory and Computation*, 15(5):3215–3222, 2019.

11 Andrew A. Peterson. Acceleration of saddle-point searches with machine learning. *The Journal of Chemical Physics*, 145(7), 2016.

12 Olli Pekka Koistinen, Freyja B. Dagbjartsdóttir, Vilhjálmur Ásgeirsson, Aki Vehtari, and Hannes Jónsson. Nudged elastic band calculations accelerated with Gaussian process regression. *The Journal of Chemical Physics*, 147(15), 2017.

13 Simon Loftager, Juan María García-Lastra, and Tejs Vegge. A density functional theory study of the ionic and electronic transport mechanisms in $LiFeBO_3$ battery electrodes. *The Journal of Chemical Physics C*, 120(33):18355–18364, 2016.

14 Alexander S. Tygesen, Mohnish Pandey, Tejs Vegge, Kristian S. Thygesen, and Juan M. García-Lastra. Role of long-range dispersion forces in modeling of Mxenes as battery electrode materials. *The Journal of Chemical Physics C*, 123:4064–4071, 2019.

15 Pál D. Mezei and O. Anatole von Lilienfeld. Non-covalent quantum machine learning corrections to density functionals. 2019.

16 D. De Fontaine. Cluster Approach to Order-Disorder Transformations in Alloys. In Henry Ehrenreich and David Turnbull, editors, *Solid State Physics*, volume 47, pages 33–176. Academic Press, 1994.

17 Xi Zhang and Marcel H F Sluiter. Cluster Expansions for Thermodynamics and Kinetics of Multicomponent Alloys. *Journal of Phase Equilibria and Diffusion*, 37(1):44–52, feb 2016.

18 J. M. Sanchez, F. Ducastelle, and D. Gratias. Generalized cluster description of multicomponent systems. *Physica A: Statistical Mechanics and its Applications*, 128(1-2):334–350, 1984.

19 Alex Zunger. First-Principles Statistical Mechanics of Semiconductor Alloys and Intermetallic Compounds. In Patrice E. A. Turchi and Antonios Gonis, editors, *Statics and Dynamics of Alloy Phase Transformations*, pages 361–419. Springer US, Boston, MA, 1994.

20 M Asta, V Ozolins, and C Woodward. A first-principles approach to modeling alloy phase equilibria. *JOM*, 53(9):16–19, sep 2001.

21 A. van de Walle. A complete representation of structure-property relationships in crystals. *Nature Materials*, 7(6):455–458, jun 2008.

22 A. Van der Ven, M. K. Aydinol, G. Ceder, G. Kresse, and J. Hafner. First-principles investigation of phase stability in Li_xCoO_2. *Physical Review B*, 58:2975–2987, 1998.

23 Anton Van der Ven, Mehmet K. Aydinol, and Gerbrand Ceder. First-Principles Evidence for Stage Ordering in Li_xCoO_2. *Journal of The Electrochemical Society*, 145(6):2149–2155, 1998.

24 C. Wolverton and Alex Zunger. First-Principles Prediction of Vacancy Order-Disorder and Intercalation Battery Voltages in Li_xCoO_2. *Physical Review Letters*, 81:606–609, 1998.

25 C. Wolverton and Alex Zunger. Cation and vacancy ordering in Li_xCoO_2. *Physical Review B*, 57:2242–2252, 1998.

26 Jinhyuk Lee, Alexander Urban, Xin Li, Dong Su, Geoffroy Hautier, and Gerbrand Ceder. Unlocking the potential of cation-disordered oxides for rechargeable lithium batteries. *Science*, 343(6170):519–522, 2014.

27 Ruiyong Chen, Shuhua Ren, Michael Knapp, Di Wang, Raiker Witter, Maximilian Fichtner, and Horst Hahn. Disordered lithium-rich oxyfluoride as a stable host for enhanced Li^+ intercalation storage. *Advanced Energy Materials*, 5(9):1–7, 2015.

28 Naoaki Yabuuchi. material design concept of lithium-excess electrode materials with rocksalt-related structures for rechargeable non-aqueous batteries. *The Chemical Record*, pages 1–19, 2018.

29 Axel van de Walle. Multicomponent multisublattice alloys, nonconfigurational entropy and other additions to the Alloy Theoretic Automated Toolkit. *Calphad*, 33(2):266–278, 2009.

30 A. van de Walle, M. Asta, and G. Ceder. The alloy theoretic automated toolkit: A user guide. *Calphad*, 26(4):539–553, 2002.

31 D Lerch, O Wieckhorst, G L W Hart, R W Forcade, and S Müller. UNCLE: a code for constructing cluster expansions for arbitrary lattices with minimal user-input. *Modelling and Simulation in Materials Science and Engineering*, 17(5):055003, 2009.

32 Jin Hyun Chang, David Kleiven, Marko Melander, Jaakko Akola, Juan Maria Garcia Lastra, and Tejs Vegge. CLEASE: A versatile and user-friendly implementation of Cluster Expansion method. *Journal of Physics: Condensed Matter*, 31:325901, 2019.

33 Anirudh Raju Natarajan and Anton Van der Ven. Machine-learning the configurational energy of multicomponent crystalline solids. *npj Computational Materials*, 4(1):1–7, 2018.

34 Lance J. Nelson, Gus L W Hart, Fei Zhou, and Vidvuds Ozoliņš. Compressive sensing as a paradigm for building physics models. *Physical Review B*, 87(3):035125, 2013.

35 Lance J. Nelson, Vidvuds Ozoliņš, C. Shane Reese, Fei Zhou, and Gus L W Hart. Cluster expansion made easy with Bayesian compressive sensing. *Physical Review B*, 88(15):155105, 2013.

36 E. Peled and S. Menkin. Review—SEI: Past, present and future. *Journal of The Electrochemical Society*, 164(7):A1703–A1719, 2017.

37 Seong Jin An, Jianlin Li, Claus Daniel, Debasish Mohanty, Shrikant Nagpure, and David L. Wood. The state of understanding of the lithium-ion-battery graphite solid electrolyte interphase (SEI) and its relationship to formation cycling. *Carbon*, 105:52–76, aug 2016.

38 Aiping Wang, Sanket Kadam, Hong Li, Siqi Shi, and Yue Qi. Review on modeling of the anode solid electrolyte interphase (SEI) for lithium-ion batteries. *npj Computational Materials*, 4(1), mar 2018.

39 Birger Horstmann, Fabian Single, and Arnulf Latz. Review on multi-scale models of solid-electrolyte interphase formation. *Current Opinion in Electrochemistry*, 13:61–69, feb 2019.

40 Dusan Strmcnik, Ivano E. Castelli, Justin G. Connell, Dominik Haering, Milena Zorko, Pedro Martins, Pietro P. Lopes, Bostjan Genorio, Thomas Østergaard, Hubert A. Gasteiger, Filippo Maglia, Byron K. Antonopoulos, Vojislav R. Stamenkovic, Jan Rossmeisl, and Nenad M. Markovic. Electrocatalytic transformation of HF impurity to H_2 and LiF in lithium-ion batteries. *Nature Catalysis*, 1(4):255–262, April 2018.

41 Martin Hangaard Hansen and Jan Rossmeisl. pH in grand canonical statistics of an electrochemical interface. *The Journal of Physical Chemistry C*, 120(51):29135–29143, December 2016.

42 Zhe Li, Oleg Borodin, Grant D. Smith, and Dmitry Bedrov. Effect of organic solvents on Li^+ ion solvation and transport in ionic liquid electrolytes: A molecular dynamics simulation study. *The Journal of Physical Chemistry B*, 119(7):3085–3096, February 2015.

43 Nicolas Lespes and Jean-Sébastien Filhol. Using implicit solvent in ab initio electrochemical modeling: Investigating Li^+/Li electrochemistry at a Li/solvent interface. *Journal of Chemical Theory and Computation*, 11(7):3375–3382, June 2015.

44 C. J. Brown. The crystal structure of ethylene carbonate. *Acta Crystallographica*, 7(1):92–96, January 1954.

45 Alexander Bagger, Ivano E. Castelli, Martin Hangaard Hansen, and Jan Rossmeisl. Fundamental atomic insight in electrocatalysis. In *Handbook of Materials Modeling*, pages 1–31. Springer International Publishing, 2018.

46 Ivano E. Castelli, Milena Zorko, Thomas Østergaard, Pedro F.B.D. Martins, Pietro P. Lopes, Byron K. Antonopoulos, Filippo Maglia, Nenad M. Markovic, Dusan Strmcnik, and Jan Rossmeisl. The role of an interface in stabilizing reaction intermediates for hydrogen evolution in aprotic electrolytes. *Chemical Science*, 11:3914–3922, 2020.

47 Thomas M. Østergaard, Livia Giordano, Ivano E. Castelli, Filippo Maglia, Byron K. Antonopoulos, Yang Shao-Horn, and Jan Rossmeisl. Oxidation of ethylene carbonate on Li metal oxide surfaces. *The Journal of Physical Chemistry C*, 122(19):10442–10449, April 2018.

48 Jerome Friedman, Trevor Hastie, and Robert Tibshirani. *The elements of statistical learning*, volume 1. Springer series in statistics New York, 2001.

49 Andrew Gelman, Hal S Stern, John B Carlin, David B Dunson, Aki Vehtari, and Donald B Rubin. *Bayesian data analysis*. Chapman and Hall/CRC, 2013.

50 Kristen A Severson, Peter M Attia, Norman Jin, Nicholas Perkins, Benben Jiang, Zi Yang, Michael H Chen, Muratahan Aykol, Patrick K Herring, Dimitrios Fraggedakis, et al. Data-driven prediction of battery cycle life before capacity degradation. *Nature Energy*, page 1, 2019.

51 Michael I Jordan and Tom M Mitchell. Machine learning: Trends, perspectives, and prospects. *Science*, 349(6245):255–260, 2015.

52 Tim Mueller, Aaron Gilad Kusne, and Rampi Ramprasad. Machine learning in materials science: Recent progress and emerging applications. *Reviews in Computational Chemistry*, 29:186–273, 2016.

53 Andreas C Damianou, Michalis K Titsias, and Neil D Lawrence. Variational inference for latent variables and uncertain inputs in Gaussian processes. *The Journal of Machine Learning Research*, 17(1):1425–1486, 2016.

54 Wojciech M Czarnecki and Igor T Podolak. Machine learning with known input data uncertainty measure. In *IFIP International Conference on Computer Information Systems and Industrial Management*, pages 379–388. Springer, 2013.

55 Mattias Teye, Hossein Azizpour, and Kevin Smith. Bayesian uncertainty estimation for batch normalized deep networks. *arXiv preprint arXiv:1802.06455*, 2018.

56 Kathryn Chaloner and Isabella Verdinelli. Bayesian experimental design: A review. *Statistical Science*, pages 273–304, 1995.

57 Rampi Ramprasad, Rohit Batra, Ghanshyam Pilania, Arun Mannodi-Kanakkithodi, and Chiho Kim. Machine learning in materials informatics: recent applications and prospects. *npj Computational Materials*, 3(1):54, 2017.

58 Edward O Pyzer-Knapp, Kewei Li, and Alan Aspuru-Guzik. Learning from the harvard clean energy project: The use of neural networks to accelerate materials discovery. *Advanced Functional Materials*, 25(41):6495–6502, 2015.

59 Keith T Butler, Daniel W Davies, Hugh Cartwright, Olexandr Isayev, and Aron Walsh. Machine learning for molecular and materials science. *Nature*, 559(7715):547, 2018.

60 Bryce Meredig, Ankit Agrawal, Scott Kirklin, James E Saal, JW Doak, Alan Thompson, Kunpeng Zhang, Alok Choudhary, and Christopher Wolverton. Combinatorial screening for new materials in unconstrained composition space with machine learning. *Physical Review B*, 89(9):094104, 2014.

61 Ghanshyam Pilania, Chenchen Wang, Xun Jiang, Sanguthevar Rajasekaran, and Ramamurthy Ramprasad. Accelerating materials property predictions using machine learning. *Scientific Reports*, 3:2810, 2013.

62 Yue Liu, Tianlu Zhao, Wangwei Ju, and Siqi Shi. Materials discovery and design using machine learning. *Journal of Materiomics*, 3(3):159–177, 2017.

63 Ephrem Chemali, Phillip J Kollmeyer, Matthias Preindl, and Ali Emadi. State-of-charge estimation of li-ion batteries using deep neural networks: A machine learning approach. *Journal of Power Sources*, 400:242–255, 2018.

64 Gozde O Sahinoglu, Milutin Pajovic, Zafer Sahinoglu, Yebin Wang, Philip V Orlik, and Toshihiro Wada. Battery state-of-charge estimation based on regular/recurrent gaussian process regression. *IEEE Transactions on Industrial Electronics*, 65(5):4311–4321, 2018.

65 Hanmin Sheng and Jian Xiao. Electric vehicle state of charge estimation: Nonlinear correlation and fuzzy support vector machine. *Journal of Power Sources*, 281:131–137, 2015.

66 Shijie Tong, Joseph H Lacap, and Jae Wan Park. Battery state of charge estimation using a load-classifying neural network. *Journal of Energy Storage*, 7:236–243, 2016.

67 Rajendra P Joshi, Jesse Eickholt, Liling Li, Marco Fornari, Veronica Barone, and Juan E Peralta. Machine learning the voltage of electrode materials in metal-ion batteries. *arXiv preprint arXiv:1903.06813*, 2019.

68 Austin D Sendek, Ekin D Cubuk, Evan R Antoniuk, Gowoon Cheon, Yi Cui, and Evan J Reed. Machine learning-assisted discovery of solid li-ion conducting materials. *Chemistry of Materials*, 31(2):342–352, 2018.

69 Omar Allam, Byung Woo Cho, Ki Chul Kim, and Seung Soon Jang. Application of DFT-based machine learning for developing molecular electrode materials in li-ion batteries. *RSC Advances*, 8(69):39414–39420, 2018.

70 Zeeshan Ahmad, Tian Xie, Chinmay Maheshwari, Jeffrey C Grossman, and Venkatasubramanian Viswanathan. Machine learning enabled computational screening of inorganic solid electrolytes for suppression of dendrite formation in lithium metal anodes. *ACS Central Science*, 4(8):996–1006, 2018.

71 Albert P Bartók, Mike C Payne, Risi Kondor, and Gábor Csányi. Gaussian approximation potentials: The accuracy of quantum mechanics, without the electrons. *Physical Review Letters*, 104(13):136403, 2010.

72 Volker L Deringer, Céline Merlet, Yuchen Hu, Tae Hoon Lee, John A Kattirtzi, Oliver Pecher, Gábor Csányi, Stephen R Elliott, and Clare P Grey. Towards an atomistic understanding of disordered carbon electrode materials. *Chemical Communications*, 54(47):5988–5991, 2018.

73 So Fujikake, Volker L Deringer, Tae Hoon Lee, Marcin Krynski, Stephen R Elliott, and Gábor Csányi. Gaussian approximation potential modeling of lithium intercalation in carbon nanostructures. *The Journal of Chemical Physics*, 148(24):241714, 2018.

74 Jörg Behler and Michele Parrinello. Generalized neural-network representation of high-dimensional potential-energy surfaces. *Physical Review Letters*, 98(14):146401, 2007.

75 Alireza Khorshidi and Andrew A Peterson. Amp: A modular approach to machine learning in atomistic simulations. *Computer Physics Communications*, 207:310–324, 2016.

76 Bin Wu, Sangwoo Han, Kang G Shin, and Wei Lu. Application of artificial neural networks in design of lithium-ion batteries. *Journal of Power Sources*, 395:128–136, 2018.

77 Neal Dawson-Elli, Seong Beom Lee, Manan Pathak, Kishalay Mitra, and Venkat R Subramanian. Data science approaches for electrochemical engineers: an introduction through surrogate model development for lithium-ion batteries. *Journal of The Electrochemical Society*, 165(2):A1–A15, 2018.

78 Koji Fujimura, Atsuto Seko, Yukinori Koyama, Akihide Kuwabara, Ippei Kishida, Kazuki Shitara, Craig AJ Fisher, Hiroki Moriwake, and Isao Tanaka. Accelerated materials design of lithium superionic conductors based on first-principles calculations and machine learning algorithms. *Advanced Energy Materials*, 3(8):980–985, 2013.

79 M Attarian Shandiz and R Gauvin. Application of machine learning methods for the prediction of crystal system of cathode materials in lithium-ion batteries. *Computational Materials Science*, 117:270–278, 2016.

80 Sungjin Kim, Adrián Jinich, and Alán Aspuru-Guzik. Multidk: A multiple descriptor multiple kernel approach for molecular discovery and its application to organic flow battery electrolytes. *Journal of Chemical Information and Modeling*, 57(4):657–668, 2017.

81 Ian Goodfellow, Yoshua Bengio, and Aaron Courville. *Deep Learning*. MIT Press, 2016. http://www.deeplearningbook.org.

82 Charles Blundell, Julien Cornebise, Koray Kavukcuoglu, and Daan Wierstra. Weight uncertainty in neural networks. *arXiv preprint arXiv:1505.05424*, 2015.

83 Danijar Hafner, Dustin Tran, Alex Irpan, Timothy Lillicrap, and James Davidson. Reliable uncertainty estimates in deep neural networks using noise contrastive priors. *arXiv preprint arXiv:1807.09289*, 2018.

84 Daniel P Tabor, Loïc M Roch, Semion K Saikin, Christoph Kreisbeck, Dennis Sheberla, Joseph H Montoya, Shyam Dwaraknath, Muratahan Aykol, Carlos Ortiz, Hermann Tribukait, et al. Accelerating the discovery of materials for clean energy in the era of smart automation. *Nat. Rev. Mater.*, 3:5–20, 2018.

85 Min-Sik Park, Yoon-Sok Kang, and Dongmin Im. A high-speed screening method by combining a high-throughput method and a machine-learning algorithm for developing novel organic electrolytes in rechargeable batteries. *ECS Transactions*, 68(2):75–81, 2015.

86 Anubhav Jain, Shyue Ping Ong, Geoffroy Hautier, Wei Chen, William Davidson Richards, Stephen Dacek, Shreyas Cholia, Dan Gunter, David Skinner, Gerbrand Ceder, et al. Commentary: The materials project: A materials genome approach to accelerating materials innovation. *Apl Materials*, 1(1):011002, 2013.

87 James E Saal, Scott Kirklin, Muratahan Aykol, Bryce Meredig, and Christopher Wolverton. Materials design and discovery with high-throughput density functional theory: the open quantum materials database (oqmd). *Jom*, 65(11):1501–1509, 2013.

88 Xiaohui Qu, Anubhav Jain, Nav Nidhi Rajput, Lei Cheng, Yong Zhang, Shyue Ping Ong, Miriam Brafman, Edward Maginn, Larry A Curtiss, and Kristin A Persson. The electrolyte genome project: A big data approach in battery materials discovery. *Computational Materials Science*, 103:56–67, 2015.

89 Leila Ghadbeigi, Taylor D Sparks, Jaye K Harada, and Bethany R Lettiere. Data-mining approach for battery materials. In *2015 IEEE Conference on Technologies for Sustainability (SusTech)*, pages 239–244. IEEE, 2015.

90 Rajan Jose and Seeram Ramakrishna. Materials 4.0: Materials big data enabled materials discovery. *Applied Materials Today*, 10:127–132, 2018.

91 Jess Wellendorff, Keld T Lundgaard, Andreas Møgelhøj, Vivien Petzold, David D Landis, Jens K Nørskov, Thomas Bligaard, and Karsten W Jacobsen. Density functionals for surface science: Exchange-correlation model development with bayesian error estimation. *Physical Review B*, 85(23):235149, 2012.

92 Vikram Pande and Venkatasubramanian Viswanathan. Robust high-fidelity DFT study of the lithium-graphite phase diagram. *Physical Review Materials*, 2(12):125401, 2018.

93 Rune Christensen, Jens S Hummelshøj, Heine A Hansen, and Tejs Vegge. Reducing systematic errors in oxide species with density functional theory calculations. *The Journal of Physical Chemistry C*, 119(31):17596–17601, 2015.

94 Albert P Bartók, Risi Kondor, and Gábor Csányi. On representing chemical environments. *Physical Review B*, 87(18):184115, 2013.

95 Giulio Imbalzano, Andrea Anelli, Daniele Giofré, Sinja Klees, Jörg Behler, and Michele Ceriotti. Automatic selection of atomic fingerprints and reference configurations for machine-learning potentials. *The Journal of Chemical Physics*, 148(24):241730, 2018.

96 Olexandr Isayev, Denis Fourches, Eugene N Muratov, Corey Oses, Kevin Rasch, Alexander Tropsha, and Stefano Curtarolo. Materials cartography: representing and mining materials space using structural and electronic fingerprints. *Chemistry of Materials*, 27(3):735–743, 2015.

97 O Anatole von Lilienfeld, Raghunathan Ramakrishnan, Matthias Rupp, and Aaron Knoll. Fourier series of atomic radial distribution functions: A molecular fingerprint for machine learning models of quantum chemical properties. *International Journal of Quantum Chemistry*, 115(16):1084–1093, 2015.

98 Michael J Willatt, Félix Musil, and Michele Ceriotti. Atom-density representations for machine learning. *The Journal of Chemical Physics*, 150(15):154110, 2019.

99 David K Duvenaud, Dougal Maclaurin, Jorge Iparraguirre, Rafael Bombarell, Timothy Hirzel, Alán Aspuru-Guzik, and Ryan P Adams. Convolutional networks on graphs for learning molecular fingerprints. In *Advances in Neural Information Processing Systems*, pages 2224–2232, 2015.

100 Kristof T Schütt, Farhad Arbabzadah, Stefan Chmiela, Klaus R Müller, and Alexandre Tkatchenko. Quantum-chemical insights from deep tensor neural networks. *Nature Communications*, 8:13890, 2017.

101 Paul C. Jennings, Steen Lysgaard, Jens Strabo Hummelshøj, Tejs Vegge, and Thomas Bligaard. Genetic algorithms for computational materials discovery accelerated by machine learning. *npj Computational Materials*, 5(1), April 2019.

102 Diederik P Kingma and Max Welling. Auto-encoding variational bayes. *arXiv preprint arXiv:1312.6114*, 2013.

103 Lars Maaløe, Marco Fraccaro, and Ole Winther. Semi-supervised generation with cluster-aware generative models. *arXiv preprint arXiv:1704.00637*, 2017.

104 Avni Jain, Jonathan A. Bollinger, and Thomas M. Truskett. Inverse methods for material design. *AIChE Journal*, 60(8):2732–2740, May 2014.

105 Benjamin Sanchez-Lengeling and Alán Aspuru-Guzik. Inverse molecular design using machine learning: Generative models for matter engineering. *Science*, 361(6400):360–365, Jul 2018.

106 Phuoc Nguyen, Truyen Tran, Sunil Gupta, Santu Rana, and Svetha Venkatesh. Hybrid generative-discriminative models for inverse materials design. *arXiv preprint arXiv:1811.06060*, 2018.

107 Muratahan Aykol, Vinay I Hegde, Santosh Suram, Linda Hung, Patrick Herring, Chris Wolverton, and Jens S Hummelshøj. Network analysis of synthesizable materials discovery. *arXiv preprint arXiv:1806.05772*, 2018.

108 Juan Maria Garcia-Lastra, Jon SG Myrdal, Rune Christensen, Kristian S Thygesen, and Tejs Vegge. DFT+U study of polaronic conduction in Li_2O_2 and Li_2CO_3: implications for Li–Air batteries. *The Journal of Physical Chemistry C*, 117(11):5568–5577, 2013.

Index

Atomic-Scale Modelling of Electrochemical Systems, First Edition.
Edited by Marko M. Melander, Tomi T. Laurila, and Kari Laasonen.